MECHANICS OF ENGINEERING MATERIALS

Mechanics of engineering materials

P. P. BENHAM

Professor Emeritus and formerly
Professor of Aeronautical Engineering,
The Queen's University of Belfast

R. J. CRAWFORD

Professor of Mechanical and Manufacturing Engineering
and Director of the School of Mechanical and Process Engineering,
The Queen's University of Belfast

Copublished in the United States with
John Wiley & Sons, Inc., New York

Longman Scientific & Technical
Longman Group UK Limited
Longman House, Burnt Mill, Harlow
Essex CM20 2JE, England
and Associated Companies throughout the world

Copublished in the United States with
John Wiley & Sons, Inc., 605 Third Avenue, New York, NY 10158

First published 1987
Seventh impression 1994

British Library Cataloguing in Publication Data

Benham, P. P.
   Mechanics of engineering materials. –
   [Rev. ed.]
   1. Materials   2. Mechanics, Applied
   I. Title   II. Crawford, R. J.
   III. Benham, P. P. Mechanics of solids
   and structures
   620.1′12   TA404.8
   ISBN 0-582-28640-9

Library of Congress Cataloging in Publication Data

Benham, P. P. (Peter Philip), 1927–
   Mechanics of engineering materials.
   Bibliography: p.
   Includes index.
   1. Strength of materials.   2. Structures, Theory of.
I. Crawford, R. J.   II. Title.
TA405.B484   1986   620.1′12   86–20064
ISBN 0–470–20704–3 (USA only)

Set in 10/12pt Linotron 202 Times Roman
Produced by Longman Singapore Publishers Pte Ltd
Printed in Singapore

Contents

Preface

The S.I. edition of *Mechanics of Solids and Structures* by P. P. Benham and F. V. Warnock was first published in 1973. It appears to have been very well received over the ensuing period. This preface is therefore written both for those who are familiar with the past text and also those who are approaching this subject for the first time. Although the subject matter is still basically the same today as it has been for decades, there are a few developing topics which have been introduced into undergraduate courses such as finite element analysis, fracture mechanics and fibre composite materials. In addition, style of presentation and illustrations in engineering texts have changed for the better and certain limitations of the previous edition, e.g. the number of problems and worked examples, needed to be rectified. Professor Warnock died in 1976 after a period of happy retirement and so it was left to the other author and the publisher to take the initiative to construct a new textbook.

In order to provide fresh thinking and reduce the time of rewriting Dr Roy Crawford, Reader in Mechanical Engineering at the Queen's University of Belfast, was invited, and kindly agreed, to join the project as a co-author. Although the present text might be regarded as a further edition of the original book the new authorship team preferred to make a completely fresh start. This is reflected in the change of title which is widely used as an alternative to *Mechanics of Solids*. The dropping of the reference to structures does not imply any reduction in that topic as will be seen in the contents.

In order that the book should not become any larger with the proposed expansion of material in some chapters, it was decided that the previous three chapters on experimental stress analysis should be omitted as there are several excellent texts in this field. The retention of that part of the book dealing with mechanical properties of materials for design was regarded as important even though there are also specialized texts in this area.

The main part (Ch. 1–18) of this new book of course still deals with the basic subject of Solid Mechanics, or Mechanics of Materials, whichever title one may prefer, being the study of equilibrium and displacement systems in engineering components and structures to enable designs to be effected in terms of stress and strain and the selection of materials. These eighteen chapters cover virtually all that is required in the three-year syllabus of a university or polytechnic degree course in engineering, or the examinations of the engineering Council, C.N.A.A. etc.

Although there is a fairly natural ordering of the material there is some scope for variation and lecturers will have their own particular detailed preferences.

As in the previous text, the first eleven chapters are concerned with forces and displacements in statically-determinate and indeterminate components and structures, and the analysis of uniaxial stress and strain due to various forms of loading such as bending, torsion, pressure and temperature change. The basic concepts of strain energy (Ch. 9) and elastic stability (Ch. 11) are also introduced. In Chapter 12 a study is made of two-dimensional states of stress and strain with special emphasis on principal stresses and the analysis of strain measurements using strain gauges. Chapter 13 combines two chapters of the previous book and brings together the topics of yield prediction and stress concentration which are of such importance in design.

Also included in these two chapters is an elementary introduction to the stress analysis and failure of fibre composite materials. These relatively new advanced structural materials are becoming increasingly used, particularly in the aerospace industry, and it is essential for engineers to receive a basic introduction to them. These thirteen chapters constitute the bulk of the syllabuses covered in first and second-year courses.

Four of the next five chapters appeared in the previous text and deal with more advanced or specialized topics such as thick-walled pressure vessels, rotors, thin plates and shells and post-yield or plastic behaviour, which will probably occupy part of final-year courses.

One essential new addition is an introductory chapter on finite element analysis. It may seem presumptuous even to attempt an introduction to such a broad subject in one chapter, but it is an attempt to provide initial encouragement and confidence to proceed to the complete texts on finite elements.

Chapters 19 to 22 cover much the same ground as in the previous text, but have been brought up to date particularly in relation to fracture mechanics. Since these chapters have such importance in relation to design, a number of worked examples have been introduced, together with problems at the end of each chapter. Bibliographies have still been included for further reading as required.

The first Appendix covers the essential material on properties of areas. The second deals with the simple principles of matrix algebra. A useful table of mechanical properties is provided in the third Appendix.

One of the recommendations of the Finniston Report to higher education was that theory should be backed up by more practical industrial applications. In this context the authors have attempted to incorporate into the worked examples and problems at the end of each chapter realistic engineering situations apart from the conventional examination-type applications of theory.

There had been a number of enquiries for a solutions manual for the previous text and this can be very helpful to both lecturer and student. Consequently this text is accompanied by another volume which contains worked solutions to nearly 300 problems. The manual should be used alongside the main text, so that steps in each solution can be referred back to the appropriate development in the relevant chapter. It is most important not to approach solutions on the basis of plucking the "appropriate formula" out of the text, inserting the numbers, and manipulating a calculator!

Every effort has been made by the authors to ensure accuracy of text and solutions, but lengthy experience demonstrates human fallibility in this respect. When errors subsequently come to light they will be corrected at the next reprinting and readers' patience and comments will be appreciated!

Some use has been made of data and diagrams from other published literature and, in addition to the individual references, the authors wish to make grateful acknowledgement to all persons and organizations concerned.

P. P. Benham
R. J. Crawford

1987

Notation

α	angle, coefficient of thermal expansion
β	angle
γ	shear strain
δ	deflection, displacement
ε	direct strain
η	efficiency, viscosity
θ	angle, angle of twist, co-ordinate
λ	lack of fit
ν	Poisson's ratio
ρ	radius of curvature, density
σ	direct stress
τ	shear stress
ϕ	angle, co-ordinate, stress function
ω	angular velocity
A	area
C	complementary energy
D	diameter
E	Young's modulus of elasticity
F	force
G	shear or rigidity modulus of elasticity, strain energy release rate
H	force
I	second moment of area, product moment of area
J	polar second moment of area
K	bulk modulus of elasticity, fatigue strength factor, stress concentration factor, stress intensity factor
L	length
M	bending moment
N	number of stress cycles, speed of rotation
P	force
Q	shear force
R	force, radius of curvature, stress ratio

S	cyclic stress
T	temperature, torque
U	strain energy
V	volume
W	weight, load
X	body force
Y	body force
Z	body force, section modulus

a	area, distance
b	breadth, distance
c	distance
d	depth, diameter
e	eccentricity, base of Napierian logarithms
g	gravitational constant
h	distance
j	number of joints
k	diameter ratio of cylinder
l	length
m	mass, modular ratio, number of members
n	number
p	pressure
q	shear flow
r	co-ordinate, radius, radius of gyration
s	length
t	thickness, time
u	displacement in the x- or r-direction
v	deflection, displacement in the y- or θ-direction, velocity
w	displacement in the z-direction, load intensity
x	co-ordinate, distance
y	co-ordinate, distance
z	co-ordinate, distance

It should be noted that a number of these symbols have also been used to denote constants in various equations.

Chapter 1

Statically-determinate force systems

INTRODUCTION

Structural and solid-body mechanics are concerned with analysing the effects of applied loads. These are *external* to the material of the structure or body and result in *internal* reacting forces, together with deformations and displacements, conforming to the principles of Newtonian mechanics. Hence a familiarity with the principles of statics, the cornerstone of which is the concept of *equilibrium of forces*, is essential.

Forces result in four basic forms of deformation or displacement of structures or solid bodies and these are *tension, compression, bending* and *twisting*.

The equilibrium conditions in these situations are discussed so that the forces may be determined for simple engineering examples.

REVISION OF STATICS

A particle is in a state of equilibrium if the resultant force and moment acting on it are zero, and hence according to Newton's law of motion it will have no acceleration and will be at rest. This hypothesis can be extended to clusters of particles that interact with each other with equal and opposite forces but have no overall resultant. Thus it is evident that solid bodies, structures, or any subdivided part, will be in equilibrium if the resultant of all external forces and moments is zero. This may be expressed mathematically in the following six equations which relate to Cartesian co-ordinate axes *x, y* and *z*.

$$\left. \begin{array}{l} \sum F_x = 0 \\ \sum F_y = 0 \\ \sum F_z = 0 \end{array} \right\} \tag{1.1}$$

1

where F_x, F_y and F_z represent the components of force vectors in the co-ordinate directions.

$$\left. \begin{array}{l} \sum M_x = 0 \\ \sum M_y = 0 \\ \sum M_z = 0 \end{array} \right\} \tag{1.2}$$

where M_x, M_y and M_z are components of moment vectors caused by the external forces acting about the axes x, y, z.

The above six equations are the necessary and sufficient conditions for equilibrium of a body.

If the forces all act in one plane, say $z = 0$, then

$$\sum F_z = \sum M_x = \sum M_y = 0$$

are automatically satisfied and the equilibrium conditions to be satisfied in a two-dimensional system are

$$\left. \begin{array}{l} \sum F_x = 0 \\ \sum F_y = 0 \\ \sum M_z = 0 \end{array} \right\} \tag{1.3}$$

Forces and moments are vector quantities and may be resolved into components, that is to say a force or a moment of a certain magnitude and direction may be replaced and exactly represented by two or more components of different magnitudes and in different directions.

Considering firstly the two-dimensional case shown in Fig. 1.1, the

Fig. 1.1

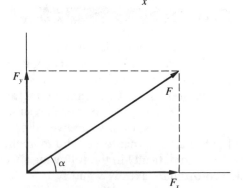

force F may be replaced by the two components F_z and F_y provided that

$$\left. \begin{array}{l} F_x = F \cos \alpha \\ F_y = F \sin \alpha \end{array} \right\} \tag{1.4}$$

Fig. 1.2

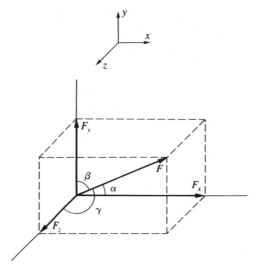

If the force F were arbitrarily oriented with respect to three axes x, y, z as in Fig. 1.2, then it could be replaced or represented by the following components:

$$\left.\begin{array}{l} F_x = F \cos \alpha \\ F_y = F \cos \beta \\ F_z = F \cos \gamma \end{array}\right\} \tag{1.5}$$

A couple or moment vector about an axis can similarly be resolved into a representative system of component vectors about other axes, as shown in Fig. 1.3 and represented by the following equations:

$$\left.\begin{array}{l} M_x = M \cos \alpha \\ M_y = M \cos \beta \\ M_z = M \cos \gamma \end{array}\right\} \tag{1.6}$$

Fig. 1.3

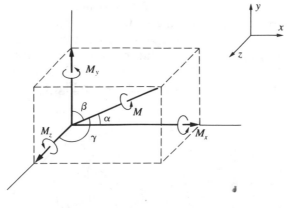

It is sometimes more convenient to replace a system of applied forces by a resultant which of course must have the same effect as those forces. Considering a two-dimensional case as illustrated in Fig. 1.4, then the most general solution is obtained by choosing any point A through which

Fig. 1.4

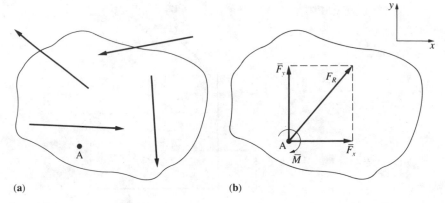

(a)　　　　　　　　　　　　　　　(b)

the resultant can act. Then the total force components in the co-ordinate directions are

$$\left.\begin{array}{l} \bar{F}_x = \sum F_x \\ \bar{F}_y = \sum F_y \end{array}\right\} \tag{1.7}$$

and the resultant force is given by

$$F_R = \sqrt{(\bar{F}_x^2 + \bar{F}_y^2)} \tag{1.8}$$

However, this is not sufficient in itself since the moment due to the forces must be represented. This is done by having a couple acting about A such that

$$\bar{M} = \sum M_z \tag{1.9}$$

In general, then, any system of forces can be replaced by a resultant force through and a couple about any chosen point.

The equivalent solution for a three-dimensional system of forces is similarly a couple and a resultant force whose direction is parallel to the axis of the couple.

One of the most useful constructions in force analysis is termed the *triangle of forces*. If a body is acted on by three forces then, for equilibrium to exist, these must act through a common point or else there will exist a couple about the point causing the body to rotate. In addition the magnitude and direction of the three force vectors must be such as to form a closed triangle as shown in Fig. 1.5.

Fig. 1.5

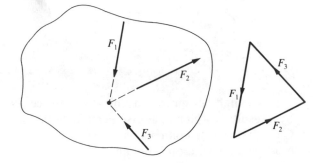

TYPES OF STRUCTURAL AND SOLID BODY COMPONENTS

Structures are made up of series of members of regular shape that have a particular function for load carrying. The shape and function are, through usage, implied in the name attached to the member. The first group is concerned with carrying loads parallel to a longitudinal axis. Examples are shown in Fig. 1.6. A member which prevents two parts of a structure from moving apart is subjected to a pull at each end, or tensile force, and is termed a *tie* (*a*). Conversely a slender member which prevents parts of a structure moving toward each other is under compressive force and is termed a *strut* (*b*). A vertical member which is perhaps not too slender and supports some of the mass of the structure is called a *column* (*c*). A *cable* (*d*) is a generally recognized term for a flexible string under tension which connects two bodies. It cannot supply resistance to bending action.

Fig. 1.6

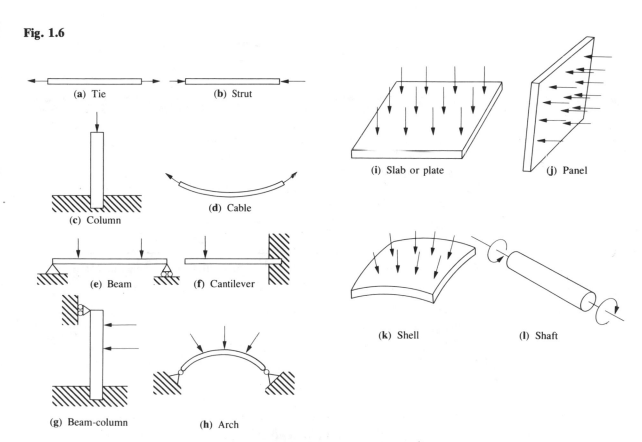

(a) Tie (b) Strut

(c) Column (d) Cable

(e) Beam (f) Cantilever

(g) Beam-column (h) Arch

(i) Slab or plate (j) Panel

(k) Shell (l) Shaft

One of the most important of structural members is that which is frequently supported horizontally and carries transverse loading. This is known as a *beam* (*e*), a common special case of which is termed a *cantilever* (*f*), where one end is fixed and provides all the necessary support. A *beam-column* (*g*), as the name implies, combines the separate functions already described. The *arch* (*h*) has the same function as the beam or beam-column, but is curved in shape.

The filling-in and carrying of load over an area or space are achieved by *flat slabs* or *plates* (*i*), by *panels* (*j*) and also by *shells* (*k*), which are curved versions of the former.

The transmission of torque and twist is achieved through a member which is frequently termed a *shaft* (*l*).

The members described above can have a variety of cross-sectional shapes depending on the particular type of loading to be carried. Some typical cross-sections are illustrated in Fig. 1.7. They are made generally by rolling steel and rolling or extruding aluminium alloy.

Fig. 1.7

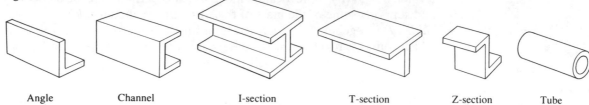

| Angle | Channel | I-section | T-section | Z-section | Tube |

TYPES OF SUPPORT AND CONNECTION FOR STRUCTURAL COMPONENTS

The applied loading on a framework, beam or column is transmitted to the supports which will provide the required reacting forces to maintain overall equilibrium. Examples of supports of various kinds suitable to react to loading in a plane (two dimensions) are shown in Fig. 1.8. In the accompanying table the possible displacement and reacting force components are indicated.

	Displacement			Reacting force		
	x	y	θ	R_x	R_y	M_z
(*a*) Built-in or fixed support				$\checkmark$	$\checkmark$	$\checkmark$
(*b*) Pin connection			$\checkmark$	$\checkmark$	$\checkmark$	
(*c*) Roller support	$\checkmark$		$\checkmark$		$\checkmark$	
(*d*) Sliding support		$\checkmark$		$\checkmark$		$\checkmark$

The separate members of a structural framework are joined together by bolting, riveting or welding, two examples of which are shown in Fig. 1.9. Now, if these joints were ideally stiff, when the members of the framework were deformed under load, the angles between the members at the joint would not change. This would also imply that the joint was capable of transmitting a couple. However, calculations for a complete structure on this basis would become rather involved and tedious.

It is found in practice that there is some degree of rotation between members at a joint due to the elasticity of the system. Furthermore, it has

Fig. 1.8

Type of support   Equivalent force system

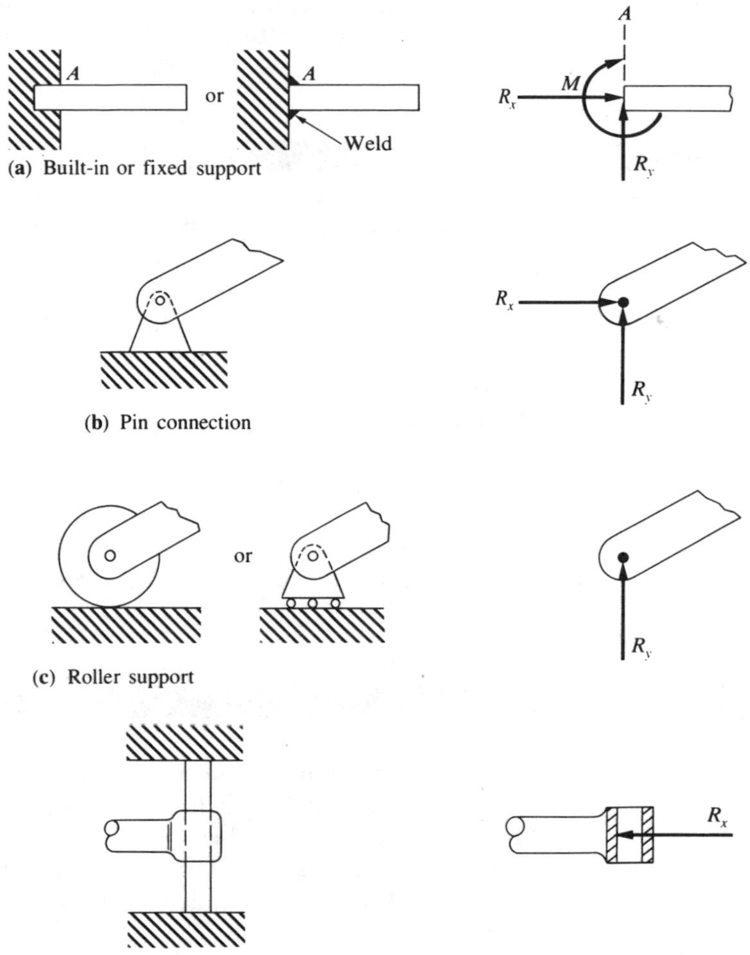

(a) Built-in or fixed support

(b) Pin connection

(c) Roller support

(d) Sliding support

been shown that it is not unreasonable, for purposes of calculation, to assume that these joints may be represented by a simple ball and socket or pin in a hole. Even with this arrangement, which of course cannot transmit a couple or bending moment (other than by friction which is ignored), deformations of the members are relatively small. Consequently changes in angle at the joints are also small, which is why this

Fig. 1.9

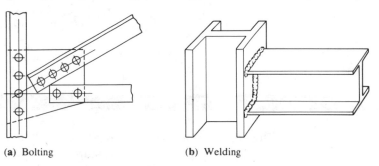

(a) Bolting   (b) Welding

approximation is not unreasonable when applied to the actual joints. Thus it is common practice when calculating the forces in the members of a framework to assume that all joints are pinned.

STATICAL DETERMINACY

In general, structural or solid body mechanics involves determination of unknown forces on the structure or body. The approach taken in this analysis depends initially on whether the system under consideration is "statically determinate" or "statically indeterminate". The former condition requires that the number of equations available from statements of equilibrium is the same as the number of unknown forces (including reactions). If the number of unknown reactions or internal forces in the structure or component is greater than the number of equilibrium equations available, then the problem is said to be *statically indeterminate*. Additional equations have to be found by considering the displacement or deformation of the body.

The above statements are quite general and apply throughout this text.

FREE-BODY DIAGRAMS

When commencing to analyse any force system acting on a component or structure it is essential firstly to have a diagram showing the forces acting. If the structure or part of it is separated from its surroundings and the appropriate reactions, required to maintain equilibrium, are inserted then a diagram of this system is called a *free-body* diagram. Examples of this are shown in Fig. 1.10.

Fig. 1.10

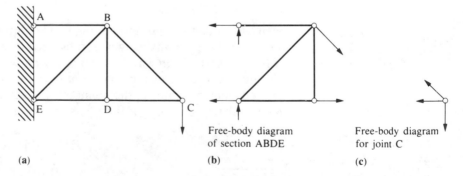

(a)

Free-body diagram of section ABDE

(b)

Free-body diagram for joint C

(c)

DETERMINACY CRITERIA FOR STRUCTURES

The principles of statical determinacy will now be considered in relation to plane and space-frame structures. There are three classes of frame or truss, in concept, although one is not of practical interest:

(*a*) *Under-stiff.* If there are more equilibrium equations than unknown forces or reactions the system is unstable and is not a structure but a mechanism.

(*b*) *Just-stiff.* This is the *statically-determinate* case for which there are the same number of equilibrium equations as unknown forces. If any member is removed then a part or the whole of the frame will collapse.

(*c*) *Over-stiff.* This is the *statically-indeterminate* case in which there are more unknown forces than available equilibrium equations. There is at least one member more than is required for the frame to be just stiff.

Figure 1.11 shows an example of each of the three conditions. Remembering that each joint is pinned, then it is clear that in (*a*) the central square of members is not stable unless there is one diagonal member inserted. The number and arrangement of the members in case (*b*) is quite correct for the "just-stiff" or statically-determinate condition. In case (*c*) in contrast to (*a*) the central square has two diagonal braces one or other of which is unnecessary and hence the frame is "over-stiff" or statically indeterminate.

Fig. 1.11

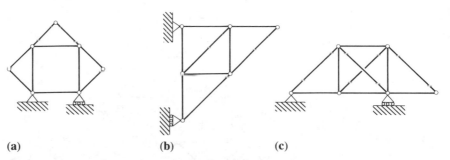

(a)   (b)   (c)

It is useful to express the three cases in the form of mathematical criteria. Let the number of joints, including support points, in a frame be j, the number of members m, and the number of reactions r. Now, for a space frame there are three equilibrium equations applicable to each joint, namely $\sum F_x = 0$, $\sum F_y = 0$, $\sum F_z = 0$; hence there are $3j$ equations to determine $m + r$ unknown forces and reactions, and the statically-determinate case is represented by

$$m + r = 3j \qquad (1.10)$$

When $m + r < 3j$ the members form a mechanism, and for $m + r > 3j$ the frame is over-stiff, or redundant, and therefore statically indeterminate.

There are six conditions for overall equilibrium; therefore the minimum value for r, when using the above criteria to allow for any general loading system, is six. Certain arrangements of loading may result in less than six reactions being required.

For frames lying only in one plane there are only two equilibrium equations at each joint and so the relationships comparable with the above are

$$\left.\begin{array}{l} m + r < 2j \\ m + r = 2j \\ m + r > 2j \end{array}\right\} \qquad (1.11)$$

The minimum value for r in these expressions must be three for general forms of loading, but may be less under certain conditions.

The above criterion for a just-stiff frame is a necessary but not a sufficient condition, since the *arrangement* of the members might still not provide the required stiffness.

EXAMPLE 1.1

Examine the plane frames illustrated in Fig. 1.12. State the class of each and where members should be inserted or removed to make each statically determinate. Also indicate any redundant reactions.

Fig. 1.12

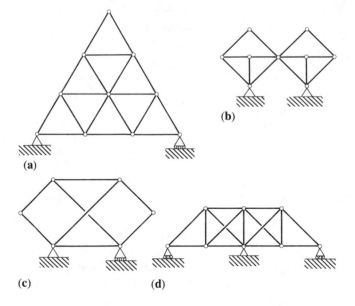

(a)

(b)

(c)   (d)

SOLUTION

(a) $m = 18$, $r = 3$, $j = 10$: hence $m + r > 2j$ and the frame is overstiff or redundant. Any member may be removed from the central hexagon structure. None of the members may be removed from the apexes.

(b) $m = 14$, $r = 4$, $j = 9$: hence $m + r = 2j$ and the frame is just stiff and statically determinate.

(c) $m = 8$, $r = 3$, $j = 6$: hence $m + r < 2j$, which constitutes a mechanism. To make statically determinate insert a member between any pair of unconnected joints.

(d) $m = 15$, $r = 4$, $j = 8$: hence $m + r > 2j$, which is redundant both in members and reactions. Hence remove either the left or right support and a diagonal member from each square.

DETERMINATION OF AXIAL FORCES BY EQUILIBRIUM STATEMENTS

PLANE PIN-JOINTED FRAMES

The members of pin-jointed plane or space frameworks can only carry axial forces and these may be determined by considering the equilibrium of various parts of the structure as "free bodies".

Equilibrium at joints

At any joint in a plane frame two equilibrium equations apply, namely

$$\sum F_x = 0 \quad \text{and} \quad \sum F_y = 0 \text{ (for a frame lying in the } z\text{-plane)}$$

hence only two unknown forces can be determined. The method therefore entails making a free-body diagram centred on each joint at where there *only two* unknowns. The forces are then resolved in the x and y directions so that the above two equations can be applied. It will probably be necessary at first to determine support reactions by considering equilibrium of the whole frame. This generally gives at least one joint where there is a known reaction and only two members having unknown forces from which to start the analysis. The method will be illustrated by the following example.

EXAMPLE 1.2

Determine the magnitude and the type of force in each member of the plane pin-jointed frame shown in Fig. 1.13.

Fig. 1.13

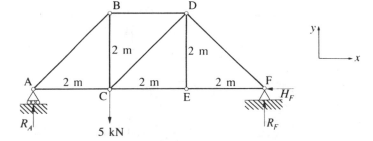

SOLUTION

In Fig. 1.13, in order to save the space and having to draw the complete frame twice, once on its supports and a second time as a free body away from the supports with the reactions in place, the possible reactions have been placed directly on to the configuration diagram.

Considering the equilibrium of the whole frame and taking moments about joint F,

$$\sum M_z = 6R_A - 5000 \times 4 = 0 \quad \text{so that} \quad R_A = 3333 \text{ N}$$

For horizontal equilibrium,

$$\sum F_x = 0 \quad \text{gives} \quad H_F = 0$$

For vertical equilibrium,

$$\sum F_y = 0 \quad \text{gives} \quad R_A + R_F - 5000 = 0$$

Hence

$$R_F = 1667 \, \text{N}$$

Fig. 1.14

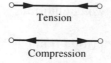

Tension

Compression

The only joints at which there are two unknowns are A and F. Start say at A and *assume initially that each member is subjected to tension which is designated as positive* and indicated by arrowheads as shown in Fig. 1.14.

Joint A   The free-body diagram for this joint is as shown in Fig. 1.15.

Fig. 1.15

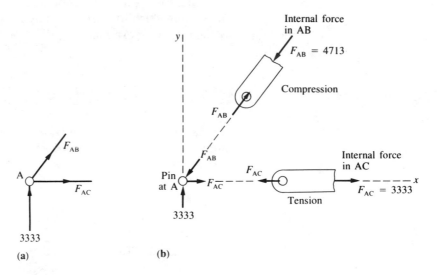

(a)   (b)

$$\sum F_y = 0 \quad \text{gives} \quad F_{AB} \sin 45° + 3333 = 0$$

Hence

$$F_{AB} = -4713 \, \text{N}$$

The negative sign shows that in this case the wrong assumption was made and that AB is in compression and not tension.

$$\sum F_x = 0 \quad \text{gives} \quad F_{AB} \cos 45° + F_{AC} = 0$$

Hence

$$F_{AC} = 4713 \cos 45° = 3333 \, \text{N}$$

The next step *has* to be at joint B (rather than C).

Joint B
(Fig. 1.16)

$$\sum F_y = F_{BA} \cos 45° + F_{BC} = 0$$

$$F_{BC} = 4713 \cos 45° = 3333 \text{ N}$$

$$\sum F_x = F_{BD} - F_{BA} \cos 45° = 0$$

$$F_{BD} = -4713 \cos 45° = -3333 \text{ N} \quad \text{(compression)}$$

Fig. 1.16

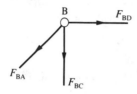

Joint C
(Fig. 1.17)

These are now only two unknowns at this joint, F_{CD} and F_{CE}.

$$\sum F_y = F_{CB} - 5000 + F_{CD} \cos 45° = 0$$

$$F_{CD} = (-3333 + 5000)\sqrt{2} = 2356 \text{ N}$$

$$\sum F_x = F_{CA} - F_{CE} - F_{CD} \cos 45° = 0$$

$$F_{CE} = 3333 - 2356 \cos 45° = 1667 \text{ N}$$

Fig. 1.17

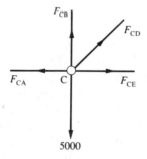

Continuing the above process for joints E and D gives

$$F_{ED} = 0 \qquad F_{EF} = 1667 \text{ N} \qquad F_{DF} = 2356 \text{ N}$$

The final force distribution in the frame is shown in Fig. 1.18.

Fig. 1.18

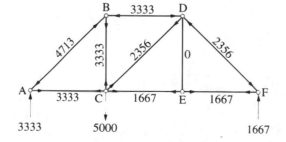

As the force in DE is zero one might ask whether that member is required. With this particular applied load, in fact, DE could be removed quite safely, since the two members CE and EF really act as one continuous member CEF in tension. However, if the applied load at C acted upwards the loads in all the members would have the same magnitude as before but would be reversed in sense and then CE and EF would be in compression. While ED was in position even without any force in it, it would keep CE and EF in line, whereas if ED were removed the joint E would move vertically due to the slightest misalignment and result in collapse.

Equilibrium of frame sections

The next step from a joint as a free body in equilibrium is to consider the equilibrium of a single member or group of members. This may have advantages in effort compared with analysing the equilibrium at every joint. In Fig. 1.19 cutting along the line XX provides either a free

Fig. 1.19

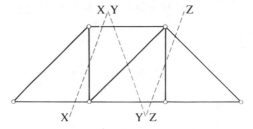

body of member AB to the left of the section or a free body of the remainder of the frame to the right of the section. Section YY splits the frame into two parts either of which could be regarded as a free body the equilibrium of which would give the forces in BD, CD and CE. In isolating part of the frame by a section it must be remembered that there are only three equilibrium equations that can be written ($\sum F_x = \sum F_y = \sum M_z = 0$); hence only three unknown forces can be found for a particular "section" and free body. Considering section XX then, as shown in Fig. 1.20, insert a force vector at the cut end of each member in the tensile sense.

Fig. 1.20

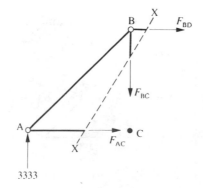

Equilibrium for the free body requires that

$$\sum F_x = 0 \qquad \sum F_y = 0 \qquad \sum M_z = 0$$

Hence

$$\sum F_y = 3333 - F_{BC} = 0$$

$$F_{BC} = 3333 \text{ N}$$

Moments about A give

$$\sum M_z = 2F_{BD} + 2F_{BC} = 0$$

$$F_{BD} = -3333 \text{ N} \quad \text{(compression)}$$

Finally,

$$\sum F_x = F_{BD} + F_{AC} = 0$$

$$F_{AC} = 3333 \text{ N}$$

The same results would have been obtained if the three equilibrium equations had been written for the remainder of the frame to the right of section XX.

OTHER SOLID BODIES AND STRUCTURES

The principle of sectioning off a "free body" and writing equilibrium statements, which was demonstrated for plane pin-jointed frames, is equally applicable in other engineering problems. Consider for example the front end of the fork-lift truck shown in Fig. 1.21(a). The individual elements of this may be analysed quite easily as illustrated in Fig. 1.21(b) so that each may be designed to withstand the forces acting on it.

Fig. 1.21 Free-body diagrams of parts of a fork-lift mechanism

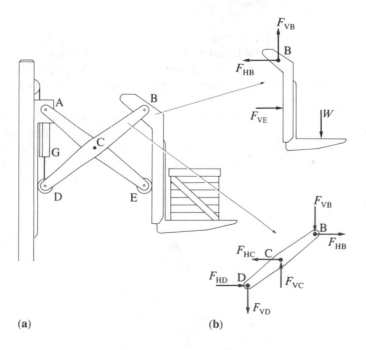

(a)

(b)

BENDING OF SLENDER MEMBERS

In the force analysis of frameworks the members were only subjected to axial force, namely tension or compression. The next step is to consider the effect of transverse loads acting on slender members. The deformation that results is termed *bending* and is of course very common in structures and machines – floor joists, railway axles, aeroplane wings, leaf springs, etc. External applied loads which cause bending give rise to internal reacting forces. These have to be determined before it is possible to calculate stress and deflection.

The transverse externally applied load on a beam or bar can take one of two forms, concentrated or distributed. The former is illustrated in Fig. 1.22(*a*) in which the load acts on the surface of the beam along a line perpendicular to the longitudinal axis. This, of course, is an idealization, and in practice a concentrated load will cover a very short length of the beam.

Fig. 1.22 (a) Point load; (b) distributed load

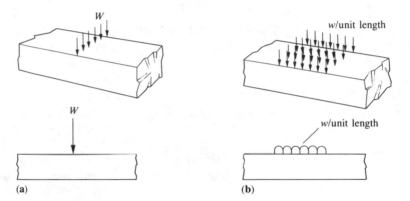

(a)   (b)

A load which is distributed is shown in Fig. 1.22(*b*) and occupies a length of beam surface. The load intensity is always taken as constant across the beam thickness but may be uniformly or non-uniformly distributed along part or the whole length of the beam. In practice the particular conditions of force and displacement at beam supports may vary considerably. Theoretical solutions of beam problems generally employ two simplified forms of support. These are termed *simply supported* and *built-in* or *fixed*. The former is illustrated in Fig. 1.23(*a*) in which the beam rests on knife-edges or rollers. When the beam is bent

Fig. 1.23 (a) Simple supports; (b) built-in support

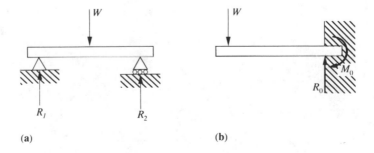

(a)   (b)

under load the support reaction is only a transverse force and there is no restraining couple. Hence the deflection at the support is zero and the beam is free to take up a slope dictated by the applied load. The built-in support shown in Fig. 1.23(*b*) reacts with a transverse force and a couple. Thus both deflection and slope are fully restrained. The particular example illustrated of a beam built-in at one end and free at the other is termed a *cantilever*.

The number and type of supports also has a further important bearing on a beam solution by making it either *statically determinate* or *statically indeterminate* (Fig. 1.24). In the former the support reactions can be found simply from force and moment equilibrium equations. This applies, for example, to beams on two simple supports or one built-in support and no other. The two equilibrium equations are insufficient to find the reactions at the supports of a statically indeterminate beam owing to the presence of redundant forces. In this case it is necessary to consider also the deflection of the beam in order to obtain additional equations to solve for the reactions.

Fig. 1.24 (a) Statically determinate; (b) statically indeterminate

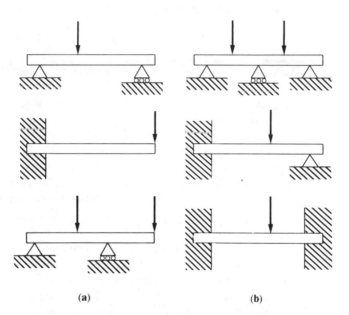

(a)   (b)

FORCE AND MOMENT EQUILIBRIUM DURING BENDING

A slender curved bar is shown in Fig. 1.25(*a*) subjected to various transverse loads. In general, to maintain equilibrium a force and a couple will be required at each end of the bar. The force can be resolved into two components, one perpendicular and the other parallel to each end cross-section. The internal forces are obtained by cutting the bar to give two free bodies and inserting at the cut sections the necessary forces and moments for equilibrium, as shown in Fig. 1.25(*b*). The couple, *M*, is

Fig. 1.25

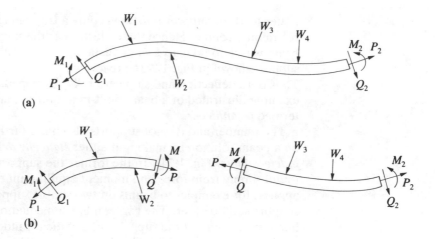

(a)

(b)

termed a *bending moment,* the transverse force Q, is called a *shear force,* and P is a *longitudinal force*. The most important stresses and deflections that occur during bending are due to M, rather than Q or P.

SIGN CONVENTION FOR BENDING

It is important that the analysis of internal forces in bending shall be consistent within and between various problems. This is best achieved by adopting a sign convention for loading, bending moment, shear force and distance/length. There is no standardized convention for bending, although there tends to be "common practice" in, for example, civil engineering structural design. There is a fair degree of uniformity amongst textbooks and the convention used in this text (as in previous editions) follows the general pattern. It is illustrated for *positive values* in Fig. 1.26. There are several points worth noting, primarily that it does not follow rigorous mathematical convention for positive directions. For example loading, W and w, are taken positive downwards since for most horizontal members applied loading is in that sense. (If loading *upwards* was taken as positive there would be a profusion of negative signs.)

Distance, x is *always* taken positive from left to right along the beam. The choice of y positive downwards is again largely convenience and practice because most deflections tend to be downwards.

Fig. 1.26

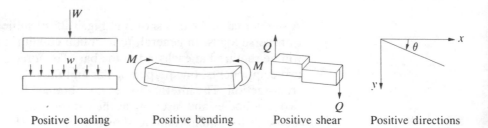

Positive loading   Positive bending   Positive shear   Positive directions

It is very important to remember that definitions of bending moment and shear force are illustrated as vector *pairs* in the positive (or negative) sense. For example in the case of shear force, Q, the right-hand vector is downwards (in the positive direction of y) but the left-hand vector although upwards constitutes part of the positive *pair*. A similar argument applies to the moment vectors, M.

The most important point to remember is *not* to change the sign convention (for some apparent convenience) in the middle of a problem solution.

SHEAR-FORCE AND BENDING-MOMENT DIAGRAMS

Both stresses and deflections during bending are directly related to S.F. (shear force) and B.M. (bending moment); it is therefore desirable to know the distribution along the member and hence where maximum or minimum values occur. To this end S.F. and B.M. are computed for a number of cross-sections along the beam and diagrams plotted showing the distribution and magnitude. A few basic examples now follow and further illustrations of shear force and bending moment distributions appear at the start of Chapter 6 on bending stresses.

CANTILEVER CARRYING A CONCENTRATED LOAD

This example is illustrated in Fig. 1.27. The first step is to choose any section XX at a distance x from the left-hand end, cut the beam at this position and draw the free-body diagram, inserting the required internal forces and moments in the positive sense according to the sign convention. It is advisable not to balance mentally the free body and put on the forces in what seems the "right" sense.

Fig. 1.27

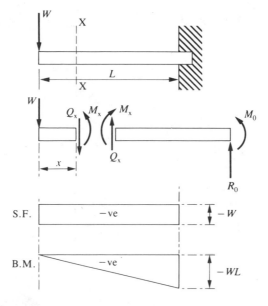

Shear force   Vertical equilibrium for the left-hand free body gives

$$W + Q_X = 0$$

where Q_X denotes the shearing force at a distance x from the left end; hence

$$Q_X = -W$$

It is fairly obvious that this value of shear force is obtained at whatever distance the beam is cut between $x = 0$ and $x = L$.

Bending moment   Taking moments about the mid-point of section XX will give a moment equilibrium equation for the left-hand free body

$$Wx + M_X = 0 \quad \text{or} \quad M_X = -Wx$$

This is a linear variation from $M = 0$ at $x = 0$ to $M = -WL$ at $x = L$, as shown in the diagram.

    The reactions at the built-in end are, from the above,

$$R_0 = W \quad \text{and} \quad M_0 = -WL$$

These could of course be found at the beginning simply from considering overall equilibrium of the beam.

    Figure 1.27 illustrates the S.F. and B.M. diagrams which, in this example, are in the negative area below each zero base line.

    The right-hand free body illustrates the equal and opposite force Q_X and moment M_X vectors at the cut section XX.

CANTILEVER CARRYING A UNIFORMLY-DISTRIBUTED LOAD

Shear force   Following the above procedure, vertical equilibrium of the free body in Fig. 1.28 is obtained as

$$wx + Q_X = 0 \quad \text{so that} \quad Q_X = -wx$$

Fig. 1.28

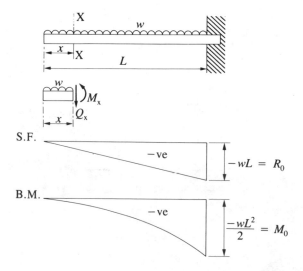

20

The shear-force diagram is therefore a linear variation from 0 to $-wL$ as shown in the figure.

Bending moment The moment equilibrium equation is obtained by taking moments about XX; thus

$$wx\frac{x}{2} + M_X = 0 \quad \text{or} \quad M_X = -\frac{wx^2}{2}$$

which gives a parabolic shape of B.M. diagram varying from $M = 0$ at $x = 0$ to $M = -wL^2/2$ at $x = L$.

The support reactions are therefore

$$R_0 = wL \quad \text{and} \quad M_0 = -\frac{wL^2}{2}$$

SIMPLY SUPPORTED BEAM CARRYING A UNIFORMLY-DISTRIBUTED LOAD

In this case, Fig. 1.29, the reaction at the left-hand end has to be determined before the values of S.F. can be expressed. From vertical equilibrium and symmetry of the whole beam,

$$R_1 = R_2 = +\frac{wL}{2}$$

Fig. 1.29

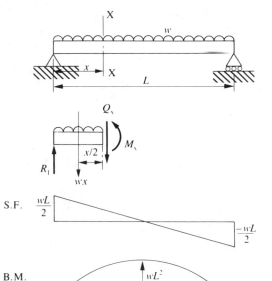

Shear force For the equilibrium of the free-body section,

$$+Q_X - \frac{wL}{2} + wx = 0$$

$$Q_X = -wx + \frac{wL}{2}$$

This equations show that the S.F. is zero at mid-span and equal to the reactions, $wL/2$, at $x = 0$ and L.

Bending moment Moment equilibrium gives

$$M_X - \frac{wL}{2}x + wx\frac{x}{2} = 0$$

$$M_X = \frac{wLx}{2} - \frac{wx^2}{2}$$

This gives a diagram which is parabolic, being zero at each support and having a maximum value of $wL^2/8$ at $x = L/2$.

EXAMPLE 1.3

Sketch the S.F. and B.M. diagrams for the simply-supported beam shown in Fig. 1.30.

SOLUTION

$$-R_A - R_B + 5000 + 10\,000 = 0$$

$$-5R_A + (5000 \times 4) - 5000 \times 2 \times 1 = 0$$

$$R_A = 2000\,\text{N} \qquad R_B = 13\,000\,\text{N}$$

Fig. 1.30

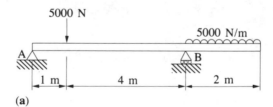

(a)

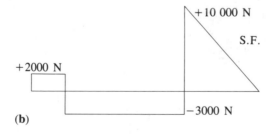

(b)

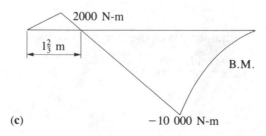

(c)

Shear force	$0 < x < 1$	$+Q - 2000 = 0$ $Q = +2000$ N

$1 < x < 5$ $+Q - 2000 + 5000 = 0$ $Q = -3000$ N

$5 < x < 7$ $+Q - 2000 + 5000 - 13\,000 + 5000(x - 5) = 0$
$Q = -5000x + 35\,000$ N

Bending moment $0 < x < 1$ $-2000x + M = 0$ $M = 2000x$ N-m

$1 < x < 5$ $-2000x + 5000(x - 1) + M = 0$

$M = 5000 - 3000x$ N-m

$5 < x < 7$ $-2000x + 5000(x - 1) - 13\,000(x - 5)$

$+ 5000(x - 5)\dfrac{(x - 5)}{2} + M = 0$

$$M = -2500x^2 + 35\,000x - 122\,500 \text{ N-m}$$

The S.F. and B.M. diagrams are shown in Fig. 1.30(*b*) and (*c*).

POINT OF CONTRAFLEXURE

In the above example it will be seen that the B.M. changes sign through a zero value and this is termed a *point of contraflexure*. Its position is determined by putting the bending moment expression equal to zero.

TORSION OF MEMBERS

Torsion is the engineering word used to describe the process of twisting a member about its longitudinal axis as illustrated in Fig. 1.31. The angle of rotation of one section relative to another at a distance *l* along the member is termed the angle of twist θ, measured in radians. The twist per unit length is thus θ/l. The forces required to cause this twist are shown at each section A and B as *F*, equal in magnitude and opposite in sense acting equidistant from the central axis of the member. The

Fig. 1.31

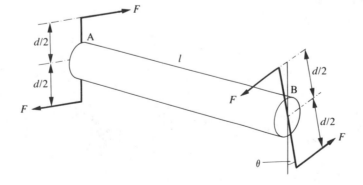

product of the forces times the distance to the axis is Fd and this couple about the axis is called a *torque T*. For equilibrium of the member the torque at each end must be equal in magnitude and opposite in sense acting about the longitudinal axis, and the torque is constant between these two sections.

The situation shown in Fig. 1.31 is described as pure torsion, i.e. the member is only subjected to torque and no other resultant lateral or longitudinal forces exist. However, in practice applied forces which set up torque frequently also induce bending and possibly end load on the component.

MEMBERS SUBJECTED TO AXIAL FORCE, BENDING MOMENT AND TORQUE

This situation was referred to at the end of the previous section and will now be considered in more detail.

Let us take a practical example with which many people have been personally involved, that of changing a wheel on a car. The wheel nuts have to be removed with a wheel brace or wrench. There are, of course, various designs to achieve the same objective which is to be able to apply enough torque to the nuts to loosen them and at a later stage to retighten them. A typical wheel brace is shown in Fig. 1.32 in which both hands apply force, one in the direction necessary to loosen the nut and the other as a reacting support. The torque set up on the nut is simply the force F times the perpendicular distance between the plane in which the force is acting and the plane containing the nut. In addition, the brace is acting as a beam with a kink in it, i.e. simply-supported at each end with a "concentrated" load at the centre. In fact each section of the brace will have to transmit some combination of bending and torsion. In this particular case once the brace has been pushed on to the nut there will not thereafter be any axial force along the axis of the brace. To calculate the bending moment, shearing force and torque in each section of the brace, free-body diagrams must be drawn. If these are done correctly it is then a simple matter to determine the equilibrium conditions for each section. The following example will illustrate the method for slightly

Fig. 1.32

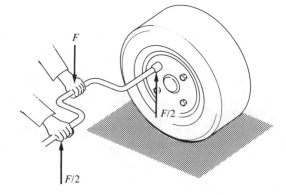

more complex internal reactions produced by the application of a single force.

EXAMPLE 1.5

A length of pipe ABCD is connected rigidly to a pump unit at A and is bent through two right angles at B and C as shown in Fig. 1.33. A force F has to be applied at D in order to connect the pipe to the next unit which is slightly out of alignment. Determine the system of forces, moments and torques which are set up in the pipe in relation to the co-ordinate axes given. AB is in the x-direction, BC in the z-direction and CD in the y-direction and the force F lies in the xz-plane.

Fig. 1.33

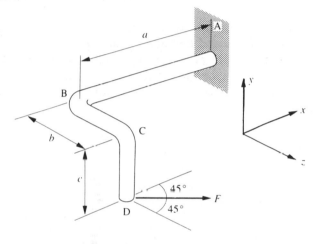

The first step is to resolve the force into the x- and z-directions, giving in

Fig. 1.34

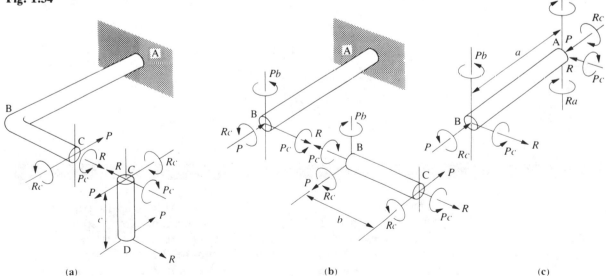

(a) (b) (c)

each case $F \cos 45$ or $F/\sqrt{2}$. To avoid confusion we shall temporarily call these P and R respectively. The next step is to draw the series of free-body diagrams as shown in Fig. 1.34, which should be self-explanatory, commencing with CD and successively following on with the segments CB and BA.

It is seen that in each free body the transfer of the forces P and R to the next end of the segment requires a moment and/or torque to maintain equilibrium about the respective axes. This process is an *essential* first step towards the analysis of bending and torsional stresses and deflections which will be derived in Chapters 5, 6 and 7.

TORQUE DIAGRAM

In the analysis of torsion problems there is an analogous diagram to the shear-force and bending-moment diagrams for bending and this is a torque diagram. An example is shown in Fig. 1.35, which illustrates the variation in torque which may occur along a shaft depending on the various inputs and outputs of power transmission.

Fig. 1.35

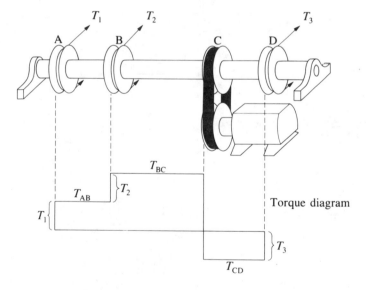

THE PRINCIPLE OF SUPERPOSITION

Consider the beam problem in Fig. 1.36(*a*). Although this would not be difficult to analyse for B.M. in its present form, imagine it split up into the two separate cases as in Fig. 1.36(*b*). These have effectively been dealt with on pages 19 and 21 respectively, and the B.M. diagrams for the two parts are as shown. Now, bending moment is always a first-order

Fig. 1.36

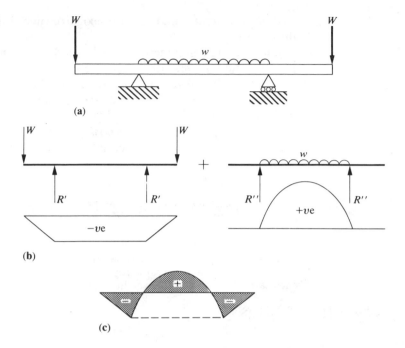

function of applied load, i.e. there are never terms in W^2 or w^2 etc.; therefore at any section along the beam the sum of the bending moments due to the loads acting separately would be exactly the same as the bending moment due to the combined load. Similarly the reactions at the supports due to the separate loads may be summed to give the reactions caused by the combined loading. The reader may also like to verify that the S.F. diagram could be derived in exactly the same manner.

The solution of the present example is obtained by summing the ordinates of bending moment at every section along the beam. However, this may be done more simply graphically than algebraically by placing the two B.M. diagrams together as shown in Fig. 1.36(c) and the resultant is obtained as the shaded area.

This most important technique is known as the *principle of superposition*. It can be used only in relation to the same types of function which have linear characteristics. It will be demonstrated at regular stages throughout subsequent chapters when analysing force, stress, strain and displacement systems.

SUMMARY

This chapter has served several purposes, the primary one being to emphasize the importance and give demonstrations of the development of *equilibrium situations* in solid bodies and structures. To this end it is also vital to be able to draw representative *free-body diagrams* with all the required force components. An appreciation of the meaning of *statical determinacy* and how to recognize such situations is also most important.

Engineering components are subjected largely to *tension, compression, bending* and *torsion,* either separately or in a variety of combinations. It is essential, therefore, to understand how these various modes of deformation arise and the associated equilibrium requirements for internal and applied forces.

Finally the *principle of superposition* can be a most valuable means of solution, as will be seen in succeeding chapters.

PROBLEMS

1.1   A maintenance cradle in a factory is supported by the pulley system in Fig. 1.37. A fitter weighing 82 kg can be raised (*a*) by pulling on the rope R himself or (*b*) by someone else pulling on the rope. Calculate the pull needed on the rope in each case. Friction and the weight of the cradle and pulleys may be ignored.

Fig. 1.37

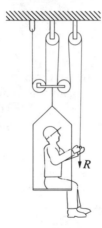

1.2   Figure 1.38 shows a typical design for a vehicle weighbridge. Each of the levers A, B and C have *a* : *b* in the ratio of 1 : 10. If a balance load of 12 N is required on the lever A then calculate the weight of the vehicle.

Fig. 1.38

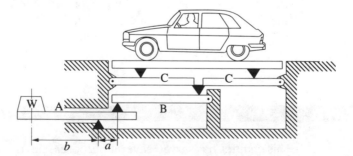

1.3   A certain design of brake drum is as shown in Fig. 1.39. If the drum rotates anticlockwise and generates a torque of 210 Nm, calculate the force needed in the cylinder C in order to stop the drum. The coefficient of friction between the drum and shoe is 0.35.

Fig. 1.39

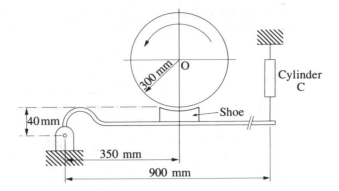

1.4   A 4 m wide gate for controlling water levels is shown in Fig. 1.40. For the levels shown calculate the magnitude and direction of the force in the hydraulic cylinder A. The density of the water is 1000 kg/m³.

Fig. 1.40

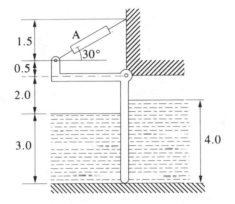

1.5   A special type of bicycle calliper brake is shown in Fig. 1.41. If it is pin-jointed at A, B, C and D and the cyclist applies a force of 300 N at A when the bicycle is stationary on a 1 in 7 slope, calculate whether or not the bicycle would move down the slope. The combined weight of the bicycle and cyclist is 102 kg and the

Fig. 1.41

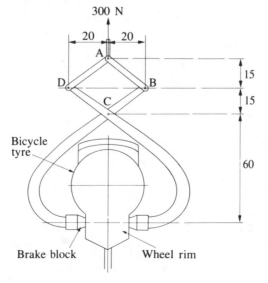

coefficient of friction between the wheel and the brake block is 0·7. The outside diameter of the tyre is 690 mm and the diameter of the line of action of the brake block is 640 mm.

1.6 A motor cycle is standing on horizontal ground with its front wheel in contact with the kerb as shown in Fig. 1.42. When the rider slowly engages gear the motor cycle starts to rise over the kerb when the rear wheel torque reaches 0.3 kNm. For the instant when the front wheel just starts to leave the road calculate (*a*) the value of the coefficient of friction between the tyre and the road to avoid wheel slip, and (*b*) the magnitude and direction of the force on the front wheel. The wheel diameters are both 600 mm and the combined weight of the rider and the motor cycle is 285 kg acting through a point midway between the wheels.

Fig. 1.42

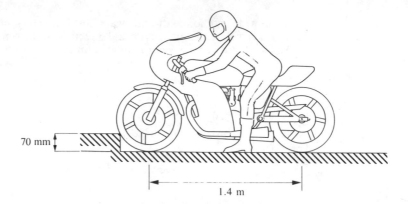

70 mm

1.4 m

1.7 At the moment of start-up, a power hacksaw is in the configuration shown in Fig. 1.43. If the torque input at the pulley drive is 150 N-m calculate the magnitude and direction of the force on the workpiece. The weight of the blade saddle is 45·9 kg and the weight of the support arm may be ignored. Friction between the blade saddle and the support arm may also be ignored.

Fig. 1.43

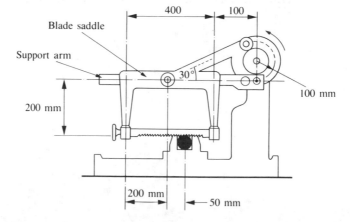

400

100

Blade saddle

Support arm

30°

100 mm

200 mm

200 mm

50 mm

1.8 When the hydraulically-operated digger shown in Fig. 1.44 is removing earth the force exerted by the bucket is horizontal with a magnitude of 15 kN. For the arm positions indicated, calculate the force in each of the three rams A, B and C. The weight of the arms of the digger should be ignored.

Fig. 1.44

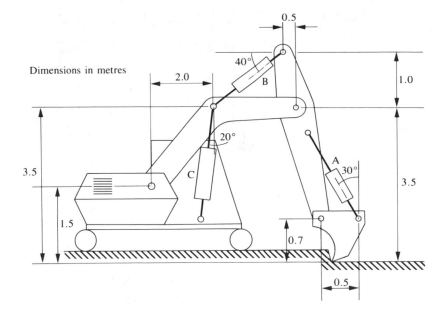

Dimensions in metres

1.9 In a fork-lift truck of the type shown in Fig. 1.45 the load of 3 kN is raised by means of the rotating screw AC. If the weight of the front fork is 102 kg (acting through its centroid G) and the weights of the arms AB and CD are 51 kg each, calculate the force in the screw AC and the reactions at A for the configuration shown.

Fig. 1.45

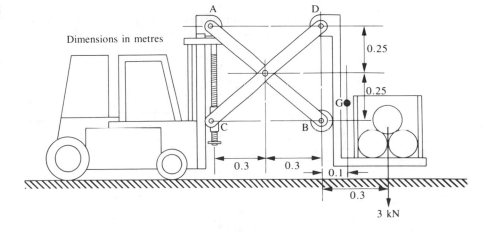

Dimensions in metres

1.10 The framework shown in Fig. 1.46 is used to support a steel car body weighing 200 kg. When the car body is suspended in (*a*) air and (*b*) totally immersed in a plating bath containing a liquid of density 1000 kg/m³, calculate the support reactions and the forces in members AB, BF and FE. Density of steel = 7·8 Mg/m³.

Fig. 1.46

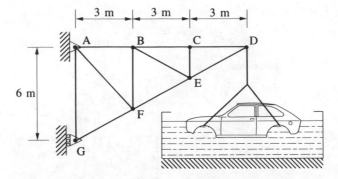

1.11   A dockyard crane may be considered to be a plane pin-jointed frame as shown in Fig. 1.47. Determine the support reactions and the forces in the members of the framework by (*a*) resolution at joints, (*b*) graphical construction.

Fig. 1.47

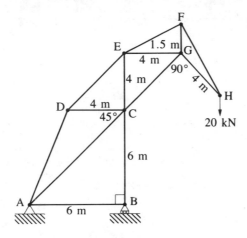

1.12   A crane with the dimensions shown in Fig. 1.48(*a*) supports a weight of 1000 kg. Use the method of sections to calculate the forces in members HC and HJ. If an

Fig. 1.48

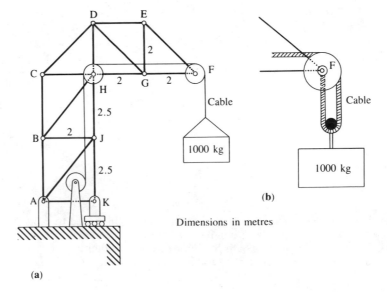

Dimensions in metres

(**a**)

alternative design is used as in (*b*) in which the cable is attached to the frame at F, calculate the force which this would cause in HJ.

1.13 Sketch the shear-force and bending-moment diagrams, inserting the principal numerical values for the beams illustrated in Fig. 1.49. Also establish any positions of contraflexure.

Fig. 1.49

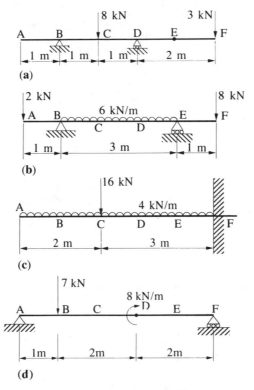

1.14 A 10 m × 6 m advertisement board is supported by three posts as in Fig. 1.50. Construct the shear-force and bending-moment diagrams for the vertical posts if the wind loading on the board is 981 N/m².

Fig. 1.50

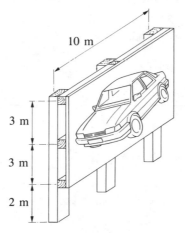

1.15 The bracket ABC is freely pivoted on the vertical rod shown in Fig. 1.51. Determine the forces, bending moments and torque transmitted on all parts of

Fig. 1.51

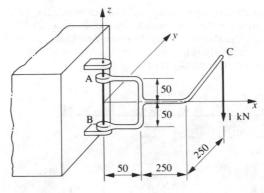

Dimensions in mm

the bracket when a 1 kN vertical force is applied at C. Also calculate the reactions on the wall at A and B.

1.16 A power transmission system is illustrated in Fig. 1.52. The belt drives at A, C and D require 15 kW, 25 kW and 10 kW of power respectively at a shaft speed of 500 rev/min. Draw a diagram to show the torque in each section of the drive shaft.

Fig. 1.52

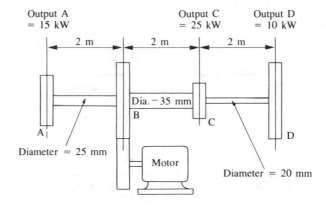

1.17 Use the principle of superposition to determine the B.M. diagram for the beam loaded as shown in Fig. 1.53.

Fig. 1.53

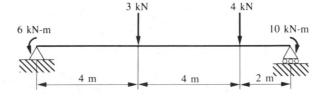

Chapter 2

Statically-determinate stress systems

INTRODUCTION

The effects of external applied forces can now be measured in terms of the internal reacting forces in a solid body or the members of a framework, as described in the previous chapter. However, at that stage no mention was made of the *cross-sectional size and shape* of the members. This aspect had no effect on the forces in the members, but conversely one should be able to describe quantitatively the way in which two members of different cross-sectional size would react to a particular value of force. This is done through the concept of *stress* and this chapter shows how stresses can be determined for simple engineering situations.

STRESS: NORMAL, SHEAR AND HYDROSTATIC

Consider the member shown in Fig. 2.1(*a*) subjected to an external force, *F*, represented by the arrow at each end parallel to the longitudinal axis. The arrow simply represents a force *resultant* on the end faces and obviously the force is not actually applied solely along the line of the arrow. Similarly the internal force reaction does not act along just a single line but is transmitted throughout the bulk of material from grain to grain. If part of the member is cut off to give a free body as in Fig. 2.1(*b*) then equilibrium will be maintained by appropriate components of internal reacting force such as dF_i acting on elements of area dA.

For equilibrium $\sum dF_i = F$ and $\sum dA = A$ the total cross-sectional area. The internal *force per unit area* is.

$$\frac{dF_i}{dA}$$

and as dA tends to zero this is termed *stress*. In general, stress varies in magnitude and direction, so that dF_i/dA is in fact the stress at a point.

Fig. 2.1

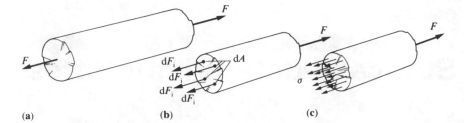

(a)   (b)   (c)

Normal stress   In the simple case in Fig. 2.1 the *average* direct stress is F/A and will be denoted by the symbol σ as in Fig. 2.1(*c*). Normal or direct stress acts perpendicular to a plane and when acting outwards from the plane is termed *tensile stress* and given a *positive* sign. Stress acting towards a plane is termed *compressive stress* and is *negative* in sign. In order to denote the direction of a stress with respect to co-ordinate axes, a suffix notation is used, so that σ_x, σ_y, σ_z represent the components of normal stress in the x-, y- and z-directions as shown in Fig. 2.2(*a*).

Fig. 2.2

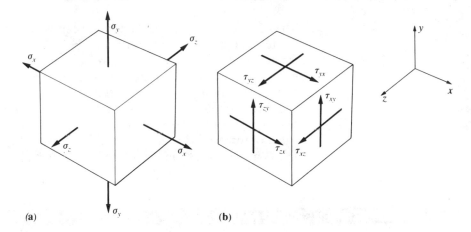

(a)   (b)

Shear stress   When a person who is running wishes to slow down he applies his "brakes". This is achieved through mounting pressure between the soles of his shoes and the ground and thus increased frictional force parallel to the ground. This concept of a force applied tangential or parallel to a surface is termed a shear force page 18 (see also page 47). If internal reacting shear force is expressed as a force per unit area then it is termed a *shear stress*. It also acts parallel to any associated plane within the material and is denoted by the symbol τ. A double suffix notation is required to define shear stresses with respect to co-ordinate axes. The first suffix gives the direction of the normal to the plane on which the stress is acting, and the second suffix indicates the direction of the shear stress component. Thus τ_{xy} is a shear stress acting on the yz-plane (the normal in the x-direction) and pointing in the y-direction. The sign convention associated with shear stress is defined as positive when the direction of the stress vector and the direction of the normal to the plane are both in the positive sense or both in the negative sense in relation to the directions of the co-ordinate axes. If the directions of shear stress and the normal to the plane are opposite in sign then the shear stress is negative. There are twelve possible shear stress components in a three-dimensional stress system as indicated in Fig. 2.2(b) (those on the obscured faces have been omitted for clarity) which reduce to six independent components as explained later.

There is a further condition of shear stresses which always exists and is explained as follows. Consider the element of unit thickness in Fig. 2.3 subjected to shearing stresses along its edges τ_{xy} and τ_{yx}. Then the shear *forces* along the sides AB and CD are $(\tau_{xy} \times 1 \times dy)$ and along AD and BC are $(\tau_{yx} \times 1 \times dx)$. Now, to maintain equilibrium of the element, the above *forces* must balance out, which may be expressed by taking moments about a z-axis through the centre of the element:

$$2(\tau_{xy} \times 1 \times dy) \times \frac{dx}{2} - 2(\tau_{yx} \times 1 \times dx) \times \frac{dy}{2} = 0$$

or

$$\tau_{xy} = \tau_{yx}$$

Fig. 2.3

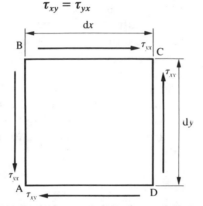

These are termed *complementary shear stresses*. Thus a shear stress on one plane is *always* accompanied by a complementary shear stress of the same sign and magnitude on a perpendicular plane.

Hydrostatic stress
This is a special state of direct stress which should be mentioned now although its importance will become more evident at later stages. Hydrostatic stress may be represented by the stress set up in a body immersed at a great depth in a fluid. The external applied pressure p being equal at all points round the body gives rise to internal reacting compressive force and hence compressive stress equal in all directions, i.e. $\sigma_x = \sigma_y = \sigma_z = \sigma = p$.

A STATICALLY-DETERMINATE STRESS SYSTEM

If the stresses within a body can be calculated purely from the conditions of equilibrium of the applied loading and the internal forces then the problem is said to be statically determinate. There are very few examples of this nature; however, they do give further illustration of the application of equilibrium, and the remainder of the chapter will be devoted to solutions in this category.

ASSUMPTIONS AND APPROXIMATIONS

Exact solutions for stress, displacement, etc., in real engineering problems are not always mathematically possible and even those that are possible can involve lengthy computation and advanced mathematical techniques which are not necessarily justifiable. This is because we seldom know the exact conditions of applied loading on a component or structure for its expected working life, and the materials used are not wholly predictable in behaviour. It therefore becomes necessary and desirable in most engineering problems to make some simplifying approximations and assumptions which, while not changing the basic nature of the problem, will allow a simpler solution and an answer which is not too far from the truth. It is important, however, that any assumptions or approximations are clearly stated at the start so that the reader may assess the validity of the answer in respect of what might be the exact solution.

Some of the problems to follow are in general not statically determinate but, with some realistic geometrical limitations, they can be solved purely from equilibrium conditions to give answers which, although not exact, are reasonably accurate for the purposes of engineering design.

TIE BAR AND STRUT (OR COLUMN)

These are the simplest examples of statically determinate stress situations, since the equilibrium condition is simply that the external force at the ends of the member must be balanced by the internal force, which is the average stress multiplied by the cross-sectional area.

The case of the tie bar is illustrated in Fig. 2.4. The bar at (a),

Fig. 2.4

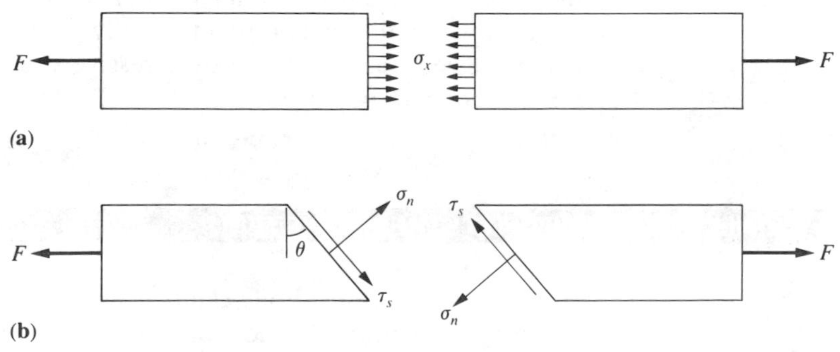

(a)

(b)

subjected to tension, has been cut perpendicular to the axis into two free bodies to show the normal stress σ_x. This multiplied by the area on which it acts must be in equilibrium with the applied force, F hence,

$$\sigma_x A = F \quad \text{or} \quad \sigma_x = \frac{F}{A} \text{ (tensile stress)}$$

In Fig. 2.4(b) it is seen that if the bar is cut into two free bodies at an angle to the axis then there will be two components of stress; one is normal to the plane σ_n, and the other is parallel to the plane τ_s. The significance of these stress components is dealt with in detail in Chapter 12.

A similar reasoning to that above applies to the compression situation in the strut (column).

EXAMPLE 2.1

A double-acting hydraulic cylinder has a piston 250 mm in diameter and a piston rod 75 mm in diameter. The water pressure is 7 MN/m² on one side of the piston and 300 kN/m² on the other side, and on the return stroke the pressures are interchanged. Determine the maximum stress in the rod.

SOLUTION

The two loading situations are shown in Fig. 2.5.
The first step is to calculate the areas on which pressures and stresses are acting.

$$\text{Full piston area} = \frac{\pi}{4} \times 0.25^2 = 0.0491 \text{ m}^2$$

$$\text{Rod area} = \frac{\pi}{4} \times 0.075^2 = 0.0044 \text{ m}^2$$

$$\text{Reduced piston area} = 0.0447 \text{ m}^2$$

In the first case the equilibrium equation is

$$0.0044\sigma_x + (0.0491 \times 7) - (0.0447 \times 0.3) = 0$$

Fig. 2.5

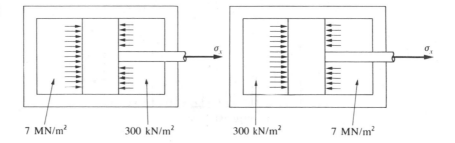

7 MN/m²   300 kN/m²   300 kN/m²   7 MN/m²

Hence

$$\sigma_x = -75 \text{ MN/m}^2 \quad \text{(compression)}$$

On the other stroke the equilibrium condition is

$$0.0044\sigma_x + (0.0491 \times 0.3) - (0.0447 \times 7) = 0$$

and

$$\sigma_x = +67.8 \text{ MN/m}^2 \quad \text{(tension)}$$

Therefore the maximum stress set up in the rod is 75 MN/m² in compression.

SUSPENSION-BRIDGE CABLES

A common form of loading on a cable, for example in suspension bridges, is shown in Fig. 2.6(a). The loading, w per unit length, is distributed uniformly on a *horizontal base,* the weight of the cable being neglected. In this particular example, the ends A and B are set at different heights above the lowest point. It is useful for the analysis to cut the cable at O, to insert a reaction at that point, and to consider the equilibrium of the right-hand part of the cable. The free-body diagram in Fig. 2.6(b) shows that equilibrium is satisfied by the triangle of forces T_B, T_0 and wx_1. The position of the lowest point O and hence the distance x_1 is not known. The distance x_1 can be determined by equilibrium of moments of the forces for either part of the cable. Thus, taking moments about B, and noting that the force wx_1 can be taken as acting through the mid-point of x_1, then

$$T_0 y_1 - wx_1 \frac{x_1}{2} = 0 \tag{2.1}$$

You should now draw the free-body diagram for the left-hand part. Moments about A gives

$$T_0 y_2 - w(l - x_1)\frac{(l - x_1)}{2} = 0 \tag{2.2}$$

Fig. 2.6

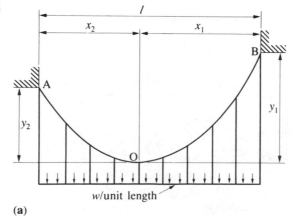

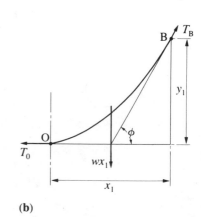

(a)　　　　　　　　　　　　　　　　　　(b)

Inspection of eqn. (2.1) shows that, for any point on the cable having co-ordinates (x, y) relative to O,

$$y = \frac{wx^2}{2T_0} \tag{2.3}$$

which is an equation for a parabola.

Returning to eqns. (2.1) and (2.2) and eliminating T_0 gives

$$\frac{y_1}{y_2} = \frac{x_1^2}{(l - x_1)^2}$$

from which

$$x_1 = \frac{l(y_1/y_2)^{1/2}}{1 + (y_1/y_2)^{1/2}} \tag{2.4}$$

If the ends A and B are at the same level then $y_2 = y_1$ and $x_1 = l/2$.

The maximum tension in the cable will naturally occur at end B since this side of the cable supports the greater proportion of the load. The force T_B can be determined from equilibrium of the triangle of forces T_0, wx_1 and T_B. Vertical equilibrium:

$$T_B \sin \phi - wx_1 = 0$$

But, from the geometry of the triangle,

$$\sin \phi = \frac{y_1}{[y_1^2 + (x_1/2)^2]^{1/2}}$$

Therefore

$$T_B = wx_1 \frac{[y_1^2 + (x_1/2)^2]^{1/2}}{y_1} \tag{2.5}$$

Similarly,

$$T_A = w(l - x_1) \frac{\left[y_2^2 + \left(\frac{l - x_1}{2}\right)^2\right]^{1/2}}{y_2} \tag{2.6}$$

where x, is given by eqn. (2.4).

The minimum tension in the cable is T_0, which is obtained from eqn. (2.1), by substituting the values of x_1 and y_1. Any required stress values are determined by dividing the appropriate tension by the cross-sectional area, a, of the cable, since it is assumed that the stress is uniformly distributed across the section and thus $\sigma = T/a$.

THIN RING OR CYLINDER ROTATING

If a cylinder or ring is rotating at constant velocity, Fig. 2.7(a), then an *inward* radial component of force is required to provide the centripetal

Fig. 2.7

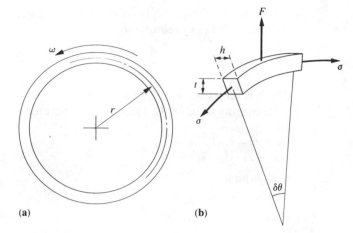

(a) (b)

acceleration on each element of material. This *inward* force may be
resolved into the tangential direction at each end of a typical element as
shown in Fig. 2.7(*b*), which is thus subjected to circumferential tensile
stress. This may be determined from a consideration of dynamical
equilibrium. Alternatively we can reduce this dynamical situation to a
static one by applying D'Alembert's principle, in which forces to
accelerate masses are replaced by equal and opposite static forces. In this
case the centripetal force is replaced by what is often termed a centrifugal
force acting radially outwards, and we can then apply statical
equilibrium.

Consider the rim of a wheel (no spokes) or a slice through a
thin-walled cylinder each rotating about a central axis at velocity ω. If
this problem is to be statically determinate then the diameter of the ring
or cylinder must be large, say >10 times the cross-sectional dimensions of
the rim. It is then possible to assume a uniform distribution of stress over
the cross-section in the circumferential direction, and in the radial and
axial directions the stresses can be taken as zero or negligible.

A small element of arc of the ring of cross-sectional area $A(=th)$
rotating at uniform velocity ω is shown in Fig. 2.7(*b*). The forces acting
on the element are the radial inertia force F and the circumferential
tensile stress σ acting over the area A. The resolved component of the
radial forces inwards is

$$2\sigma th \sin\frac{\delta\theta}{2} \simeq 2\sigma th\frac{\delta\theta}{2} \quad \text{for small values of } \delta\theta$$

Mass of element $= \rho thr\delta\theta$

where ρ is the mass per unit volume.

Radial centrifugal force, $F = (\rho thr\,\delta\theta)\omega^2 r$

For radial equilibrium, the resolved inward radial force must balance the
outward force F, hence

$$2\sigma th\frac{\delta\theta}{2} - F = 0$$

$$\sigma th\,\delta\theta - \rho thr^2\omega^2\delta\theta = 0$$

and

$$\sigma = \rho \omega^2 r^2$$
$$= \rho v^2 \tag{2.7}$$

where v is the tangential velocity of the central point of the element cross-section.

It should be noted that the tensile stress is independent of the shape and area of the cross-section of the rim.

An important practical example of the above effect of centrifugal action is the stress set up in the blades of gas turbine rotors which rotate at very high speed.

THIN SHELLS UNDER PRESSURE

Thin-walled shells are used extensively in engineering in two principal forms: (a) made of reinforced concrete to form part of a civil engineering structure and stressed partly by self-weight and partly by the environment, namely wind, snow, etc.; (b) made of metal and used generally in engineering as storage containers for liquid, powder, gas, etc. Stresses will arise due to, say, uniform internal liquid or gas pressure, e.g. in a steam boiler, or pressure due to weight of liquids or solids contained.

In general, a shell of arbitrary wall thickness subjected to pressure contains a three dimensional stress system. The stresses are perpendicular to the thickness and in two principal orthogonal directions tangential to the surface geometry. Each of these stresses has a non-uniform distribution through the thickness of material, and the solution is statically indeterminate. If, however, the wall thickness is less than about one-tenth of the principal radii of curvature of the shell, the variation of tangential stresses through the wall thickness is small and the radial stress may be neglected. The solution can then be treated as statically determinate.

In the present treatment the following additional simplifying assumptions will be made: that the shell acts as a membrane and does not provide bending resistance so that there are only uniform direct stresses present, which are often called *membrane* stresses; that the shell is formed by a surface of revolution and there are no discontinuities or sharp bends.

Consider first a general axi-symmetrical shell, Fig. 2.8(a) from which is cut an element bounded by two meridional lines and two lines perpendicular to the meridians as at (b). The notation is as follows:

$\sigma_1 = $ tensile stress in meridional direction, meridional stress
$\sigma_2 = $ tensile stress along a parallel circle, hoop stress
$r_1 = $ meridional radius of curvature
$r_2 = $ radius of curvature perpendicular to the meridian
$t = $ wall thickness

Fig. 2.8

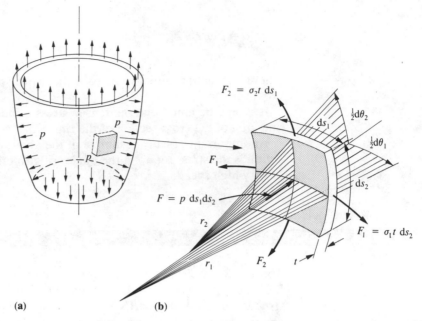

(a)　　　　　　(b)

The forces on the edges of the element are, in the meridional direction, $\sigma_1 t\, ds_2$ and, in the perpendicular direction, $\sigma_2 t\, ds_1$. These forces have components inwards towards their centres of curvature. The radial components are $2\sigma_1 t\, ds_2 \sin(d\theta_1/2)$ and $2\sigma_2 t\, ds_1 \sin(d\theta_2/2)$. For small values of $d\theta$ the total radial force becomes

$$\sigma_1 t\, ds_2 d\theta_1 + \sigma_2 t\, ds_1 d\theta_2$$

or

$$\sigma_1 t\, ds_2 \frac{ds_1}{r_1} + \sigma_2 t\, ds_1 \frac{ds_2}{r_2}$$

The radial force due to the pressure p acting over the surface area of the element is

$$p\, ds_1 ds_2$$

Thus, for equilibrium of the element,

$$\sigma_1 t\, ds_2 \frac{ds_1}{r_1} + \sigma_2 t\, ds_1 \frac{ds_2}{r_2} = p\, ds_1 ds_2$$

and, simplifying,

$$\frac{\sigma_1}{r_1} + \frac{\sigma_2}{r_2} = \frac{p}{t} \tag{2.8}$$

In general σ_1 and σ_2 are both different and unknown and so another equilibrium equation has to be formulated relevant to the particular problem, in addition to eqn. (2.8), in order to solve for the stresses.

Thin sphere　　The simplest case to which eqn. (2.8) may be applied is the sphere. The symmetry about any axis implies that $\sigma_1 = \sigma_2 = \sigma$, and of course

Fig. 2.9

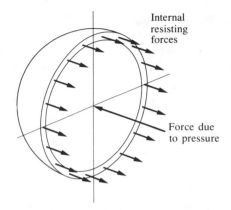

Internal resisting forces

Force due to pressure

$r_1 = r_2 = r$; therefore the circumferential stress in any direction is $\sigma = pr/2t$. This stress may also be obtained by cutting the sphere across any diameter (Fig. 2.9) and considering the equilibrium of either of the free-body hemispheres in the same manner as shown in the next section for the cylinder.

Thin cylinder

This particular problem will be considered from first principles rather than using eqn. (2.8) in order to demonstrate the further use of free-body diagrams.

Axial equilibrium

The force acting on each closed end of the cylinder owing to the internal pressure p, Fig. 2.10(b), is obtained from the product of the pressure and the area on which it acts. Thus

Axial force $= p\pi r^2$

The part of the vessel shown in the free-body diagram, Fig. 2.10(b), is in axial equilibrium simply under the action of the axial force above and the axial stress, σ_x, in the material. The radial pressure shown has no axial resultant force. The cross-sectional area of material is approximately $2\pi rt$, and therefore the reacting force is $\sigma_x \times 2\pi rt$, and, for equilibrium,

$$2\pi rt\sigma_x = \pi r^2 p$$

or

$$\sigma_x = \frac{pr}{2t} \tag{2.9}$$

Fig. 2.10

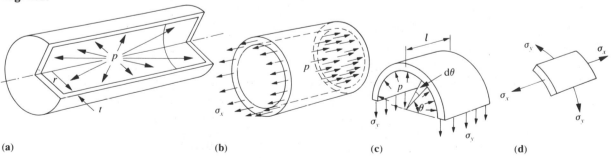

(a)

(b)

(c)

(d)

Circumferential
equilibrium

Considering the equilibrium of part of the vessel, if the cylinder is cut across a diameter as in the free-body diagram in Fig. 2.10(c) the internal pressure acting outwards must be in equilibrium with the circumferential stress, σ_y, as shown. Consider the length of cylinder l and the small arc of shell subtending an angle dθ shown in the diagram. The radial component of force on the element is $p \times l \times r$ dθ; hence the vertical component is $p \times l \times r$ dθ sin θ. Therefore the total vertical force due to pressure is

$$\int_0^\pi prl \sin \theta \, d\theta = 2prl$$

It is useful to note that the vertical force can also be found by considering the pressure acting on the projected area at the diameter which is $p(2rl)$.

The internal force required for equilibrium is obtained from the stress σ_y acting on the two ends of the strip of shell. Hence the internal force is $\sigma_y(2tl)$. For equilibrium, $2tl\sigma_y = 2rlp$, or

$$\sigma_y = \frac{pr}{t} \tag{2.10}$$

Comparing eqns. (2.9) and (2.10) it is seen that the circumferential stress is twice the axial stress. Figure 2.10(d) shows a small element of the shell subjected to the axial and circumferential (hoop) stresses.

EXAMPLE 2.2

A concrete dome is 250 mm thick, has a radius of 30 m and subtends an angle of 120° at the support ring. Calculate the stresses at the support due to self-weight. The density for concrete is 2.3 Mg/m³. The dome is illustrated in Fig. 2.11.

Fig. 2.11

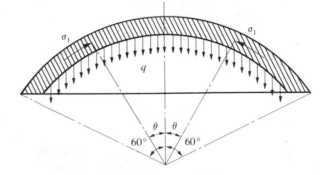

SOLUTION

Consider the vertical equilibrium of a circular segment containing an angle 2θ; then if q is the force per unit area due to weight of concrete,

$$(2\pi r \sin \theta \times t\sigma_1 \sin \theta) + 2\pi r(r - r \cos \theta)q = 0$$

$$\sigma_1 = -\frac{qr}{t} \frac{(1 - \cos \theta)}{\sin^2 \theta} = -\frac{qr}{t} \frac{1}{(1 + \cos \theta)}$$

Since self-weight, unlike applied pressure, does not act radially everywhere, eqn. (2.8) has to be modified to take account of the radially resolved component of weight, giving

$$\frac{\sigma_1}{r_1} + \frac{\sigma_2}{r_2} = -\frac{q}{t}\cos\theta$$

Substituting for σ_1,

$$\sigma_2 = \frac{qr}{t}\left[\frac{1}{1+\cos\theta} - \cos\theta\right]$$

The stresses at the supports are obtained by putting $\theta = 60°$ and $q = 2.3 \times 9.81 \times 250/1000 = 5.65 \text{ kN/m}^2$.

$$\sigma_1 = -\frac{5.65 \times 30}{0.25} \times \frac{1}{1\frac{1}{2}} = -452 \text{ kN/m}^2$$

$$\sigma_2 = \frac{5.65 \times 30}{0.25}(\tfrac{2}{3} - \tfrac{1}{2}) = 113 \text{ kN/m}^2$$

SHEAR IN A COUPLING

A pinned coupling for a tow bar is illustrated in Fig. 2.12(a). Tensile force P is transmitted across the joint being carried by shearing action on sections XX and YY of the pin.

Assuming that there is a uniform shear stress ι over the cross-sectional area A of the pin, and since two cross-sections are in shear, as shown in Fig. 2.12(b),

Total internal shear force $= 2\tau A$

For equilibrium,

$$F = 2\tau A$$

so the shear stress at the two sections of the pin is

$$\tau = \frac{F}{2A} \tag{2.11}$$

Fig. 2.12

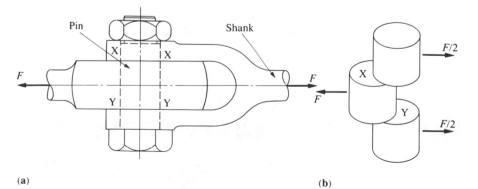

(a)   (b)

TORSION OF A THIN CIRCULAR TUBE

In general the twisting or torsion of solid and hollow circular-section members is statically indeterminate and is discussed fully in Chapter 5 However, as with thin-walled shells, if the radius of a circular tube is large (say >10 times) compared with the wall thickness then the stress can reasonably be taken as uniform through the thickness and the problem can be treated as statically determinate.

In Fig. 2.13 a tube of mean radius r and wall thickness t is subjected to opposing torques at each end. The deformation action is that of twisting, and considering the equilibrium of part of the tube, by cutting it at some section away from the end effects and perpendicular to the axis, an internal reacting torque will be required for equilibrium and to maintain the twist. This reaction torque will take the form of a "uniform" shearing stress acting on the cut face shown. The shear stress does not vary around the tube as this would imply a variation in the complementary longitudinal shear stress and thus a resultant axial force when there is none.

Fig. 2.13

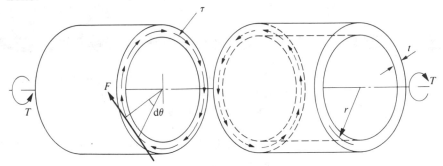

Considering a small element of the tube wall subtending an angle $d\theta$, it has an area of $tr\,d\theta$. The tangential force F shown in Fig. 2.13 is given as the shear stress acting over the element wall, so that $F = \tau tr\,d\theta$.

This force exerts a moment about the central axis of Fr and the total reacting torque is obtained by summing all such moments around the periphery, giving

$$\int_0^{2\pi} \tau tr^2 d\theta = 2\pi tr^2 \tau$$

and this is in equilibrium with the applied torque. Hence $T = 2\pi tr^2\tau$ and the magnitude of the shear stress is

$$\tau = \frac{T}{2\pi tr^2} \qquad (2.12)$$

JOINTS

All branches of engineering have to be able to connect together pieces of material which may then have to carry working loads through the connection or joint. The most common ways of constructing joints are by holding the segments firmly together with bolts, rivets or welds. The forces which have to be transmitted through a joint generally subject the connectors, e.g. bolts, to shear, and the holes in the material through which the connectors pass to tensile and compressive stresses. The determination of most of these stresses may be regarded as statically indeterminate, owing to the complex nature of load transfer through rows of bolts or rivets. Hence for a thorough treatment the reader is referred to the texts specifically concerned with engineering design. A few simple mechanical joints can be treated as statically determinate and two examples illustrate the equilibrium application as follows.

EXAMPLE 2.3

A steel bracket is bolted to a column by three similar bolts as shown in Fig. 2.14. The design case is for the 36 kN load carried in the off-centre position. Using a factor of safety of 4 determine a suitable minimum diameter for the bolts. It may be assumed that the bracket and plate to which it is attached are essentially rigid.

Fig. 2.14

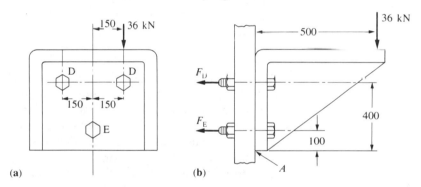

(a)   (b)

Allowable stresses for the bolt material are 600 MN/m^2 in tension and 350 MN/m^2 in shear.

SOLUTION

The design load is obtained by multiplying the nominal applied load by the factor of safety, giving $36 \times 4 = 144$ kN. The bolts are subjected to two types of reacting force simultaneously, tension and shear. Tension is caused by the applied force inducing rotation about the bottom edge at A, thus stretching the bolts. Shear is caused by the off-centre loading inducing rotation in the plane of the joint. These two loading effects may be represented as in Fig. 2.15 where the load of 144 kN has been transferred to the central axis of the bolt group together with a couple of $144 \times 0.15 = 21.6$ kN-m acting about the centre of area of the bolt group.

Taking moments about the lower edge A of the bracket, Fig. 2.14(b),

$$144 \times 500 - 2F_D \times 400 - F_E \times 100 = 0$$

At this point we have to make a guess at the proportions of F_D and F_E

Fig. 2.15

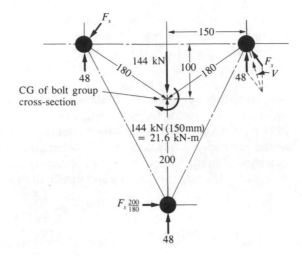

since the problem is not strictly statically determinate. We can assume that $F_D/F_E = 400/100 = 4$ on the basis that for rigid plate and bracket the upper bolts will stretch four times the amount of the lower bolt, from which $F_D = 87.27$ and $F_E = 21.81$ kN.

The required minimum bolt diameter d to resist the tensile force is

$$\frac{\pi}{4}d^2 \times 600 \times 10^6 = 87\,270$$

$$d = 13.6\,\text{mm}$$

If we neglect any assistance from friction in resisting the twisting moment at the joint plane, then the bolts are subjected to shearing action in two ways: (i) due to the vertical component of force of 144 kN which we will assume is divided equally between the three bolts giving 48 kN on each. We will assume that the rotational couple (Fig. 2.15) of 21·6 kN-m is reacted by shear forces acting on each bolt in proportion to the perpendicular distance from the centre of area of the bolt group, giving forces F_s, F_s and $(200/180)F_s$. Equilibrium of moments about C gives

$$2F_s \times 0.18 + 1.1F_s \times 0.2 = 21.6$$

$$F_s = 37.1\,\text{kN}$$

The greatest vector sum of the pairs of shear forces for each bolt is clearly that for the top right-hand one, for which the resultant is 81.5 kN. The bolt diameter (minimum) to resist shear is therefore

$$\frac{\pi}{4}d^2 \times 350 \times 10^6 = 81\,500$$

$$d = 17.2\,\text{mm}$$

It is seen that this diameter is larger than that found to resist tension, therefore this latter must be regarded as the minimum for which a standard bolt size is selected. (At a later stage in the book, in Chapter 12, it will be shown that tensile stress and shear stress acting simultaneously must not be considered in isolation.)

EXAMPLE 2.4

A closed-ended cylindrical steel pressure vessel is 2 m in internal diameter. It is made up from 10 mm thick steel sheet formed into two semi-cylinders, which are riveted along longitudinal single-lap seams (as shown in Fig. 2.16) with 12 mm diameter rivets at 36 mm pitch. The dished ends are also single-lap riveted on to the cylinder at each end and the circumferential joints are made with 8 mm diameter rivets at 30 mm pitch. If the maximum shear stress in any rivet is not to exceed 60 MN/m² calculate the maximum allowable internal pressure and the resulting longitudinal and circumferential stresses in the wall of the cylinder.

Fig. 2.16

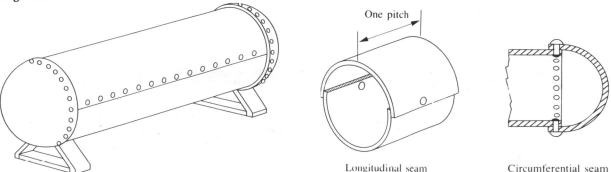

Longitudinal seam   Circumferential seam

SOLUTION

Longitudinal seam

Maximum shear force per rivet

$$60 \times 10^6 \times \frac{\pi}{4} \times \frac{12^2}{10^6} = 6.8 \text{ kN}$$

Separating force at joints due to pressure over one pitch length is

$$0.036 \times 2 \times p$$

The equilibrium condition is

$$0.036 \times 2 \times p = 2 \times 6.8$$

so that

$$p = 189 \text{ kN/m}^2$$

Circumferential seam

Maximum shear force per rivet is

$$60 \times 10^6 \times \frac{\pi}{4} \times \frac{8^2}{10^6} = 3.02 \text{ kN}$$

Number of rivets required

$$\pi \times \frac{2000}{30} = 209$$

Total force in shear allowable $= 630 \text{ kN}$

This must be in equilibrium with the longitudinal force due to pressure, which is

$$\frac{\pi}{4} \times 2^2 \times p$$

Hence

$$\pi p = 630 \quad \text{and} \quad p = 201 \text{ kN/m}^2$$

Therefore the maximum allowable pressure is 189 kN/m².

$$\text{Circumferential stress, } \sigma_1 = \frac{pr}{t} = \frac{189 \times 1}{0 \cdot 01} = 18.9 \text{ MN/m}^2$$

$$\text{Axial stress, } \sigma_2 = \frac{18.9}{2} = 9.45 \text{ MN/m}^2$$

SUMMARY

The purpose of this chapter has been to introduce the concept of *stress* which represents the internal reacting force per unit area within the material. Engineering design of components to establish size and shape depends on allowable values of stress which the material can tolerate when subjected to applied loads or forces. In a few cases it is possible to establish the stress for a given size or decide on the required size for an allowable stress by consideration *only* of the *equilibrium* of the system of external applied forces and internal reacting forces. These design examples are said to be *statically determinate*.

PROBLEMS

2.1 A steel rod of varying cross-section is loaded as shown in Fig. 2.17. Determine where the maximum stress occurs.

Fig. 2.17

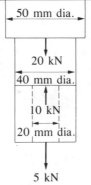

2.2 It is required to make a large concrete foundation block which, when supporting a compressive load together with its self-weight, will have the same compressive stress at all cross-sections. Determine a suitable profile.

2.3 The two parabolic cables of a suspension bridge are subjected to a horizontal uniformly-distributed load of 80 kN/m as shown in Fig. 2.18. Calculate the required area of the cables at each end if their maximum permissible stress is 200 MN/m². What is the compressive load in the vertical columns?

Fig. 2.18

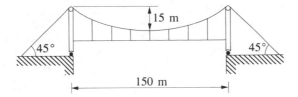

2.4 A suspension footbridge spanning a ravine is constructed with twin cables and carries a horizontal uniformly-distributed loading of 2 kN/m of span, which is 300 m. The lowest point of the cables is 50 m below one cliff support, which is 10 m below the higher cliff support. Determine a suitable cross-sectional area for each cable using a safety factor of 2 and a tensile stress of 300 MN/m².

2.5 A cable is freely suspended from two points which are at the same horizontal level. If the cable is subjected to a uniformly-distributed loading of w per unit length (self-weight, snow, birds), derive an expression for the maximum tension in the cable.

2.6 A small boat is anchored as shown in Fig. 2.19. When the tide causes a horizontal force of 1 kN on the boat the steel rope is tangential at the anchor point. If the rope diameter is 20 mm calculate (*a*) the distance between the boat and the anchor point, and (*b*) the maximum stress in the rope. The density of the steel is 7800 kg/m³ and the density of the water is 1000 kg/m³.

Fig. 2.19

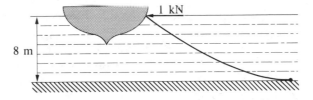

2.7 A chemical reaction process is carried out in a thin-walled steel cylinder of internal diameter 400 mm with closed ends rotated about a longitudinal axis at a speed of 5000 rev/min. Whilst it is rotating, it is subjected to an internal pressure of 4 MN/m². If the maximum allowable tensile stress in any direction in the material is 175 MN/m², calculate a suitable shell thickness. Density of steel = 7.83 Mg/m³.

2.8 The pipeline reducer shown in Fig. 2.20 has a uniform wall thickness of 3 mm. If the pipeline carries a fluid at a pressure of 0.7 MN/m², calculate the axial and hoop stresses in the reducer at a point halfway along its length. Assuming that the

Fig. 2.20

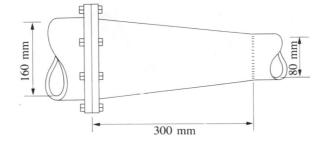

coupling at the large end of the reducer takes all the axial thrust, calculate the stress in each of the six 10 mm diameter retaining bolts.

2.9 A thin spherical steel vessel is made up of two hemispherical portions bolted together at flanges. The inner diameter of the sphere is 300 mm and the wall thickness is 6 mm. Assuming that the vessel is a homogeneous sphere, what is the maximum working pressure for an allowable tensile stress in the shell of 150 MN/m^2?

If twenty bolts of 16 mm diameter are used to hold the flanges together, what is the tensile stress in the bolts when the sphere is under full pressure?

2.10 A conical storage tank has a wall thickness of 20 mm and an apex angle of 60°. If the vessel is filled with water to a depth of 3 m, calculate the maximum meridional and circumferential stresses. The water loading is 9.81 kN/m^3.

2.11 In the mechanical digger shown in Fig. 2.21 the combined weight of the bucket and its contents is 1000 kg and the centre of gravity is at G. For the position shown in which the arm NKM is horizontal, calculate the shear stress in the pins at N and M. The inset sketch shows the joint arrangement in each case, and the pin diameters at N and M are 10 mm and 15 mm respectively.

Fig. 2.21

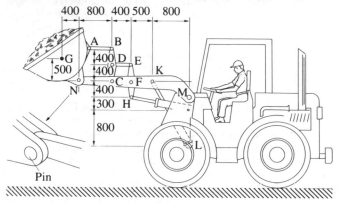

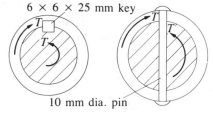

Pin

2.12 The hub of a pulley may be fastened to a 25 mm diameter shaft either by a square key or by a pin, as shown in Fig. 2.22. Determine the torque that each connection can transmit if the average shear stress in the key or pin is not to exceed 70 MN/m^2.

Fig. 2.22

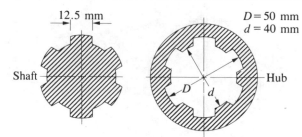

6 × 6 × 25 mm key

10 mm dia. pin

2.13 A splined shaft connection as shown in Fig. 2.23 is 50 mm long and is used to permit axial movement of the shaft relative to the hub during torque

Fig. 2.23

12.5 mm

$D = 50$ mm
$d = 40$ mm

Shaft

Hub

transmission. In order to facilitate axial movement in the connection it is to be designed so that the side pressure on the splines does not exceed 7 MN/m². Calculate the power that could be transmitted by the shaft at 2000 rev/min and the shear stress in the splines at this power.

2.14   A thin-walled circular tube of 50 mm mean radius is required to transmit 300 kW at 500 rev/min. Calculate a suitable wall thickness so that the shear stress does not exceed 80 MN/m².

2.15   A torque tube consists of two sections which are riveted together, as in Fig. 2.24, by 50 rivets of 4 mm diameter pitched uniformly and the radius of the mating surface of the tubes is 100 mm. If the limiting shear stress for the rivets is 180 MN/m² determine the maximum torque that can be transmitted through the joint.

Fig. 2.24

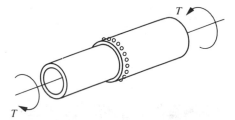

Chapter 3

Stress–strain relations

INTRODUCTION

As explained previously a statically-indeterminate problem cannot be solved from the conditions of equilibrium alone; additional equations are required to find all the unknowns. These equations are obtained by studying the *geometry of deformation* of the component or structure and the *load–deformation* or *stress–strain relationship* for the material. These topics are dealt with in this chapter, but the detailed application in worked examples is carried out in Chapter 4.

DEFORMATION

Deformations may occur in a material for a number of reasons, such as external applied loads, change in temperature, tightening of bolts, irradiation effects, etc. Bending, twisting, compression, torsion and shear or combinations of these are common modes of deformation. In some materials, e.g. rubber, plastics, wood, the deformations are quite large for relatively small loads, and readily observable by eye. In metals, however, the same loads would produce very small deformations requiring the use of sensitive instruments for measurement.

Stress values do not always provide the limiting factor in design, for although a component may be safe and employ material economically with regard to stress, the deformations accompanying that stress might be dangerous or inconvenient. For example, too high a deflection of an aeroplane wing can result, among other things, in a detrimental change in aerodynamic characteristics. A lathe bed which was not sufficiently rigid would not permit of the required tolerances in machining. A perfectly safe sag in a dance-hall floor might upset the poise of the dancers.

In this and succeeding chapters there will be many problems in which the analysis of displacements will be considered specifically in addition to the determination of stress magnitude.

As explained in Chapter 2 the effect of a force applied to bodies of different size can be compared in terms of stress, i.e. the force per unit area. Likewise the deformation of different bodies subjected to a particular load is a function of size, and therefore comparisons are made by expressing deformation as a non-dimensional quantity given by the change in dimension per unit of original dimension, or in the case of shear as a change in angle between two initially perpendicular planes. The non-dimensional expression of deformation is termed *strain*.

Direct or normal strain

Consider the bar shown in Fig. 3.1 subjected to axial tensile loading F. If the resulting extension of the bar is δl and its unloaded length is l_0, then the direct tensile strain is

$$\varepsilon = \frac{\delta l}{l_0}$$

Fig. 3.1

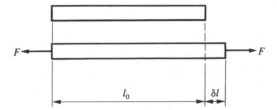

If two bars identical in material, length l_0 and cross-sectional area were each subjected to a tensile force F then the extension in each would be the same, say δl, and the strain would be $\delta l/l_0$. The bars are now joined end on end and the same tensile force F is applied. The overall extension of the combined bar will be $2\delta l$, but since the original length is now $2l_0$ the strain is $2\delta l/2l_0$ or $\delta l/l_0$, i.e. the same as for the separate bars.

Similarly, if the bar had been compressed by an amount δl, then the compressive strain would be

$$\varepsilon = -\frac{\delta l}{l_0}$$

Strain is defined as positive for an increase in dimension and negative for a reduction in dimension.

The same suffix notation is used for strains as for stresses. ε_x is the strain of a line measured in the x-direction, and ε_y the strain of a line in the y-direction.

Shear strain

An element which is subjected to shear stress experiences deformation as shown in Fig. 3.2. The tangent of the angle through which two adjacent sides rotate relative to their initial position is termed *shear strain*. In many cases the angle is very small and the angle itself is used, expressed

Fig. 3.2

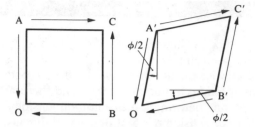

in radians, instead of the tangent, so that

$$\gamma = \angle AOB - \angle A'OB' = \phi$$

When $\angle A'OB' < \angle AOB$, then γ is defined as a positive shear strain, and when $\angle A'OB' > \angle AOB$, γ is termed a negative shear strain.

Volumetric strain

The term "hydrostatic stress" was used in Chapter 2 to describe a state of tensile or compressive stress equal in all directions within or external to a body. Hydrostatic stress causes a change in volume of the material which, if expressed per unit of original volume, gives a volumetric strain denoted by ε_v.

Volumetric strain may be expressed in terms of the three co-ordinate linear strains, and for some problems this is a necessary relationship which can be derived as follows. Let a cuboid of material have sides initially of lengths, x, y and z. If under some load system the sides change in length by dx, dy and dz then the new volume is

$$(x + \mathrm{d}x)(y + \mathrm{d}y)(z + \mathrm{d}z)$$

Neglecting products of small quantities,

New volume $= xyz + zy\,\mathrm{d}x + xz\,\mathrm{d}y + xy\,\mathrm{d}z$

Original volume $= xyz$

Change in volume $= zy\,\mathrm{d}x + xz\,\mathrm{d}y + xy\,\mathrm{d}z$

$$\text{Volumetric strain, } \varepsilon_v = \frac{zy\,\mathrm{d}x + xz\,\mathrm{d}y + xy\,\mathrm{d}z}{xyz}$$

$$= \frac{\mathrm{d}x}{x} + \frac{\mathrm{d}y}{y} + \frac{\mathrm{d}z}{z}$$

$$= \varepsilon_x + \varepsilon_y + \varepsilon_z \tag{3.1}$$

i.e. volumetric strain is given by the sum of the three linear co-ordinate strains.

ELASTIC LOAD–DEFORMATION BEHAVIOUR OF MATERIALS

Studies of material behaviour made by Robert Hooke in 1678 showed that up to a certain limit the extension δl of a bar subjected to an axial

Fig. 3.3 (a) Linear elasticity; (b) exceeding the elastic limit; (c) non-linear elasticity

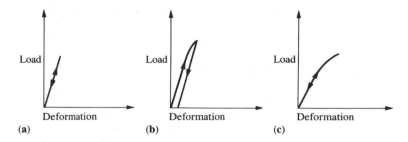

(a) (b) (c)

tensile loading F was often directly proportional to F, as in Fig. 3.3(a). This behaviour in which $\delta l \propto F$ is known as *Hooke's law*. It is similarly found that for many materials uniaxial compressive load and compressive deformation are proportional up to a certain limit of load. A cylindrical bar which is twisted about its axis by opposing torques applied at each end is also found to have a linear torque–twist relationship up to a certain point. The maximum load up to which Hooke's law is applied is termed the *limit of proportionality*. If in each of the above cases at any particular load the same deformation exists both with increasing and decreasing load, and if after completely unloading the body it returns to exactly its original size, then it is said to exhibit the property of *elasticity*. This behaviour exists only over a certain range of load and deformation, the end point being termed the *elastic limit*. In general, the limit of proportionality is a shade lower than the elastic limit. If the elastic limit is exceeded it is found that some permanent deformation remains after removal of the load, as illustrated in Fig. 3.3(b).

Metals generally obey a linear load–deformation law up to their elastic limits, exhibiting what is termed *linear elasticity*; however, there are some materials, principally non-metallic, which have an elastic range as defined above, but exhibit a non-linear load–deformation relationship, Fig. 3.3(c).

ELASTIC STRESS–STRAIN BEHAVIOUR OF MATERIALS

If the load in Fig. 3.3(a) is divided by the original cross-sectional area of the bar, A, and the extensions on the abscissa are divided by the original length of the bar, l_0, a graph of stress against strain is obtained. Since A and l_0 are constants the stress–strain behaviour is also linear in the elastic range.

The slope of the line is constant and may be expressed as

$$\frac{W}{A} \bigg/ \frac{\delta l}{l_0} = \frac{\sigma}{\varepsilon} = E$$

where E is a constant for the material, and is called *Young's modulus of elasticity*. Since ε is non-dimensional, E has the dimensions of stress, i.e. force per unit area. Some typical values of E are given for a few materials in Table 3.1.

A relationship between shear stress and shear strain may be derived from a torsion test on a cylindrical bar in which applied torque and

Table 3.1

Material	E (GN/m²)	G (GN/m²)
Steels	190–207	77–83
Copper	110–120	37–46
Aluminium	69–70	24–28
Glass	50–80	20–35

angular twist are measured. The connections between torque and shear stress and twist and shear strain will be derived in Chapter 5. It is sufficient to state here that shear stress is proportional to shear strain within the elastic limit. Hence $\tau/\gamma = $ constant $= G$.

The constant of proportionality, G, is known as the *modulus of rigidity,* or *shear modulus,* and has the dimensions of force per unit area. Typical values of G for various materials are given in Table 3.1.

It can also be demonstrated experimentally that volumetric strain is proportional to hydrostatic stress within the elastic range. The constant relating those two quantities is termed the *bulk modulus* and is denoted by the symbol K. Thus

$$\frac{\sigma}{\varepsilon_v} = K$$

It will be shown in Chapter 12 that the elastic constants, E, G and K are related to one another.

EXAMPLE 3.1

Calculate the overall change in length of the tapered rod shown in Fig. 3.4. It carries a tensile load of 10 kN at the free end, and at the step

Fig. 3.4

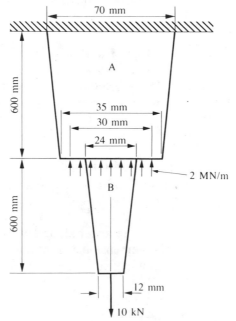

change in section a compressive load of 2 MN/m evenly distributed around a circle of 30 mm diameter. $E = 208\,\text{GN/m}^2$.

SOLUTION

Firstly consider the general case of a tapered rod fixed at one end and subjected to a tensile load W at the other end as in Fig. 3.5. The mean

Fig. 3.5

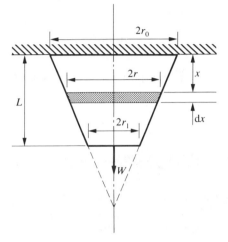

radius of any arbitrary slice at a distance x from the upper end is

$$r = r_0 - (r_0 - r_1)\frac{x}{L}$$

So the cross-sectional area of the slice is

$$A_x = \pi\left[r_0 - (r_0 - r_1)\frac{x}{L}\right]^2$$

If the slice extends an amount du under load then its strain is

$$\frac{du}{dx} = \frac{W}{A_x}\frac{1}{E}$$

The total extension of the rod is

$$u = \int_0^L \frac{W}{A_x E}\,dx$$

$$= \frac{W}{\pi E}\int_0^L \frac{dx}{\left[r_0 - (r_0 - r_1)\dfrac{x}{L}\right]^2}$$

$$= \frac{WL}{E\pi r_0 r_1}$$

Returning now to the stepped taper rod in Fig. 3.4, the extension of the lower part will be

$$u_B = \frac{10\,000 \times 0.6}{208 \times 10^9 \times \pi \times 0.024 \times 0.012} = +0.0319\,\text{mm}$$

The compressive load on the upper part will be treated as an axial concentrated load of magnitude

$$2 \times \pi \times 0.03 = 0.06\pi \text{ MN} = 188.5 \text{ kN}$$

Resultant load acting on part A $= -188.5 + 10 = -178.5$ kN

$$\text{Compression of A} = -\frac{178.5 \times 10^3 \times 0.6}{208 \times 10^9 \times \pi \times 0.07 \times 0.035}$$

$$= -0.0669 \text{ mm}$$

Therefore

$$\text{Overall deformation of rod} = -0.0669 + 0.0319$$

$$= -0.035 \text{ mm}$$

LATERAL STRAIN AND POISSON'S RATIO

If a bar is subjected to say longitudinal tensile stress then it will extend in the direction of the stress and contract in the transverse or lateral directions Fig. 3.6. If the member were subjected to uniaxial compressive stress then an expansion would occur in the lateral directions. It is found that the lateral strain is proportional to the longitudinal strain, and the constant of proportionality is termed *Poisson's ratio* denoted by the symbol v. Hence

$$\text{Lateral strain} = -v \times \text{Direct strain (due to stress)}$$

For most metals v is in the range from 0.28 to 0.32. It is important to remember that lateral strain can occur without being accompanied by lateral stress.

Fig. 3.6

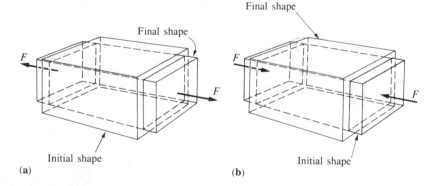

(a)   (b)

THERMAL STRAIN

The effect of a change of temperature on a piece of material is a small change in size and hence strain. This can, in some circumstances, induce considerable stresses. The dependence of size on temperature variation is measured in terms of the basic quantity known as the *coefficient of linear thermal expansion* per unit temperature per unit length, denoted by α.

A rod of length l_0 has its temperature changed from T_0 to T and the accompanying change in length is

$$\delta l = \alpha l_0 (T - T_0)$$

This may be expressed as a thermal strain:

$$\varepsilon_T = \frac{\delta l}{l_0} = \alpha(T - T_0)$$

Increasing temperature causes expansion and thus a positive strain, while decreasing temperature results in contraction and negative strain. An important feature about this behaviour is that if there is no restraint on the material there can be strain unaccompanied by stress. However, if there is any restriction on free change in size then a *thermal stress* will result.

The total strain in a body experiencing thermal stress may be divided into two components, the strain associated with the stress, ε_σ, and the strain resulting from temperature change, ε_T. Thus

$$\varepsilon = \varepsilon_\sigma + \varepsilon_T$$

Hence

$$\varepsilon = \frac{\sigma}{E} + \alpha(T - T_0)$$

which is a more general form of the simple uniaxial stress–strain law.

GENERAL STRESS–STRAIN RELATIONSHIPS

Consider an element of material as in Fig. 3.7(*a*) subjected to a uniaxial stress, σ_x; the corresponding strain system is shown at (*b*). In the

Fig. 3.7 (a) Stress; (b) strain

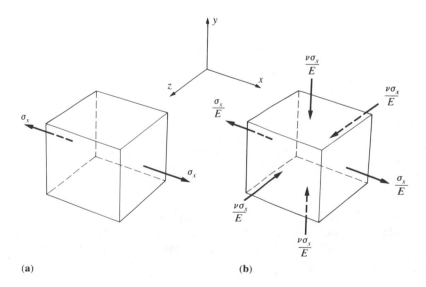

(a) (b)

x-direction the strain is ε_x and in the y- and z-directions the strains are $-v\varepsilon_x$ and $-v\varepsilon_x$, respectively. These strains may be written in terms of stress as $\varepsilon_x = \sigma_x/E$ and $\varepsilon_y = \varepsilon_z = -v\sigma_x/E$, the negative sign indicating contraction.

The element in Fig. 3.8(a) is subjected to triaxial stresses σ_x, σ_y and σ_z. The total strain in the x-direction is therefore composed of a strain

Fig. 3.8 (a) Stress; (b) strain

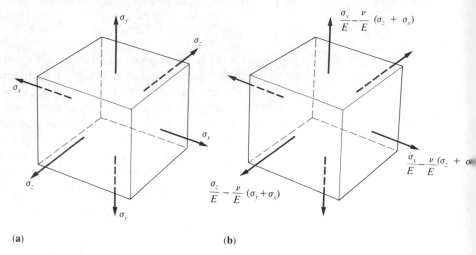

(a)

(b)

due to σ_x, a lateral strain due to σ_y and a further lateral strain due to σ_z. Using the principle of superposition, the resultant strain in the x-direction is therefore the sum of the separate strains as shown at (b); hence

$$\varepsilon_x = \frac{\sigma_x}{E} - \frac{v\sigma_y}{E} - \frac{v\sigma_z}{E}$$

or

$$\varepsilon_x = \frac{\sigma_x}{E} - \frac{v}{E}(\sigma_y + \sigma_z)$$

Similarly

$$\varepsilon_y = \frac{\sigma_y}{E} - \frac{v}{E}(\sigma_z + \sigma_x) \left.\right\}$$

(3.2)

and

$$\varepsilon_z = \frac{\sigma_z}{E} - \frac{v}{E}(\sigma_x + \sigma_y)$$

There is no lateral strain associated with shear strain; hence the shear-stress/shear-strain relationship is the same for both uniaxial and complex strain systems.

PLANE STRESS AND PLANE STRAIN

In many practical situations the stress component in the z-direction is zero and this is referred to as a *plane stress* condition. The above equations may be applied with $\sigma_z = 0$, but note that the strain in the z-direction is not zero.

If the strain in the z-direction is zero, then this condition is referred to as *plane strain* and the above equations may be applied using $\varepsilon_z = 0$. However, zero strain in the z-direction does not imply zero stress in that direction. This may be confirmed quite simply by considering the sample of material in Fig. 3.9 subjected to tensile stresses in the x- and y-directions. Due to the Poisson's ratio effect there would be a change in dimensions in the z-direction. To keep $\varepsilon_z = 0$ it would be necessary to have a stress in the z-direction.

Fig. 3.9

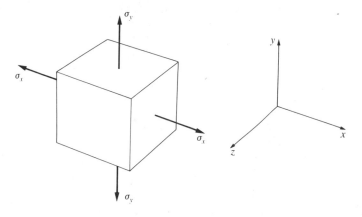

THERMAL STRAINS If in addition to strain due to stress there is also thermal strain due to change in temperature, then the most general form of the six stress–strain relationships is

$$\left. \begin{array}{l} \varepsilon_x = \dfrac{\sigma_x}{E} - \dfrac{v}{E}(\sigma_y + \sigma_z) + \alpha(T - T_0) \\[2mm] \varepsilon_y = \dfrac{\sigma_y}{E} - \dfrac{v}{E}(\sigma_z + \sigma_x) + \alpha(T - T_0) \\[2mm] \varepsilon_z = \dfrac{\sigma_z}{E} - \dfrac{v}{E}(\sigma_x + \sigma_y) + \alpha(T - T_0) \end{array} \right\} \tag{3.3}$$

$$\gamma_{xy} = \frac{\tau_{xy}}{G}, \qquad \gamma_{yz} = \frac{\tau_{yz}}{G}, \qquad \gamma_{zx} = \frac{\tau_{zx}}{G} \tag{3.4}$$

STRAINS IN A STATICALLY-DETERMINATE PROBLEM

The stresses in a thin-walled cylinder under internal pressure were found in Chapter 2 as a statically determinate problem, and now that stress–strain relationships have been developed the strains in the cylinder can be found.

From eqns. (2.9) and (2.10) from equilibrium,

$$\sigma_x = \frac{pr}{2t} \quad \text{and} \quad \sigma_y = \frac{pr}{t}$$

as shown in Fig. 3.10, and σ_z is negligible in comparison; therefore, from

Fig. 3.10

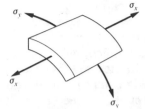

the stress–strain equations (3.3) the axial strain is

$$\varepsilon_x = \frac{pr}{2tE} - \frac{vpr}{tE} = \frac{pr}{2tE}(1 - 2v) \tag{3.5}$$

and the circumferential or hoop strain is

$$\varepsilon_y = \frac{pr}{tE} - \frac{vpr}{2tE} = \frac{pr}{2tE}(2 - v) \tag{3.6}$$

Taking a value for v of 0.3 it is found that the ratio of the hoop to the axial strain is

$$\frac{\varepsilon_y}{\varepsilon_x} = \frac{1.7}{0.4} = 4.25$$

whereas the ratio for hoop to axial stresses was 2.0.

In the thin sphere there is only circumferential stress and strain:

$$\varepsilon = \frac{\sigma}{E} - \frac{v\sigma}{E}$$

and since

$$\sigma = \frac{pr}{2t}$$

$$\varepsilon = \frac{pr}{2tE}(1 - v) \tag{3.7}$$

ELASTIC STRAIN ENERGY

When a piece of material is deformed in simple tension, compression, bending or torsion, etc., within its elastic range, work is done by the applied loading. On removal of the loading the material returns to its undeformed state due to the release of stored energy. This is termed *elastic strain energy* and has the same magnitude as the external work done. It is the release of strain energy in a stretched rubber band that enables a pellet to be projected from a catapult. The release of strain energy in a metal loaded and then unloaded in the elastic range is less obvious. The best illustration of it is a metal spring, the performance of which depends on the energy stored when the wire is bent and twisted during loading. The term *resilience* often associated with springs has in fact a more general meaning as the strain energy stored per unit volume.

STRAIN ENERGY FROM NORMAL STRESS

Consider the load-extension diagram in Fig. 3.11: the external work done during a small increment of extension dx, is w dx and the total area under the curve up to the proportional limit at point A is $\frac{1}{2}W\,\delta l$. Dividing by the area a and length l of the specimen,

$$\frac{1}{2}\frac{W}{a}\frac{\delta l}{l}$$

Fig. 3.11

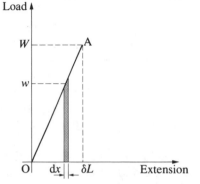

which is the stored strain energy, denoted by U; hence

$$U = \frac{1}{2}\frac{W}{a}\frac{\delta l}{l} = \frac{1}{2}\sigma\varepsilon \quad \text{per unit volume}$$

Since $\varepsilon - \sigma/E$,

$$U = \frac{1}{2}\frac{\sigma^2}{E} \quad \text{per unit volume} \tag{3.8}$$

STRAIN ENERGY FROM SHEAR STRESS

If a piece of material is subject to pure shear then the strain energy stored per unit volume is represented by the area under the shear-stress/shear-strain curve shown in Fig. 3.12.

Fig. 3.12

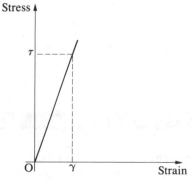

Hence $U = \frac{1}{2}\tau\gamma$ per unit volume, and since $\gamma = \tau/G$,

$$U = \frac{\tau^2}{2G} \quad \text{per unit volume} \tag{3.9}$$

This expression only applies if the shear stress is uniform over the element of material.

PLASTIC STRESS–STRAIN BEHAVIOUR OF MATERIALS

With relatively few exceptions in the design of structures and machines, stresses and strains are limited to the elastic range of a material as described on page 59. However, it is important to appreciate how a material behaves beyond the elastic range in what is termed the *plastic range* of stress and strain. As there is some elementary analysis involving yielding and plasticity in Chapters 13 and 16, the basic concept will be briefly introduced here. A more detailed treatment will be found in Chapter 18.

When a metal specimen is subjected to uniaxial tension to fracture, the measurements of stress and strain will result in a diagram typically as shown in Fig. 3.13(*a*) or (*b*). When the material has passed through the elastic range and enters the plastic range it is said to be *yielding*. Stress continues to increase with strain, but at a slower rate than in the elastic range, until a maximum value of nominal stress (load divided by original cross-sectional area) is reached which is termed the *tensile strength*. Thereafter the specimen enters a failure range terminating in complete fracture at one cross-section. The plastic range of strain is generally between 50 and 300 times the elastic range of strain, and demonstrates what is described as *ductility*, namely the ability to deform plastically.

Fig. 3.13

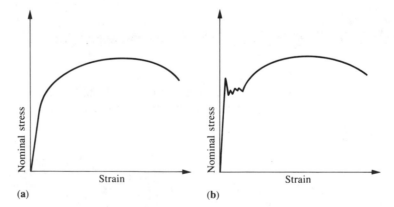

(a) (b)

VISCOELASTIC STRESS–STRAIN BEHAVIOUR OF MATERIALS

Some materials, notably plastics and rubbers, do not exhibit a linear elastic range like metals, but show an interdependence of stress and strain with time. In addition these materials have a "memory" in the sense that current strain or stress is always dependent on the loading history and after unloading considerable recovery of residual strain can occur at zero load. The above behaviour is termed *viscoelasticity*.

The simplest representation of viscoelasticity is the combined features of a Hookean solid and a Newtonian liquid. The former provides an elastic component, and the latter a viscous component (Fig. 3.14(*a*) and (*b*)). From the latter, stress is proportional to strain rate, $\dot{\varepsilon}$, and the constant of proportionality is η, the *coefficient of viscosity*. Hence the simplest possible relationship between stress, strain and time for linear viscoelastic behaviour is

$$\sigma = E\varepsilon + \eta\dot{\varepsilon} \tag{3.10}$$

Fig. 3.14

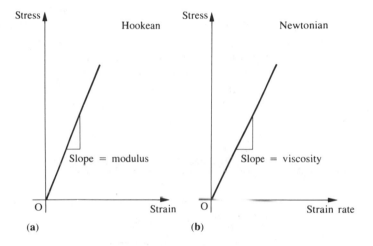

(a) (b)

Because of the dependence on time and the behaviour known as *creep* (*see* Ch. 22) experimentally determined stress–strain relationships demonstrate a strain-rate dependence and non-linearity as shown in Fig. 3.15. As a result, the modulus of a viscoelastic material is not as constant as it is for metals, but depends on the magnitude and nature of the loading applied to the material. Design analysis for viscoelastic materials,

Fig. 3.15

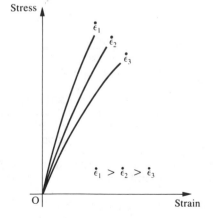

e.g. the thermoplastic components, although still following the basic steps of equilibrium of forces and geometry of deformation, which are independent of the nature of the material, is somewhat more complex owing to the time-dependent stress–strain relationship. This matter is dealt with further in Chapter 22, which specifically relates to creep and viscoelasticity.

SUMMARY

In this chapter the concept of *strain* was introduced together with the relationships that exist between *stress* and *strain* for engineering materials. The "stage" is now set for the analysis of most engineering design problems involving a knowledge of *strength* and *stiffness*. The next three chapters apply these basic concepts to the design of important widely-used engineering components. An early appreciation of the importance of *strain energy* is essential for all engineers, since it has a major part to play through energy theorems in design analysis and further in relation to the mechanics of yielding and fracture.

PROBLEMS

3.1 State whether the components, illustrated in Fig. 3.16 are in a state of plane stress or plane strain.
(*a*) A grinding wheel rotating at high speed.
(*b*) A plate of steel being cold-rolled.
(*c*) A long thick-walled cylinder containing a fluid pressurized by two end pistons.

Fig. 3.16

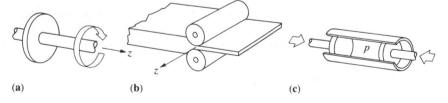

(a)   (b)   (c)

3.2 A 60 mm diameter mild-steel sphere has parallel flats machined on it 20 mm each side of the central axis. If a compressive load of 5 MN is applied perpendicular to the flats, calculate the decrease in length along the loading axis. The modulus of steel is 207 GN/m².

3.3 A copper band 20 mm wide and 2 mm thick is a snug fit on a 100 mm diameter steel bar which may be assumed to be rigid. Determine the stress in the copper if its temperature is lowered by 50 °C. $\alpha = 18 \times 10^{-6}/°C$, $E = 105$ GN/m².

3.4 What are the shear strain and angle of twist per unit length for the tube in Problem 2.14. $G = 85$ GN/m².

3.5 Express the stresses σ_x, σ_y, σ_z in terms of the three co-ordinate strains and the elastic constants. Obtain similar expressions for the cases of plane stress, $\sigma_z = 0$ and plane strain, $\varepsilon_z = 0$.

3.6 Show that the volumetric strain, ε_{vol}, in an element subjected to triaxial stresses

σ_x, σ_y and σ_z, is given by

$$\varepsilon_{vol} = \frac{1 - 2v}{E}(\sigma_x + \sigma_y + \sigma_z)$$

3.7 Determine the maximum strain and change in diameter of the cylinder in Problem 2.7 if $E = 208\ \text{GN/m}^2$ and $v = 0.3$.

3.8 A rectangular steel plate of uniform thickness has a strain gauge rosette bonded to one surface at the centre as shown in Fig. 3.17. It is placed in a test rig which can apply a biaxial force system along the edges of the plate. If the measured strains are $+0.0005$ and $+0.0007$ in the x- and y-directions, determine the corresponding stresses set up in the plate and the strain through the thickness. $E = 208\ \text{GN/m}^2$, $v = 0.3$.

Fig. 3.17

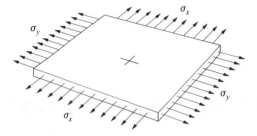

3.9 A trapeze artist weighs 50 kg and is balanced at the centre of a 3 mm diameter wire tightrope of 20 m length. There is an initial stress of 100 MN/m² in the tightrope before the artist balances on it. Determine the strain energy stored in the wire. $E = 208\ \text{GN/m}^2$.

3.10 Determine the shear-strain energy stored in the torsion tube of Fig. 2.13. $G = 85\ \text{GN/m}^2$.

Chapter 4

Statically indeterminate stress systems

INTRODUCTION

The subject has now been developed sufficiently so that statically indeterminate problems involving uniform normal tensile or compressive stress can be solved. The principles of solution are fundamental to the mechanics of solids and structures and require the writing of:

1. Equation(s) of *equilibrium of forces* (external (applied); internal (as a function of stress and area)).
2. Equation(s) describing the *geometry of deformation* or *compatibility of displacements*.
3. Relationships between *load–deformation* or *stress–strain* for the material(s).

It is of the utmost importance to remember the above principles and apply them in a logical manner to all future problems. This chapter is devoted to illustrating the application of these principles to a variety of problems encountered in design. The solutions have been deliberately worked in letter symbols rather than numerical values to enable the steps to be followed more easily.

INTERACTION OF DIFFERENT MATERIALS

If two or more different materials are integrated into the design of an engineering component, one contributing factor to the statical indeterminacy of the problem can be the different elastic constants.

Two simple examples will now be studied involving two different materials and also two different component configurations. It is important to note the differences in equilibrium and deformation relationships that arise in these two component arrangements.

EXAMPLE 4.1

A bi-metallic rod is subjected to a compressive force, F, as shown in Fig. 4.1. Determine the overall change in length.

Fig. 4.1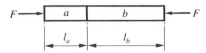

SOLUTION

The two materials will be denoted by subscripts a and b.

Equilibrium

It should be clear that the same force is carried through each material, so that

$$F_a = F_b = F \tag{4.1}$$

Geometry of deformation

The overall change in length is the sum of the changes in the two parts of the rod, so that

$$\delta = \delta_a + \delta_b \tag{4.2}$$

Stress–strain relations

Since it is a simple uniaxial stress system

$$\frac{\sigma_a}{\varepsilon_a} = E_a \tag{4.3}$$

$$\frac{\sigma_b}{\varepsilon_b} = E_b \tag{4.4}$$

Let the cross-sectional areas be A_a and A_b. Then equs. (4.3) and (4.4) can be re-expressed as

$$\frac{F_a}{A_a} = E_a \frac{\delta_a}{l_a} \tag{4.5}$$

and

$$\frac{F_b}{A_b} = E_b \frac{\delta_b}{l_b} \tag{4.6}$$

Substituting the values of δ_a and δ_b into eqn. (4.2) gives

$$\delta = \frac{F_a l_a}{A_a E_a} + \frac{F_b l_b}{A_b E_b}$$

Thus, using eqn. (4.1) and putting $A_a = A_b = A$

$$\delta = \frac{F}{A} \left(\frac{l_a}{E_a} + \frac{l_b}{E_b} \right)$$

EXAMPLE 4.2

Two components of different materials are arranged concentrically and loaded through rigid end plates as shown in Fig. 4.2. Determine the force carried by each component.

Fig. 4.2

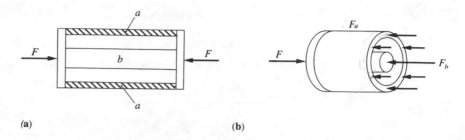

(a)   (b)

SOLUTION

Equilibrium   In this case the load is *shared* in some unknown proportions between the two parts, so that

$$F_a + F_b = F \tag{4.7}$$

Geometry of deformation   If the unloaded lengths, l, are initially the same, then they will remain the same under load; hence

$$\delta_a = \delta_b = \delta \tag{4.8}$$

Stress–strain relations   Again, for a simple uniaxial stress situation,

$$\frac{\sigma_a}{\varepsilon_a} = E_a \tag{4.9}$$

$$\frac{\sigma_b}{\varepsilon_b} = E_b \tag{4.10}$$

From eqns. (4.8), (4.9) and (4.10),

$$F_a = E_a A_a \frac{\delta}{l}$$

$$F_b = E_b A_b \frac{\delta}{l}$$

Substituting in the equilibrium equation (4.7),

$$E_a A_a \frac{\delta}{l} + E_b A_b \frac{\delta}{l} = F$$

Thus

$$\delta = \frac{Fl}{E_a A_a + E_b A_b} \tag{4.11}$$

$$F_a = \frac{F E_a A_a}{E_a A_a + E_b A_b} \tag{4.12}$$

and

$$F_b = \frac{F E_b A_b}{E_a A_a + E_b A_b} \tag{4.13}$$

INTERACTION OF DIFFERENT STIFFNESS COMPONENTS

Situations similar to those described on p. 73 can arise if an assembly is made of one type of material but certain component parts have different stiffnesses (load per unit deformation of an elastic member) which interact. For example, consider a flexible mounting for a machine of weight W consisting of a *rigid* (does not bend) rectangular plate supported on four coil springs each of stiffness K placed symmetrically with respect to the corners of the plate as shown in Fig. 4.3. After the machine has been located centrally on the plate, the springs are further loaded by means of four bolts each of stiffness k, having their lower ends fixed firmly into concrete and their upper ends passing freely through the plate. The nuts are tightened onto the plate and this has two effects because of the interaction between the bolts and the springs. The former are stretched slightly and the latter compressed. Various operational situations could be conceived in which the machine generated vertical reciprocating forces. These would alternately increase and decrease the interacting forces in the bolts and springs. A design criterion would consist of tightening the nuts sufficiently so that a tensile force of a particular magnitude was always maintained in the bolts.

Fig. 4.3

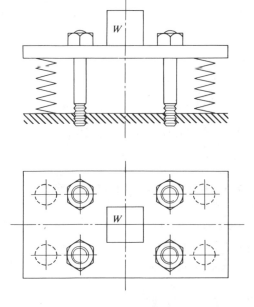

The following worked example will illustrate the form of solution.

EXAMPLE 4.3

A rigid member AB, the weight of which can be neglected, is supported horizontally at the pin joints A and C and by the spring at B is shown in

Fig. 4.4

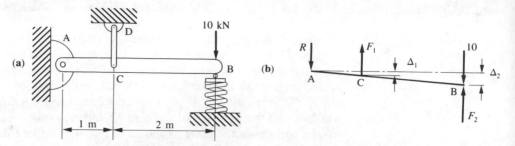

Fig. 4.4(a). The stiffness of member CD is 2 kN/m and of the spring is 5 kN/m. Calculate the force in CD, which is initially unstressed, and the reaction at A when a vertical load of 10 kN is applied at B.

SOLUTION

Equilibrium

Let the reaction at A be R and the forces in CD and in the spring be F_1 and F_2 respectively.

Vertical $10 + R - F_1 - F_2 = 0$ $\hspace{3cm}$ (4.14)

Moments $3R - 2F_1 = 0$ $\hspace{3.5cm}$ (4.15)

Geometry of deformation

The deformation is as shown in Fig. 4.4(b), so that from similar triangles

$$\frac{\Delta_1}{\Delta_2} = \frac{1}{3} \hspace{3cm} (4.16)$$

Load-deformation relations

$$\frac{F_1}{\Delta_1} = 2 \text{ kN/m} \hspace{3cm} (4.17)$$

$$\frac{F_2}{\Delta_2} = 5 \text{ kN/m} \hspace{3cm} (4.18)$$

From eqns. (4.16), (4.17) and (4.18),

$$F_1 = \frac{2}{15} F_2$$

From eqns. (4.14) and (4.15), eliminating R,

$$F_1 + 3F_2 = 30$$

Hence the force in CD is

$$F_1 = 1.276 \text{ kN}$$

and the reaction at A is

$$R = \frac{2}{3} F_1 = 0.851 \text{ kN}$$

RESTRAINT OF THERMAL STRAIN

When the temperature of a piece of material is changed its size will also change, and when expressed non-dimensionally this is termed *thermal* strain (*see* page 62). If the thermal strain is not restricted in any manner, then at the new steady temperature a previously unstressed component will remain unstressed. Clearly if there is any form of restraint to free change in size then *thermal stress* will result. The following two examples will illustrate this situation.

EXAMPLE 4.4

The bimetallic component illustrated in Fig. 4.2 consists of a steel rod of cross-sectional area 600 mm^2 coaxially surrounded by a copper tube of cross-sectional area 1200 mm^2. It is not subjected to any external load but its temperature is changed from 20 °C to 100 °C. Determine the axial stresses set up in the copper and the steel.

$$E_s = 205 \text{ GN/m}^2, E_c = 115 \text{ GN/m}^2, \alpha_s = 11 \times 10^{-6}/°C, \alpha_c = 16 \times 10^{-6}/°C.$$

SOLUTION

Equilibrium Since there is no applied external force, the sum of the internal forces in the copper and steel must be zero. Therefore

$$F_c + F_s = 0 \tag{4.19}$$

or

$$\sigma_c A_c + \sigma_s A_s = 0 \tag{4.20}$$

Geometry of deformation Since the two materials are initially stress-free and their ends are tied together the total strain must be the same for each. Therefore

$$\varepsilon_c = \varepsilon_s \quad \text{or} \quad (\varepsilon_\sigma + \varepsilon_T)_c = (\varepsilon_\sigma + \varepsilon_T)_s \tag{4.21}$$

where ε_σ = strain due to stress, ε_T = strain due to temperature change.

Stress–strain relation

$$\varepsilon_c = \frac{\sigma_c}{E_c} + \alpha_c(T - T_0) \tag{4.22}$$

$$\varepsilon_s = \frac{\sigma_s}{E_s} + \alpha_s(T - T_0) \tag{4.23}$$

Equating ε_c and ε_s from eqns. (4.22) and (4.23),

$$\frac{\sigma_c}{E_c} + \alpha_c(T - T_0) = \frac{\sigma_s}{E_s} + \alpha_s(T - T_0)$$

Using eqn. (4.20) to eliminate σ_s gives

$$\frac{\sigma_c}{E_c} + \alpha_c(T - T_0) = -\frac{1}{E_s}\frac{A_c}{A_s} + \alpha_s(T - T_0)$$

$$\sigma_c\left(\frac{1}{E_c} + \frac{A_c}{E_s A_s}\right) = (T - T_0)(\alpha_s - \alpha_c)$$

or

$$\sigma_c = \frac{A_s E_s E_c(T - T_0)(\alpha_s - \alpha_c)}{A_s E_s + A_c E_c}$$

and

$$\sigma_s = -\frac{A_c E_c E_s (T - T_0)(\alpha_s - \alpha_c)}{A_s E_s + A_c E_c}$$

The negative sign for σ_s does not necessarily indicate a compressive stress but simply that it is opposite in sign to σ_c. The type of stress in each material is determined by the numerical values of the quantities, T_0, T, α_s and α_c. Substituting the numerical values in the above equations gives

$$\sigma_c = -21.7\,\text{MN/m}^2$$

and

$$\sigma_s = -\frac{\sigma_c A_c}{A_s} = -\frac{-21.7 \times 1200}{600} = +43.4\,\text{MN/m}^2$$

Thus for an *increase* in temperature, α_c being greater than α_s, the copper is prevented from expanding as much as if it were free and is put into compression. The steel is forced to expand more than it would if free and is therefore in tension.

If there was the situation of both an applied load as in Fig. 4.2 and a change in temperature then the solution could conveniently be obtained by using the principle of superposition and adding together the two separate results.

EXAMPLE 4.5

A copper ring having an internal diameter of 150 mm and external diameter of 154 mm is to be shrunk onto a steel ring, of the same width, having internal and external diameters of 140 mm and 150.05 mm respectively.

What change in temperature is required in the copper ring so that it will just slide on to the steel ring?

What will be the uniform circumferential stress in each ring and also the interface pressure when assembled and back at room temperature?

Assume that there is no stress in the width direction. $E_s = 205\,\text{GN/m}^2$, $E_c = 100\,\text{GN/m}^2$, $\alpha_c = 18 \times 10^{-6}$ per deg C.

SOLUTION

The circumferential length of the copper ring has to be increased by heating till it is fractionally larger than the circumferential length of the steel ring.

Minimum required change in circumference $= \pi d_s - \pi d_c$

$$= \pi \times 0.05$$

Change in circumference due to heating $= \pi d_c \times \alpha(T - T_0)$

$$= \pi \times 150 \times 18 \times 10^{-6}(T - T_0)$$

Therefore

$$T - T_0 = \frac{0.05}{2700 \times 10^{-6}} = 18.5\,°C$$

When the assembly has returned to ambient temperature assume that the circumferential stresses in the copper and steel are uniformly distributed over each cross-section.

Equilibrium Let the width of each ring be w; then, with thicknesses of 2 and 5 mm respectively,

$$(w \times 2)\sigma_c + (w \times 5)\sigma_s = 0$$

$$\sigma_c = -2.5\sigma_s \qquad (4.24)$$

Geometry of deformation The circumferential strains in the copper and steel must be the same at the mating surface, so

$$\varepsilon_s = \varepsilon_c \qquad (4.25)$$

Stress–strain relations $$\varepsilon_s = \sigma_s/E_s \qquad (4.26)$$

and

$$\varepsilon_c = \sigma_c/E_c + \alpha_c\Delta T \qquad (4.27)$$

since it is only the copper ring that has the thermal strain component. From eqns. (4.25), (4.26) and (4.27),

$$\frac{\sigma_s}{E_s} = \frac{\sigma_c}{E_c} + \alpha_c\Delta T$$

Using eqn. (4.24),

$$\sigma_s\left(\frac{1}{E_s} + \frac{2.5}{E_c}\right) = -18 \times 10^{-6} \times 18.5$$

The negative sign is due to ΔT being a reduction in temperature. Substituting for E_s and E_c,

$$\sigma_s = -11.15\,\text{MN/m}^2$$

and

$$\sigma_c = +27.9\,\text{MN/m}^2$$

The radial pressure at the interface between the two rings may be treated as a thin cylinder under internal or external pressure, so that

$$p = \sigma_c t/r = 27.9 \times 2/75 = 745\,\text{kN/m}^2$$

VOLUME CHANGES

The following problem analyses the change in volume of a vessel subjected to pressure and makes use of the relationship between hydrostatic stress and volume strain.

EXAMPLE 4.6

A thin spherical steel shell has a mean diameter of 3 m, a wall thickness of 6 mm, and is just filled with water at 20 °C and atmospheric pressure. Find the rise in gauge pressure if the temperature of the water and shell rises to 50 °C, and then determine the volume of water that would escape if a small leak developed at the top of the vessel.

Steel: Young's modulus, $E = 200 \text{ GN/m}^2$
Coefficient of linear expansion $= 11 \times 10^{-6}$ per deg C
Poisson's ratio $= 0.3$
Water: Bulk modulus, $K = 2.2 \text{ GN/m}^2$
Coefficient of volumetric expansion $= 0{\cdot}207 \times 10^{-3}$ per deg C

SOLUTION

Equilibrium

Let the gauge pressure in the sphere after rise in temperature be p; then from Chapter 2 the equilibrium condition is

$$\sigma = \frac{pr}{2t} = 125p \tag{4.28}$$

Geometry of deformation

If there is to be a pressure at all then the water and sphere must remain in overall contact, and hence

Change in volume of sphere = Change in volume of water

or

$$\varepsilon_{v\,sphere} = \varepsilon_{v\,water} \tag{4.29}$$

since the original volume is the same for each.

Stress–strain relations

For the water the total volumetric strain is the sum of that due to pressure and that due to thermal strain:

$$\varepsilon_{v\,water} = -(p/K) + \alpha_v(T - T_0) \tag{4.30}$$

For the sphere the total volumetric strain is a function of strain due to stress (from pressure) and thermal strain:

$$\varepsilon_{v\,sphere} = \varepsilon_{v\,stress} + \varepsilon_{v\,thermal} \tag{4.31}$$

From eqn. (4.30),

$$\varepsilon_{v\,water} = -\frac{p}{2{\cdot}2 \times 10^9} + (0{\cdot}207 \times 10^{-3} \times 30)$$
$$= -(0{\cdot}445p \times 10^{-9}) + (6210 \times 10^{-6})$$

The change in the internal capacity or volume of the sphere is

$$\tfrac{4}{3}\pi(r + \delta r)^3 - \tfrac{4}{3}\pi r^3$$

which gives, neglecting products of the small quantity δr,

$$\tfrac{4}{3}\pi \times 3r^2 \delta r$$

Expressing this as a volumetric strain,

$$\frac{\frac{4}{3}\pi \times 3r^2 \delta r}{\frac{4}{3}\pi r^3} = 3\frac{\delta r}{r}$$

It will now be shown that $\delta r/r$ is the linear or hoop strain in the material of the sphere.

Change in circumference $= 2\pi(r + \delta r) - 2\pi r = 2\pi\delta r$

Therefore

$$\text{Hoop strain} = \frac{2\pi\delta r}{2\pi r} = \frac{\delta r}{r}$$

So that

Volumetric strain of vessel $= 3 \times$ hoop strain

Now, the hoop strain is given by eqn. (3.3):

$$\varepsilon = \frac{\sigma}{E} - \frac{v\sigma}{E} + \alpha(T - T_0)$$

$$= \frac{125p}{200 \times 10^9}(1 - 0.3) + (11 \times 10^{-6} \times 30)$$

Therefore the total volumetric strain is

$$\varepsilon_{v\,sphere} = 3\left(\frac{125 \times 0.7p}{200 \times 10^9} + 330 \times 10^{-6}\right)$$

Hence, using eqn. (4.29),

$$(-0.445p \times 10^9) + (6210 \times 10^{-6}) = 3\left[\frac{125 \times 0.7p}{200 \times 10^9} + (330 \times 10^{-6})\right]$$

from which

$$p = 2.97 \text{ MN/m}^2$$

The volume of water which escapes through the leak is simply the difference of the *free* thermal expansions of the water and the vessel, since obviously there is no pressure present to affect the issue.

Volume of water escaping
$$= [(6210 \times 10^{-6}) - (3 \times 330 \times 10^{-6})] \times (4/3)\pi \times 1500^3$$
$$= 7.4 \times 10^{-3} \text{ m}^3$$

CONSTRAINED MATERIAL

Examples 4.1 and 4.2 introduced the concept of two materials reacting against each other, but in a simple *uniaxial* load and stress situation. There are some engineering components in which material is constrained on a two- or three-dimensional basis, so that more complex stress and strain distributions result. The following example, although somewhat contrived, provides an illustration of the use of the general stress–strain relationships.

EXAMPLE 4.7

A cylindrical block of concrete is encompassed by a close-fitting thin steel tube of inside radius r and wall thickness t as shown in Fig. 4.5. If the concrete is subjected to a uniform axial compressive stress σ_x, determine the required ratio of r/t so that the axial strain in the tube is equal to that in the concrete. The moduli and Poisson's ratio for concrete and steel are E_c, ν_c and E_s, ν_s respectively.

Fig. 4.5

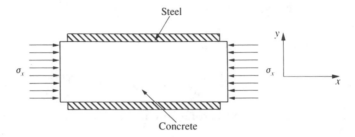

SOLUTION

The physical nature of the problem is that as the concrete is compressed it expands laterally on to the tube. This action sets up circumferential tensile stress in the tube which in turn results in axial contraction. It is required to make the latter the same as the compression of the concrete by a suitable choice of r/t. Superscripts c and s will be used to anotate the stresses and strains in the concrete and steel. Using eqns. (3.2),

$$\text{Axial strain in concrete, } \varepsilon_x^c = \frac{\sigma_x^c}{E_c} - \frac{\nu_c}{E_c}\sigma_y^c - \frac{\nu_c}{E_c}\sigma_z^c \qquad (4.32)$$

($\sigma_y^c = \sigma_z^c$ from symmetry).

Circumferential strain in concrete,

$$\varepsilon_y^c = \frac{\sigma_y^c}{E_c}(1 - \nu_c) - \frac{\nu_c\sigma_x^c}{E_c} \qquad (4.33)$$

Since the hoop stress, σ_y^s, in the tube is derived from the pressure σ_y^c exerted by the concrete substituted into eq. (2.10), and $\sigma_x^s = \sigma_z^s = 0$.

$$\text{Circumferential strain in tube, } \varepsilon_y^s = \frac{\sigma_y^s}{E_s} + \frac{\sigma_y^c r}{tE_s} \qquad (4.34)$$

$$\text{Axial strain in tube, } \varepsilon_x^s = -\nu_s\varepsilon_y^s = -\frac{\nu_s\sigma_y^s}{E_s} = -\frac{\nu_s\sigma_y^c r}{tE_s} \qquad (4.35)$$

The geometry of deformation conditions are that the circumferential strains in the concrete and steel are equal:

$$\varepsilon_y{}^c = \varepsilon_y{}^s$$

and similarly the axial strains are equal:

$$\varepsilon_x{}^c = \varepsilon_x{}^s$$

From eqns. (4.33) and (4.34),

$$\frac{\sigma_y{}^c}{E_c}(1 - v_c) - \frac{v_c\sigma_x{}^c}{E_c} = \frac{\sigma_y{}^c r}{tE_s}$$

$$\sigma_y{}^c\left(\frac{1 - v_c}{E_c} - \frac{r}{tE_s}\right) = \frac{v_c\sigma_x{}^c}{E_c} \qquad (4.36)$$

and from eqns. (4.32) and (4.35),

$$\frac{\sigma_x{}^c}{E_c} - \frac{2v_c}{E_c}\sigma_y{}^c = -\frac{v_s\sigma_y{}^c r}{tE_s}$$

$$\sigma_y{}^c\left(\frac{2v_c}{E_c} - \frac{v_s r}{tE_s}\right) = \frac{\sigma_x{}^c}{E_c} \qquad (4.37)$$

From eqns. (4.36) and (4.37),

$$\frac{r}{tE_s}(1 - v_s v_c) = \frac{1 - v_c}{E_c} - \frac{2v_c{}^2}{E_c}$$

Hence

$$\frac{r}{t} = \frac{E_s}{E_c}\frac{1 - v_c - 2v_c{}^2}{1 - v_s v_c} \qquad (4.38)$$

Taking $E_s = 15\,E_c$, $v_s = 0.29$ and $v_c = 0.25$,

$$\frac{r}{t} = \frac{15[1 - 0.25 - (2 \times 0.25^2)]}{1 - (0.29 \times 0.25)} = \frac{15 \times 0.625}{0.9275} = 10.1$$

MAXIMUM STRESS DUE TO A SUDDENLY-APPLIED LOAD

In Chapter 3, page 67 dealing with the elastic strain energy stored
under uniaxial stress, the load was applied gradually so that the work
done was the *average* load times the distance moved at the point where
the load was applied. Now suppose that a bar fixed at the top with a
flange at the lower end, as in Fig. 4.6(a), has a load W suddenly released
on to the flange. Let the momentary maximum extension, strain and
stress in the bar be $\delta l'$, ε' and σ' respectively. In addition, if the masses
of the bar and flange are small compared with the load W, then a
reasonable approximation to the behaviour is made by neglecting the
effect of the former. Because the *full* load moves through the extension
$\delta l'$,

Work done $= W\delta l'$

Fig. 4.6

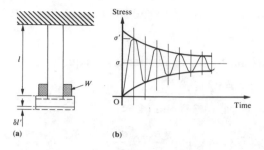

(a) (b)

The strain energy stored per unit volume momentarily is $\frac{1}{2}\sigma'\varepsilon'$, so that

$$U = \tfrac{1}{2}\sigma'\varepsilon'al = \tfrac{1}{2}\sigma'a\delta l'$$

Since the work done is equal to the strain energy,

$$W\delta l' = \tfrac{1}{2}\sigma'a\delta l'$$

Thus $\sigma' = 2W/a$; but $W/a = \sigma$, the stress due to a gradually applied load, so that

$$\sigma' = 2\sigma \tag{4.39}$$

or the momentary maximum stress due to a suddenly-applied load is twice the stress for a gradually applied load. The bar will subsequently oscillate about the statical equilibrium position while the stresses and deformations rapidly die away, as shown in Fig. 4.6(b), to the value obtained for a gradually-applied load. However, the momentary stress intensification by a factor of 2 might have serious consequences on a component.

MAXIMUM STRESS DUE TO IMPACT

An extension of the above problem is the case where the load W is dropped on to the flange from a height h, causing a momentary extension of the bar $\delta l'$. The total potential energy is $W(h + \delta l')$, and the momentarily stored strain energy is

$$U' = \tfrac{1}{2}\sigma'a\delta l'$$

Then neglecting the mass of the bar and flange and assuming no losses of energy during impact,

$$\tfrac{1}{2}\sigma'a\delta l' = W(h + \delta l')$$

or

$$\tfrac{1}{2}\sigma'\delta l' = \frac{Wh}{a} + \frac{W}{a}\delta l' \tag{4.40}$$

Now $\delta l' = (\sigma'/E)l$, and $W/a = \sigma$ is the final steady stress; substituting into eq. (4.40),

$$\sigma'^2 - 2\sigma\sigma' - 2\sigma\frac{Eh}{l} = 0$$

Therefore

$$\sigma' = \sigma + \left(\sigma^2 + 2\sigma\frac{Eh}{l}\right)^{1/2} \qquad (4.41)$$

It will be seen from this equation that if $h = 0$ then $\sigma' = 2\sigma$, which is the result obtained in the previous section.

The true situation for the stress during impact of one body on another is more complicated than is indicated in this approximate analysis. In practice the deformation and stress imposed on the bar at the point of impact take time to propagate along the length of the bar. The *stress wave*, as it is called, on reaching the fixed end of the bar will be reflected towards the point of initiation and thus a complex state of stress will arise.

EXAMPLE 4.8

The lower part of a child's pogo stick is illustrated in Fig. 4.7 when the child and stick are just about to descend to the ground and the spring is undeformed. Determine the momentary maximum stress in the steel compression tube on impact with the ground, and compare this with the final steady stress. It may be assumed that the outer sleeve and supports are rigid and that the ground does not deform. The weights of the various parts may be neglected. $E = 208\ \text{GN/m}^2$; spring stiffness $= 18\ \text{kN/m}$.

Fig. 4.7

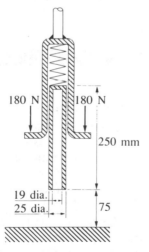

180 N   180 N

250 mm

19 dia.
25 dia.

75

SOLUTION

Let δ be the compression of the spring, and x the compression of the tube. If the force in the spring on impact is momentarily F then the strain energy stored in the spring and tube is

$$\frac{F\delta}{2} + \frac{\sigma^2}{2E} \times \text{volume}$$

$$\text{Tube volume} = 250\frac{\pi}{4}(25^2 - 19^2) \times 10^{-9} = 51.8 \times 10^{-6}\,\text{m}^3$$

$$\text{Potential energy lost on impact} = 2 \times 180(0.075 + \delta + x)$$

and

$$\delta = \frac{F}{18\,000}\,\text{m} \qquad x = \frac{Fl}{AE} = \frac{F \times 250 \times 10^{-3}}{208 \times 10^{-6} \times 208 \times 10^9}\,\text{m}$$

Equating potential and strain energies,

$$360\left(0.075 + \frac{F}{18\,000} + \frac{250F \times 10^{-6}}{208 \times 208}\right)$$

$$= \frac{F^2}{36\,000} + \frac{F^2 \times 51.8 \times 10^{-6}}{2 \times (208 \times 10^{-6})^2 \times 208 \times 10^9}$$

$$(F^2 \times 0.0278 \times 10^{-3}) - (F \times 20 \times 10^{-3}) - 27 = 0$$

$$F^2 - 720F - (970 \times 10^3) = 0$$

from which $F = 1410\,\text{N}$.

$$\sigma_{max} = \frac{1410}{208 \times 10^{-6}} = 6.78\,\text{MN/m}^2$$

$$\text{Final steady stress, } \sigma = \frac{360}{208 \times 10^{-6}} = 1.73\,\text{MN/m}^2$$

SUMMARY

The importance of this chapter centres on the three requirements of principle set out in the Introduction, page 72. Because a problem cannot be solved by equilibrium statements alone, it is then described as statically indeterminate and we have to assess the geometry of deformation and also link stress and strain through the modulus and Poisson's ratio for the material. The bulk of the chapter has been devoted to a series of illustrative examples which set out the above three steps as appropriate to each case. It is therefore very important to work through each example to achieve a full understanding of the formulation of the equations, since from that point the solution is merely manipulative computation. The foregoing principles will be extensively used in the next two chapters on torsion and bending and thus this chapter must be thoroughly understood.

PROBLEMS

4.1 A composite shaft consists of a brass bar 50 mm in diameter and 200 mm long, to each end of which are concentrically friction-welded steel rods of 20 mm diameter and 100 mm length. During a tensile test to check the welds, on the composite bar at a particular stage the overall extension is measured as 0·15 mm. What are the axial stresses in the two parts of the bar? $E_{\text{brass}} = 120 \text{ GN/m}^2$, $E_{\text{steel}} = 208 \text{ GN/m}^2$.

4.2 A spring-loaded buffer stop is illustrated in Fig. 4.8. The spring, which has a stiffness of 6 kN/mm, is located on the end of a steel tube of inner and outer diameters 21 mm and 29 mm respectively and 150 mm length. Determine accurately the total axial displacement of the system under a load of 30 kN. What would be the simple approximate solution? $E = 207 \text{ GN/m}^2$.

Fig. 4.8

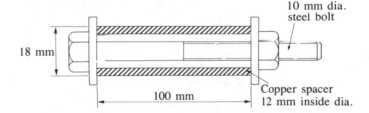

4.3 The steel bolt shown in Fig. 4.9 has a thread pitch of 1·6 mm. If the nut is initially tightened up by hand so as to cause no stress in the copper spacing tube calculate the stresses in the copper and the bolt if a spanner is then used to turn the nut through 90°. $E_c = 100 \text{ GN/m}^2$, $E_s = 208 \text{ GN/m}^2$.

Fig. 4.9

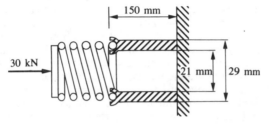

4.4 A hydraulic cylinder of 80 mm inside diameter and 4 mm wall thickness is welded to rigid end plates as shown in Fig. 4.10. The end plates are tied together by four rods of 8 mm diameter symmetrically arranged around the cylinder. Calculate the stresses in the rods and the cylinder at the cylinder design pressure of 20 MN/m². The cylinder and tie rods are made from steel with a Poisson's ratio value of 0.3.

Fig. 4.10

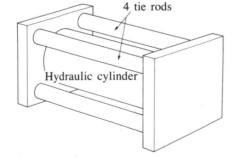

4.5 A rigid chute 2 m in length is supported horizontally, at a height of 0.33 m above a hopper, by a spring at one end of stiffness 30 kN/m and a second spring at

Fig. 4.11

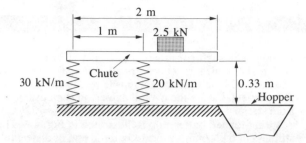

mid-length of stiffness 20 kN/m (Fig. 4.11). Determine the position on the chute which a component of 2.5 kN reaches when the unsupported end of the chute just touches the edge of the hopper.

4.6 An elastic packing piece is bolted between a rigid rectangular plate and a rigid foundation by two bolts pitched 300 mm apart and symmetrically placed on the long centre-line of the plate, which is 450 mm long. The tension in each bolt is initially 120 kN, the extension of each bolt is 0.015 mm and the compression of the packing piece is 0.6 mm. If one bolt is further tightened to a tension of 150 kN, determine the tension in the other bolt.

4.7 A bimetallic temperature-sensitive component consists of a short steel tube of outside diameter 70 mm and inside diameter 60 mm, surrounding a solid copper rod of 50 mm diameter. At 20 °C the rod and cylinder have exactly the same length. If a 100 kN load is placed on top of the rod and cylinder, calculate the forces in the two materials if the whole assembly is heated to 60 °C. Calculate also the temperature at which the copper would take all the force. $E_s = 208$ GN/m^2, $E_c = 104$ GN/m^2, $\alpha_s = 12 \times 10^{-6}$ per deg C, $\alpha_c = 18.5 \times 10^{-6}$ per deg C.

4.8 A steel tube of 150 mm internal diameter and 8 mm wall thickness in a chemical plant is lined internally with a well-fitting copper sleeve of 2 mm wall thickness. If the composite tube is initially unstressed, calculate the circumferential stresses set up, assumed to be uniform through the wall thickness, in a unit length of each part of the tube due to an increase in temperature of 100 °C. Neglect any temperature effect in the axial direction. For steel $\alpha = 11 \times 10^{-6}$ per deg C, $E = 208$ GN/m^2; for copper $\alpha = 18 \times 10^{-6}$ per deg C, $E = 104$ GN/m^2.

4.9 A part of an aeroplane structure may be represented by the simplified form shown in Fig. 4.12. The vertical member is made of titanium ($E = 105$ GN/m^2) and the horizontal members are stainless steel ($E = 200$ GN/m^2). All sections are 25 mm thick. If the titanium member is subjected to a temperature rise of 100 °C calculate the maximum bending moment set up in the steel members. The latter

Fig. 4.12

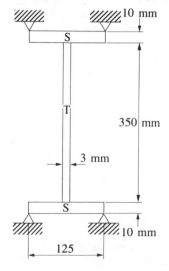

may be regarded as simply supported at each end and the weight of all members may be neglected. The coefficient of thermal expansion of titanium is 9×10^{-6} per deg C.

4.10 For a hydraulic test a steel tube of 80 mm internal diameter, 2 mm wall thickness and 1.2 m in length is fitted with end plugs and filled with oil at a pressure of $2 \, \text{MN/m}^2$. Determine the volume of oil leakage which would cause the pressure to fall to $1.5 \, \text{MN/m}^2$. Bulk modulus for the oil $= 2.8 \, \text{GN/m}^2$; for the steel $E = 208 \, \text{GN/m}^2$, $v = 0.29$.

4.11 A drop-weight shearing device consists of a vertical rod with a cross-sectional area of $125 \, \text{mm}^2$ which has an end collar which supports a spring of stiffness 150 kN/m, as shown in Fig. 4.13. If the shear tool of 10 kg mass is dropped through a height of 500 mm on to the spring, calculate (*a*) the initial instantaneous extension of the rod, (*b*) the maximum stress in the rod, and (*c*) the initial instantaneous compression of the spring. $E = 208 \, \text{GN/m}^2$.

Fig. 4.13

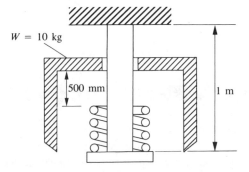

4.12 A drop-hammer used for forging metal is illustrated in Fig. 4.14. If the hammer is dropped through a height of 1 m on to the workpiece, calculate the resulting compression of the workpiece. Compare the force transmitted to the foundation for the system shown with that transmitted if the workpiece were resting on the foundation before the hammer was dropped 1 m on to it. Press: Mass of hammer $= 12\,000 \, \text{kg}$, mass of anvil $= 5000 \, \text{kg}$, spring stiffness $= 15 \, \text{MN/m}$. Workpiece: 30 mm dia. $\times$ 30 mm tall, modulus, $E = 208 \, \text{GN/m}^2$.

Fig. 4.14

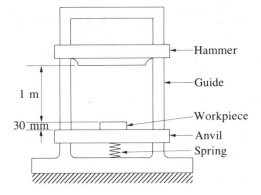

Chapter 5

Torsion

INTRODUCTION

One of the common engineering modes of deformation is that of torsion, in which a solid or tubular member is subjected to torque about its longitudinal axis resulting in twisting deformation. A design analysis is required in order to estimate shear stress distribution and angular twist for solid and hollow shafts of circular cross-section and thin-walled closed and open noncircular cross-sections. Engineering examples of the above are obtained in shafts transmitting power in machinery and transport, structural members in aeroplanes, springs, etc.

TORSION OF A THIN-WALLED CYLINDER

The thin cylinder of mean radius r, thickness t and length L, shown in Fig. 5.1, is subjected to an axial torque T at each end which causes the cylinder to twist about its longitudinal axis. In Chapter 2 it was shown that as a statically determinate problem only circumferential uniform shear stress in the wall of the cylinder $\tau_{z\theta}$ was set up as a reaction to the torque T as shown in Fig. 5.1.

Equilibrium Reiterating the steps on page 48, the shear stress $\tau_{z\theta}$ acting on an element of wall $tr\,d\theta$ gives a shear force

$$F = \tau_{z\theta} tr\,d\theta$$

This will provide a reacting moment about the central axis

$$Fr = \tau_{z\theta} tr^2 d\theta$$

hence the total reacting torque will be

$$\int_0^{2\pi} \tau_{z\theta} tr^2 d\theta$$

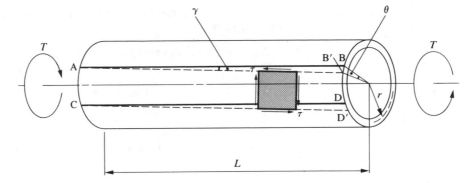

Fig. 5.1

which is in equilibrium with the applied torque T. Therefore

$$T = \tau_{z\theta} tr^2 2\pi$$

or

$$\tau_{z\theta} = \frac{T}{2\pi r^2 t} \tag{5.1}$$

It should also be noted that the circumferential shear stress $\tau_{z\theta}$ is associated with a complementary shear stress $\tau_{\theta z}$ in the longitudinal direction in the wall. As there are no other shear stresses present, for simplicity the $z\theta$ suffices will be omitted.

Geometry of deformation

The rotation of one end of the cylinder relative to the other through an angle θ results in a change in angle γ between a cross-section and a longitudinal generator on the cylinder as in Fig. 5.1. The angle γ is the shear strain associated with the shear stress. The displacement of B to B' may be expressed both as $r\theta$ and γl and therefore

$$\gamma l = r\theta$$

and

$$\gamma = \frac{r\theta}{l} \tag{5.2}$$

Stress–strain relationship

The stress–strain relationhship in shear is expressed as

$$\frac{\tau}{\gamma} = G \tag{5.3}$$

From eqns. (5.1), (5.2) and (5.3) we can define the interrelationships for the cylinder as

$$\tau = \frac{Gr\theta}{l} = \frac{T}{2\pi r^2 t} \tag{5.4}$$

TORSION OF A SOLID CIRCULAR SHAFT

In the case of the thin-walled cylinder, in the previous section the shear stress was assumed to be constant throughout the wall, but in the case of a solid cylinder the shear stress varies over the cross-section. Firstly we shall require that:

(a) The shaft is straight and of uniform cross section over its length.
(b) The torque is constant along the length of the shaft.

Further we note that the longitudinal and transverse symmetry of the shaft in relation to the applied torque enables the following deductions to be made:

1. Cross-sections which are plane before twisting remain plane during twisting.
2. Radial lines remain radial during twisting.
3. Deformation is by rotation of one plane relative to the next and planes remain normal to the axis of the shaft.

RELATION BETWEEN STRESS, STRAIN AND ANGLE OF TWIST

Geometry of deformation

The cylindrical shaft of length L and outer radius r_0 subjected to torque T may be regarded as being built up of a large number of thin-walled tubes just fitting inside each other. They are all twisted through the same angle of rotation θ, therefore for any arbitrary tubes of radius r_p and r_q experiencing shear strain γ_p, and γ_q from (5.2) we may write

$$\theta = \frac{\gamma_p L}{r_p} = \frac{\gamma_q L}{r_q}$$

or

$$\frac{\gamma_p}{r_p} = \frac{\gamma_q}{r_q} = \text{constant}$$

which demonstrates that at the centre of the shaft where r is zero, γ is zero, and that at the surface γ is maximum and the variation is linear. Therefore

$$\frac{\gamma}{r} = \frac{\theta}{L} \tag{5.5}$$

Stress–strain relation

The shear stress – shear strain relation in terms of the shear modulus is

$$\frac{\tau}{\gamma} = G \tag{5.6}$$

From eqns. (5.5) and (5.6) we have

$$\frac{\tau}{r} = \frac{G\theta}{L} \tag{5.7}$$

Thus shear stress has a linear distribution across the shaft diameter, being zero at the centre and maximum at the other surface, as shown in Fig. 5.2.

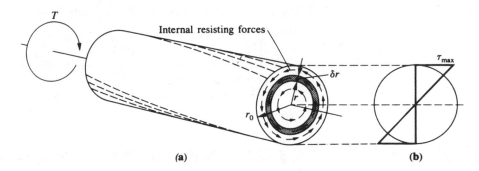

Fig. 5.2 (a) Section of shaft; (b) stress distribution

T

Internal resisting forces

δr

r_0

τ_{max}

(a)

(b)

RELATION BETWEEN TORQUE AND SHEAR STRESS

Consider any one of the thin tubes of radius r and thickness δr on which the shear stress is τ.

Equilibrium

Force per unit length $= \tau \, \delta r$
Torque per unit length of tube about
shaft axis $= \tau r \, \delta r$
Resisting torque on whole tube $= \tau r \, \delta r 2\pi r$

Resisting torque for whole cross-section $= \int_0^{r_0} \tau 2\pi r^2 dr$

This is equal to the applied torque; therefore

$$T = \int_0^{r_0} \tau 2\pi r^2 dr \tag{5.8}$$

Using, eqn. (5.7) to substitute for τ,

$$T = \int_0^{r_0} \frac{G\theta}{L} 2\pi r^3 dr = \frac{G\theta}{L} \int_0^{r_0} 2\pi r^3 dr \tag{5.9}$$

The integral function

$$\int_0^{r_0} 2\pi r^3 dr = \frac{\pi r_0^4}{2}$$

is the polar second moment of area of the section (see Appendix A) denoted by J. Therefore,

$$T = \frac{G\theta}{L} J$$

or

$$\frac{T}{J} = \frac{G\theta}{L} \tag{5.10}$$

and from eqn. (5.7),

$$\frac{T}{J} = \frac{G\theta}{L} = \frac{\tau}{r} \tag{5.11}$$

The quantities concerned and their units are

T = Torque, N–m
J = Polar second moment of area, m^4 (or mm^4)
τ = Shear stress, N/m^2 at radius r, m
G = Shear modulus, N/m^2
θ = Angle of twist, rad, over length L, m

EXAMPLE 5.1

Calculate the size of shaft which will transmit 40 kW at 2 rev/s. The shear stress is to be limited to 50 MN/m^2 and the twist of the shaft is not to exceed 1° for each 2 m length of shaft. The shear modulus G is 77 GN/m^2.

SOLUTION

Firstly converting the power to be transmitted into a torque,

$$T = \frac{40\,000}{2\pi \times 2} = 3183 \text{ N-m}$$

From eqn. (5.11)

$$\tau_{max} = \frac{Tr_0}{J} = \frac{T}{\frac{1}{2}\pi r_0^3}$$

$$r_0^3 = \frac{2 \times 3183 \times 10^9}{\pi \times 50 \times 10^6} = 40.6 \times 10^3 \text{ mm}^3$$

$$r_0 = 34.4 \text{ mm } on \ a \ stress \ basis$$

Considering the twist criterion, from eqn. (5.10),

$$r_0^4 = \frac{2TL}{\pi G\theta}$$

$$= \frac{2 \times 3183 \times 2 \times 57.3 \times 10^{12}}{\pi \times 77 \times 10^9 \times 1}$$

$$= 302 \times 10^4 \text{ mm}^4$$

$$r_0 = 41.7 \text{ mm } on \ a \ twist \ basis.$$

This is the governing criterion, and therefore shaft diameter = 83.4 mm.

TORSION OF A HOLLOW CIRCULAR SHAFT

The above analysis for the solid shaft is similarly applicable to the hollow shaft (thick-walled tube). Thus the torsion relationship equation (5.11) also expresses the conditions of equilibrium and compatibility for a hollow circular shaft. However, the radial boundaries are now $r = r_1$ and $r = r_2$, the outer and inner radii respectively, and thus the polar second moment of area is

$$J = \int_{r_2}^{r_1} 2\pi r^3 dr$$

$$= \frac{\pi}{2}(r_1{}^4 - r_2{}^4)$$

The shear stress varies linearly from Tr_2/J at the bore to Tr_1/J at the outer surface, as shown in Fig. 5.3.

Fig. 5.3

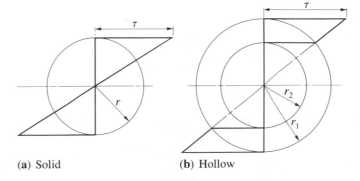

(a) Solid   (b) Hollow

The hollow shaft is more efficient in its use of stressed material than the solid shaft because the core of a solid shaft has relatively low stresses as compared with the outer layers. However, hollow shafts are not used widely in practice owing to the cost of machining, unless saving of weight is at a premium or it is necessary to pass services down the centre of the shaft.

EXAMPLE 5.2

Compare the torque that can be transmitted by a hollow shaft with that of a solid shaft, as shown in Fig. 5.3, of the same material, weight, length, and allowable stress.

SOLUTION

Let r_1 and r_2 be the outer and inner radii of the hollow shaft and let r be the radius of the solid shaft. Then, for the same maximum shear stress τ,

$$T_{hollow} = \frac{\tau}{r_1}\frac{\pi}{2}(r_1{}^4 - r_2{}^4)$$

and

$$T_{solid} = \frac{\tau}{r}\frac{\pi}{2}r^4$$

Eliminating τ we have

$$\frac{T_{hollow}}{T_{solid}} = \frac{r_1{}^4 - r_2{}^4}{r_1 r^3} \tag{5.12}$$

Since, for shafts of the same weight, $\pi r^2 = \pi(r_1{}^2 - r_2{}^2)$ per unit length, eqn. (5.12) can be simplified to

$$\frac{T_{hollow}}{T_{solid}} = \frac{r_1{}^2 + r_2{}^2}{r_1 r} = \frac{r_1}{r}\left(1 + \frac{1}{n^2}\right) \tag{5.13}$$

where $n = r_1/r_2$

Now $r^2 = r_1{}^2 - r_2{}^2$, and putting r_2 in terms of r_1/n gives

$$r^2 = r_1{}^2 - \left(\frac{r_1}{n}\right)^2 \quad \text{or} \quad \left(\frac{r_1}{r}\right) = \frac{n}{\sqrt{(n^2 - 1)}}$$

Therefore

$$\frac{T_{hollow}}{T_{solid}} = \frac{n^2 + 1}{n\sqrt{(n^2 - 1)}} \tag{5.14}$$

It is common practice to take $n = 2$, which gives

$$\frac{T_{hollow}}{T_{solid}} = \frac{5}{2\sqrt{3}} = 1.44$$

Thus the hollow shaft can carry 44% greater torque than the solid shaft for the same weight, etc.

TORSION OF NON-UNIFORM AND COMPOSITE SHAFTS

In certain shaft arrangements, the complete shaft is continuous but not of uniform diameter (Fig. 5.4), or the arrangement may be such that one

Fig. 5.4

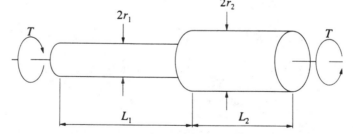

shaft is hollow with another shaft arranged coaxially (Fig. 5.5). In each case it is necessary to investigate the conditions of both the torque and the angle of twist in each part of the system in order to obtain a sufficient number of equations for the solution.

Fig. 5.5

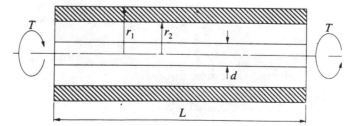

CONTINUOUS SHAFT HAVING TWO DIFFERENT DIAMETERS

In this case (Fig. 5.4) the total torque T is transmitted by each portion of the shaft; thus:

Equilibrium

$$T = T_1 = T_2 \tag{5.15}$$

Geometry of deformation

The total deformation θ is due to θ_1 over length L_1, plus θ_2 over length L_2, so that

$$\theta = \theta_1 + \theta_2 \tag{5.16}$$

Substituting for T_1 and T_2 in eqn. (5.15),

$$T = \frac{\tau_1}{r_1} J_1 = \frac{\tau_2}{r_2} J_2$$

where τ_1 and τ_2 are the surface shear stresses

$$\frac{\tau_1}{r_1} \frac{\pi r_1^4}{2} = \frac{\tau_2}{r_2} \frac{\pi r_2^4}{2}$$

$$\frac{\tau_1}{\tau_2} = \left(\frac{r_2}{r_1}\right)^3 \tag{5.17}$$

Considering the ratio of the angle of twist per unit length, we have

$$\theta_1 = \frac{\tau_1 L_1}{r_1 G} \quad \text{and} \quad \theta_2 = \frac{\tau_2 L_2}{r_2 G}$$

Hence

$$\frac{\theta_1/L_1}{\theta_2/L_2} = \frac{\tau_1}{\tau_2} \frac{r_2}{r_1} = \left(\frac{r_2}{r_1}\right)^4 \tag{5.18}$$

The total twist of the shaft is obtained using (5.16),

$$\theta = \frac{1}{G}\left(\frac{\tau_1 L_1}{r_1} + \frac{\tau_2 L_2}{r_2}\right)$$

or, in terms of the torque,

$$\theta = \frac{T}{G}\left(\frac{L_1}{J_1} + \frac{L_2}{J_2}\right) \tag{5.19}$$

CONCENTRIC SHAFTS

In Fig. 5.5 the shafts have a common axis and are joined at the ends so that the total torque T is made up of that carried by the hollow shaft and that carried by the solid shaft, these being T_1 and T_2 respectively:

Equilibrium

$$T = T_1 + T_2$$

$$= \frac{\tau_1}{r_1} \frac{\pi}{2}(r_1^4 - r_2^4) + \frac{\dot{\tau}}{r} \frac{\pi}{2} r^4 \tag{5.20}$$

Geometry of deformation

Both shafts twist through the same angle θ since their ends are rigidly connected; hence

$$\theta = \theta_1 = \theta_2$$

or

$$\frac{T_1 L}{GJ_1} = \frac{T_2 L}{GJ_2}$$

when both shafts have the same shear modulus G. Therefore

$$\frac{T_1}{T_2} = \frac{J_1}{J_2} = \frac{r_1^4 - r_2^4}{r^4} \qquad (5.21)$$

Substituting for T_1 and T_2 in terms of the maximum shear stresses τ_1 and τ,

$$\frac{\tau_1 \dfrac{\pi}{2} \dfrac{(r_1^4 - r_2^4)}{r_1}}{\tau \dfrac{\pi}{2} r^3} = \frac{r_1^4 - r_2^4}{r^4}$$

Hence

$$\frac{\tau_1}{\tau} = \frac{r_1}{r} \qquad (5.22)$$

so that the ratio of the maximum shear stresses is the same as the ratio of the outer diameters.

EXAMPLE 5.3

A solid alloy shaft of 50 mm diameter is to be friction-welded concentrically to the end of a hollow steel shaft of the same external diameter (Fig. 5.6). Find the internal diameter of the steel shaft if the angle of twist per unit length is to be 75% of that of the alloy shaft.

Fig. 5.6

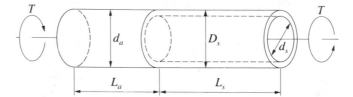

What is the maximum torque that can be transmitted if the limiting shear stresses in the alloy and the steel are 50 MN/m² and 75 MN/m² respectively? $G_{steel} = 2.2 G_{alloy}$.

SOLUTION

Equilibrium $\qquad T_{alloy} = T_{steel} = T \qquad (5.23)$

Geometry of deformation $\qquad \dfrac{\theta_s}{L_s} = 0.75 \dfrac{\theta_a}{L_a} \qquad (5.24)$

Hence

$$\frac{T_s}{J_s G_s} = 0.75 \frac{T_a}{J_a G_a}$$

Since

$$T_s = T_a \quad \text{and} \quad G_s = 2.2 G_a$$

$$J_a = 2.2 \times 0.75 J_s$$

$$\frac{\pi d_a^{\,4}}{32} = 2.2 \times 0.75 \frac{\pi}{32} (D_s^{\,4} - d_s^{\,4})$$

$$50^4 = (2.2 \times 0.75 \times 50^4) - (2.2 \times 0.75 \times d_s^{\,4})$$

$$d_s^{\,4} = \frac{0.65 \times 50^4}{2.2 \times 0.75} \qquad d_s = 39.6\,\text{mm}$$

The torque that can be carried by the alloy is

$$T = \frac{\pi d^3}{16}\tau = \frac{\pi \times 50^3}{16 \times 10^9} \times 50 \times 10^6 = 1227\,\text{N-m}$$

The torque that can be carried by the steel is

$$T = \frac{\pi}{16}\frac{(50^4 - 39.6^4)}{50 \times 10^9} \times 75 \times 10^6 = 1120\,\text{N-m}$$

Hence the maximum allowable torque is 1120 N-m.

EXAMPLE 5.4

A composite shaft of circular cross-section 0.5 m long is rigidly fixed at each end, as shown in Fig. 5.7. A 0.3 m length of the shaft is 50 mm in diameter and is made of bronze to which is joined the remaining 0.2 m length of 25 mm diameter made of steel. If the limiting shear stress in the steel is 55 MN/m² determine the maximum torque that can be applied at the joint. What is then the maximum shear stress in the bronze? $G_{steel} = 82\,\text{GN/m}^2$; $G_{bronze} = 41\,\text{GN/m}^2$.

Fig. 5.7

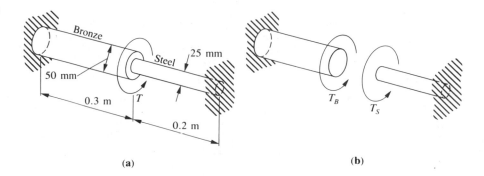

(a)   (b)

SOLUTION

Equilibrium

The free-body diagram Fig. 5.7(b) shows that the torque T is shared between the two parts of the shaft. Therefore

$$T = T_s + T_b$$

The torque that can be carried by the steel is

$$T_s = \frac{2 \times 55 \times 10^6}{0.025} \times \frac{\pi \times 0.025^4}{32} = 169 \text{ N-m}$$

Geometry of deformation

The angle of twist must be the same for each part at the joint, therefore

$$\theta_s = \theta_b$$

and the angle of twist for the steel is given by

$$\theta_s = \frac{169}{\frac{\pi}{32} \times 0.025^4} \times \frac{0.2}{82 \times 10^9} = 0.0108 \text{ rad}$$

Therefore $\theta_b = 0.0108$ rad and so

$$T_b = \frac{41 \times 10^9 \times 0.0108}{0.3} \times \frac{\pi \times 0.05^4}{32} = 906 \text{ N-m}$$

The total torque that can be applied at the joint is

$$T_b + T_s = 906 + 169 = 1075 \text{ N-m}$$

The maximum shear stress in the bronze is

$$\tau_b = \frac{41 \times 10^9 \times 0.0108}{0.3} \times \frac{25}{10^3} = 36.8 \text{ MN/m}^2$$

TORSION OF A TAPERED SHAFT

Suppose a twisting moment T is applied to the tapered shaft of length L, shown in Fig. 5.8. The resisting moment of all cross-sections of the shaft

Fig. 5.8

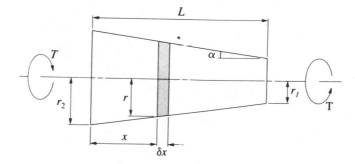

must be equal to T. Let

τ_1 = Maximum shear stress at end of radius r_1
τ_2 = Maximum shear stress at end of radius r_2
τ = maximum shear stress at cross-section of radius r

Then

$$\frac{\pi\tau_1 r_1^3}{2} = \frac{\pi\tau_2 r_2^3}{2} = \frac{\pi\tau r^3}{2} = T$$

or

$$\tau_1 r_1^3 = \tau_2 r_2^3 = \tau r^3 \tag{5.25}$$

Let δx be a small length of the shaft at distance x from the larger end and having a mean radius r and angle of twist $\delta\theta$. Then

$$\delta\theta = \frac{T}{JG}\,\delta x = \frac{2T}{G\pi}\frac{\delta x}{r^4} \tag{5.26}$$

But

$$r = r_2 - x\tan\alpha = r_2 - x\frac{r_2 - r_1}{L}$$

$$= r_2 - ax \tag{5.27}$$

where $a = (r_2 - r_1)/L$. Substituting eqn. (5.27) in eqn. (5.26),

$$\delta\theta = \frac{2T}{G\pi}(r_2 - ax)^{-4}\,\delta x$$

or, in the limit,

$$d\theta = \frac{2T}{G\pi}(r_2 - ax)^{-4}dx$$

The total angle of twist θ for the length L is given by

$$\theta = \int_0^L d\theta = \int_0^L \frac{2T}{G\pi}(r_2 - ax)^{-4}dx$$

$$= \frac{2T}{G\pi}\frac{1}{3a}\left[(r_2 - ax)^{-3}\right]_0^L$$

$$= \frac{2T}{G\pi}\cdot\frac{L}{3(r_2 - r_1)}\left[\frac{1}{r_1^3} - \frac{1}{r_2^3}\right]$$

Therefore

$$\theta = \frac{2TL}{G\pi}\frac{(r_1^2 + r_1 r_2 + r_2^2)}{3r_1^3 r_2^3} \tag{5.28}$$

In the special case when $r_1 = r_2$, i.e. the shaft is parallel,

$$\theta = \frac{2T}{G\pi}\frac{L}{r_1^4} = \frac{TL}{GJ} \tag{5.29}$$

which is the result already obtained.

TORSION OF A THIN TUBE OF NON-CIRCULAR SECTION

The torsion of a solid shaft of non-circular section is a complex problem, but in the case of a thin hollow shaft, or tube, a simple theory can be developed even if the tube thickness is not constant.

The thin-walled tube shown in Fig. 5.9 is assumed to be of constant cross-section throughout its length. The wall thickness is variable but at any point is taken to be t. Assume the applied torque T to act about the longitudinal axis XX and to introduce shearing stresses over the end of the tube, these stresses have a direction parallel to that of a tangent to the tube at a given point. A shearing stress of magnitude τ at any point in the circumference has a complementary shear stress of the same magnitude acting in a longitudinal direction.

Fig. 5.9

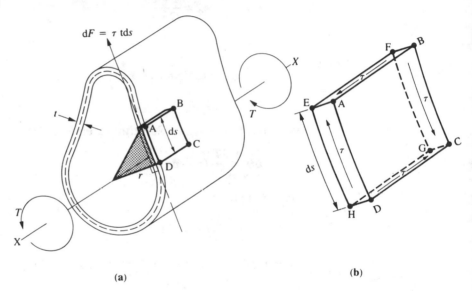

(a) (b)

Consider the small portion ABCD of the tube and assume that the shearing stress τ is constant throughout the wall thickness t. The shearing force along the thin edge AB is τt per unit length, and for longitudinal equilibrium of ABCD, this force must be equal to that on the thin edge CD.

Now, since ABCD was an arbitrary choice, it follows that τt is constant for all parts of the tube. The value $\tau t = q$ is called the *shear flow* and is an internal shearing force per unit length of the circumference of the section of the thin tube.

The force, dF, acting in a tangential direction on an element of the perimeter of length ds is $\tau t\,ds$, and if r is the mean radius at the point, then the moment of this force about the axis XX is $\tau tr\,ds$ and the total torque on the cross-section of the tube is

$$T = \oint \tau tr \, ds \tag{5.30}$$

The integration extends over the whole circumference. Since $\tau t = q$ is constant, we may write

$$T = \tau t \oint r \, ds = q \oint r \, ds \tag{5.31}$$

Now, $r \, ds$ is twice the shaded area shown, and $\oint r \, ds$ for the whole circumference is therefore equal to $2A$, where A is the area enclosed by the centre-line of the wall of the tube (shown dotted); therefore

$$T = 2Aq \tag{5.32}$$

Thus at any point the shearing stress is given by

$$\tau = \frac{q}{t} = \frac{T}{2At} \tag{5.33}$$

Owing to the variation of shear stress around the circumference of the tube it is not possible to predict the deformations, and hence the angle of twist, in the simple manner used for the circular shaft or tube. The angle of twist θ can be determined, however, from the strain energy stored in the tube. Considering an axial strip along which the shear stress is constant, then from eqn. (3.9) the shear strain energy per unit volume is

$$U_s = \frac{\tau^2}{2G}$$

If the strip is of length l, thickness t and width ds, then the energy stored in the strip is

$$U_s = \frac{\tau^2}{2G} \times lt \, ds$$

Therefore the energy stored in the complete tube is

$$U_s = \oint \frac{\tau^2}{2G} lt \, ds$$

Substituting for τ from eqn. (5.33),

$$U_s = \oint \frac{T^2}{8A^2t^2G} lt \, ds$$

$$= \frac{T^2 l}{8A^2 G} \oint \frac{ds}{t} \tag{5.34}$$

But the stored energy is equal to the work $\frac{1}{2}T\theta$, since in the elastic range the torque is proportional to the angle of twist θ; therefore

$$\tfrac{1}{2}T\theta = \frac{T^2 l}{8A^2 G} \oint \frac{ds}{t}$$

so that

$$\theta = \frac{Tl}{4A^2 G} \oint \frac{ds}{t} \tag{5.35}$$

If the tube is of constant thickness t around the circumference s, then

$$\theta = \frac{Tls}{4A^2Gt}$$

or, substituting for T from eqn. (5.33)

$$\theta = \frac{\tau ls}{2AG} \qquad\qquad (5.36)$$

The light-alloy stabilizing strut of a high-wing monoplane is 2 m long and has the cross-section shown in Fig. 5.10. Determine the torque that can be sustained and the angle of twist if the maximum shear stress is limited to 28 MN/m². $G = 27$ GN/m².

Fig. 5.10

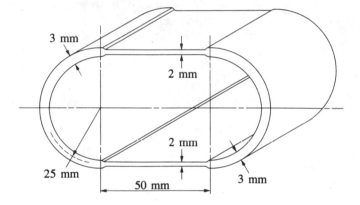

SOLUTION

The area enclosed by the median line of the wall thickness is

$$A = (\pi \times 25^2) + (50 \times 50) = 4460 \text{ mm}^2$$

The allowable torque is obtained using the minimum wall thickness:

$$T = 2At\tau$$

$$= 2 \times \frac{4460}{10^6} \times \frac{2}{10^3} \times 28 \times 10^6 = 500 \text{ N-m}$$

The angle of twist is obtained from eqn. (5.35):

$$\theta = \frac{Tl}{4A^2G} \oint \frac{ds}{t}$$

Therefore

$$\theta = \frac{500 \times 2}{4 \times 4460^2 \times 10^{-12} \times 27 \times 10^9} \left[\frac{100}{2} + \frac{50\pi}{3} \right]$$

$$= 0.0476 \text{ rad} = 2.73°$$

TORSION OF A THIN RECTANGULAR STRIP

An approximate solution for the torsion of a strip of rectangular cross-section whose thickness is small compared with the width, Fig. 5.11(a), can be obtained by considering the strip to be built up of a series of thin-walled concentric tubes which all twist by the same amount. One of these tubes is shown in Fig. 5.11(b), and the area enclosed by the median line is

$$A = (b - 2h)2h + \pi h^2$$

Fig. 5.11

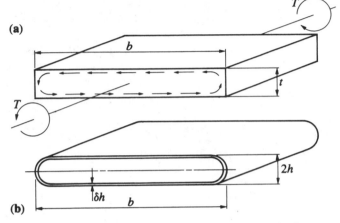

(a)

(b)

If b is large compared with h then the terms in h^2 can be ignored and

$$A \approx 2bh$$

If the tube is subjected to a torque δT then, from eqn. (5.33), the shear stress is

$$\tau = \frac{\delta T}{4bh\,\delta h} \tag{5.37}$$

As the tube becomes infinitely thin,

$$\frac{dT}{dh} = 4bh\tau$$

and the torque carried by the strip is

$$T = \int_0^{t/2} 4bh\tau\,dh \tag{5.38}$$

From eqn. (5.36),

$$\tau = \frac{G\theta}{l}\frac{2A}{s} = \frac{G\theta}{l}\frac{4bh}{2b} \tag{5.39}$$

Substituting in eqn. (5.38) for τ,

$$T = \int_0^{t/2} \frac{G\theta}{l} 8bh^2\,dh$$

$$= \frac{1}{3}bt^3\frac{G\theta}{l} \tag{5.40}$$

The quantity $bt^3/3$ is termed the *torsion constant*, but it is *not* the polar second moment of area for the section.

Eqn. (5.39) shows that the shear stress parallel to the long edge of the cross-section is proportional to the distance h from the central axis. The maximum shear stress occurs at the outer surface and is

$$\tau_{max} = \frac{tG\theta}{l} \tag{5.41}$$

EXAMPLE 5.6

Determine the angle of twist per unit length and the maximum shear stress in the aluminium channel section shown in Fig. 5.12 when subjected to a pure torque of 20 N-m. Shear modulus = 27 GN/m².

Fig. 5.12

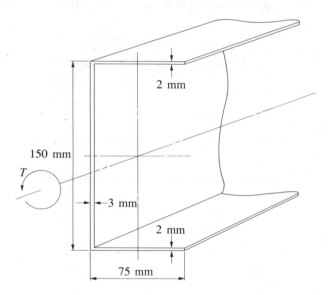

150 mm

T

2 mm

3 mm

2 mm

75 mm

SOLUTION

The channel may be analysed as three rectangular strips, the two flanges and the web, and the above solution will be used for each part. Let the proportions of the torque carried by the flanges and web be T_1 and T_2 respectively; then from eqn. (5.40),

$$T_1 = \frac{1}{3} \times \frac{75}{10^3} \times \left(\frac{2}{10^3}\right)^3 \times 27 \times 10^9 \frac{\theta}{l}$$

$$= 5.4 \frac{\theta}{l}$$

$$T_2 = \frac{1}{3} \times \frac{150}{10^3} \times \left(\frac{3}{10^3}\right)^3 \times 27 \times 10^9 \frac{\theta}{l}$$

$$= 36.5 \frac{\theta}{l}$$

But $2T_1 + T_2 = 20$; therefore

$$47.3\frac{\theta}{l} = 20$$

and

$$\frac{\theta}{l} = 0.422 \text{ rad/m} = 24.2°/\text{m}$$

The maximum shear stress in the flanges, from eqn. (5.41), is

$$\tau_{max} = \frac{2}{10^3} \times 27 \times 10^9 \times 0.422 = 22.8 \text{ MN/m}^2$$

and in the web is

$$\tau_{max} = \frac{3}{10^3} \times 27 \times 10^9 \times 0.422 = 34.2 \text{ MN/m}^2$$

EFFECT OF WARPING DURING TORSION

In the previous section and the above example dealing with a thin-walled open section subjected to torsion, the pure torque was supposed to be applied to each end of the member in such a way that there was no axial restraint. Owing to the variation in transverse shear stress, for example in the flanges of the above channel or the I-section shown in Fig. 5.13, there is also a variation in longitudinal complementary shear stress which results in axial movement of one flange with respect to the other. Therefore, cross-sections which were initially plane do not remain so during torsion and there is warping of any cross-section. If one or more sections of a member are constrained in some manner to remain plane during torsion then warping is restrained.

Fig. 5.13

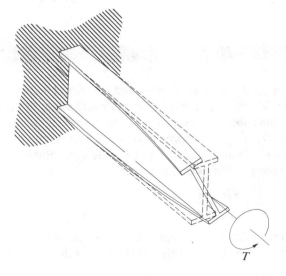

T

Resisting torque in the section is supplied in two ways, by simple torsion and also by torque set up through the restraint of warping. Thus an applied torque will cause a smaller angle of twist than when the section is free to warp, and torsional stiffness may be considerably increased if warping is restrained. Non-circular closed and solid-section members also exhibit warping, but the effect is much smaller in comparison with the open section. Further discussion of warping behaviour is beyond the scope of this book.

TORSION OF SOLID RECTANGULAR AND SQUARE CROSS SECTIONS

The solution on page 105 is only applicable for a rectangular section in which the longer side is much greater than the other. Unfortunately the solution for torsion of general rectangular and square sections, although of some engineering interest, is very complex and beyond the scope of this text. (The reader who is interested in the full solution should consult a text such as *Theory of Elasticity* by Timoshenko and Goodier.) The result in terms of torque, maximum shear stress at the centre of the long side and angle of twist may be expressed as

$$T = \alpha b h^2 \tau_{max} = \beta b h^3 G \frac{\theta}{l} \tag{5.42}$$

where b is the longer and h the shorter side, and α, β are factors dependent on the geometry, as given in Table 5.1.

Table 5.1

b/h	1	1.5	2	2.5	3	4	6	10	∞
α	0.208	0.231	0.246	0.256	0.267	0.282	0.299	0.312	0.333
β	0.141	0.196	0.229	0.249	0.263	0.281	0.299	0.312	0.333

SUMMARY

The analysis of the thin-walled circular tube highlights the basic steps of equilibrium of internal and external forces, the geometry of deformation of the tube and the elastic shear-stress/shear-strain relationship for the material. A full appreciation of the foregoing makes the understanding of the solution for the solid and hollow circular shafts relatively simple, leading to the basic relationship

$$\frac{T}{J} = \frac{G\theta}{L} = \frac{\tau}{r}$$

from which we see that shear stress and strain are linearly distributed across the section, being maximum at the outer surface. In the case of composite shafts the additional assessed information to the above is the manner in which the applied torque or the angle of twist is shared between different components.

Non-circular thin tubes and other open sections play an important part in engineering, and, whereas the stresses can be determined from the equilibrium state, we have to evaluate the shear strain energy to determine the amount of twist.

Finally the effect of warping of the cross-section and its restraint have a considerable effect on the stiffness of the torsion member and this must be taken into account.

PROBLEMS

5.1 A hollow steel shaft with an external diameter of 150 mm is required to transmit 1 MW at 300 rev/min. Calculate a suitable internal diameter for the shaft if its shear stress is not to exceed 70 MN/m^2.

Compare the torque-carrying capacity of this shaft with a solid steel shaft having the same weight per unit length and limiting shear stress. $G = 80$ GN/m^2.

5.2 The gearbox in Fig. 5.14 is required to supply an output torque of 300 N-m at a shaft speed of 100 rev/min. The gear ratios are such that shaft 1 rotates at three times the speed of shaft 2, and shaft 2 rotates at five times the speed of shaft 3. Calculate the input power required from the motor and the diameters of shafts 2 and 3 if the allowable shear stress of the shaft material is 400 MN/m^2. Neglect losses in the system and assume a safety factor of 2.

Fig. 5.14

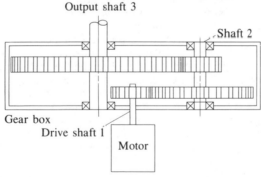

5.3 A steel shaft has to transmit 1 MW at 240 rev/min so that the maximum shear stress does not exceed 55 MN/m^2 and there is not more than 2° twist on a length of 30 diameters. Determine the required diameter of shaft. $G = 80$ GN/m^2.

5.4 A shaft carries five pulleys, A, B, C, D and E, and details of shaft diameters, lengths and pulley torques are given in Table 5.2. Determine in which sections the maximum shear stress and angle of twist occur. $G = 80$ GN/m^2.

Table 5.2

Pulley	Torque (N-m)	Direction	Shaft	Length (m)	Diameter (mm)
A	60	Clockwise	AB	1	38
B	900	Anticlockwise	BC	$\frac{1}{2}$	100
C	300	Clockwise	CD	$\frac{1}{2}$	75
D	640	Clockwise	DE	$1\frac{1}{2}$	75
E	100	Anticlockwise			

5.5 A hollow steel drive shaft with an outside diameter of 50 mm and an inside diameter of 40 mm is fastened to a coupling using six 8 mm diameter bolts on a pitch circle diameter of 150 mm. If the shear stress in the shaft or bolt material is not to exceed 170 MN/m² calculate the maximum power which the system could transmit at a rotational speed of 100 rev/min. What would be the effect of only using three of the bolts?

5.6 A torsional vibration damper consists of a hollow steel shaft fixed at one end and to the other end is attached a solid circular steel shaft which passes concentrically along the inside of the hollow shaft, as shown in Fig. 5.15. Determine the maximum torque T that can be applied to the free end of the solid shaft so that the angle of twist where the torque is applied does not exceed 5°. Local effects where the two parts are connected may be ignored. Shear modulus, $G = 80$ GN/m².

Fig. 5.15

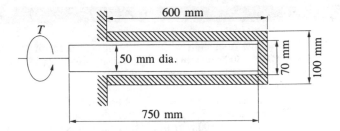

5.7 Two shafts are connected end-to-end by means of a coupling in which there are twelve bolts on a pitch circle diameter of 250 mm. The maximum shear stress is limited to 36 MN/m² in the shafts and 16 MN/m² in the bolts. If one shaft is solid, 50 mm diameter, and the other hollow, 60 mm external diameter, calculate the internal diameter of the latter and the bolt diameter so that both shafts and coupling are equally strong.

5.8 A compound drive shaft consists of a solid circular steel bar B surrounded for part of its length by an aluminium tube A as shown in Fig. 5.16. The contacting surfaces are smooth and the collar and the shaft are both rigidly fixed to a machine at D. Pin C fills a hole drilled completely through a diameter of the collar and shaft. The shearing deformation in the pin and the bearing deformation between the pin and shaft can be neglected. Calculate the maximum torque T which can be applied to the steel shaft as shown, without exceeding an average shearing stress of 8 MN/m² on the cross sectional area of the pin at C, the interface between the shaft and collar. $G_{steel} = 80$ GN/m²; $G_{aluminium} = 28$ GN/m².

Fig. 5.16

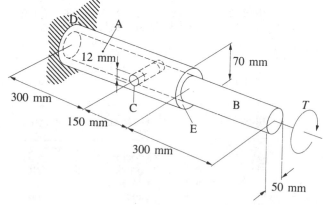

5.9 An aluminium-alloy strut for a light aircraft having the cross-section shown in Fig. 5.17 is 3 m in length. If the shear stress is not to exceed 30 MN/m² and the applied torque is 134 N-m, determine the required thickness t of metal. What is the angle of twist? $G = 28 \, \text{GN/m}^2$.

Fig. 5.17

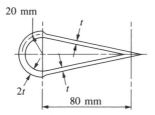

20 mm

t

$2t$

t

80 mm

5.10 Compare the torsional strength and stiffness of thin-walled tubes of circular cross-section of mean radius R and thickness t, with and without a longitudinal slot.

5.11 A torsional member used for stirring a chemical process is made of a circular tube to which are welded four rectangular strips as shown in Fig. 5.18. The tube has inner and outer diameters of 94 mm and 100 mm respectively, each strip is 50 mm by 18 mm, and the stirrer is 3 m in length. If the maximum shearing stress in any part of the cross-section is limited to 56 MN/m², neglecting any stress concentration, calculate the maximum torque which can be carried by the stirrer and the resulting angle of twist over the full length. $\alpha = 0.264$, $\beta = 0.258$, $G = 83 \, \text{GN/m}^2$.

Fig. 5.18

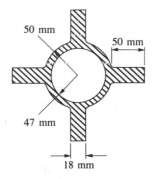

50 mm

50 mm

47 mm

18 mm

5.12 An I-beam has a width of 100 mm and depth of 150 mm, as shown in Fig. 5.19. Calculate the maximum pure torque which could be applied to this beam if the yield shear stress is 240 MN/m². Assume a safety factor of 3.

Fig. 5.19

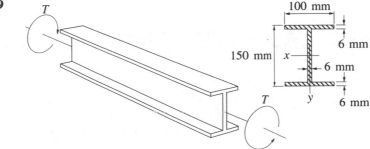

T

T

100 mm

6 mm

150 mm x

6 mm

y 6 mm

Chapter 6

Bending: stress

One of the most common modes of deformation of engineering structures and components is that of bending. The simplest form is the slender member, often termed a beam, subjected to transverse loading, which is the subject of this chapter. In Chapter 17 the more complex states of bending in plates and shells are studied.

Beams are designed both for strength and limited deflection, and this chapter is concerned with the former. The types of stresses that are set up in bending are normal tension and compression and transverse shear. These stresses derive directly from the internal reactions of shear force and bending moment introduced in Chapter 1 in the applications of statical equilibrium. The importance of being able to establish shear-force and bending-moment distributions cannot be overemphasized, since the correct determination of a design based on allowable stresses and deflection depends upon that first step. Because of this, the chapter commences with a number of illustrations of shear force and bending-moment distributions. This is then followed by analyses of bending stresses and shear stresses for various beam configurations.

SHEAR FORCE AND BENDING-MOMENT DISTRIBUTIONS

The reader should refer back to Chapter 1, pages 16–23, if necessary, to revise the basic notation, sign conventions and equilibrium principles for establishing shear-force and bending-moment distributions. The following worked examples should help to reinforce the basic understanding of the procedure.

SIMPLY-SUPPORTED BEAM CARRYING CONCENTRATED LOADS

The left-hand reaction is first required and as the beam is statically determinate the reactions can be found from the following two equilibrium equations:

$$-R_A - R_D + W_1 + W_2 = 0$$

$$-R_A L + W_1(L - a) + W_2(L - b) = 0$$

from which

$$R_A = \frac{W_1(L - a) + W_2(L - b)}{L}$$

The two concentrated loads cause mathematical discontinuities and so separate free-body diagrams and equilibrium equations have to be expressed for each different section, as shown in Fig. 6.1 and in the

Fig. 6.1

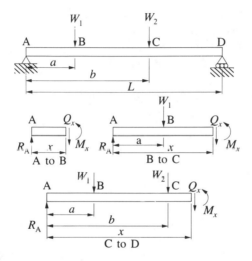

following equations respectively:

Shear force

AB	$0 < x < a$	$+Q - R_A = 0$	$Q = R_A$
BC	$a < x < b$	$+Q - R_A + W_1 = 0$	
		$Q = +R_A - W_1$	
CD	$b < x < L$	$+Q - R_A + W_1 + W_2 = 0$	
		$Q = R_A - W_1 - W_2$	

Bending moment

AB	$0 < x < a$	$M - R_A x = 0$	$M = R_A x$
BC	$a < x < b$	$M - R_A x + W_1(x - a) = 0$	
		$M = R_A x - W_1(x - a)$	
CD	$b < x < L$	$M - R_A x + W_1(x - a) + W_2(x - b) = 0$	
		$M = R_A x - W_1(x - a) - W_2(x - b)$	

The diagrams resulting from the above equations are shown in Fig. 6.2. It

Fig. 6.2

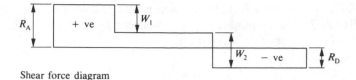

Shear force diagram

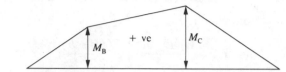

Bending moment diagram

should be noted that

(*a*) The S.F. diagram *changes* in value at a support or concentrated load by the amount of the reaction or load.
(*b*) The B.M. is *always zero* at the ends of a beam which are either unsupported or on a simple support.

SIMPLY-SUPPORTED BEAM WITH AN APPLIED COUPLE

Apart from transverse loads, bending can be caused by a couple applied to the beam at a cross-section, as shown in Fig. 6.3. The reactions at the supports are obtained from moment equilibrium of the whole beam, from which

$$R_A L = R_B L = \bar{M} \quad \text{or} \quad R_A = R_B = \frac{\bar{M}}{L}$$

Fig. 6.3

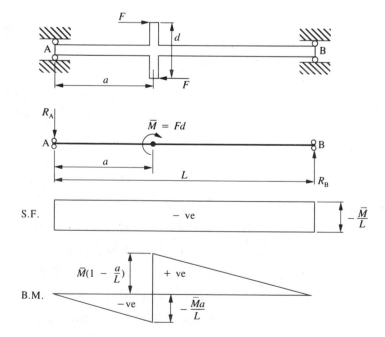

Shear force   The shear force is constant along the length of the beam and of value $Q = -\bar{M}/L$.

Bending moment

For $x < a$ $M + R_A x = 0$;   hence $M = -\dfrac{\bar{M}x}{L}$

For $x > a$ $M + R_A x - \bar{M} = 0$;   hence $M = \bar{M} - \dfrac{\bar{M}x}{L}$

$$= \frac{\bar{M}}{L}(L - x)$$

The S.F. and B.M. diagrams were derived from these equations.

EXAMPLE 6.1

Sketch a B.M. diagram for the curved member shown in Fig. 6.4(*a*), giving the principal numerical values.

Fig. 6.4

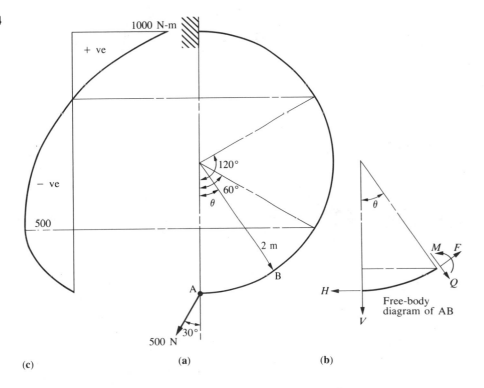

(c)   (a)   (b)

SOLUTION

The determination of bending-moment distribution along a slender curved member follows exactly the same procedure as for straight beams.

It is more convenient to resolve the applied load into vertical and horizontal components. These are:

$$V = 500 \cos 30° = 250\sqrt{3} \quad \text{and} \quad H = 500 \sin 30° = 250$$

A portion of the bar is cut off to give a free body and the forces and couple are inserted as shown in Fig. 6.4(b). The B.M. required to balance H is

$$250 \times 2(1 - \cos \theta)$$

and to balance V,

$$-250\sqrt{3} \times 2 \sin \theta$$

Hence

Bending moment
$$M = 500(1 - \cos \theta) - 500\sqrt{3} \sin \theta$$

At

$$\theta = 0 \qquad M = 0$$
$$\theta = \pi/2 \qquad M = -365 \text{ N-m}$$
$$\theta = \pi \qquad M = +1000 \text{ N-m}$$

Also, if $M = 0 = 500(1 - \cos \theta - \sqrt{3} \sin \theta)$, then

$$\cos \theta = -\tfrac{1}{2} \quad \text{and} \quad \theta = 120°$$

which gives a point of contraflexure, where the bending moment changes sign.

When $dM/d\theta = 0$,

$$\sin \theta - \sqrt{3} \cos \theta = 0 \qquad \theta = 60°$$

$$M_{60°} = 500\left(1 - \tfrac{1}{2} - \sqrt{3}\,\frac{\sqrt{3}}{2}\right) = -500 \text{ N-m}$$

The B.M. diagram is plotted for convenience along the vertical axis as shown in Fig. 6.4(c).

EXAMPLE 6.2

Sketch the S.F. and B.M. diagrams for the cantilever shown in Fig. 6.5.

SOLUTION

Even though there is no loading on AB and therefore no shear force or bending moment, the distance x will still be measured from A rather than B.

Shear force
$$0 < x < 2 \qquad Q = 0$$
$$2 < x < 5 \qquad Q = -5000 - 2000(x - 2)$$
$$5 < x < 8 \qquad Q = -5000 - 6000 - 3000 = -14\,000 \text{ N}$$

Bending moment
$$0 < x < 2 \qquad M = 0$$

$$2 < x < 5 \qquad M = -5000(x - 2) - 2000(x - 2)\frac{(x - 2)}{2}$$

$$= -1000(x - 2)(x + 3)\text{N-m (parabolic distribution)}$$

$$5 < x < 8 \qquad M = -5000(x - 2) - 6000(x - 3\tfrac{1}{2}) - 3000(x - 5)$$

$$= -14\,000x + 46\,000 \text{ N-m (linear distribution)}$$

The diagrams and principal values are shown in Fig. 6.5.

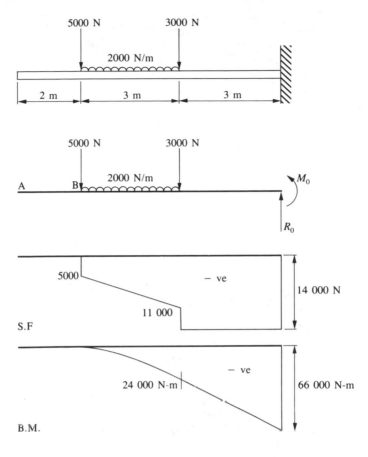

Fig. 6.5

The beam shown in Fig. 6.6(a) carries distributed loading which varies in an arbitrary manner, f(w). Consider the free body, Fig. 6.6(b), of a small slice of length dx for which the loading may be regarded as uniform, w.

Vertical equilibrium

$$-Q + w\,dx + \left(Q + \frac{dQ}{dx}\,dx\right) = 0$$

$$w + \frac{dQ}{dx} = 0$$

$$w = -\frac{dQ}{dx} \tag{6.1}$$

Moment equilibrium Taking moments about the mid-point of the right-hand edge,

$$-M - Q\,dx + w\,dx\,\frac{dx}{2} + \left(M + \frac{dM}{dx}\,dx\right) = 0$$

Fig. 6.6

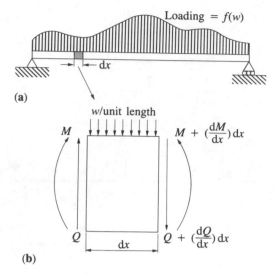

(a)

(b)

Neglecting $(dx)^2$,

$$-Q + \frac{dM}{dx} = 0$$

$$Q = \frac{dM}{dx} \tag{6.2}$$

When $Q = 0$, $dM/dx = 0$, i.e. the S.F. is zero when the *slope* of a B.M. diagram is zero and not simply when M is a maximum, although this also is implied. It is possible to have $M = M_{max}$ when $dM/dx \neq 0$ (see Fig. 6.4).

Also, from eqns. (6.1) and (6.2), we note that

$$w = -\frac{d^2M}{dx^2} \tag{6.3}$$

From eqn. (6.1) it follows that, between any two sections denoted by 1 and 2,

$$\int_1^2 dQ = \int_1^2 -w\, dx$$

or

Shear force $$Q_2 - Q_1 = \int_1^2 -w\, dx + A \tag{6.4}$$

Thus the *change* in shear force between any two cross-sections may be obtained from the area under the load distribution curve between those sections.

From eqn. (6.2), considering two sections, 1 and 2,

$$\int_1^2 dM = \int_1^2 Q\, dx$$

or

Bending moment
$$M_2 - M_1 = \int_1^2 Q \, dx + B \qquad (6.5)$$

Thus the *change* in bending moment between any two sections is found from the area under the shear force diagram between those sections.

EXAMPLE 6.3

Use the relationships developed on page 118 to find the position of zero shear force and the value of the maximum bending moment for the simply-supported beam shown in Fig. 6.7. It carries a distributed load which varies linearly in intensity from zero at the left to w Newtons per metre at the right-hand end. Plot the S.F. and B.M. diagrams.

Fig. 6.7

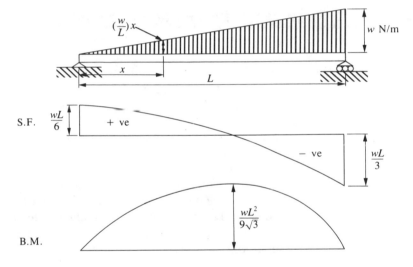

SOLUTION

By similar triangles, the load intensity at x from the left end is $(w/L)x$. From eqn. (6.4) the shear force is given by

Shear force
$$Q = -\int \frac{wx}{L} \, dx + A = -\frac{wx^2}{2L} + A$$

and from eqn. (6.5),

Bending moment
$$M = \int Q \, dx + B = -\frac{wx^3}{6L} + Ax + B$$

Now, $M = 0$ when $x = 0$ and L. Hence

$$B = 0 \quad \text{and} \quad A = \frac{wL}{6}$$

119

Thus

$$Q = -\frac{wx^2}{2L} + \frac{wL}{6}$$

and when $Q = 0$, $x = L/\sqrt{3}$.

Now,

$$M = -\frac{wx^3}{6L} + \frac{wLx}{6}$$

Since $\mathrm{d}M/\mathrm{d}x = Q = 0$ is a possible solution in this case for M_{max},

$$M_{max} = -\frac{w}{6L}\left(\frac{L}{\sqrt{3}}\right)^3 + \frac{wL}{6}\frac{L}{\sqrt{3}} = \frac{wL^2}{9\sqrt{3}}$$

The S.F. and B.M. diagrams are as shown in Fig. 6.7.

DEFORMATIONS IN PURE BENDING

ASSUMPTIONS

This is a statically-indeterminate problem, and hence all three basic principles stated in the Introduction of Chapter 4 will need to be employed. However, the steps will be considered in a different order since at this stage there is insufficient information about the stress distribution to enable an equilibrium equation to be formulated. The approach initially will be to consider a prismatic beam of symmetrical cross-section subjected to pure bending, from which the *geometry of deformation* can be studied and strain distribution determined. The *stress–strain relations* will give the stress distribution which can be related to forces and moments through an *equilibrium condition*.

Before commencing the analysis it is necessary to make some assumptions and these are as follows:

1. Transverse sections of the beam which are plane before bending will remain plane during bending.
2. From consideration of symmetry during bending, transverse sections will be perpendicular to circular arcs having a common centre of curvature.
3. The radius of curvature of the beam during bending is large compared with the transverse dimensions.
4. Longitudinal elements of the beam are subjected only to simple tension or compression, and there is no lateral stress.
5. Young's modulus for the beam material has the same value in tension and compression.

Longitudinal deformation

Consideration of the beam subjected to pure bending shown in Fig. 6.8 indicates that the lower surface stretches and is therefore in tension and the upper surface shortens and thus is in compression. Hence there must

Fig. 6.8

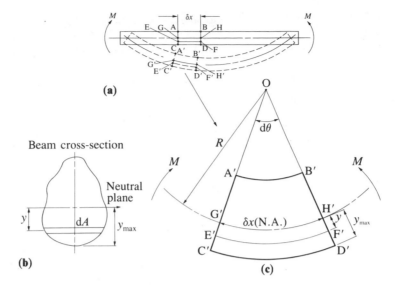

(a)

Beam cross-section

(b)

(c)

be an xz-plane in between in which longitudinal deformation is zero. This is termed the *neutral plane*, and a transverse axis lying in the neutral plane is the *neutral axis*. Consider the deformations between two sections AC and BD, a distance δx apart, of an initially straight beam. A longitudinal fibre EF at distance y below the neutral axis will have initially the same length as the fibre GH at the neutral axis. During bending EF stretches to become E'F', but GH being at the neutral axis is unstrained when it becomes G'H'. Therefore, if R is the radius of curvature of G'H',

$$G'H' = GH = \delta x = R \, \delta\theta$$

$$E'F' = (R + y) \, \delta\theta$$

and the longitudinal strain in fibre E'F' is

$$\varepsilon_x = \frac{E'F' - EF}{EF}$$

But $EF = GH = G'H' = R \, \delta\theta$; therefore

$$\varepsilon_x = \frac{(R + y) \, \delta\theta - R \, \delta\theta}{R \, \delta\theta}$$

Hence

$$\varepsilon_x = \frac{y}{R} \tag{6.6}$$

and since $R = \mathrm{d}x/\mathrm{d}\theta$, therefore also

$$\varepsilon_x = y \, \frac{\mathrm{d}\theta}{\mathrm{d}x} \tag{6.7}$$

From eqn. (6.6) it will be seen that strain is distributed linearly across the section, being zero at the neutral surface and having maximum values

121

at the outer surfaces. It is important to note here that eqn. (6.6) is entirely independent of the type of material, whether it is in an elastic or plastic state and linear or non-linear in stress and strain.

Transverse deformation

Regarding deformations in the y- and z-directions, it is apparent that changes in length of the beam will result in changes in the transverse dimensions. For example, the fibres in compression will be associated with an increase in thickness, whereas the region in tension will show a decrease in beam thickness. The transverse strains will be

$$\varepsilon_z = \varepsilon_y = -\frac{\nu\sigma_x}{E}$$

and a beam of initially rectangular cross-section will take up the shape shown in Fig. 6.9. This can be easily demonstrated by bending an eraser. The neutral surface, instead of being plane, will be curved. This behaviour is termed *anticlastic curvature*. The deformations are extremely small and do not affect the solution for longitudinal strains.

Fig. 6.9

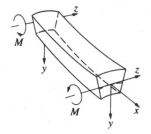

STRESS–STRAIN RELATIONSHIP

Consider the segmental length of beam illustrated in Fig. 6.10 the cross-section of which, for simplicity at this stage of analysis, has symmetry about the vertical axis y, but not about a horizontal axis z.

As it has been assumed that $\sigma_y = \sigma_z = 0$ then the stress–strain relationship that is applicable for *linear-elastic bending* is

$$\varepsilon_x = \frac{\sigma_x}{E}$$

Fig. 6.10

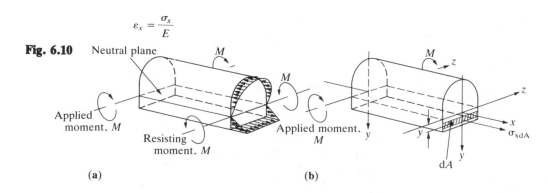

(a)　　　　　　　(b)

and, from eqn. (6.6),

$$\varepsilon_x = \frac{\sigma_x}{E} = \frac{y}{R}$$

or

$$\frac{\sigma_x}{y} = \frac{E}{R} \qquad\qquad (6.8)$$

Thus bending stress is also distributed in a linear manner over the cross-section, being zero where y is zero, that is at the neutral plane, and being maximum tension and compression at the two outer surfaces where y is maximum, as shown in Fig. 6.10(a).

EQUILIBRIUM OF FORCES AND MOMENTS

POSITION OF THE NEUTRAL PLANE AND AXIS

Consider a small element of area, dA, at a distance y from some arbitrary location of the neutral surface, Fig. 6.10(b).

The force on the element in the x-direction is

$$\mathrm{d}F_x = \sigma_x \mathrm{d}A$$

Therefore the total longitudinal force on the cross-section is

$$F_x = \int_A \sigma_x \mathrm{d}A$$

where A is the total area of the section.

Since there is no external axial force in pure bending, the internal force resultant must be zero; therefore

$$F_x = \int_A \sigma_x \mathrm{d}A = 0$$

Using eqn. (6.8) to substitute for σ_x,

$$\frac{E}{R} \int_A y \, \mathrm{d}A = 0$$

Since E/R is not zero, the integral must be zero, and as this is the first moment of area about the neutral axis, it is evident that the centroid of the section must coincide with the neutral axis. (The first moment of area of a section about its centroid is zero; see Appendix A.)

INTERNAL RESISTING MOMENT

Returning to the element in Fig. 6.10(b), the moment of the axial force about the neutral surface is d$F_x y$. Therefore the total *internal resisting moment* is

$$\int_A y \, \mathrm{d}F_x = \int_A y \sigma_x \mathrm{d}A$$

This must balance the external applied moment M, so that for equilibrium

$$\int_A y\sigma_x \, dA = M \tag{6.9}$$

or, substituting for σ_x,

$$M = \frac{E}{R} \int_A y^2 \, dA$$

Now $\int_A y^2 \, dA$ is the *second moment of area* (see Appendix A) of the cross-section about the neutral axis and will be denoted by I. Thus

$$M = \frac{EI}{R}$$

or

$$\frac{M}{I} = \frac{E}{R} \tag{6.10}$$

THE BENDING RELATIONSHIP

Combining eqns. (6.9) and (6.10) gives the fundamental relationship between bending stress, moment and geometry of deformation:

$$\frac{M}{I} = \frac{\sigma}{y} = \frac{E}{R} \tag{6.11}$$

THE GENERAL CASE OF BENDING

The bending relationship, eqn. (6.11), is only an exact solution for the case of pure bending; however, in practice many beam problems involve bending moment and shearing force which vary along the length. In these cases it has been shown that eqn. (6.11), even if not exact, provides a solution which is quite accurate for engineering design, except for cross-sections at support points and concentrated loads where stress concentration may occur (see Chapter 13).

SECTION MODULUS

For the outer surfaces of the beam,

$$M = \hat{\sigma}_t \frac{I}{y_{tmax}} = \hat{\sigma}_c \frac{I}{y_{cmax}}$$

where the subscripts denote tension and compression. The quantities

I/y_{tmax} and I/y_{cmax} are functions of geometry only; they are termed the *section moduli* and are denoted by Z_t and Z_c. Thus

$$\sigma_{tmax} = \frac{M}{Z_t} \quad \text{and} \quad \sigma_{cmax} = \frac{M}{Z_c} \tag{6.12}$$

EXAMPLE 6.4

A pulley block is attached to one end of a horizontal solid cylindrical bar of 2 m length having the other end fixed to a structure by welding. The maximum concentrated load to be raised on the pulley block will be 1 kN and to allow for the weight of the block and any possible overload a safety factor of 2 will be used. If the maximum design bending stress for the material is 150 MN/m² calculate a suitable diameter for the bar.

SOLUTION

The bending moment on this cantilever bar will increase linearly from zero at the free end to a maximum value at the fixed end. It is this value which we must use for the design. Using the safety factor, the design load is $1 \times 2 = 2$ kN and the maximum bending moment is therefore $2 \times 2 = 2$ kN-m.

Let the diameter of the bar be d; then

$$I = \frac{\pi d^4}{64}; \quad y_{max} = \frac{d}{2}; \quad \sigma_{max} = 150 \text{ MN/m}^2$$

Using eqn. (6.11),

$$\frac{4000}{\pi d^4/64} = \frac{150 \times 10^6}{d/2}$$

$$d^3 = \frac{4000 \times 32}{150 \times 10^6 \times \pi} \text{ m}^3$$

$$d = 0.065 \text{ m} = 65 \text{ mm}$$

EXAMPLE 6.5

A tee-section bar has dimensions as shown in Fig. 6.11(*b*). The bar is used as a simply-supported beam of span 1.5 m, the flange being horizontal as shown in Fig. 6.11(*a*). Calculate the uniformly-distributed

Fig. 6.11

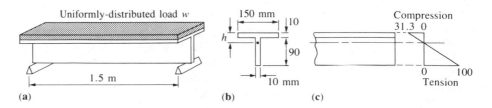

(a)　　　　　(b)　　　　(c)

load which can be applied if the maximum tensile stress is not to exceed $100 \, MN/m^2$. What is then the greatest bending stress in the flange?

SOLUTION

The first step is to find the position of the centre of area, which will also give the neutral axis. Taking first moments of area about the top surface,

$$(240 \times 10)h = (1500 \times 5) + (900 \times 55)$$

$$h = 23.8 \, mm$$

To find the second moment of area we can use the parallel axis theorem (see Appendix A):

$$I_{xx} = \frac{150 \times 10^3}{12} + (150 \times 10 \times 18.8^2) + \frac{10 \times 90^3}{12} + (10 \times 90 \times 31.2^2)$$

$$= 2028 \times 10^3 \, mm^4$$

$$M_{max} = \frac{w \times 1.5^2}{8} = 0.281w \, \text{N-m}$$

The maximum tensile stress occurs on the bottom edge of the web where $y = y_{max} = 76.2 \, mm$. Therefore

$$\frac{100 \times 10^6}{0.0762} = \frac{0.281w}{2028 \times 10^{-9}}$$

$$w = \frac{202.8}{0.0762 \times 0.281} = 9.5 \, \text{kN/m}$$

Since stress is proportional to distance from the neutral axis, as shown in Fig. 6.11(c),

$$\frac{\hat{\sigma}_c}{y_c} = \frac{\hat{\sigma}_t}{y_t}$$

Therefore the greatest compressive stress in the flange is

$$\hat{\sigma}_c = 100 \times 10^6 \times \frac{23.8}{76.2} = 31.3 \, MN/m^2$$

BEAMS MADE OF DISSIMILAR MATERIALS

In some circumstances it may be necessary or desirable to construct a beam such that the cross-section contains two different materials. Usually the object is for one material to act as a reinforcement to the other, perhaps weaker, material. There would be a number of reasons (cost, weight, size, etc.) why the whole beam could not be made from the stronger material. The positioning of the reinforcement material might not be symmetrical with respect to the centroid of the cross-section and it could be embedded within or fixed in some manner to the outside of the main bulk material.

The arguments which were applied to the analysis of simple bending of a homogeneous beam also apply to the composite beam since the two materials constrain each other to deform in the same manner, e.g. to an arc of a circle for pure bending.

SYMMETRICAL SECTIONS

Consider a beam cross-section consisting of a central part of, say, plastic or timber with reinforcing plates firmly bonded (no sliding) to the upper and lower surfaces along the length of the beam as shown in Fig. 6.12. The section is symmetrical about the centroid and neutral surface.

Fig. 6.12

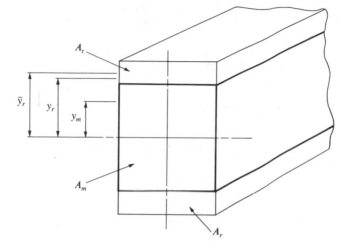

Equilibrium

(a) Since there is no *external* end load, longitudinal equilibrium requires that

$$\int_{Am} \sigma_m \mathrm{d}A_m + \int_{Ar} \sigma_r \mathrm{d}A_r = 0 \tag{6.13}$$

where the subscripts m and r refer to the main and reinforcing materials, respectively, as shown in Fig. 6.12.

(b) The sum of the internal resisting moments must be in equilibrium with the external applied moment

$$M_m + M_r = M$$

Hence, using eqn. (6.9) for each of the materials,

$$\int_{Am} \sigma_m y_m \mathrm{d}A_m + \int_{Ar} \sigma_r y_r \mathrm{d}A_r = M \tag{6.14}$$

The above equations are independent of whether the material is in an elastic or a plastic condition.

Geometry of deformation

Both materials deform to a circular arc and, for a linear strain distribution, from eqn. (6.6)

$$\frac{1}{R} = \frac{\varepsilon_m}{y_m} = \frac{\varepsilon_r}{y_r} \tag{6.15}$$

These relationships are purely a function of geometry and therefore are independent of the material and its properties.

Stress–strain relations

For linear elastic materials under uniaxial stress

$$\left.\begin{array}{l} \sigma_m = E_m\varepsilon_m \\ \sigma_r = E_r\varepsilon_r \end{array}\right\} \tag{6.16}$$

From eqns. (6.15) and (6.16), $\sigma_m = y_m E_m/R$ and $\sigma_r = y_r E_r/R$. Substitution in the equilibrium equation (6.14) gives

$$\left(\frac{E_m}{R}\right)\int_{A_m} y_m{}^2\mathrm{d}A_m + \left(\frac{E_r}{R}\right)\int_{A_r} y_r{}^2\mathrm{d}A_r = M$$

but the integrals are the second moments of area of the core and reinforcing plates, I_m and I_r, respectively, about the neutral surface. Therefore

$$\frac{E_m I_m + E_r I_r}{R} = M$$

Substituting for $1/R$ gives

$$\frac{\sigma_m}{E_m y_m} = \frac{\sigma_r}{E_r y_r} = \frac{M}{E_m I_m + E_r I_r}$$

or

$$\left.\begin{array}{l} \sigma_m = \dfrac{M E_m y_m}{E_m I_m + E_r I_r} \\[2mm] \sigma_r = \dfrac{M E_r y_r}{E_m I_m + E_r I_r} \end{array}\right\} \tag{6.17}$$

EXAMPLE 6.6

A timber beam of depth 100 mm and width 50 mm is to be reinforced with steel plates on each side, as shown in Fig. 6.13. The composite section will be subjected to a maximum bending moment of 6 kN-m. If the maximum stress in the timber is not to exceed 12 MN/m², calculate

Fig. 6.13

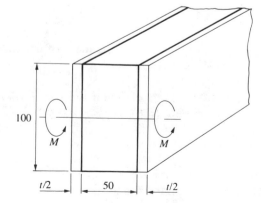

the required thickness for the steel plates, which are 100 mm in depth. What is the maximum stress in the steel? $E_{steel} = 205 \text{ GN/m}^2$, $E_{timber} = 15 \text{ GN/m}^2$.

SOLUTION

$$I_{timber} = \frac{50 \times 100^3}{12} = 4.17 \times 10^6 \text{ mm}^4$$

$$I_{steel} = \frac{t}{12} \times 100^3 = 0.0833 \times 10^6 t \text{ mm}^4$$

From eqn. (6.17) for limiting stress in the timber, at the top or bottom surface,

$$12 \times 10^6 = \frac{6000 \times 0.05 \times 15 \times 10^9 \times 10^{12}}{(15 \times 10^9 \times 4.17 \times 10^6) + (205 \times 10^9 \times 0.0833 \times 10^6 t)}$$

from which $t = 18.3$ mm; and since this is the total thickness of steel, the plates are each 9.15 mm thick.

The maximum stress in steel plates also occurs at the top and bottom edges,

$$\sigma = \frac{6000 \times 0.05 \times 205 \times 10^9 \times 10^{12}}{(15 \times 10^9 \times 4.17 \times 10^6) + (205 \times 10^9 \times 0.0833 \times 10^6 \times 18.3)}$$

$$= 164 \text{ MN/m}^2$$

EQUIVALENT SECTIONS

In beam sections which are symmetrical geometrically, but unsymmetrical with respect to location of the different materials, as, for example, in Fig. 6.14(a), the neutral axis no longer coincides with the centroid of the section. The problem can still be solved using eqns. (6.13)–(6.16), but the arithmetic is more tedious since the neutral axis has first to be found.

A more convenient approach, which is also valid, is to transform the composite section into an equivalent (from the view of resisting forces

Fig. 6.14

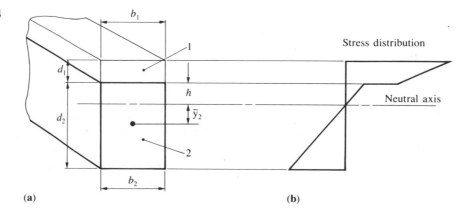

(a)   (b)

and moments) section of only one of the two (or more) materials. The solution is then of course a simple routine.

Consider the beam section shown in Fig. 6.14(a), in which the unknown neutral axis has been placed at a distance h from the interface between the two materials. Assuming no sliding, the strain at the interface must be the same for each material; therefore, at the interface,

$$\varepsilon = \frac{\sigma_1}{E_1} = \frac{\sigma_2}{E_2} \tag{6.18}$$

Also, rearranging eqns. (6.17),

$$M = \frac{\sigma_1}{h}\left(I_1 + \frac{E_2}{E_1} I_2\right) = \frac{\sigma_2}{h}\left(I_2 + \frac{E_1}{E_2} I_1\right) \tag{6.19}$$

where I_1 and I_2 are the respective second moments of area about the neutral axis. Now,

$$\frac{E_2}{E_1} I_2 = \frac{E_2}{E_1}\left(\frac{b_2 d_2^{\,3}}{12} + b_2 d_2 \bar{y}_2^{\,2}\right) = b_1'\left(\frac{d_2^{\,3}}{12} + d_2 \bar{y}_2^{\,2}\right)$$

where $b_1' = (E_2/E_1)b_2$, so exactly the same resisting moment will exist if material 2 is replaced by material 1 having the *same* depth d_2 but a new width of $(E_2/E_1)b_2$. This forms the equivalent section shown in Fig. 6.15(a), which is entirely made of material 1 and is shown for the case of

Fig. 6.15

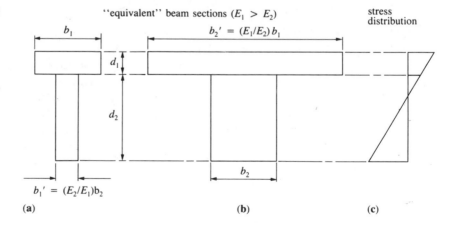

"equivalent" beam sections ($E_1 > E_2$)

stress distribution

$E_1 > E_2$. In the converse manner, material 1 can be replaced by material 2 if the width is made $(E_1/E_2)b_1$ to give the equivalent section as in Fig. 6.15(b). The neutral axis can be found quite simply (from either equivalent section) as it now coincides with the centroid. The stress distribution in either equivalent section, Fig. 6.15(c), can be determined and transposed to that actually occurring as in Fig. 6.14(b) by using eqn. (6.18) for the condition at the interface.

EXAMPLE 6.7

A timber beam of rectangular section 100 mm × 50 mm and simply supported at the ends of a 2 m span has a 30 mm × 10 mm steel strip securely fixed to the top surface as shown in Fig. 6.16(a) to protect the timber from trolley wheels. When the trolley is exerting a force at mid-span of 2 kN, determine the stress distribution at that section. $E_{\text{steel}} = 20E_{\text{wood}}$.

Fig. 6.16

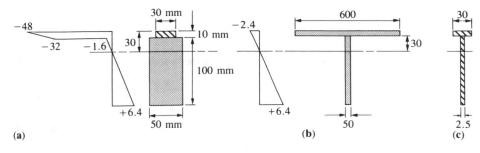

(a) (b) (c)

SOLUTION

Firstly consider the equivalent section made entirely of timber as in Fig. 6.16(b). The 30 mm width of steel becomes $30 \times 20 = 600$ mm width of timber equivalent. The position of the centroid and hence of the neutral axis from the top surface is given by

$$\bar{y}(6000 + 5000) = (6000 \times 5) + (5000 \times 60)$$

$$\bar{y} = 30 \text{ mm}$$

$$I = \frac{600 \times 10^3}{12} + (6000 \times 25^2) + \frac{50 \times 100^3}{12} + (5000 \times 30^2)$$

$$= 1246.6 \times 10^4 \text{ mm}^4$$

Maximum bending moment,

$$M = WL/4 = (2000 \times 2)/4 = 1000 \text{ N-m}$$

At lower surface,

$$\sigma = \frac{1000 \times 0.08}{1246.6 \times 10^{-8}} = 6.4 \text{ MN/m}^2$$

At interface in timber, multiplying the stress by the ratio of the respective y-values,

$$\sigma = 6.4 \times -20/80 = -1.6 \text{ MN/m}^2$$

Since $E_{\text{steel}} = 20E_{\text{wood}}$, then from eqn. (6.18), $\sigma_{\text{steel}} = 20\sigma_{\text{wood}}$. Hence at interface in steel,

$$\sigma = -1.6 \times 20 = -32 \text{ MN/m}^2$$

At top surface,

$$\sigma = 6.4 \times -\frac{30}{80} \times 20 = -48 \text{ MN/m}^2$$

The distribution in the equivalent section is shown in Fig. 6.16(b) and the actual distribution in Fig. 6.16(a).

If, next, the equivalent section is made out of steel then this is represented in Fig. 6.16(c) where the width of the timber is made into an equivalent width of steel by dividing 50 mm by 20, the modulus ratio, to give 2.5 mm.

The centroid is at

$$\bar{y}(300 + 250) = (300 \times 5) + (250 \times 60)$$

$$\bar{y} = 30 \, \text{mm}$$

which is the same as before since the proportions are the same. As only width dimensions have been changed by a factor of 20 compared to the previous equivalent section, there is no need to recalculate I, since

$$I_s = \frac{I_w}{20} = \frac{1246.6 \times 10^4}{20} = 623.3 \times 10^3 \, \text{mm}^4$$

At top surface, $\sigma = \dfrac{1000 \times (-0.03)}{623.3 \times 10^{-9}} = -48 \, \text{MN-m}^2$

At interface in steel, $\sigma = -48 \times \dfrac{20}{30} = -32 \, \text{MN/m}^2$

At interface in timber, $\sigma = -32 \times \dfrac{1}{20} = -1.6 \, \text{MN/m}^2$

and

At lower surface, $= 48 \times \dfrac{80}{30} \times \dfrac{1}{20} = +6.4 \, \text{MN/m}^2$

REINFORCED CONCRETE SECTIONS

Perhaps the most common example of a composite beam is one using steel bars to reinforce concrete. The steel is always embedded in the concrete on the tension side of the beam owing to the weakness of concrete in tension, but reinforcement may also be included on the compression side to reduce the amount of concrete required if it were not reinforced.

Consider the case illustrated in Fig. 6.17 and make the conventional assumption that the concrete takes all the compression and the reinforcing bars take all the tension. All the required relationships have been derived above, and it is only necessary now to solve for the unknown quantities as required.

Let the distance of the neutral axis from the outer surface in compression be h, Fig. 6.17(b), and the ratio of the elastic moduli

$$\frac{E_{\text{steel}}}{E_{\text{concrete}}} = m$$

Fig. 6.17

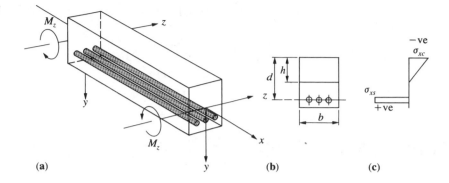

(a)   (b)   (c)

From eqns. (6.15) and (6.16),

$$\sigma_c = \frac{y_c E_c}{R} \quad \text{and} \quad \sigma_s = \frac{y_s E_s}{R}$$

Substituting the stresses into the equation (6.13) for longitudinal equilibrium, and noting that the stresses in the steel and concrete are of opposite sign, gives

$$-\frac{E_c}{R}\int_{A_c} y_c dA_c + \frac{E_s}{R}\int_{A_s} y_s dA_s = 0 \tag{6.20}$$

For the steel reinforcement the tensile stress is considered constant over the cross-sectional area A_s and concentrated at $y = (d - h)$ as in Fig. 6.17(c), so that eqn. (6.20) becomes

$$-E_c \frac{bh^2}{2} + E_s(d - h)A_s = 0 \tag{6.21}$$

and, substituting $E_s/E_c = m$ and rearranging gives

$$h = \left[\left(\frac{mA_s}{b}\right)^2 + \frac{2mA_s d}{b}\right]^{1/2} - \frac{mA_s}{b} \tag{6.22}$$

which gives the position of the neutral axis.

Substituting σ_c and σ_s for σ_m and σ_r in eqn. (6.14) for equilibrium of moments,

$$\frac{E_c}{R}\int_{A_c} y_c^2 dA_c + \frac{E_s}{R}\int_{A_s} y_s^2 dA_s = M$$

or

$$\frac{E_c}{R}\frac{bh^3}{3} + \frac{E_s}{R}A_s(d - h)^2 = M$$

Now,

$$\frac{1}{R} = \frac{\sigma_c}{y_c E_c} = \frac{\sigma_s}{y_s E_s}$$

so that substituting for $1/R$ gives

$$\left.\begin{aligned}\sigma_c &= \frac{My_c}{\frac{1}{3}bh^3 + mA_s(d-h)^2} \\ \sigma_s &= \frac{M(d-h)m}{\frac{1}{3}bh^3 + mA_s(d-h)^2}\end{aligned}\right\}$$ (6.23)

EXAMPLE 6.8

A reinforced concrete T-beam, Fig. 6.18 has a flange 1.5 m wide and 100 mm deep. The reinforcement is placed in the web 380 mm from the upper edge of the flange. The beam is designed so that the neutral axis coincides with the lower edge of the flange. The limits of stress are for steel 110 MN/m² and for concrete 4 MN/m². The modular ratio $E_{\text{steel}}/E_{\text{concrete}}$ is 15. Calculate (a) the area of the reinforcement, (b) the moment of resistance of the beam, (c) the actual maximum stress in the steel and the concrete.

Fig. 6.18

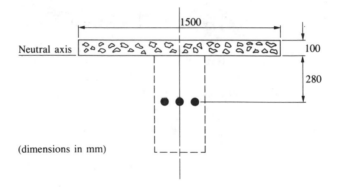

(dimensions in mm)

SOLUTION

The area of steel can be calculated using eqn. (6.21):

$$-\frac{1.5 \times 0.1^2}{2} + 15(0.38 - 0.1)A_s = 0$$

whence

$$A_s = 1790 \text{ mm}^2$$

The denominator in eqns. (6.23) is given as

$$\tfrac{1}{3} \times 1.5 \times 0.1^3 + [15 \times 0.00179(0.38 - 0.1)^2] = 2.61 \times 10^{-3} \text{ m}^4$$

Assuming the steel reaches the maximum stress, then from eqn. (6.23),

$$M = \frac{2.61 \times 10^{-3} \times 110 \times 10^6}{0.28 \times 15} = 68.3 \text{ kN-m}$$

Alternatively, if the concrete reaches the limiting stress,

$$M = \frac{2.61 \times 10^{-3} \times 4 \times 10^6}{0.1} = 104.5 \text{ kN-m}$$

Therefore

Maximum moment of resistance $= 68.3$ kN-m

Actual maximum stress in concrete, $\sigma_c = \dfrac{68.3 \times 10^3 \times 0.1}{2.61 \times 10^{-3}}$

$$= 2.62 \text{ MN/m}^2$$

COMBINED BENDING AND END LOADING

A number of situations arise in practice where a member is subjected to a combination of bending and longitudinal load. Problems of this type can be most easily dealt with by superposition of the individual components of stress to give resultant values.

Consider the rectangular-section beam shown in Fig. 6.19, which is subjected to bending moments M about the z-axis, and an axial load P in the x-direction.

Fig. 6.19

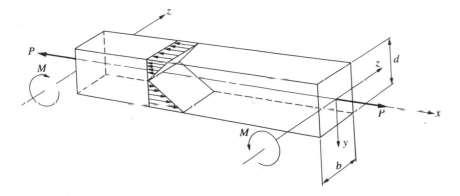

If the end load acted alone, there would be a longitudinal stress

$$\sigma_x = \frac{P}{A}$$

If the moment acted alone, the longitudinal bending stress would be

$$\sigma_x = \pm \frac{My}{I}$$

By superposition the resultant stress due to P and M is

$$\sigma_x = \frac{P}{A} \pm \frac{My}{I}$$

$$= \frac{P}{bd} \pm \frac{12My}{bd^3} \tag{6.24}$$

Fig. 6.20

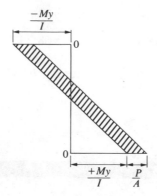

The distribution of σ_x over the cross-section is shown in Figs. 6.19 and 6.20 by the shaded portions. An interesting feature is that the neutral surface no longer passes through the centroid of the cross-section since

$$\int_A \sigma_x \,\mathrm{d}A \neq 0$$

as in the case of simple bending.

Note that y in eqn. (6.24) is a distance from the centroid of the section and *not* from the new neutral axis.

ECCENTRIC END LOADING

If the end load P does not act at the centroid of the cross-section then it will itself set up bending moments about the principal axes. For example, in Fig. 6.21(*a*) a short beam is subjected to a compressive load P, which is eccentric from the z- and y-axes by amounts m and n respectively. The equivalent equilibrium system is with the load P acting at the centroid and moments Pm and Pn acting about the z- and y-axes, as in Fig. 6.21(*b*).

The direct stress due to P alone is therefore

$$\sigma_x' = -\frac{P}{A}$$

Fig. 6.21

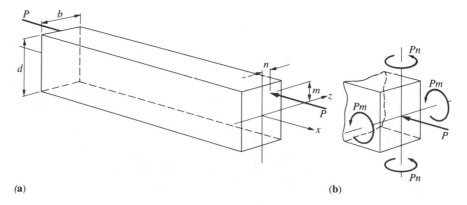

(a)   (b)

Due to bending about the y-axis,

$$\sigma''_x = \pm \frac{Pnz}{I_{yy}}$$

Due to bending about the z-axis,

$$\sigma'''_x = \pm \frac{Pmy}{I_{zz}}$$

Therefore the resultant longitudinal stress is by superposition

$$\sigma_x = \sigma'_x + \sigma''_x + \sigma'''_x$$

$$= -\frac{P}{A} \pm \frac{Pnz}{I_{yy}} \pm \frac{Pmy}{I_{zz}} \tag{6.25}$$

$$= \frac{P}{A}\left(-1 \pm \frac{nz}{r_{yy}^2} \pm \frac{my}{r_{zz}^2}\right) \tag{6.26}$$

where r_{yy} and r_{zz} are the radii of gyration about the relevant axes. For the rectangular section shown in Fig. 6.21, $r^2 = I/A$; therefore

$$r_{yy}^2 = \frac{b^2}{12} \quad \text{and} \quad r_{zz}^2 = \frac{d^2}{12}$$

and

$$\sigma_x = \frac{P}{bd}\left(-1 \pm \frac{12nz}{b^2} \pm \frac{12my}{d^2}\right) \tag{6.27}$$

where z lies between $\pm b/2$ and y between $\pm d/2$.

In some materials which are strong in compression but weak in tension, such as concrete, it is necessary to limit the eccentricities m and n so that no tensile stress is set up. The condition for no tension is that the compressive stress due to P is greater than or equal to the maximum tensile bending stresses set up by the moments Pm and Pn. Therefore, from eqn. (6.27),

$$\frac{6n}{b} + \frac{6m}{d} \leqslant 1 \tag{6.28}$$

This relationship defines the locus of maximum eccentricity as shown by the shaded area in Fig. 6.22. When P is applied within the shaded area

Fig. 6.22

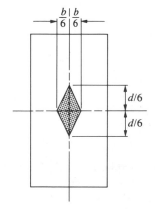

137

then no tensile stress will be set up anywhere in the cross-section. The limits on the z- and y-axes are $\pm b/6$ and $\pm d/6$, which has resulted in what is known as the *middle-third rule* for no tension. Typical distributions for various amounts of eccentricity along *one* principal axis are shown in Fig. 6.23.

Fig. 6.23

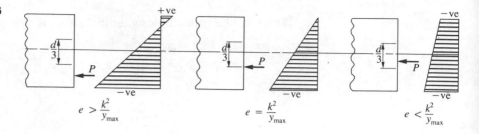

$$e > \frac{k^2}{y_{max}} \qquad e = \frac{k^2}{y_{max}} \qquad e < \frac{k^2}{y_{max}}$$

EXAMPLE 6.9

In a tensile test within the elastic range on a specimen of circular cross-section an extensometer is being used which will only measure deformation on one side of the specimen. Determine how much eccentricity of loading will give rise to a 5% difference between the surface stress derived from the extensometer and the average stress over the cross-section.

SOLUTION

Let the average stress be σ; then for a 5% error due to non-axial loading the resultant stress on one edge of the specimen will be 0.95σ, and at the opposite end of the diameter 1.05σ. From eqn. (6.26),

$$0.95\sigma = \sigma\left(1 - \frac{my_{max}}{r_z^2}\right)$$

Therefore

$$m = \frac{0.05r_z^2}{y_{max}}$$

For a circular cross-section $I = Ar^2 = \pi d^4/64$, and hence $r_z^2 = d^2/16$. Therefore

$$m = \frac{0.05d^2/16}{d/2} = 0.00625d$$

Thus in a tensile test on a specimen of 10 mm diameter, the eccentricity of loading must be *less* than 0.063 mm to avoid surface stresses being more than 5% greater than the average direct stress.

EXAMPLE 6.10

A slotted machine link 6 mm thick, illustrated in Fig. 6.24(*a*) is subjected to a tensile load of 40 kN acting along the centre line of the end faces. Find the stress distribution for a section through the slot such as AA.

Fig. 6.24

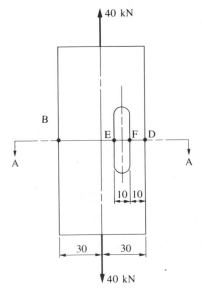

Section AA
(dimensions in mm)
(a)

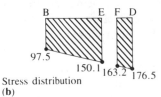

Stress distribution
(b)

SOLUTION

We must first find the centroid C of the section AA and by taking the moment of area about the left side:

$$[(40 \times 6) + (10 \times 6)]\bar{x} = (40 \times 6 \times 20) + (10 \times 6 \times 55)$$

so that

$$\bar{x} = 27 \text{ mm}$$

Hence the load is acting eccentrically with respect to the centroid of the slot cross-section AA by an amount

$$e = 30 - 27 = 3 \text{ mm}$$

This eccentricity gives rise to a moment

$$M = 40\,000 \times 0.003 = 120 \text{ N-m}$$

which will cause tensile bending stress along the edge at D and compressive bending stress along the edge at B.

For bending only, the neutral axis passes through C and the greatest bending stresses occur at B and D; thus $y_{max} = 27$ and 33 mm and the second moment of area is $91.28 \times 10^3 \text{ mm}^4$. Therefore

$$\sigma_B = -\frac{120 \times 0.027}{91.28 \times 10^{-9}} = -35.5 \text{ MN/m}^2$$

$$\sigma_D = +\frac{120 \times 0.033}{91.28 \times 10^{-9}} = +43.5 \text{ MN/m}^2$$

$$\text{Direct stress} = \frac{40\,000}{0.05 \times 0.006} = +133 \text{ MN/m}^2$$

The resultant stresses will be

At B	$+133 - 35.5 = 97.5 \text{ MN/m}^2$
At E	$+133 + 17.1 = 150.1 \text{ MN/m}^2$
At F	$+133 + 30.2 = 163.2 \text{ MN/m}^2$
At D	$+133 + 43.5 = 176.5 \text{ MN/m}^2$

The distribution of stress is shown in Fig. 6.24(b).

SHEAR STRESSES IN BENDING

The presence of shear force indicates that there must be shear stress on transverse planes in the beam. It is not possible to make use of the conditions of geometry of deformation and the stress–strain relationships except in the development of an exact solution. However, from the assumptions about the validity of the bending-stress distribution, it is possible to estimate the transverse and longitudinal shear-stress distributions in the beam by using only the condition of equilibrium.

Firstly, consider the bending-stress distribution in the short section of beam of length dx shown in Fig. 6.25. The bending moment increases from M on GK to $M + (dM/dx)\, dx$ on HJ, therefore the bending stress on any arbitrary fibre must increase from

$$\sigma = \frac{My}{I} \quad \text{on} \quad \text{GK} \tag{6.29}$$

Fig. 6.25

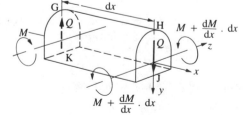
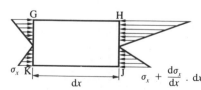

to

$$\sigma + \frac{d\sigma}{dx}dx = \left(M + \frac{dM}{dx}dx\right)\frac{y}{I} \quad \text{on} \quad \text{HJ} \tag{6.30}$$

at each value of y.

Now consider the strip of beam below an xz-plane ABCD, Fig. 6.26, at a distance y_1, from the neutral surface. Each fibre in this strip will have

Fig. 6.26

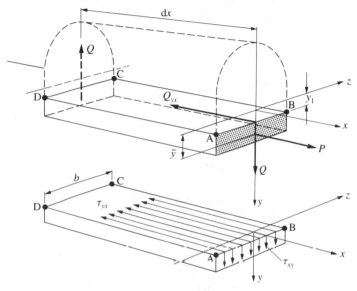

an increase in bending stress $d\sigma$ along the length dx, so that taken over the shaded area A there will be a resultant axial force equal to

$$\int_A d\sigma \, dA = P \tag{6.31}$$

One way of maintaining equilibrium of the strip is by means of a shear force Q_{yx} acting on the surface of the plane ABCD. Therefore

$$Q_{yx} = \int_A d\sigma \, dA \tag{6.32}$$

Now subtracting eqn. (6.29) from eqn. (6.30) gives

$$d\sigma = dM\frac{y}{I}$$

Hence, from eqn. (6.32),

$$Q_{yx} = \int_A dM\frac{y}{I}dA \tag{6.33}$$

If the width of the strip at the plane $y = y_1$ is b and this is small compared with the depth of the beam, then the shear stress on ABCD is almost uniformly distributed and can therefore be expressed as

$$\tau_{yx} = \frac{Q_{yx}}{b\,dx}$$

Note that this is a *positive* shear stress, being the product of negative directions of the normal to the plane and the stress direction.

Substituting for Q_{yx} in eqn. (6.33) gives

$$\tau_{yx} = \frac{\mathrm{d}M}{\mathrm{d}x} \frac{1}{bI} \int_A y \, \mathrm{d}A \qquad (6.34)$$

But, from eqn. (6.2), $\mathrm{d}M/\mathrm{d}x = Q$ the vertical shear force on the section and the integral is the first moment of area A about the neutral surface; therefore eqn. (6.34) becomes

$$\tau_{yx} = Q \frac{A\bar{y}}{bI}$$

Using the principle of complementary shear stresses, Fig. 6.26(*b*), it is then evident that the vertical shear stress, τ_{xy}, is also given by

$$\tau_{xy} = Q \frac{A\bar{y}}{bI} \qquad (6.35)$$

Again we note that τ_{xy} is *positive* and so is correctly related to the *positive* shear force Q.

This solution is only exact for a constant shear force along the beam; however, if the cross-section is small compared with the span, the error introduced by a varying shear force is quite small.

EXAMPLE 6.11

A beam of rectangular cross-section, depth d, thickness b, is simply supported over a span of length l, and carries a concentrated load W at mid-span. Determine the distribution and maximum value of the transverse shear stress.

SOLUTION

Although changing sign at the centre of the span, the shear force Q is constant in magnitude along the whole span and is equal to $W/2$.

Considering the cross-section shown in Fig. 6.27, the transverse shear stress on some arbitary line AB at a distance y from the neutral surface is given by eqn. (6.35), where A is the shaded area below AB and $\bar{y}$ is the

Fig. 6.27

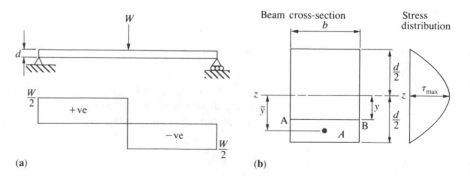

(a) (b)

distance of the centroid of A from the neutral surface. Therefore

$$A\bar{y} = b\left(\frac{d}{2} - y\right) \times \frac{1}{2}\left(\frac{d}{2} + y\right) = \frac{b}{2}\left[\left(\frac{d}{2}\right)^2 - y^2\right]$$

The vertical shear stress is therefore

$$(\tau_{xy})_{AB} = \frac{Q}{bI} \times \frac{b}{2}\left[\left(\frac{d}{2}\right)^2 - y^2\right]$$

$$= \frac{W}{4I}\left[\left(\frac{d}{2}\right)^2 - y^2\right] \tag{6.36}$$

The above expression shows that the distribution of vertical shear stress down the depth of the section is parabolic. The shear stress is zero at the outer fibres where $y = \pm d/2$, as it must be since the complementary shear stress in the longitudinal direction must be zero at a free surface. The maximum value is at the neutral surface where $y = 0$; therefore

$$\tau_{xy\,max} = \frac{Wd^2}{16I} = \frac{3W}{4bd}$$

If uniformly distributed the shear stress would be given by the shear force divided by the area, or

$$\tau_{xy\,mean} = \frac{W}{2bd}$$

Hence the maximum shear stress is 1.5 times the mean value.

EXAMPLE 6.12

A box beam is built up of plate material riveted together as shown by the cross-section, Fig. 6.28. It is simply supported at each end of a 3 m span and carries a concentrated load of 12 kN at 1 m from one end. Estimate a suitable rivet diameter if the rivets are to be pitched at about 100 mm intervals. The shear stress is not to exceed 50 MN/m² in each rivet.

Fig. 6.28

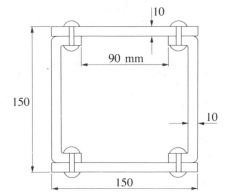

SOLUTION

The maximum shear force will be equal to the larger of the two reactions, $12 \times 2/3 = 8$ kN.

Now, the force tending to shear the rivets is due to the variation of bending stress along the length of the beam and is given by eqn. (6.34) and the next equation if slightly rearranged as follows:

$$\text{Shear force per unit length} = \tau_{yx}b = \frac{QA\bar{y}}{I}$$

If the rivet pitch is p and the cross-sectional area a, then the allowable shear force per unit length is $\hat{\tau}a/p$. Hence

$$\frac{\hat{\tau}a}{p} = \frac{QA\bar{y}}{I}$$

$$\text{Required area, } a = \frac{QA\bar{y}p}{\hat{\tau}I}$$

$$I \approx \frac{0.15 \times 0.15^3}{12} - \frac{0.13 \times 0.11^3}{12} - (2 \times 0.09 \times 0.01 \times 0.06^2)$$
$$\approx 0.2132 \times 10^{-4} \, \text{m}^4$$

At the interface between the outer plates and the side plates where shearing would occur,

$$A\bar{y} = 150 \times 10 \times 70 = 105 \times 10^3 \, \text{mm}^3$$

Therefore

$$a = \frac{8000 \times 105 \times 10^{-6} \times 0.1}{50 \times 10^6 \times 0.2132 \times 10^{-4}} = 7.89 \times 10^{-5} \, \text{m}^2$$
$$= 78.9 \, \text{mm}^2$$

As there are two rivets at each interface resisting shear, the diameter of each is

$$d = \left(\frac{78.9}{2} \times \frac{4}{\pi}\right)^{1/2} = 7.1 \, \text{mm}$$

The best compromise is to use 7 mm rivets at 97 mm pitch, giving 31 pitches in the length of the beam.

BENDING AND SHEAR STRESSES IN I-SECTION BEAMS

The I-section beam shown in Fig. 6.29(a) is widely used in the construction of buildings, bridges, etc. The shape is efficient to resist both bending and shear. The latter is carried almost entirely by the web, and the flanges are located where the bending stress is highest. The practical I-section is usually idealized for ease of calculation into the rectangular shapes shown at (b). The second moment of area of the section is derived

Fig. 6.29

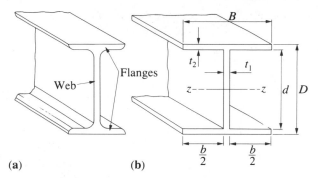

(a)   (b)

in Appendix A in two ways; and one of these gives

$$I = \frac{BD^3 - bd^3}{12}$$

the bending stress distribution being

$$\sigma = \frac{12My}{BD^3 - bd^3} \qquad (6.37)$$

which is illustrated in Fig. 6.30.

Fig. 6.30

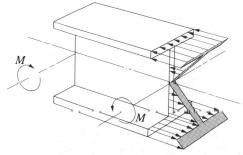

It can be shown that for a typical I-section the flanges carry approximately 80% of the total bending moment on the cross-section. In designing an I-section against failure, consideration must be given, in addition to the strength of the tension flange, to the avoidance of buckling (Chapter 13) in the compression flange.

SHEAR STRESSES IN WEB AND FLANGES

The shear-stress distributions are rather more complex than in a rectangular section. Firstly, we will examine the distribution of vertical shear τ_{xy} parallel to the axis yy.

For the web, from eqn. (6.35),

$$\tau_{xy} = \frac{Q}{It_1} \int_y^{D/2} y \, \mathrm{d}A$$

Because the section has a different width for the flange and the web this integral has to be expressed in two parts:

$$\tau_{xy} = \frac{Q}{It_1} \left[\int_y^{d/2} t_1 y \, \mathrm{d}y + \int_{d/2}^{D/2} B y \, \mathrm{d}y \right]$$

$$= \frac{Q}{2I} \left[\left(\frac{d}{2} \right)^2 - y^2 \right] + \frac{QB}{8It_1} (D^2 - d^2) \qquad (6.38)$$

Fig. 6.31

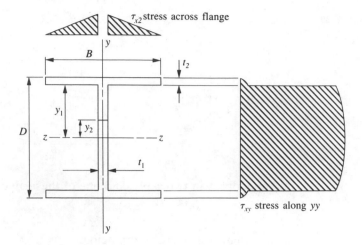

This expression gives a parabolic distribution superimposed on a constant value as shown in Fig. 6.31.

The maximum value occurs at the neutral axis, where $y = 0$, and is

$$\tau_{xy} = \frac{Q}{8I}\left[d^2 + \frac{B}{t_1}(D^2 - d^2)\right] \tag{6.39}$$

In that part of the flange directly above and below the web the vertical shear stress can be expressed as

$$\tau_{xy} = \frac{Q}{BI}\int_{y'}^{D/2} By \, dy = \frac{Q}{2I}\left[\left(\frac{D}{2}\right)^2 - y'^2\right]$$

However, in those parts of the flange on each side of the web the top and bottom surfaces are "free" from load, and therefore the longitudinal and complementary vertical shear stresses must be zero. Thus the distribution is parabolic as for a rectangular section.

It is evident that in the flanges the vertical shear stress and its complementary component contribute little to balancing the longitudinal variation in bending stress. However, this may be achieved, as illustrated in Fig. 6.32, by means of a shear force Q_{zx} lying in an xy-plane which cuts off a segment of the flange. A complementary shear force Q_{xz} then occurs in a yz-plane. As in the analysis on page 141 the net end load is equal

Fig. 6.32

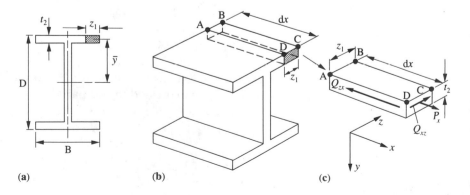

(a)　　　　　(b)　　　　　(c)

to the shear force Q_{zx}, which is the shear stress (positive) τ_{zx} multiplied by the area $t_2\,dx$, i.e.

$$P_x = Q_{zx} = \tau_{zx}t_2dx$$

Also

$$P_x = \int_A d\sigma_x dA$$

Therefore

$$\tau_{zx}t_2dx = \int_A d\sigma_x dA$$

$$= \frac{dM}{I}\int_A y\,dA$$

$$\tau_{zx} = \frac{Q}{t_2I}A\bar{y} \tag{6.40}$$

where Q is the (positive) *vertical* shear force on the section and $\bar{y}$ is still measured from the neutral axis to the centroid of the area.

$$A\bar{y} = zt_2\frac{D - t_2}{2}$$

From eqn. (6.40) and the complementary (positive) shear-stress condition,

$$\tau_{zx} = \tau_{xz} = Qz\frac{D - t_2}{2I} \tag{6.41}$$

This is a linear distribution of shear stress in the z-direction, being zero at the outer edges of the flanges and a maximum at the joint with the web. The distribution is shown in Fig. 6.31, in which the maximum value is

$$\tau_{xz} = \frac{Q(D - t_2)(B - t_1)}{4I} \tag{6.42}$$

EXAMPLE 6.13

The vertical steel column of 5 m height and rolled I-section shown in Fig. 6.33 is built in at the lower end and subjected to a transverse force of

Fig. 6.33

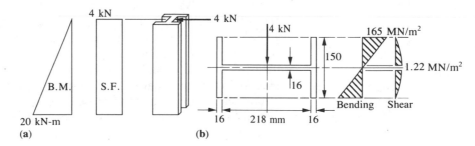

B.M.

S.F.

20 kN-m

(a)

4 kN

4 kN

4 kN

150

16

16   218 mm   16

165 MN/m²

1.22 MN/m²

Bending   Shear

(b)

4 kN at the free end. Calculate the bending and shear stress distributions at the fixed-end cross-section.

SOLUTION   At the base

Bending moment $= 4 \times 5 = 20$ kN-m

Shear force $= 4$ kN

Second moment of area $= \dfrac{32 \times 150^3}{12} + \dfrac{218 \times 16^3}{12}$

$$= 9.075 \times 10^6 \text{ mm}^4$$

$y_{max} = \pm 0.075$ m

Therefore

$$\sigma_{max} = \pm \frac{20 \times 10^3 \times 0.075}{9.075 \times 10^{-6}} = \pm 165 \text{ MN/m}^2$$

In the flanges the shear stress in the y-direction is given by

$$\tau_{xy} = \frac{4000}{2 \times 0.016 \times 9.075 \times 10^{-6}} \int_y^{75} 2 \times 0.016 y \, dy$$

$$= \frac{440 \times 10^6}{2} (0.075^2 - y^2)$$

This is a parabolic distribution varying from zero at the outer surfaces to

$$\tau_{xy} = 220 \times 10^6 (0.075^2 - 0.008^2) = 1.22 \text{ MN/m}^2$$

at the section where the flanges join the web.

In the web itself, since the width, which appears in the denominator for shear stress above, is 250 mm, it is evident that the shear stress in it can be neglected compared with that in the flanges.

A consideration of the geometry of the section indicates that shear stresses of the τ_{xz} type are also insignificant.

The distributions of bending and shear stresses are shown in Fig. 6.33.

ASYMMETRICAL OR SKEW BENDING

The previous analysis has been concerned with bending about an axis of symmetry. However, many occasions arise in practice where bending will occur either of a section which does not have any axes of symmetry or of a symmetrical section about an asymmetrical axis. In order to express the conditions of equilibrium in asymmetrical bending a knowledge of first and second moments of area about arbitrary axes through the centroid of the section is required. It is therefore necessary to study certain properties of areas before embarking on the analysis of stress distribution in asymmetrical bending. The reader is here referred to Appendix A.

Fig. 6.34

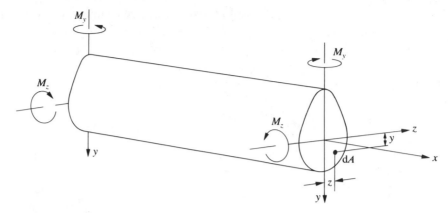

Asymmetrical pure bending of a beam is shown in Fig. 6.34 for positive external moments M_y and M_z applied about an arbitrary set of centroidal axes. Considering first bending in the xy-plane only, the equilibrium equations are

$$\int_A \sigma_x dA = 0 \quad \text{(neutral surface through the centroid)}$$

$$\int_A \sigma_x y \, dA = M_z \quad \text{and} \quad \int_A \sigma_x z \, dA = M_y$$

But $\sigma_r = yE/R_y$, where R_y is the radius of curvature in the xy-plane; hence

$$M_z = \frac{E}{R_y} \int_A y^2 dA = \frac{EI_z}{R_y}$$

and

$$M_y = \frac{E}{R_y} \int_A yz \, dA = \frac{EI_{yz}}{R_y}$$

I_{yz} is called the *product moment of area* (*see* Appendix A).

For bending only in the xz-plane a procedure similar to the above gives

$$M_y = \frac{E}{R_z} \int z^2 dA = \frac{EI_y}{R_z}$$

$$M_z = \frac{E}{R_z} \int zy \, dA = \frac{EI_{yz}}{R_z}$$

For simultaneous bending in the xy- and xz-planes the above relationships may be superimposed to give

$$\left. \begin{aligned} M_y &= \frac{EI_y}{R_z} + \frac{EI_{yz}}{R_y} \\[2mm] M_z &= \frac{EI_z}{R_y} + \frac{EI_{yz}}{R_z} \end{aligned} \right\} \tag{6.43}$$

The radii of curvature are obtained from the above equations as

$$\frac{1}{R_y} = \frac{M_z I_y - M_y I_{yz}}{E(I_y I_z - I_{yz}^2)}$$

$$\frac{1}{R_z} = \frac{M_y I_z - M_z I_{yz}}{E(I_y I_z - I_{yz}^2)}$$

The resultant bending stress is therefore the sum of the components for bending in each of the xy- and xz-planes:

$$\sigma_x = \frac{yE}{R_y} + \frac{zE}{R_z}$$

$$= \frac{y(M_z I_y - M_y I_{yz}) + z(M_y I_z - M_z I_{yz})}{I_y I_z - I_{yz}^2} \tag{6.44}$$

If either M_z or M_y is in the opposite sense, i.e. negative to that shown in Fig. 6.34, then the appropriate signs must be changed in eqn. (6.44) and elsewhere.

The neutral surface, where $\sigma_x = 0$, is defined by the plane

$$y(M_z I_y - M_y I_{yz}) + z(M_y I_z - M_z I_{yz}) = 0 \tag{6.45}$$

If bending is about the principal axes of the section for which $I_{yz} = 0$, then

$$\sigma_x = \frac{M_z y}{I_z} + \frac{M_y z}{I_y} \tag{6.46}$$

If there is only a single external applied moment M about an axis ss inclined at θ to one of the principal axes, as in Fig. 6.35, then the moment vector M can be resolved into components M_y and M_z about the y- and z-axes, so that

$$-M_y = M \sin \theta \quad \text{and} \quad M_z = M \cos \theta$$

Fig. 6.35

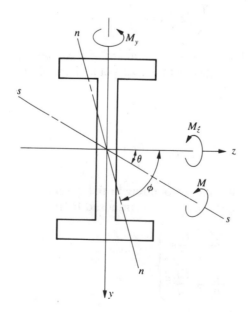

and substituting in eqn. (6.46),

$$\sigma_x = \frac{-Mz \sin \theta}{I_y} + \frac{My \cos \theta}{I_z} \qquad (6.47)$$

The neutral plane will no longer be perpendicular to the plane of bending, as in the symmetrical problem, but can be determined by putting $\sigma_x = 0$ above; then

$$\frac{-z \sin \theta}{I_y} + \frac{y \cos \theta}{I_z} = 0$$

and

$$\frac{y}{z} = \frac{I_z}{I_y} \tan \theta = \tan \phi$$

where ϕ is the inclination of *nn*, the neutral surface, to the *z*-axis.

The neutral surface is perpendicular to the plane of bending if either $I_z = I_y$ or $\theta = 0$.

EXAMPLE 6.14

The angle section shown in Fig. 6.36 is subjected to a bending moment of 2 kN-m about the *z*-axis. Determine the bending stress distribution.

Fig. 6.36

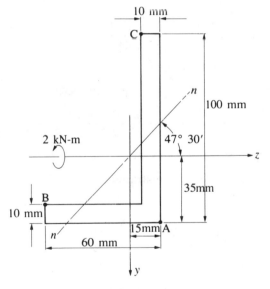

SOLUTION

Considering the first moments of area about vertical and horizontal edges respectively,

$$\bar{z} = \frac{(90 \times 10 \times 5) + (60 \times 10 \times 30)}{(90 + 60)10} = 15 \text{ mm}$$

$$\bar{y} = \frac{(90 \times 10 \times 55) + (60 \times 10 \times 5)}{(90 + 60)10} = 35 \text{ mm}$$

which gives the co-ordinates of the centroid.

$$I_z = \frac{10 \times 90^3}{12} + (10 \times 90 \times 20^2) + \frac{60 \times 10^3}{12} + (60 \times 10 \times 30^2)$$

$$= 151.2 \times 10^4 \, \text{mm}^4$$

$$I_y = \frac{90 \times 10^3}{12} + (90 \times 10 \times 10^2) + \frac{10 \times 60^3}{12} + (10 \times 60 \times 15^2)$$

$$= 41.3 \times 10^4 \, \text{mm}^4$$

$$I_{zy} = [90 \times 10 \times 10 \times (-20)] + [60 \times 10 \times (-15) \times 30]$$

$$= -45 \times 10^4 \, \text{mm}^4$$

The position of the neutral axis may now be found using eqn. (6.45), from which, with $M_y = 0$,

$$y(2000 \times 41.3 \times 10^{-8}) + z(-2000 \times -45 \times 10^{-8}) = 0$$

$$\frac{y}{z} = -1.09 \quad \text{and} \quad \phi = -47°30'$$

The maximum tensile stress will occur at A, and using eqn. (6.44), in which M_y will be zero,

$$\sigma_A = \frac{2000[(0.035 \times 41.3 \times 10^{-8}) - (+0.015 \times -45 \times 10^{-8})]}{[(41.3 \times 151.2) - 45^2]10^{-16}}$$

$$= 101 \, \text{MN/m}^2$$

At B,

$$\sigma_B = \frac{2000[(0.025 \times 41.3 \times 10^{-8}) - (-0.045 \times -45 \times 10^{-8})]}{[(41.3 \times 151.2) - 45^2]10^{-16}}$$

$$= -47 \, \text{MN/m}^2$$

The maximum compressive stress occurs at C, where

$$\sigma_C = \frac{2000[(-0.065 \times 41.3 \times 10^{-8}) - (+0.005 \times -45 \times 10^{-8})]}{[(41.3 \times 151.2) - 45^2]10^{-16}}$$

$$= -116.5 \, \text{MN/m}^2$$

The distribution is illustrated in Fig. 6.37.

Fig. 6.37

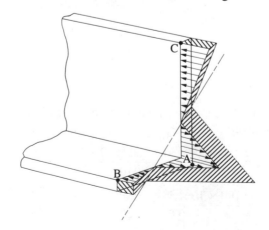

SHEAR STRESS IN THIN-WALLED OPEN SECTIONS AND SHEAR CENTRE

There are a number of beam sections widely used, particularly in aircraft construction, in which the thickness of material is small compared with the overall geometry and there is only one or no axis of symmetry. These members are termed *thin-walled open sections,* and some common shapes are shown in Fig. 6.38.

Fig. 6.38

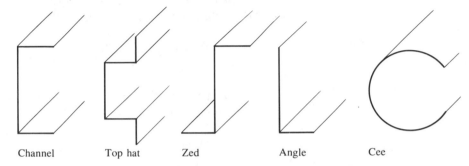

Channel   Top hat   Zed   Angle   Cee

The arguments applied to the shear stress distribution in the flanges of the I-section (page 145) may also be applied in the above cases; however, there is one important difference owing to the lack of symmetry in the latter.

If the external applied forces, which set up bending moments and shear forces, act through the centroid of the section, then in addition to bending, twisting of the beam will generally occur. To avoid twisting, and cause only bending, it is necessary for the forces to act through a particular point, which may not coincide with the centroid. The position of this point is a function only of the geometry of the beam section; it is termed the *shear centre*.

Before deriving a general theory for the bending of open-walled sections a simple example will be studied and the existence of a shear centre established. Referring to Fig. 6.39, in which the channel section is

Fig. 6.39

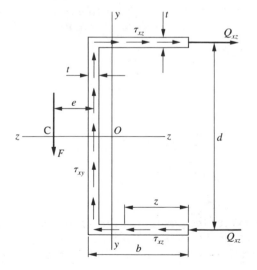

loaded by a vertical force F, the y- and z-axes are principal axes and hence $I_{yz} = 0$. For the shear stress in the flanges the analysis is similar to that, on page 145 eqn. (6.40), for the I-section.

$$\tau_{xz} = \frac{Q}{tI_z} \int_A y \, dA$$

where Q is the vertical shear force on the section caused by the applied force F. For a length z of the flange,

$$\tau_{xz} = \frac{Q}{tI_z} \int_0^z \frac{d}{2} t \, dz = \frac{Q}{tI_z} \times \frac{tzd}{2} \tag{6.48}$$

The shear stress varies linearly with z from zero at the left to a maximum at the centre line of the web:

$$\tau_{xz\,max} = \frac{Qbd}{2I_z}$$

The average shear stress is $Qbd/4I_z$, and therefore the horizontal shear force in the top and bottom flange is

$$Q_{xz} = \frac{Qb^2td}{4I_z}$$

The couple about the x-axis of these shear forces which would cause twisting of the section is

$$Q_{xz}d = \frac{Qb^2d^2t}{4I_z}$$

Twisting of the section is avoided if there is an opposing couple of equal magnitude. Let the vertical force F act through a point C, the shear centre, at a distance e from the middle of the web as shown in Fig. 6.39, so that Fe balances $Q_{xz}d$. Now Q must equal F for equilibrium, therefore

$$Qe = Q_{xz}d$$

hence

$$Qe = \frac{Qb^2d^2t}{4I_z}$$

or

$$e = \frac{b^2d^2t}{4I_z} \tag{6.49}$$

which locates the position of the shear centre and is only a function of the geometry of the section. The vertical shear stress τ_{xy} in the web may be found in the same way as was that for the I-section.

GENERAL CASE OF BENDING OF A THIN-WALLED OPEN SECTION

The analysis of the channel section was relatively simple, since there was one axis of symmetry about which bending was made to occur, and had much in common with the analysis of shear stresses in the I-section. The more general case is that of a geometrically asymmetrical open section subjected to bending which is not about a principal axis.

The method will be illustrated for a cantilever of arbitrarily curved open section, as shown in Fig. 6.40(a), subjected to a transverse load, W, through the centroid of the section at the free end. The resulting components of applied force and shear force in the co-ordinate directions will be W_z, W_y and Q_z, Q_y respectively acting through the centroid. The force components will give rise to bending moments M_y and M_z, which will vary along the length of the member.

Fig. 6.40

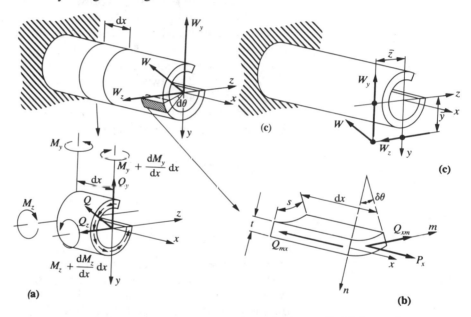

(a)

(b)

(c)

It will be assumed that shear stress through the wall thickness is negligible and that it is the longitudinal and peripheral shear stresses which are of importance. The equilibrium condition in the x-direction for the strip, Fig. 6.40(b), cut from the section subjected to bending in the xy-plane alone, due to M_z, is

$$Q_{mx} = P_x = \int_{A'} d\sigma_x dA$$

where $d\sigma_x$ is the increase in bending stress along the strip and Q_{mx} is the shear force on the edge of the strip. The orthogonal axes m and n are tangential and radial to the wall at the particular strip under examination. Now, from eqn. (6.44), putting $M_y = 0$,

$$d\sigma_x = dM_z \frac{yI_y - zI_{yz}}{I_y I_z - I_{yz}^2}$$

Therefore

$$Q_{mx} = \frac{dM_z}{I_y I_z - I_{yz}^2} \int_{A'} (y I_y - z I_{yz}) \, dA$$

or

$$\tau_{mx} = \tau_{xm} = \frac{Q_{mx}}{t \, dx} = \frac{dM_z}{dx} \frac{1}{t(I_y I_z - I_{yz}^2)} \left[I_y \int_{A'} y \, dA - I_{yz} \int_{A'} z \, dA \right]$$

$$\tau_{xm} = \frac{Q_y}{t(I_y I_z - I_{yz}^2)} \left[I_y \int_{A'} y \, dA - I_{yz} \int_{A'} z \, dA \right] \tag{6.50}$$

For bending in the xz-plane alone, due to M_y,

$$\tau'_{xm} = \frac{Q_z}{t(I_y I_z - I_{yz}^2)} \left[I_z \int_{A'} z \, dA - I_{yz} \int_{A'} y \, dA \right] \tag{6.51}$$

The integrals in the above expressions are the first moments of the area A about the y- and z-axes. The resultant value of the shear stress at a particular point is given by superposition of the above equations.

The shear stresses τ_{xm} and τ'_{xm} will each give rise to a torque about the x-axis and hence twisting of the beam. In order that there shall be bending only and no twist, it is therefore necessary that the force components W_y and W_z act through the shear centre having co-ordinates $\bar{y}$ and $\bar{z}$ as in Fig. 6.40(c). These co-ordinates can be found by considering torsional equilibrium of the section so that

$$W_y \bar{z} = \oint n \tau_{xm} t \rho \, d\theta \tag{6.52}$$

and

$$W_z \bar{y} = \oint n \tau'_{xm} t \rho \, d\theta \tag{6.53}$$

where ρ is the radius of curvature of the element and n is the perpendicular distance of τ_{xm} or τ'_{xm} from the x-axis. (τ_{xm} and τ'_{xm} are given by eqns. (6.50) and (6.51)). Since the shear forces Q_y and Q_z equal the components of applied load W_y and W_z in the above equations and therefore cancel, it will be seen that the position of the shear centre is only a function of the geometry of the cross-section.

In conclusion the load W must act through the shear centre given by the point of intersection of the lines of action of W_y and W_z to avoid twisting of the section.

EXAMPLE 6.15

A thin-walled tube of circular cross-section has inner and outer diameters of 50 and 70 mm respectively. If it is slit longitudinally on one side, at what position must a vertical force of 8 kN be applied so that there is only bending and no twisting of the section? Calculate the maximum shear stress in the section.

SOLUTION

Referring to Fig. 6.41, since the slit is narrow the I for the section may be taken as $(\pi/64)(70^4 - 50^4)$.

$$I = 872 \times 10^3 \text{ mm}^4$$

Fig. 6.41

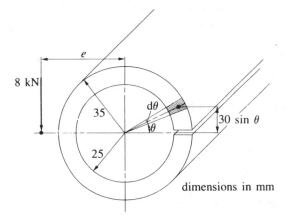

dimensions in mm

For the element shown, $y = 30 \sin \theta$ mm, therefore

$$d\sigma = \frac{dM \times 0.03 \sin \theta}{872 \times 10^{-9}}$$

Net end load $= d\sigma tr \, d\theta$

$$= \frac{dM \times 0.03 \sin \theta}{872 \times 10^{-9}} 0.01 \times 0.03 \, d\theta$$

Total end load $= \dfrac{9 \times 10^{-6}}{872 \times 10^{-9}} dM \displaystyle\int_0^\theta \sin \theta \, d\theta$

For equilibrium with the horizontal shear stress,

$$\tau \times 0.01 \, dx = 10.32 \, dM(1 - \cos \theta)$$
$$\tau = 1032Q(1 - \cos \theta)$$

The maximum shear stress occurs when $\theta = \pi$ so that

$$\hat{\tau} = 1032 \times 8000 \times 2 = 16.5 \text{ MN/m}^2$$

Torque set up by shear-stress distribution

$$= \int_0^{2\pi} \tau \times 0.01 \times 0.03^2 d\theta$$

$$= 1032 \times 8000 \times 0.01 \times 0.03^2 \int_0^{2\pi} (1 - \cos \theta) \, d\theta$$

Let the shear centre be at a distance e from the centre of the tube. Then equilibrium of torques gives

$$8000e = 1032 \times 8000 \times 9 \times 10^{-6} \times 2\pi$$
$$e = 58.4 \text{ mm}$$

BENDING OF INITIALLY CURVED BARS

The theory of bending developed so far has been related to initially straight bars and beams. The analysis will now be extended to include members which are initially curved.

The geometry of curved bars has an important bearing on the bending stress distribution. If the depth of the cross-section is small compared with the radius of curvature, then the stress distribution is linear as for straight beams. On the other hand, if the depth of section is of the same order as the radius of curvature, then a non-linear stress distribution occurs during bending.

Similar assumptions are made for curved beams as for straight beams, plane cross-sections remaining plane, etc., although a few of the assumptions are not strictly accurate for the case of a bar with a small radius of curvature.

Consider the curved bar shown unloaded in Fig. 6.42(a) and subjected to pure bending M (Fig. 6.42(b)) with initial and final radii of the neutral

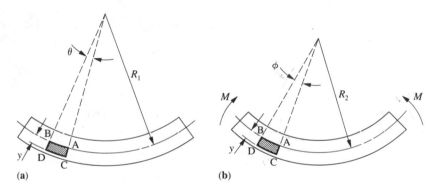

Fig. 6.42 (a) Unloaded; (b) loaded

(a)　　　　(b)

axis R_1 and R_2 respectively. The strain in a small element CD at a distance y from the neutral axis is derived as for the straight beam and is

$$\varepsilon_{CD} = \frac{(R_2 + y)\phi - (R_1 + y)\theta}{(R_1 + y)\theta}$$

but for an element AB at the neutral surface there is no change in length, so that $R_1\theta = R_2\theta$. Therefore

$$\varepsilon_{CD} = \frac{y(\phi - \theta)}{(R_1 + y)\theta}$$

Making the substitution $\phi = R_1\theta/R_2$ gives

$$\varepsilon_{CD} = \frac{y[(R_1/R_2) - 1]}{R_1 + y} = \frac{y(R_1 - R_2)}{R_2(R_1 + y)} \tag{6.54}$$

For the slender beam, y can be neglected compared with R_1 and

$$\varepsilon = y\left(\frac{1}{R_2} - \frac{1}{R_1}\right) \tag{6.55}$$

For R_1 infinite, i.e. a straight beam, the expression reduces to that found previously.

By using the same concept as for the straight beam it can be shown that for no applied end load the centroidal axis and the neutral axis coincide, and for equilibrium of the bending moment with the internal resisting moment

$$\frac{M}{I} = \frac{\sigma}{y} = E\left(\frac{1}{R_2} - \frac{1}{R_1}\right) \tag{6.56}$$

BEAMS WITH A SMALL RADIUS OF CURVATURE

For the beam in which y is not negligible compared with R_1, the strain at distance y from the neutral axis is given by eqn. (6.54). This is no longer a linear distribution of strain across the section as for the slender beam, and hence *the distribution of stress is non-linear,* as indicated in Fig. 6.43 and *the centroidal and neutral axes no longer coincide,* as will now be shown.

Fig. 6.43

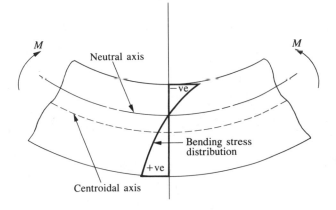

Assuming that there is no applied end load,

$$\int_A \sigma \, dA = 0$$

and from eqn. (6.54)

$$\sigma = E\varepsilon = \frac{Ey(R_1 - R_2)}{R_2(R_1 + y)} \tag{6.57}$$

Therefore

$$\frac{E(R_1 - R_2)}{R_2} \int_A \frac{y}{R_1 + y} \, dA = 0$$

since the integral must be zero, and this is not the first moment of area about the centroid, therefore the centroidal and neutral axes do not coincide.

For equilibrium of internal and external moments,

$$M = \int_A \sigma y \, dA$$

$$M = \frac{E(R_1 - R_2)}{R_2} \int_A \frac{y^2}{R_1 + y} \, dA \tag{6.58}$$

The integral term may be re-expressed as

$$\int_A \frac{y^2}{R_1 + y} \, dA = \int_A y \, dA - R_1 \int_A \frac{y}{R_1 + y} \, dA$$

$$= \int_A y \, dA$$

since the second integral is zero as shown above. Now let the distance between the centroidal and neutral axes be n, then, for the cross-section of a curved beam shown in Fig. 6.44, $y = y' + n$; hence

$$\int_A y \, dA = \int_A y' \, dA + \int_A n \, dA = nA$$

Fig. 6.44

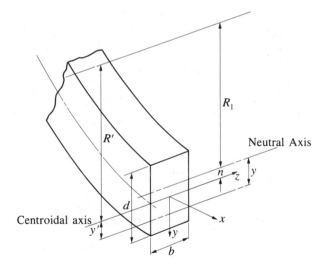

since $\int_A y' \, dA$ is zero, being the first moment of area about the centroid. Thus

$$\int_A \frac{y^2}{R_1 + y} \, dA = nA$$

Substituting into eqn. (6.58),

$$\frac{M}{nA} = \frac{E(R_1 - R_2)}{R_2}$$

and further substituting from eqn. (6.57),

$$\frac{M}{nA} = \frac{\sigma}{y}(R_1 + y)$$

or

$$\frac{\sigma}{y} = \frac{M}{nA(R_1 + y)} \tag{6.59}$$

which is in a form similar to the bending-stress relationship for slender beams, the second moment of area term, I, being replaced by $nA(R_1 + y)$.

In order to determine the magnitude of bending stress it is first necessary to find the values of n and R_1 for the particular shape of cross-section. From the condition that

$$\int_A \frac{y}{R_1 + y}\, dA = 0$$

it follows that

$$\int_A \frac{y' + n}{R' + y'}\, dA = 0$$

or

$$\int_A \left(\frac{R' + y'}{R' + y'} - \frac{R'}{R' + y'} + \frac{n}{R' + y'} \right) dA = 0$$

$$\int_A dA - \int_A \frac{R'}{R' + y'}\, dA + \int_A \frac{n}{R' + y'}\, dA = 0$$

Hence

$$n = R' - \frac{A}{\displaystyle\int_A \frac{dA}{R' + y'}} \tag{6.60}$$

and

$$R_1 = R' - n \tag{6.61}$$

For rectangular and circular sections the values of n and R_1 are obtained from eqns. (6.60) and (6.61).

Rectangular section

$$n = R' - \frac{d}{\log_e\left[\dfrac{R' + d/2}{R' - d/2}\right]} \tag{6.62}$$

An accurate value of the denominator is necessary and may be expressed as

$$\log_e\left\{\frac{R' + d/2}{R' - d/2}\right\} = \frac{d}{R'}\left[1 + \frac{1}{3}\left(\frac{d}{2R'}\right)^2 + \frac{1}{5}\left(\frac{d}{2R'}\right)^4 + \ldots\right]$$

Circular section of radius r

$$n = R' - \frac{r^2}{2[R' - (R'^2 - r^2)^{1/2}]}$$

$$= \frac{R'}{2}\left[\frac{1}{2}\left(\frac{r}{R'}\right)^2 + \frac{1}{8}\left(\frac{r}{R'}\right)^4 + \dots\right] \tag{6.63}$$

EXAMPLE 6.16

A crane hook as illustrated in Fig. 6.45(a) is designed to carry a maximum force of 12 kN. Calculate the maximum tensile and compressive stresses set up on the cross-section AB shown at (b).

Fig. 6.45

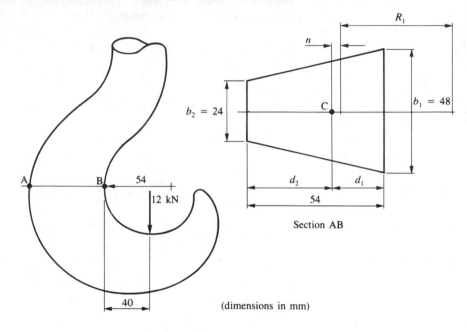

Section AB

(dimensions in mm)

SOLUTION

The position of the centroid is given by

$$d_1 = \frac{[48 + (2 \times 24)]}{72}\frac{54}{3} = 24 \text{ mm}$$

Hence

$$d_2 = 54 - 24 = 30 \text{ mm}$$

The value of n is given as follows:

$$n = R' - \frac{\frac{1}{2}(b_1 + b_2)d}{\left[b_2 + \frac{b_1 - b_2}{d}(R' + d_2)\right]\log_e\left(\frac{R' + d_2}{R' - d_1}\right) - (b_1 - b_2)}$$

$$= 78 - \frac{36 \times 54}{72\log_e 2 - 24} = 3 \text{ mm}$$

and

$$R_1 = 78 - 3 = 75 \text{ mm}$$
$$A = 36 \times 54 = 1944 \text{ mm}^2$$
$$M = -12\,000 \times 0.064 = -768 \text{ N-m}$$

(negative since curvature is reduced)

$$\hat{y}_1 = -24 + 3 = -21 \text{ mm}$$
$$\hat{y}_2 = +30 + 3 = +33 \text{ mm}$$

Direct stress on section $AB = \dfrac{12\,000}{0.001944} = +6.17 \text{ MN/m}^2$

Bending stress at $B = \dfrac{-768 \times -0.021}{0.003 \times 0.001\,944(0.075 - 0.021)}$

$$= +51.1 \text{ MN/m}^2$$

Total stress at $B = +57.27 \text{ MN/m}^2$

Bending stress at $A = \dfrac{-768 \times 0.033}{0.003 \times 0.001\,944(0.075 + 0.033)}$

$$= -40.1 \text{ MN/m}^2$$

Total stress at $A = -33.93 \text{ MN/m}^2$

SUMMARY

Many important practical problems of members subjected to bending have been analysed in this chapter and if this summary is to be succinct it must pick out the most important principles. Firstly, little progress can be made unless there is a clear understanding of the calculation of *shear-force and bending-moment distributions*. Then the *properties of areas* are always present in order to find the position of the neutral axis and the second moment of area of the cross-section. From this point in order to analyse a wide range of situations involving bending the same basic procedure is required for every case: (i) to write the *equilibrium statement* relating applied forces and internal reactions (as a function of stress), (ii) with appropriate assumptions to decide on *the geometry of deformation* of a suitable free body of the beam, and (iii) to link (i) and (ii) through the elastic *stress–strain relationship*. The final solution will also entail the use of the specific *boundary conditions* of the problem.

PROBLEMS

6.1 Draw the bending-moment and shear-force diagrams for the beam loaded as shown in Fig. 6.46 and insert the principal values on each diagram.

Fig. 6.46

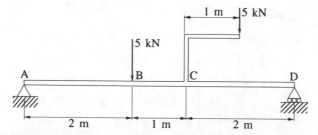

6.2 A gear wheel is mounted on a 25 mm diameter shaft which is driven through a 600 mm diameter pulley as in Fig. 6.47. During operation the gear wheel is subjected to a horizontal force of 3 kN and a vertical force of 5 kN. Calculate the value of the maximum bending moment on the shaft and sketch the bending-moment diagram.

Fig. 6.47

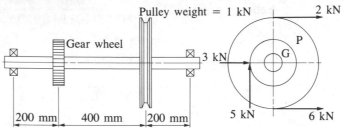

6.3 The portal frame shown in Fig. 6.48 is pinned at A, C and E. Sketch the bending-moment diagrams for the sides and top of the frame and insert the principal values.

Fig. 6.48

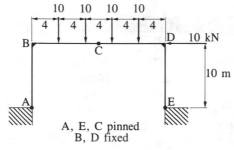

6.4 A dam 12 m in length retains water at a depth of 3 m as shown in Fig. 6.49. Sketch the bending-moment and shear-force diagrams for the dam. The density of water may be taken as 1000 kg/m³.

Fig. 6.49

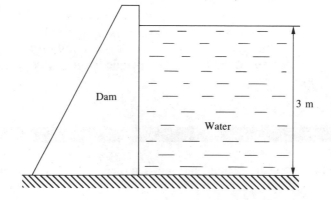

6.5 A floor joist 6 m in length is simply-supported at each end and carries a varying distributed load of grain over the whole span. The loading distribution is represented by the equation $w = ax^2 + bx + c$, where w is the load intensity in kN/m at a distance x along the beam, and a, b and c are constants. The load intensity is zero at each end and has its maximum value of 4 kN/m at mid-span.

Apply the differential equations of equilibrium relating load, shear force and bending moment to obtain the maximum values and distributions of shear force and bending moment.

6.6 A horizontal beam, the weight of which may be ignored, is loaded as shown in Fig. 6.50. The distributed load varies linearly from zero at the left-hand end to 6 kN/m at a distance of 3 m from the left-hand end. Sketch the shearing-force and bending-moment diagrams for the loading shown and insert the principal values of each on these diagrams.

Fig. 6.50

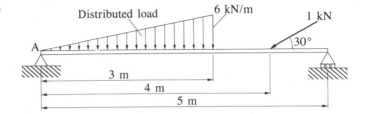

6.7 A cantilevered balcony which projects 2 m out from a wall is constructed of timber joists at 0.33 m spacing supporting a boarded floor 12 mm thick. The design loading on the floor is 4.5 kN/m² and the self-weight of a joist and its associated boarding may be neglected.

If each joist is to be 120 mm deep, determine the required width so that the maximum tensile bending stress does not exceed 10 MN/m². For the purpose of calculating the second moment of area, it may be assumed that the neutral axis of the combined joist cross-section and its associated strip of boarding is at the mid-depth of the whole section.

Also find where the neutral axis is correctly located when the joist width is determined.

6.8 A channel which carries water is made of sheet metal 3 mm thick and has a cross-section of 400 mm width and 200 mm depth. If the maximum allowable depth of water is 150 mm determine the maximum simply-supported span. The maximum bending stress (tension or compression) is not to exceed 35 MN/m², and self-weight may be neglected. Loading due to water, 9.81 kN/m³.

6.9 The roof of a petrol station is made up of two main beams and eleven cross-beams (purlins) as shown in Fig. 6.51. The main beams have an I-section 200 mm wide, 600 mm deep with a web thickness of 10 mm and a flange thickness of 15 mm. The purlins also have an I-section 125 mm wide, 250 mm deep with a web thickness of 6 mm and a flange thickness of 10 mm. Calculate the maximum stresses in each type of beam if there is a snow loading of 120 kg/m² on the roof. The density of the beam steel is 7850 kg/m³ and you should allow for the weight of the beams.

6.10 A tapered shaft of length l is built in at the larger end of diameter d_2, and is free at the smaller end of diameter d_1. A force W is applied at the free end perpendicular to the axis of the shaft. Show that the maximum bending stress at any section distant x from the free end of the shaft is given by

$$\frac{Wx}{(\pi/32)\{d_1 + (d_2 - d_1)(x/l)\}^3}$$

Hence determine the distance x at which point the greatest value of the maximum bending stress occurs.

Fig. 6.51

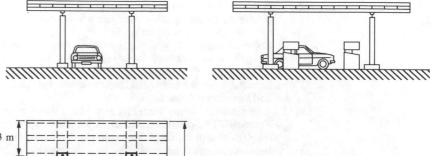

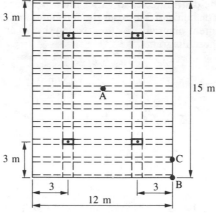

6.11 A small trailer has a suspension system as shown in Fig. 6.52. If the weight of the trailer is 4 kN and its centre of gravity is 0.5 m forward of the wheels, calculate the bending moments and torques in sections AB and BC. Calculate also the maximum bending and shear stresses in these sections. Ignore any effects at corners or changes in section.

Fig. 6.52

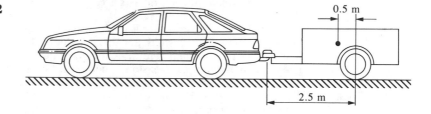

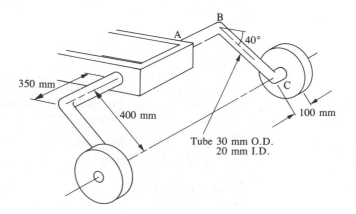

6.12 A timber beam 80 mm wide by 160 mm deep is to be reinforced with two steel plates 5 mm thick. Compare the resisting moments for the same value of the maximum bending stress in the timber when the plates are: (i) 80 mm wide and fixed to the top and bottom surfaces of the beam, and (ii) 160 mm deep and fixed to the vertical sides of the beam. E for steel $= 20 \times E$ for timber.

6.13 A composite beam is to be made up of a U-shaped steel sheet with a wooden board glued on top as in Fig. 6.53. What depth must the wood be to cause the neutral axis in pure bending to be at the horizontal diameter of the semicircle? Calculate the maximum bending moment which may be applied to this beam if the maximum stresses in the steel and wood are not to exceed 280 MN/m² and 7 MN/m², respectively. The moduli for steel and wood are 210 GN/m² and 7 GN/m² respectively.

Fig. 6.53

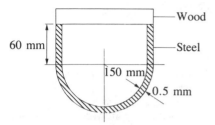

6.14 A reinforced-concrete beam has a rectangular cross-section 500 mm deep and 250 mm wide. The area of steel reinforcement is 1100 mm², and it is placed at 50 mm above the tension face. Calculate the resisting moment of the section and the stress in the steel if the compressive stress in the concrete is not to exceed 4.2 MN/m² and the modular ratio is 15.

6.15 A horizontal beam of rectangular cross-section 100 mm deep and 50 mm wide is simply supported at each end of a 1.5 m span. Vertical loads of 5 kN are applied at 0.5 m and 1 m from one end, and a horizontal tension of 40 kN is applied at the ends 25 mm below the upper surface. Determine and plot the distribution of longitudinal stress across the section at mid-span.

What eccentricity of end load is required so that there is just no resultant compressive stress at the outer surface?

6.16 The cross-section through a concrete dam is illustrated in Fig. 6.54. Calculate the required width of the base AB so that there is just no tensile stress at B. What is the resultant compressive stress at A? Loading due to water $= 9.81$ kN/m³. Weight of concrete $= 22.7$ kN/m³.

Fig. 6.54

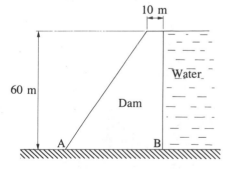

6.17 A concrete cooling tower may be assumed to consist of two truncated cones as in Fig. 6.55. If the estimated maximum horizontal wind pressure is 1.5 kN/m², calculate the wall thickness of the tower in order to avoid tensile stresses in the concrete. The density of the concrete is 2400 kg/m³. The volume of a truncated cone $= (1/3)\pi h \{R^2 + r^2 + Rr\}$.

Fig. 6.55

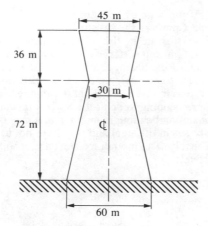

6.18 Derive an expression for the shear-stress distribution across a solid circular section rod of radius R, subjected to bending. Calculate the ratio of the maximum shear stress to the average shear stress on this section.

6.19 A channel-section beam is 50 mm wide and 50 mm deep with a 5 mm wall thickness. It is simply-supported over a length of 1 m and carries a uniformly-distributed load of 50 kN/m over its whole length. It also has a line load of 50 kN at mid-span. Sketch the shear stress distribution across the beam section 0.25 m from one of the supports and indicate the important values.

6.20 A circular tube of mean radius r and wall thickness $t (\ll r)$ is subjected to a transverse shear force Q during bending. Show, by derivation, that the maximum shear stress, τ, occurs at the neutral axis and is equal to $Q/\pi rt$.

6.21 A wooden beam section can be made up by one of the two methods shown in Fig. 6.56. If the shear force in the beam is constant at 3 kN, calculate which design is preferable to keep the shearing forces in the nails to a minimum. If the nails can withstand a shear force of 400 N, determine the maximum permissible spacing of the nails along the beam for the design selected.

Fig. 6.56

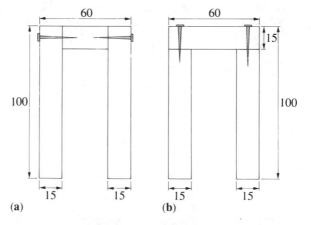

(a)   (b)

6.22 A beam is made up of four 50 mm × 100 mm pieces of timber glued to a 25 mm × 500 mm web of the same wood as shown in Fig. 6.57. Calculate the maximum allowable shear force and bending moment that this section can carry. The maximum shearing stresses in the wood and glued joints must not exceed 500 and 250 kN/m² respectively, and the maximum permissible direct stress is 1 MN/m².

Fig. 6.57

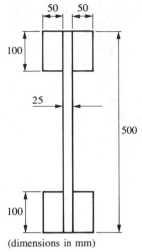

(dimensions in mm)

6.23 A Z-section beam is 2 m long and is supported as a cantilever with a 1 kN load at the free end. The direction of the 1 kN relative to the beam section is as shown in Fig. 6.58. Calculate the magnitude and position of the maximum tensile and compressive stresses on the section.

Fig. 6.58

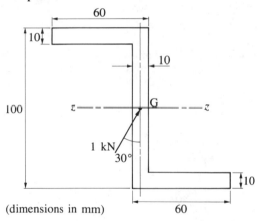

(dimensions in mm)

Fig. 6.59

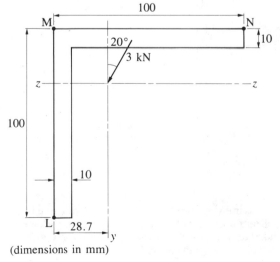

(dimensions in mm)

6.24 A 100 mm × 100 mm angle section as shown in Fig. 6.59 is built in at one end of its 1 m length and subjected to a point load of 3 kN at the free end. The point load is applied at an angle of 20° to the vertical axis as indicated. Calculate the stresses at L, M and N and the position of the neutral axis.

6.25 A channel-section beam 1 m long is built in at one end and subjected to a point load of 1 kN at the free end. If the direction of the load is as in Fig. 6.60, calculate the direction of the neutral axis and the maximum values of the tensile and compressive stresses on the section.

Fig. 6.60

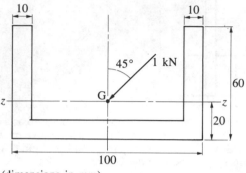

(dimensions in mm)

6.26 Determine the location of the shear centre for the beam cross-section shown in Fig. 6.61. Also calculate maximum values of the horizontal and vertical shear stresses in a flange and the web respectively for a vertical load of 500 kN applied at the shear centre.

Fig. 6.61

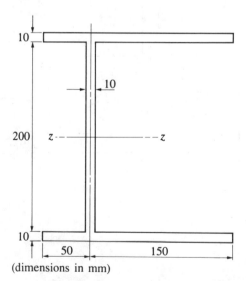

(dimensions in mm)

6.27 A proving ring, used to calibrate a testing machine, has a mean diameter of 500 mm and a rectangular section 76 mm wide and 12.7 mm thick. If the maximum permitted stress under diametral compression is 55 MN/m², determine the maximum calibration load.

6.28 A chain coupling is made up of a 20 mm diameter steel rod bent into an 'S' shape as shown in Fig. 6.62. If the coupling is subjected to a tensile load of 1 kN as indicated, calculate the maximum stress in the steel.

Fig. 6.62

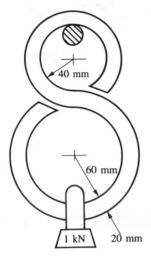

Chapter 7

Bending: slope and deflection

INTRODUCTION

In Chapter 6 the stresses during bending were investigated. In this chapter the problem will be approached from an equally important direction, namely with regard to stiffness. The total deflection of a beam is due to a very large extent to the deflection caused by bending, and to a very much smaller extent to the deflection caused by shear. In practice it is usual to put a limit on the allowable deflection, in addition to the stresses. It is important that we should be able to calculate the deflection of a beam of given section, since for given conditions of span and load it would be possible to adopt a section which would meet a strength criterion but would give an unacceptable deflection.

Various methods are available for determining the slope and deflection of a beam due to elastic bending, and examples of the use of each method will be found in this chapter. That part of the total deflection caused by shear will be discussed in Chapter 9.

THE CURVATURE–BENDING-MOMENT RELATIONSHIP

In Fig. 7.1, θ is the angle which the tangent to the curve at C makes with the x-axis, and $(\theta - d\theta)$ that which the tangent at D makes with the same axis. The normals to the curve at C and D meet at O. The point O is the centre of curvature and R is the radius of curvature of the small portion CD of the deflection curve of the neutral axis.

Numerically $ds = R\,d\theta$ and $1/R = d\theta/ds$. Using the sign convention on page 18 it will be seen that positive increments of ds from left to right are associated with a negative change in $d\theta$. Thus, when signs are taken into account, the above equation becomes

$$\frac{1}{R} = -\frac{d\theta}{ds} \tag{7.1}$$

Fig. 7.1

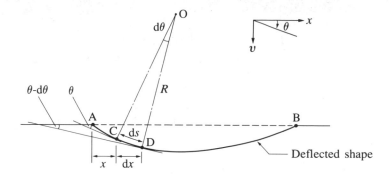

Deflections of the neutral axis are denoted by the symbol v, measured positive downwards, and are relatively small, giving a flat form of deflection curve; therefore no error is introduced in assuming that $ds \approx dx$, that $\theta \approx \tan\theta = dv/dx$, and hence that $d\theta/ds = d^2v/dx^2$. Therefore

$$\frac{1}{R} = -\frac{d^2v}{dx^2} \tag{7.2}$$

On page 124 it was shown that, when elastic bending occurs,

$$\frac{1}{R} = \frac{M}{EI}$$

Therefore

$$\frac{d^2v}{dx^2} = -\frac{M}{EI} \tag{7.3}$$

This is the differential equation of the deflection curve.

The exact relationship between the radius of curvature and geometry of the deformed beam is also obtained from $1/R = -d\theta/ds$ by simple calculus, and is given by

$$\frac{1}{R} = \pm\frac{d^2v/dx^2}{[1 + (dv/dx)^2]^{3/2}}$$

For small deflections, this again may be reduced to the form $1/R = \pm d^2v/dx^2$, since $(dv/dx)^2$ is small compared with unity. Therefore

$$M = \pm EI\frac{d^2v}{dx^2}$$

The correct sign is chosen by observing that in the beam shown M is positive and the rate of change of slope, d^2v/dx^2, is negative; hence

$$M = -EI\frac{d^2v}{dx^2}$$

or

$$EI\frac{d^2v}{dx^2} = -M \tag{7.4}$$

SLOPE AND DEFLECTION BY THE DOUBLE INTEGRATION METHOD

The first integration of eqn. (7.3) gives the slope of the beam at a distance x along its length when the origin is taken at A. Therefore

$$\frac{dv}{dx} = \int \frac{-M}{EI}\,dx + C \tag{7.5}$$

The second integration gives the deflection of the beam at the above point, or

$$v = \int \frac{dv}{dx}\,dx = \iint \frac{-M}{EI}\,dx\,dx + Cx + C_1 \tag{7.6}$$

C and C_1, the constants of integration, can be evaluated from the known conditions of slope and deflection at certain points, usually at the supports. Equations (7.5) and (7.6) are widely used for determining the slope and deflection of a beam at a given point. Examples of their use will be found in the following paragraphs. In each case the sign convention used to obtain eqn. (7.3) will be adopted.

BEAM SIMPLY SUPPORTED WITH DISTRIBUTED LOADING

The bending moment at D is, from Fig. 7.2,

$$M_D = \frac{wL}{2}x - \frac{wx^2}{2}$$

and

$$EI\frac{d^2v}{dx^2} = -\frac{wL}{2}x + \frac{wx^2}{2} \tag{7.7}$$

and

$$EI\frac{dv}{dx} = -\frac{wL}{4}x^2 + \frac{wx^3}{6} + C$$

Fig. 7.2

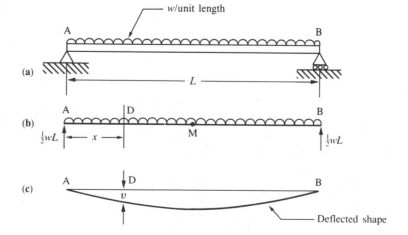

(a)   (b)   (c)   Deflected shape

From symmetry $\mathrm{d}v/\mathrm{d}x = 0$, at $x = \frac{1}{2}L$; therefore $C = wL^3/24$. Hence

$$\frac{\mathrm{d}v}{\mathrm{d}x} = -\frac{w}{2EI}\left[\frac{Lx^2}{2} - \frac{x^3}{3} - \frac{L^3}{12}\right] \tag{7.8}$$

The slope at the ends of the beam is given by $wL^3/24EI$ at A, where $x = 0$, and $-wL^3/24EI$ at B, where $x = L$.

$$v = -\frac{w}{2EI}\left[\frac{Lx^3}{6} - \frac{x^4}{12} - \frac{L^3x}{12}\right] + C_1$$

At $x = 0$, $v = 0$; therefore $C_1 = 0$ and

$$v = -\frac{w}{12EI}\left[Lx^3 - \frac{x^4}{2} - \frac{L^3x}{2}\right] \tag{7.9}$$

The maximum deflection occurs at mid-span M, where $x = \frac{1}{2}L$. Therefore

$$v_{max} = -\frac{w}{12EI}\left[\frac{L^4}{8} - \frac{L^4}{32} - \frac{L^4}{4}\right] = \frac{5}{384}\frac{wL^4}{EI} \tag{7.10}$$

SIMPLY-SUPPORTED BEAM WITH UNIFORM BENDING MOMENT

The beam shown in Fig. 7.3 is acted on by the terminal couples $\bar{M}$. The bending moment at any point along the beam is $\bar{M}$, and with the previous sign convention,

$$EI\frac{\mathrm{d}^2v}{\mathrm{d}x^2} = -\bar{M} \tag{7.11}$$

Therefore

$$EI\frac{\mathrm{d}v}{\mathrm{d}x} = -\bar{M}x + C$$

and

$$EIv = -\frac{\bar{M}x^2}{2} + Cx + C_1$$

Fig. 7.3

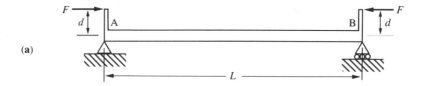

(a)

(b) $\bar{M} = Fd$ $\bar{M} = Fd$

(c)

Deflected shape

At $x = 0$, $v = 0$; therefore $C_1 = 0$; and at $x = L$, $v = 0$; therefore $C = \frac{1}{2}\bar{M}L$, and the slope dv/dx is given by

$$\frac{dv}{dx} = \frac{1}{EI}\left[-\bar{M}x + \frac{\bar{M}L}{2}\right] \tag{7.12}$$

from which the slope at the ends is $+\bar{M}L/2EI$ at A, $-\bar{M}L/2EI$ at B and at mid-span is zero. The deflection at any point is given by

$$v = \frac{1}{EI}\left[-\frac{\bar{M}x^2}{2} + \frac{\bar{M}L}{2}x\right] \tag{7.13}$$

The maximum deflection occurs at mid-span, and hence

$$v_{max} = \frac{\bar{M}L^2}{8EI} \tag{7.14}$$

CANTILEVER WITH UNIFORMLY-DISTRIBUTED LOADING

The bending moment at D in Fig. 7.4 is

$$M_D = R_A x - M_A - \frac{wx^2}{2}$$

where $R_A = wL$ and $M_A = \frac{1}{2}wL^2$, the fixing moment at the support. Therefore

$$EI\frac{d^2v}{dx^2} = -M = -wLx + \frac{wL^2}{2} + \frac{wx^2}{2} \tag{7.15}$$

$$EI\frac{dv}{dx} = -\frac{wL}{2}x^2 + \frac{wL^2}{2}x + \frac{wx^3}{6} + C$$

At $x = 0$, $dv/dx = 0$; therefore $C = 0$. Hence

$$\frac{dv}{dx} = \frac{1}{EI}\left[-\frac{wL}{2}x^2 + \frac{wL^2}{2}x + \frac{wx^3}{6}\right] \tag{7.16}$$

Fig. 7.4

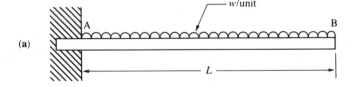

(a)

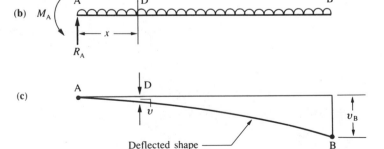

(b) M_A

(c)

At the free end, $x = L$. Therefore

$$\text{Slope at free end} = \frac{wL^3}{6EI} \tag{7.17}$$

$$v = \frac{1}{EI}\left[-\frac{wL}{6}x^3 + \frac{wL^2}{4}x^2 + \frac{wx^4}{24}\right] + C_1$$

and at $x = 0$, $v = 0$; therefore $C_1 = 0$. Hence

$$v = \frac{1}{EI}\left[-\frac{wL}{6}x^3 + \frac{wL^2}{4}x^2 + \frac{wx^4}{24}\right] \tag{7.18}$$

The deflection at the free end B, where $w = L$, is given by

$$v_B = \frac{1}{EI}\left[-\frac{wL^4}{6} + \frac{wL^4}{4} + \frac{wL^4}{24}\right] = \frac{1}{8}\frac{wL^4}{EI} \tag{7.19}$$

CANTILEVER CARRYING A CONCENTRATED LOAD

The bending moment at D in Fig. 7.5 is given by

$$M_D = -M_A + R_A x$$

where $M_A = Wl$, the fixing moment at the support, and $R_A = W$. Therefore

$$M = -Wl + Wx$$

$$EI\frac{d^2v}{dx^2} = -M = Wl - Wx \tag{7.20}$$

and

$$EI\frac{dv}{dx} = Wlx - \frac{Wx^2}{2} + C$$

Fig. 7.5

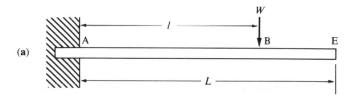

(a)

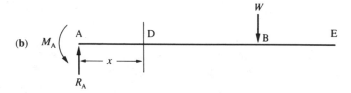

(b)

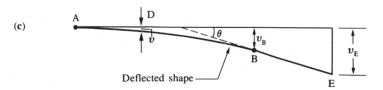

(c)

At $x = 0$, $\mathrm{d}v/\mathrm{d}x = 0$; therefore $C = 0$. Hence

$$\frac{\mathrm{d}v}{\mathrm{d}x} = \frac{Wlx}{EI} - \frac{Wx^2}{2EI} \tag{7.21}$$

At $\quad x = l, \qquad \frac{\mathrm{d}v}{\mathrm{d}x} = \frac{Wl^2}{2EI} \tag{7.22}$

$$v = \frac{Wlx^2}{2EI} - \frac{Wx^3}{6EI} + C_1$$

and at $x = 0$, $v = 0$; therefore $C_1 = 0$, and

$$v = \frac{W}{2EI}\left(lx^2 - \frac{x^3}{3}\right) \tag{7.23}$$

which is the equation for the deflection curve for the beam.

For the deflection under the load we substitute $x = l$ in eqn. (7.23) and

$$v_B = \frac{1}{3}\frac{Wl^3}{EI} \tag{7.24}$$

Deflection at free end E = deflection at B + slope at B $\times (L - l)$. Therefore

$$v_E = \frac{1}{3}\frac{Wl^3}{EI} + \frac{Wl^2}{2EI}(L - l)$$

$$= \frac{Wl^2}{2EI}\left(L - \frac{l}{3}\right) \tag{7.25}$$

CANTILEVER WITH CONCENTRATED LOAD AT FREE END

The slope and deflection under the load are obtained by substituting L for l in eqns. (7.22) and (7.24), and thus

$$v_{max} = \frac{1}{3}\frac{WL^3}{EI} \quad \text{and} \quad \theta = \frac{WL^2}{2EI} \tag{7.26}$$

DISCONTINUOUS LOADING: MACAULAY'S METHOD

It was seen on pages 22 and 113, when considering the bending moment distribution for a beam with discontinuous loading, that a bending moment expression has to be written for each part of the beam. This means that in deriving slope and deflection a double integration would have to be performed on each bending moment expression and two constants would result for each section of the beam. A further example of discontinuous loading is shown in Fig. 7.6(a); in this case there would be three bending moment equations and thus six constants of integration. There are apparently only two boundary conditions, those of zero deflection at each end. However, at the points of discontinuity, B and C, both slope and deflection must be continuous from one section to

Fig. 7.6

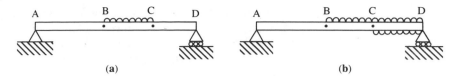

(a)   (b)

the next, so that

$$\text{At B} \quad \left(\frac{\mathrm{d}v}{\mathrm{d}x}\right)_{AB} = \left(\frac{\mathrm{d}v}{\mathrm{d}x}\right)_{BC} \quad \text{and} \quad v_{AB} = v_{BC}$$

$$\text{At C} \quad \left(\frac{\mathrm{d}v}{\mathrm{d}x}\right)_{BC} = \left(\frac{\mathrm{d}v}{\mathrm{d}x}\right)_{CD} \quad \text{and} \quad v_{BC} = v_{CD}$$

The above four plus the two end conditions enable the six constants of integration to be determined. The derivation of the deflection curve by the above approach is rather tedious; it is therefore an advantage to use the mathematical technique termed a *step function,* commonly known as Macaulay's method when applied to beam solutions. This approach requires one bending moment expression to be written down for a point close to the right-hand end to cover the bending-moment conditions for the whole length of beam, and hence, on integration, only two unknown constants have to be determined.

The step function is a function of x of the form $f_n(x) = [x - a]^n$ such that for $x < a$, $f_n(x) = 0$ and for $x > a$, $f_n(x) = (x - a)^n$. Note the change in the form of brackets used: the square brackets are particularly chosen to indicate the use of a step function, the curved brackets representing normal mathematical procedure. The important features when using the step function in analysis are that, if on substitution of a value for x the quantity inside the square brackets becomes negative, it is omitted from further analysis. Square bracket terms must be integrated in such a way as to preserve the identity of the bracket, i.e.

$$\int [x - a]^2 \, \mathrm{d}(x - a) = \tfrac{1}{3}[x - a]^3$$

Also, for mathematical continuity, distributed loading which does not extend to the right-hand end, as in Fig. 7.6(a), must be arranged to

Fig. 7.7

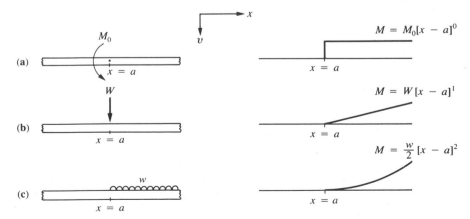

(a) $M = M_0[x - a]^0$

(b) $M = W[x - a]^1$

(c) $M = \dfrac{w}{2}[x - a]^2$

179

continue to $x = l$, whether starting from $x = 0$ or $x = a$. This may be effected by the superposition of loadings which cancel each other in the required portions of the beam as shown in Fig. 7.6(b).

An applied couple M_0 must be expressed as a step function in the form $M_0[x - a]^0$ so that the bracket can be integrated correctly (*see* example on page 182).

The three common step functions for bending moment are shown in Fig. 7.7 and several illustrative examples now follow.

BEAM SIMPLY SUPPORTED WITH CONCENTRATED LOAD

Taking moments about one end, the reactions at the supports in Fig. 7.8 are

$$R_1 = \frac{W(L - a)}{L} \quad \text{and} \quad R_2 = \frac{Wa}{L}$$

When $x > a$, the bending moment at D is $M = R_1x - W[x - a]$. Hence

$$EI\frac{d^2v}{dx^2} = -M = -R_1x + W[x - a] \tag{7.27}$$

When $x < a$, the term $W[x - a]$ becomes negative and is inapplicable, and $M = R_1x$.

Integrating eqn. (7.27),

$$EI\frac{dv}{dx} = -\frac{R_1x^2}{2} + \frac{W}{2}[x - a]^2 + C \tag{7.28}$$

and

$$EIv = -\frac{R_1x^3}{6} + \frac{W}{6}[x - a]^3 + Cx + C_1 \tag{7.29}$$

If we omit the term inside the brackets on the right-hand side of eqns. (7.28) and (7.29) when $x < a$, the equations are then of the correct form

Fig. 7.8

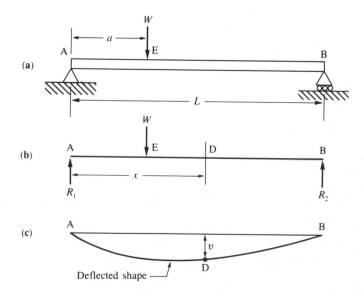

(a)

(b)

(c)

Deflected shape

for the portion AE of the beam, and since the second term on the right-hand side of these equations vanishes for a value of $x = a$, then when these equations are used for the whole beam, both dv/dx and v will be continuous at the point E.

When $x = 0$, $v = 0$, and since the term inside the brackets is omitted, $C_1 = 0$. For $x = L$, $v = 0$; therefore

$$0 = -\frac{R_1 L^3}{6} + \frac{W}{6}(L-a)^3 + CL$$

and

$$C = \frac{R_1 L^2}{6} - \frac{W}{6L}(L-a)^3$$

$$= \frac{W(L-a)L}{6} - \frac{W}{6L}(L-a)^3$$

$$= \frac{Wa}{6L}(L-a)(2L-a) \tag{7.30}$$

Substituting the values of C and R_1 in eqn. (7.29) and rearranging,

$$v = \frac{Wx}{6EI}\frac{L-a}{L}(2aL - a^2 - x^2) + \frac{W}{6EI}[x-a]^3 \tag{7.31}$$

This equation gives the deflection at any point along the beam if the last term on the right-hand side is rejected when it becomes negative, i.e. for $x < a$. For the particular case when $x = a$, the deflection under the load is given by

$$v_E = \frac{Wa^2(L-a)^2}{3EIL} \tag{7.32}$$

It should be noted that the maximum deflection occurs where $dv/dx = 0$ which is not, in fact, where the load is applied.

If W is placed at mid-span so that $a = \frac{1}{2}L$, the deflection under the load is also the maximum deflection and is

$$v = \frac{WL^3}{48EI} \tag{7.33}$$

BEAM WITH DISTRIBUTED LOAD ON PART OF THE SPAN

As explained above (Fig. 7.6(b)), for mathematical continuity the loading must be continued to the right-hand end, and to maintain the same equilibrium upward loading must be inserted from D to B as shown in Fig. 7.9.

At point E between D and B,

$$M = R_1 x - \frac{w}{2}[x-a]^2 + \frac{w}{2}[x-(a+b)]^2 \tag{7.34}$$

the second and third terms being rejected when $x < a$ and the third term when $x < (a+b)$.

$$EI\frac{d^2v}{dx^2} = -R_1 x + \frac{w}{2}[x-a]^2 - \frac{w}{2}[x-(a+b)]^2$$

Fig. 7.9

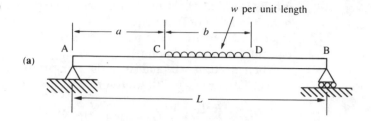

(a)

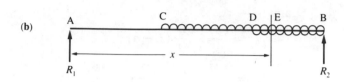

(b)

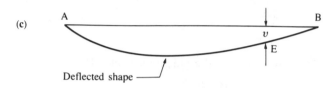

(c)

and

$$EI\frac{dv}{dx} = -\frac{R_1 x^2}{2} + \frac{w}{6}[x-a]^3 - \frac{w}{6}[x-(a+b)]^3 + C \qquad (7.35)$$

$$EIv = -\frac{R_1 x^3}{6} + \frac{w}{24}[x-a]^4 - \frac{w}{24}[x-(a+b)]^4 + Cx + C_1 \qquad (7.36)$$

The values of C and C_1 are found from the conditions of $v = 0$ when $x = 0$ and $x = L$, the terms inside the brackets being rejected when negative.

BEAM SIMPLY-SUPPORTED WITH AN APPLIED BENDING MOMENT

From moment equilibrium in Fig. 7.10,

$$R_1 = R_2 = \frac{\bar{M}}{L}$$

The bending moment at D, where $x > a$, is

$$M_D = R_1 x - \bar{M} = \frac{\bar{M}}{L}x - \bar{M}$$

A convenient way of dealing with a couple by Macaulay's method is to introduce a term $[x-a]^0$ which is in fact unity, but allows for subsequent integration in the correct manner:

$$EI\frac{d^2v}{dx^2} = -\frac{\bar{M}}{L}x + \bar{M}[x-a]^0$$

and on integration we may write

$$EI\frac{dv}{dx} = -\frac{\bar{M}x^2}{2L} + \bar{M}[x-a]^1 + C$$

Fig. 7.10

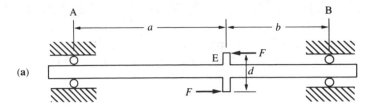

(a)

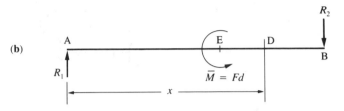

(b)

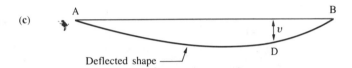

(c)

Deflected shape

where the second term is integrated with respect to $(x - a)$. Therefore

$$EIv = -\frac{\bar{M}x^3}{6L} + \frac{\bar{M}}{2}[x - a]^2 + Cx + C_1 \qquad (7.37)$$

When $x < a$, the bracketed term on the right-hand side of the equation becomes negative and is rejected.

At $x = 0$, $v = 0$; therefore $C_1 = 0$; and at $x = L$, $v = 0$; therefore

$$0 = -\frac{\bar{M}L^2}{6} + \frac{\bar{M}}{2}[L - a]^2 + CL$$

and

$$C = \frac{\bar{M}}{6L}\{-2L^2 + 6aL - 3a^2\} \qquad (7.38)$$

Hence

$$v = \frac{1}{EI}\left\{-\frac{\bar{M}}{6L}x^3 + \frac{\bar{M}}{2}[x - a]^2 - \frac{\bar{M}}{6L}(2L^2 - 6aL + 3a^2)x\right\} \qquad (7.39)$$

At E, where $x = a$, the deflection is given by

$$EIv_E = -\frac{\bar{M}}{6L}a^3 - \frac{\bar{M}}{6L}[2L^2 - 6aL + 3a^2]a$$

or, putting $L = (a + b)$,

$$EIv_E = -\frac{\bar{M}}{6L}[a^3 - a^3 - 2ab(a - b)]$$

$$v_E = \frac{\bar{M}}{3EI}\frac{(a - b)ab}{L} \qquad (7.40)$$

183

EXAMPLE 7.1

A simply-supported beam is subjected to the loading shown in Fig. 7.11. Calculate the deflection at a section 1.8 m from the left-hand end. $E = 70 \text{ GN/m}^2$, $I = 832 \text{ cm}^4$.

Fig. 7.11

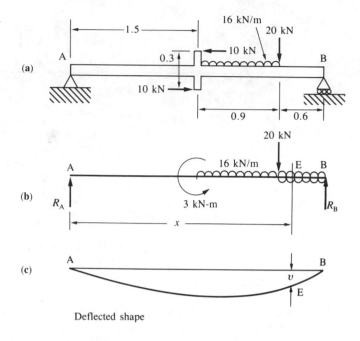

Deflected shape

SOLUTION

This example combines the features of the cases above, and so, to satisfy the Macaulay conditions, the distributed load must be extended to B and an equivalent negative load inserted to restore the correct resultant load distribution. Then

$$M = R_A x - 3[x - 1.5]^0 - 16\frac{[x - 1.5]^2}{2} - 20[x - 2.4] + 16\frac{[x - 2.4]^2}{2}$$

$M = 0$ when $x = 3$; therefore $R_A = 10$ kN.

$$EI\frac{dv}{dx} = -\frac{10x^2}{2} + 3[x - 1.5] + \frac{16[x - 1.5]^3}{6} + \frac{20[x - 2.4]^2}{2}$$

$$-\frac{16[x - 2.4]^3}{6} + C$$

$$EIv = -\frac{10x^3}{6} + \frac{3[x - 1.5]^2}{2} + \frac{16[x - 1.5]^4}{24} + \frac{20[x - 2.4]^3}{6}$$

$$-\frac{16[x - 2.4]^4}{24} + Cx + C_1$$

When $x = 0$, $v = 0$; therefore $C_1 = 0$; and when $x = 3$, $v = 0$; therefore

$$0 = -\frac{10(3)^3}{6} + \frac{3(1.5)^2}{2} + \frac{16(1.5)^4}{24} + \frac{20(0.6)^3}{6} - \frac{16(0.6)^4}{24} + 3C$$

from which $C = 12.54$. When $x = 1.8$ the third and fourth bracketed terms are omitted, and

$$EIv = -\frac{10 \times 1.8^3}{6} + \frac{3 \times 0.3^2}{2} + \frac{16 \times 0.3^4}{24} + (12.54 \times 1.8)$$

$$= 13 \text{ kN-m}^3$$

$$v = \frac{13 \times 10^3 \times 10^3}{70 \times 10^9 \times 832 \times 10^{-8}} = 22.3 \text{ mm}$$

EXAMPLE 7.2

Calculate the position and magnitude of the maximum deflection for the beam shown in Fig. 7.12. $EI = 1000 \text{ kN-m}^2$.

Fig. 7.12

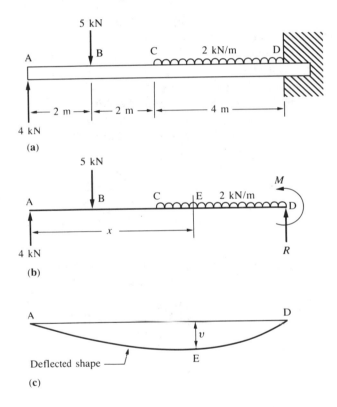

(a)

(b)

(c)

SOLUTION

$$EI\frac{d^2v}{dx^2} = -4x + 5[x-2] + \frac{2}{2}[x-4]^2$$

$$EI\frac{dv}{dx} = -\frac{4x^2}{2} + \frac{5}{2}[x-2]^2 + \frac{2}{6}[x-4]^3 + C$$

$$EIv = -\frac{4x^3}{6} + \frac{5}{6}[x-2]^3 + \frac{2}{24}[x-4]^4 + Cx + C_1$$

The boundary conditions are, $\dfrac{dv}{dx} = 0$ at $x = 8$, and $v = 0$ at $x = 8$, therefore

$$0 = -(2 \times 64) + (5 \times 18) + \frac{64}{3} + C$$

$$C = +16.7$$

$$0 = -\left(4 \times \frac{8^3}{6}\right) + (5 \times 36) + \frac{16^2}{12} + (16.7 \times 8) + C_1$$

$$C_1 = -6.6$$

At the left-hand end the deflection is obtained when $x = 0$; therefore

$$EIv = -6.6 \text{ kN-m}^3$$

$$v = -\frac{6.6 \times 10^3}{1000} = -6.6 \text{ mm}$$

This may not be the maximum deflection and we must check elsewhere in the span. However, it is not sufficient merely to equate dv/dx to zero since bracketed terms would then be included which might not be appropriate, depending on where v_{max} occurred. The best way is to make a sensible guess as to the section where the maximum deflection is likely to occur and to determine the slope at each end of that section. The slopes will be of opposite sign if the guess was correct. If not, then an adjacent section must be treated in the same way. For example, assuming $dv/dx = 0$ occurs between B and C, then

$$\text{At B } EI\frac{dv}{dx} = -\left(4 \times \frac{2^2}{2}\right) + 16.7 = +8.7$$

$$\text{At C } EI\frac{dv}{dx} = -\left(4 \times \frac{4^2}{2}\right) + \left(\frac{5}{2} \times 2^2\right) + 16.7 = -5.3$$

and the assumption was correct, therefore using the condition that zero slope occurs between B and C,

$$\frac{dv}{dx} = 0 = -\frac{4x^2}{2} + \frac{5}{2}(x - 2)^2 + 16.7$$

Thus

$$x^2 - 20x + 53.4 = 0$$

from which $x = 3.17$ m.

The deflection at this point is given by

$$EIv = -\left(\frac{4 \times 3.17^3}{6}\right) + \left(\frac{5}{6} \times 1.17^3\right) + (16.7 \times 3.17) - 6.6$$

$$= +26.44 \text{ kN-m}^3$$

$$v = +\frac{26.44 \times 10^3}{1000} = +26.44 \text{ mm}$$

Hence the maximum deflection occurs at 3.17 m from A and is downwards.

SUPERPOSITION METHOD

This principle can be applied to give the total deflection of a beam which carries individual loads W_1, W_2, W_3, etc., or distributed loads w_1, w_2, w_3, etc. Let the bending moments at a section of the beam caused by each load when acting separately on the beam be M_1, M_2, M_3, etc., and the corresponding deflections be v_1, v_2, v_3, etc. Then the total bending moment is

$$M = M_1 + M_2 + M_3 + \ldots \tag{7.41}$$

But

$$M = -EI \frac{d^2v}{dx^2}$$

Therefore

$$v = -\frac{1}{EI} \int\int M \, dx \, dx$$

$$= -\frac{1}{EI} \int\int (M_1 + M_2 + M_3 + \ldots) \, dx \, dx$$

$$= -\frac{1}{EI} \int\int M_1 \, dx \, dx + \int\int M_2 \, dx \, dx + \int\int M_3 \, dx \, dx + \ldots$$

$$= v_1 + v_2 + v_3 + \ldots \tag{7.42}$$

Thus the deflection at a section of a beam subjected to complex loading can be obtained by the summation of the deflections caused at that section by the individual components of the loading.

EXAMPLE 7.3

Use the principle of superposition to determine the deflections at the ends and centre of the beam shown in Fig. 7.13. $EI = 500 \text{ kN-m}^2$.

Fig. 7.13

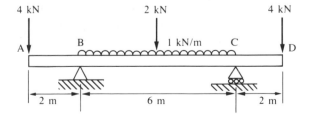

SOLUTION

Splitting the problem into three components shown in Figs. 7.14(a), (b) and (c), the respective deflections at the centre are

(a) $\delta_1 = \dfrac{Wl^3}{48EI} = +\dfrac{2000 \times 6^3 \times 10^3}{48 \times 500 \times 10^3} = +18 \text{ mm}$

(b) $\delta_2 = \dfrac{5wl^4}{384EI} = +\dfrac{5 \times 1000 \times 6^4 \times 10^3}{384 \times 500 \times 10^3} = +33.8 \text{ mm}$

(c) This may be treated as a beam subjected to couples at B and C of 8 kN-m magnitude:

$$\delta_3 = \dfrac{Ml^2}{8EI} = -\dfrac{8000 \times 6^2 \times 10^3}{8 \times 500 \times 10^3} = -72 \text{ mm}$$

Resultant deflection $= +18 + 33.8 - 72 = -20.2 \text{ mm}$

Fig. 7.14

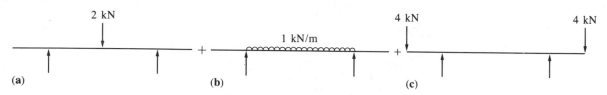

(a) (b) (c)

To find the deflection at A or D it is necessary to know the slope in each case at B or C. Then

(a) $\delta_{1A} = \theta_{1B} l_{AB} = \dfrac{Wl_{BC}^2}{16EI} l_{AB} = -\dfrac{2000 \times 6^2 \times 2 \times 10^3}{16 \times 500 \times 10^3}$

$\qquad = -18 \text{ mm}$

(b) $\delta_{2A} = \theta_{2B} l_{AB} = \dfrac{wl_{BC}^3}{24EI} l_{AB} = -\dfrac{1000 \times 6^3 \times 2 \times 10^3}{24 \times 500 \times 10^3}$

$\qquad = -36 \text{ mm}$

(c) $\delta_{3A} = \theta_{3B} l_{AB} + \dfrac{W_A l_{AB}^3}{3EI}$

$\qquad = \dfrac{Ml_{BC} l_{AB}}{2EI} + \dfrac{W_A l_{AB}^3}{3EI}$

$\qquad = \dfrac{10^3}{500 \times 10^3} \left[\dfrac{8000 \times 6 \times 2}{2} + \dfrac{4000 \times 8}{3} \right]$

$\qquad = 117.3 \text{ mm}$

Resultant deflection at A or D $= -18 - 36 + 117.3$

$\qquad\qquad\qquad = +63.3 \text{ mm}$

DEFLECTIONS DUE TO ASYMMETRICAL BENDING

The stresses which arise due to asymmetrical bending were analysed on page 148. The deflection of the centroid of a section subjected to asymmetrical bending may be calculated using the standard deflection formulae developed in this chapter. Since bending occurs about the neutral axis the direction of the deflection of the centroid will be perpendicular to the neutral axis. For example, for a cantilever of length L, carrying a load W at the free end, the maximum deflection of the centroid of the cross-section will be given by

$$\delta = \frac{W_{NA}L^3}{3EI_{NA}}$$

where W_{NA} is the component of the load perpendicular to the neutral axis and I_{NA} is the second moment of area about the neutral axis.

The most convenient way to obtain I_{NA} is to use the co-ordinates I_z, I_{yz} and I_y, $-I_{yz}$ to construct a Mohr's circle for moments of area as illustrated in Appendix A. This then allows the second moment of area at any angle to the y- or z-direction to be determined.

EXAMPLE 7.4

Calculate the maximum deflection of the centroid of the beam section shown in Example 6.14, if the beam is a cantilever of length 1 m. Young's modulus for the beam material is 210 GN/m².

SOLUTION

The beam deflection at the free end is perpendicular to the N.A. and is given by

$$\delta = \frac{W_{NA}L^3}{3EI_{NA}}$$

The solution to Example 6.14 shows that the direction of the neutral axis is 47.45° anticlockwise from the z-axis.

The *vertical* end load on the beam is 2 kN, so

$$W_{NA} = (2000 \cos 47.45)\,\text{N}$$

The second moment of area about the neutral axis is obtained from the Mohr's circle construction shown in Fig. 7.15. This gives $I_{NA} = 0.46 \times 10^{-6}\,\text{m}^4$, therefore

$$\delta = \frac{(2000 \cos 47.45)(1)^3}{3 \times 210 \times 10^9 \times 0.47 \times 10^{-6}} = 4.57\,\text{mm}$$

This deflection is downwards to the right perpendicular to the neutral axis. The vertical and horizontal deflections of the centroid are given by

Vertical deflection = 4.57 cos 47.45 = 3.1 mm↓

Horizontal deflection = 4.57 sin 47.45 = 3.4 mm→

Fig. 7.15 I_v and I_u are second moments of area about the *principal axes* (i.e. those where $I_{yz} = 0$)

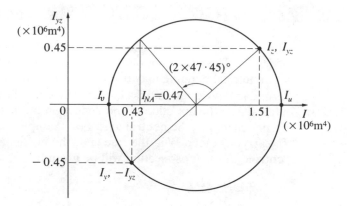

BEAMS WITH VARYING LOAD DISTRIBUTION

The deflection of an irregularly loaded beam can be found if both the bending moment M and term $\int M \, dx$ can be integrated with respect to x. It has been shown that

$$\frac{d^2v}{dx^2} = -\frac{M}{EI}$$

which can be written in the alternative form

$$\frac{d}{dx}\left(\frac{dv}{dx}\right) = -\frac{M}{EI}$$

Thus

$$\frac{dv}{dx} = -\frac{1}{EI}\int_0^x M \, dx + C \tag{7.43}$$

when EI has a constant value, and

$$v = -\frac{1}{EI}\int_0^x \int_0^x M \, dx \, dx + Cx + C_1 \tag{7.44}$$

Thus, if M and $\int M \, dx$ can be integrated with respect to x, the value of v can be calculated since the support conditions allow the determination of C and C_1.

If the analytical process cannot be carried out, a graphical method may be utilized as shown by the following example.

A beam is simply supported at each end and carries an irregular load distribution $f(w)$ as shown in Fig. 7.16(a).

The curve AMB, Fig. 7.16(b), represents the bending-moment diagram on AB. In diagram (c) the ordinate GH represents the area of the moment diagram on the portion of the span x, i.e. the shaded area AMC. Hence the curve ED is obtained by plotting ordinates so obtained. Therefore

$$GH = \int_0^x M \, dx \quad \text{and} \quad DF = \int_0^L M \, dx$$

Fig. 7.16

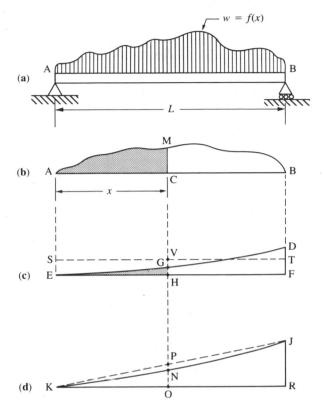

In diagram (d), the ordinate ON represents the shaded area EGH in curve (c), or

$$ON = \int_0^x GH \, dx = \int_0^x \int_0^x M \, dx \, dx$$

and

$$JR = \int_0^L \int_0^L M \, dx \, dx$$

At $x = 0$, the deflection is $v = 0$; hence from eqn. (7.44), $C_1 = 0$. At B, $x = L$, $v = 0$, and the ordinate JR in (d) represents $\int_0^L \int_0^L M \, dx \, dx$. Therefore, from eqn. (7.44),

$$0 = -\frac{1}{EI} JR + CL$$

or

$$C = \frac{JR}{EIL}$$

Join KJ. Then

$$\frac{OP}{x} = \frac{JR}{L}$$

Therefore

$$OP = \frac{x}{L} JR = CEIx$$

$$PN = OP - ON = CEIx - \int_0^x \int_0^x M \, dx \, dx$$

$$= EI\left[Cx - \frac{1}{EI} \int_0^x \int_0^x M \, dx \, dx \right]$$

$$= EIv$$

from eqn. (7.44). Thus the ordinates at a given point, measured between the straight line KPJ and the curve KNJ, represent EI times the deflection at the point.

If the straight line ST is drawn at a height equal to $JR/L = CEI$, then

$$VG = VH - GH$$

$$= CEI - \int_0^x M \, dx$$

$$= EI\left[C - \frac{1}{EI} \int_0^x M \, dx \right]$$

$$= EI \frac{dv}{dx}$$

from eqn. (7.43). Thus the ordinates, measured at a given point between the straight line ST and the curve EGD, represent EI times the slope at the point. The slope changes sign where ST intersects EGD.

EXAMPLE 7.5

A horizontal beam AB, simply-supported at each end, carries a load which increases, at a uniform rate, from zero at one end. Determine the position of, and the value of, the maximum deflection.

SOLUTION

Let w be the intensity of loading at unit distance from A, then the intensity at C, Fig. 7.17, is xw and that at B is LW.

Fig. 7.17

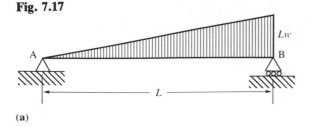

(a)

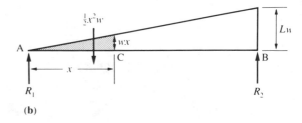

(b)

Taking moments about B,

$$R_1 L - Lw \cdot \frac{L}{2} \cdot \frac{L}{3} = 0$$

$$R_1 = \frac{L^2 w}{6}$$

Bending moment at $C = R_1 x - \frac{x^3 w}{6}$

$$= \frac{L^2 wx}{6} - \frac{x^3 w}{6}$$

$$EI\frac{d^2 v}{dx^2} = -\frac{L^2 wx}{6} + \frac{x^3 w}{6}$$

$$EI\frac{dv}{dx} = -\frac{L^2 w}{6}\frac{x^2}{2} + \frac{x^4 w}{24} + C$$

$$EIv = -\frac{L^2 w}{36}x^3 + \frac{x^5 w}{120} + Cx + C_1$$

The boundary conditions are, where $x = 0$, $v = 0$; therefore $C_1 = 0$; and at $x = L$, $v = 0$; therefore

$$0 = -\frac{L^5 w}{36} + \frac{L^5 w}{120} + CL$$

$$C = \frac{7}{360}L^4 w$$

and

$$EI\frac{dv}{dx} = -\frac{L^2 w}{12}x^2 + \frac{x^4 w}{24} + \frac{7}{360}L^4 w$$

$$= \frac{1}{360}[-30L^2 wx^2 + 15x^4 w + 7L^4 w]$$

At maximum deflection $dv/dx = 0$. Therefore

$$x^2 = \frac{30L^2 \pm \sqrt{(900L^4 - 420L^4)}}{30} = \frac{30L^2 - 21.9L^2}{30}$$

$$= 0.27L^2$$

so that

$$x = 0.52L$$

$$EIv = -\frac{L^2 w}{36}x^3 + \frac{w}{120}x^5 + \frac{7}{360}L^4 wx$$

$$= \frac{w}{360}[-10L^2 x^3 + 3x^5 + 7L^4 x]$$

$$v_{max} = \frac{w}{360}[-10L^2(0.52L)^3 + 3(0.52L)^5 + 7L^4(0.52L)]\frac{1}{EI}$$

$$= \frac{2.354 L^5 w}{360 EI}$$

EXAMPLE 7.6

A beam 4 m long is simply-supported at its ends and carries a varying distributed load over the whole span. The equation to the loading curve is $w = ax^2 + bx + c$, where w is the load intensity in kN/m run, at a distance x along the beam, measured from an origin at the left-hand support, and a, b and c are constants. The load intensity is zero at each end of the beam and reaches a maximum value of 100 kN/m at the centre of the span. Calculate the slope of the beam at each support and the deflection at the centre. $E = 208 \text{ GN/m}^2$; $I = 405 \times 10^{-6} \text{ m}^4$.

SOLUTION

The loading conditions are such that at $x = 0$, $w = 0$; therefore $c = 0$. Also, at $x = 4$, $w = 0$; therefore

$$0 = 16a + 4b \quad \text{and} \quad b = -4a$$

At $x = 2$, $w = 100$; therefore

$$100 = 4a + 2b \quad \text{and} \quad 100 = 4a - 8a$$

hence

$$a = -25 \quad \text{and} \quad b = 100$$

Therefore

Loading distribution, $w = -25x^2 + 100x$ kN/m.

$$\text{Total load on beam} = \int_0^4 w \, dx = \int_0^4 (-25x^2 + 100x) \, dx = 267 \text{ kN}$$

Therefore the support reactions are each 133.5 kN, by symmetry.
The shear force distribution is given by

$$Q = -\int w \, dx = -\int (-25x^2 + 100x) \, dx = +25\frac{x^3}{3} - \frac{100x^2}{2} + A$$

At $x = 0$, $Q = R_1 = 133.5$; therefore $A = +133.5$.

$$\text{Bending moment, } M = \int Q \, dx = \int \left(+\frac{25x^3}{3} - \frac{100x^2}{2} + 133.5 \right) dx$$

$$= +\frac{25x^4}{12} - \frac{100x^3}{6} + 133.5x + B$$

At $x = 0$, $M = 0$; therefore $B = 0$.

$$\text{Slope} = -\frac{1}{EI} \int M \, dx = -\frac{1}{EI} \int \left(+\frac{25x^4}{12} - \frac{100x^3}{6} + 133.5x \right) dx$$

$$= \frac{1}{EI} \left[-\frac{25x^5}{60} + \frac{100x^4}{24} - \frac{133.5x^2}{2} + C \right]$$

At $x = 2$, $\theta = 0$; therefore $C = 213$.
When $x = 0$ and 4 m, $\theta = \pm 0.002\,53$ rad.

The deflection is given by

$$v = \frac{1}{EI} \int \theta \, dx = \frac{1}{EI} \int \left(-\frac{25x^5}{60} + \frac{100x^4}{24} - \frac{133.5x^2}{2} + 213 \right) dx$$

$$= \frac{1}{EI} \left[-\frac{25x^6}{360} + \frac{100x^5}{120} - \frac{133.5x^3}{6} + 213x + D \right]$$

At $x = 0$, $v = 0$; therefore $D = 0$. At mid-span, $x = 2$ and $v = 3.2$ mm.

BEAMS OF UNIFORM STRENGTH

The bending equation gives $\sigma = My/I$. The condition, therefore, of uniform strength is that My/I shall be constant; that is, the section modulus shall be proportional to the bending-moment. The value of I/y may be varied, in the case of rectangular beams, by altering the depth or altering the breadth.

BEAM HAVING CONSTANT DEPTH

Taking the case of a beam supported at each end and considering one-half of the moment diagram, the maximum deflection is obtained from

$$v = \int_0^{L/2} \frac{M}{EI} x \, dx \tag{7.45}$$

Let M_1 be the bending-moment and I_1 the second moment of area at mid-span, and M and I the bending moment and second moment of area at a point distant x from one end. Since the depth is constant,

$$\frac{M_1}{I_1} = \frac{M}{I}$$

Therefore

$$v = \frac{M_1}{EI_1} \int_0^{L/2} x \, dx = \frac{M_1 L^2}{8EI_1} \tag{7.46}$$

For an isolated load at mid-span, $M_1 = \frac{1}{4}WL$, and therefore

$$v = \frac{WL^3}{32EI_1}$$

For a uniformly-distributed load, $M_1 = \frac{1}{8}wL^2$ and

$$v = \frac{wL^4}{64EI_1}$$

The deflection of a cantiliver may be obtained by substituting L for $\frac{1}{2}L$ in eqn. (7.46), when

$$v = \frac{M_1 L^2}{2EI_1}$$

I_1 being the second moment of area at the constraint and M_1 the bending-moment at the same point. For a cantilever with a uniform load, $M_1 = \frac{1}{2}wL^2$ and

$$v = \frac{wL^4}{4EI_1}$$

For a cantilever with a concentrated load at the free end, $M_1 = WL$ and

$$v = \frac{WL^3}{2EI_1}$$

BEAM HAVING CONSTANT BREADTH

$$\sigma = \frac{M_1 y_1}{I_1} = \frac{My}{I} \quad \text{or} \quad \frac{M_1}{\frac{1}{6}bD_1{}^2} = \frac{M}{\frac{1}{6}bd^2}$$

Therefore

$$\frac{D_1}{d} = \sqrt{\frac{M_1}{M}}$$

and

$$\frac{M}{I} = \frac{M_1}{I_1}\frac{D_1}{d} = \frac{M_1}{I_1}\sqrt{\frac{M_1}{M}}$$

where D_1 is the depth at mid-span and d the depth at a point distant L from the support. In the case of a beam resting on supports, then, as before, considering one-half of the span,

$$v = \int_0^{L/2} \frac{M}{EI}x \, \mathrm{d}x$$

$$= \frac{M_1^{3/2}}{EI_1} \int_0^{L/2} M^{-1/2}x \, \mathrm{d}x \tag{7.47}$$

For a concentrated load at mid-span, $M = \frac{1}{2}Wx$ and $M_1 = \frac{1}{4}WL$. Therefore

$$v = \frac{WL^3}{24EI_1}$$

Other cases may be solved by substituting the value of M in eqn. (7.47) and integrating.

DEFLECTION OF A BEAM UNDER IMPACT LOADING

Let a load W strike a beam of span L simply supported at its ends, at mid-span. If h is the distance fallen by W and δ is the deflection produced, then the work done is $W(h + \delta)$.

If W_1 is the equivalent static load applied at mid-span to produce the deflection δ then the work done by W_1 is given by $\frac{1}{2}W_1\delta$; therefore

$$\tfrac{1}{2}W_1\delta = W(h + \delta)$$

But the central deflection is given by eqn. (7.33),

$$\delta = \frac{W_1 L^3}{48EI}$$

Thus

$$\frac{48EI\delta^2}{2L^3} = W(h + \delta)$$

$$\delta^2 - \frac{WL^3\delta}{24EI} - \frac{WL^3h}{24EI} = 0$$

$$\delta = \frac{WL^3}{48EI} + \frac{1}{2}\sqrt{\left[\left(\frac{WL^3}{24EI}\right)^2 + \frac{WL^3h}{6EI}\right]} \tag{7.48}$$

SLOPE AND DEFLECTION BY THE MOMENT-AREA METHOD

In some problems, the slope and deflection of a beam can be determined more simply and more quickly by a semi-graphical solution known as the *moment-area method*. It is also particularly valuable for finding the deflection of beams of varying cross-section.

SLOPE RELATED TO AREA OF BENDING MOMENT DIAGRAM

In Fig. 7.18 a portion of length AB of a beam has a bending moment diagram of area A represented by CDEF. The distance of the centre of area G of the diagram from any chosen vertical line HH is $\bar{x}$. An exaggerated view of the deflected beam is shown below the bending-moment diagram.

Consider a small piece of the beam of length δx over which the bending moment may be assumed to be constant and equal to M. The *change* of slope over the small piece δx is given by $\delta\theta$, where $\delta\theta$ is the small angle included between tangents drawn at each extremity of δx. Let R be the radius of curvature of the small length δx when deflected; now $\delta\theta$ is also the angle subtended at the centre of curvature of the element δx, therefore

$$R\,\delta\theta = \delta x$$

and substituting for R we get

$$\delta\theta = \frac{1}{EI} M\,\delta x$$

and integrating between the limits L and l, the distances from the chosen line HH,

$$\theta = \int_l^L \frac{1}{EI} M\,\mathrm{d}x = \frac{1}{EI}\int_l^L M\,\mathrm{d}x \tag{7.49}$$

when the cross-section of the beam is constant.

Now $\int_l^L M\,\mathrm{d}x$ is the area A of the bending-moment diagram CDEF

Fig. 7.18 (a) Loading; (b) bending-moment diagram; (c) deflected shape

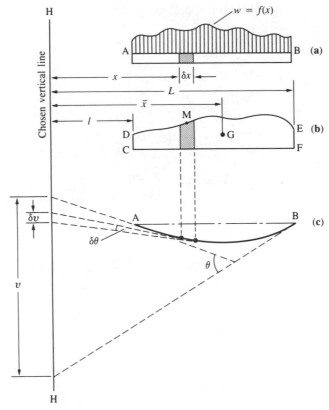

over the length AB. Therefore

$$\theta = \frac{A}{EI} \tag{7.50}$$

θ being the *change of slope* over the length AB.

Thus we have the important relationship that, over any portion of a loaded beam, the change of slope is equal to the area of the bending-moment diagram divided by EI. If the beam varies in cross-section, then I must be retained within the integral in eqn. (7.49).

DEFLECTION RELATED TO AREA OF BENDING MOMENT DIAGRAM

The next stage is to develop an expression for the deflection of the segment BD. The intercept δv on HH can be represented as

$$\delta v \approx x\, \delta\theta = x\frac{M}{EI}\, \delta x$$

Therefore

$$v = \frac{1}{EI}\int_{l}^{L} Mx\, \mathrm{d}x$$

where I is constant. Now

$\int_{l}^{L} Mx\, \mathrm{d}x$ is the first moment of the area A of the bending moment diagram on AB about the axis HH; and since $\bar{x}$ is the distance of the

centre of area of A from HH, then

$$v = \frac{A\bar{x}}{EI} \qquad (7.51)$$

Thus the distance between the intercepts, on any chosen line, of the tangents drawn to the ends of any portion of a loaded beam is equal to the product of the area of the bending-moment diagram, over that portion of the beam, and the distance of the centre of area of this diagram from the chosen line, divided by EI.

In Fig. 7.19, ACB is the deflected form of a loaded beam of length l greatly exaggerated and we wish to find the deflection of any arbitrary point P. The slopes at A and B are θ_A and θ_B respectively.

Fig. 7.19

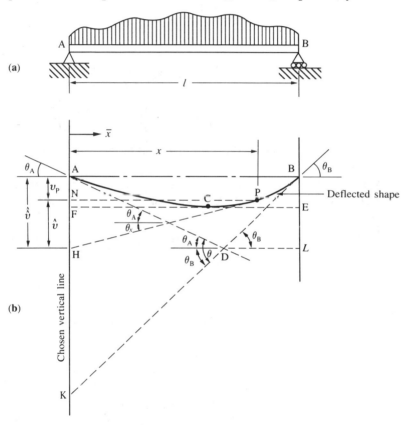

The reference intercept lines are AK and BL at each end of the beam, and FE is a horizontal tangent at C.

The deflection at P is v_P, where $v_P = \hat{v} - \check{v}$. Now

$$\hat{v} = \frac{A_{PN}\bar{x}_P}{EI}$$

where A_{PN} is the area of the bending moment diagram on PN and

$$\check{v} \approx PN \cdot \theta_x$$

$$\therefore v_P = \frac{A_{PN}\bar{x}_P}{EI} - PN \cdot \theta_x$$

since $\theta_A + \theta_x = \theta_P$. Therefore $\theta_x = \theta_P - \theta_A$, and

$$v_P = \frac{A_{PN}\bar{x}_P}{EI} - PN(\theta_P - \theta_A)$$

Now

$$\theta_A = \theta - \theta_B = \frac{A}{EI} - \theta_B$$

and

$$\theta_B l = AK = \frac{A\bar{x}}{EI}$$

therefore

$$\theta_A = \frac{A}{EI} - \frac{A\bar{x}}{EIl} = \frac{A}{EIl}(l - \bar{x})$$

Hence

$$v_P = \frac{A_{PN}\bar{x}_P}{EI} - PN\left\{\frac{A_{PN}}{EI} - A\frac{(l - \bar{x})}{EIl}\right\}$$

$$v_P = \frac{A_{PN}\bar{x}_P}{EI} - \frac{PNA_{PN}}{EI} + PN\frac{A(l - \bar{x})}{EIl} \qquad (7.52)$$

The method is illustrated in the following examples.

BEAM SIMPLY-SUPPORTED WITH POINT LOAD AT MID-SPAN

The moment diagram and the deflected form of the beam are shown in Fig. 7.20. At a point distant x from the left-hand support, the deflection is given by

$$v_x = v_1 + v_2 = v_1 + x\theta_2 \quad (\theta_2 \text{ is actually very small})$$

$$v_1 = \frac{A_x\bar{x}}{EI}$$

$$v_1 = \frac{1}{EI}\left(\frac{Wx}{2} \cdot \frac{x}{2}\right)\frac{2x}{3} = \frac{Wx^3}{6EI}$$

$$\theta_2 = \theta - \theta_1 = \frac{A\bar{x}}{EIL} - \frac{A_x}{EI}$$

$$= \frac{1}{EIL}\left(\frac{WL^2}{8}\right)\frac{L}{2} - \frac{1}{EI} \cdot \frac{Wx^2}{2}$$

$$v_2 = x\theta_2 = \frac{W}{4EI}\left(\frac{L^2}{4} - x^2\right)x$$

$$v_x = v_1 + v_2 = \frac{Wx^3}{6EI} + \frac{W}{4EI}\left(\frac{L^2x}{4} - x^3\right)$$

$$= \frac{Wx}{4EI}\left(\frac{L^2}{4} - \frac{x^2}{3}\right) \qquad (7.53)$$

At mid-span, where $x = \frac{1}{2}L$, and $v = v_{max}$, eqn. (7.53) gives

$$v_{max} = \frac{WL^3}{48EI}$$

Fig. 7.20 (a) Loading;
(b) bending-moment dia-
gram; (c) deflected shape

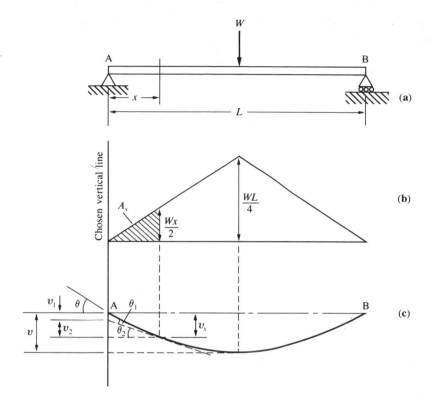

Fig. 7.21 (a) Loading;
(b) bending-moment dia-
gram; (c) deflected shape

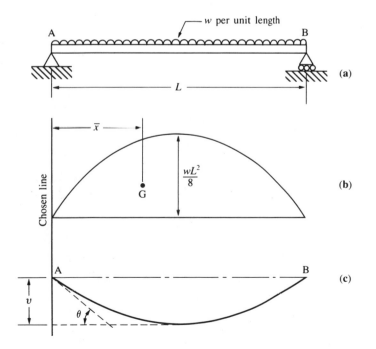

BEAM SIMPLY-SUPPORTED WITH DISTRIBUTED LOADING

Referring to Fig. 7.21, and again considering one-half of the span,

$$A = \frac{2}{3}\frac{wL^2}{8}\frac{L}{2} = \frac{wL^3}{24} \quad \text{and} \quad \bar{x} = \frac{5}{8}\frac{L}{2} = \frac{5}{16}L$$

Therefore

$$v = \frac{5}{384}\frac{wL^4}{EI} \tag{7.54}$$

CANTILEVER CARRYING DISTRIBUTED LOADING

The area of the bending moment diagram, Fig. 7.22, is given by

$$A = \frac{1}{3}\frac{wL^2}{2}L = \frac{wL^3}{6} \quad \text{and} \quad \bar{x} = \frac{3}{4}L$$

$$v = \frac{wL^3}{6}\frac{3}{4}L\frac{1}{EI} = \frac{1}{8}\frac{wL^4}{EI} \tag{7.55}$$

Fig. 7.22 (a) Loading; (b) bending-moment diagram; (c) deflected shape

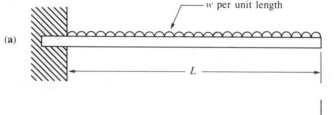

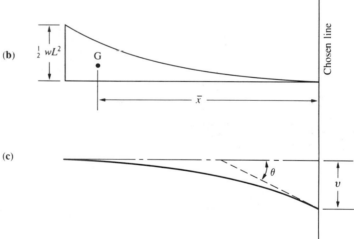

MOMENT APPLIED TO OVERHANGING END OF BEAM

The moment diagram and the deflected form of the beam are shown in Fig. 7.23. The deflection at the free end is given by

$$v = v_2 - v_1\frac{a+b}{a}$$

and v_1 is obtained by considering the portion of span a:

$$v_1 = \frac{Ma}{2}\frac{a}{3}\frac{1}{EI}$$

$$= \frac{Ma^2}{6EI} \tag{7.56}$$

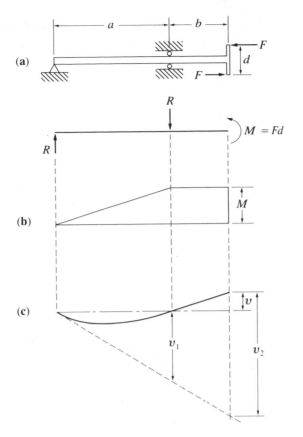

Fig. 7.23 (a) Loading; (b) bending-moment diagram; (c) deflected shape

Also, v_2 is found by considering the whole of the moment diagram:

$$v_2 = \left[\frac{Mb^2}{2} + \frac{Ma}{2} \left(\frac{a}{3} + b \right) \right] \frac{1}{EI}$$

$$= \left[\frac{Mb^2}{2} + \frac{Ma^2}{6} + \frac{Mab}{2} \right] \frac{1}{EI} \tag{7.57}$$

Therefore

$$v = \frac{1}{EI} \left[\frac{Mb^2}{2} + \frac{Ma^2}{6} + \frac{Mab}{2} - \frac{Ma^2}{6} - \frac{Mab}{6} \right]$$

$$= \frac{1}{EI} \left[\frac{Mb^2}{2} + \frac{Mab}{3} \right]$$

$$= \frac{Mb}{6EI} [2a + 3b] \tag{7.58}$$

where $M = Ra$.

EXAMPLE 7.7

A rectangular section timber beam 4 m long, 75 mm wide and 100 mm deep is simply supported at each end. It is subjected to loads of 2 kN and

4 kN at 2 m and 3 m from the left-hand end respectively. Use the moment-area method to find the deflection resulting at 2.5 m. $E = 14 \, \text{GN/m}^2$.

SOLUTION

Taking moments about each end, the reactions are found to be

$$R_A = \tfrac{1}{4}[(2 \times 2) + (4 \times 1)] = 2 \, \text{kN}$$
$$R_B = 4 \, \text{kN}$$
$$M_C = 2 \times 2 = 4 \, \text{kN-m}$$
$$M_D = (2 \times 3) - (2 \times 1) = 4 \, \text{kN-m}$$

The bending-moment diagram is shown in Fig. 7.24. Applying the relationship $v = A\bar{x}/EI$, to a reference line at the left-hand end,

$$v_1 = \frac{1}{EI}\left[\left(\frac{4 \times 2}{2} \times \frac{4}{3}\right)_X + (4 \times 1 \times 2.5)_Y + \left(\frac{4 \times 1}{2} \times 3\tfrac{1}{3}\right)_Z\right]$$

$$= \frac{22}{EI} = \frac{22\,000 \times 12 \times 10^3}{14 \times 10^9 \times 75 \times 100^3 \times 10^{-12}} = 250 \, \text{mm}$$

From similar triangles,

$$v_2 = \frac{1.5}{4} v_1 = 93.7 \, \text{mm}$$

Taking a reference line at 2.5 m from A, then

$$v_3 = \frac{1}{EI}\left[(4 \times \tfrac{1}{2} \times \tfrac{1}{4}) + \left(\frac{4 \times 1}{2} \times \frac{5}{6}\right)\right]$$

$$= \frac{2.17}{EI} = 24.7 \, \text{mm}$$

Fig. 7.24

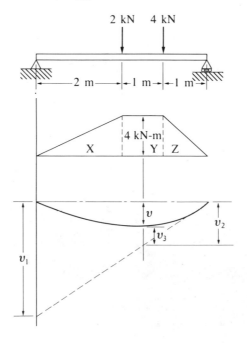

Therefore

Deflection at 2.5 m from left-hand end $= v_2 - v_3 = 69$ mm

All solutions for slope and deflection of beams depend on the relationship between bending-moment distribution and curvature (eqn. 7.4). It is therefore vital that the correct bending-moment expression can be stated. Although the double integration is relatively simple, the constants of integration can only be found by applying the correct boundary conditions at the supports. Equally, the successful application of Macaulay's method for loading discontinuities does depend on

Table 7.1
Principal slope and deflection for beams with basic loading

		Slope	Deflection
(a)		$-\dfrac{Ml}{EI}$ at B	$-\dfrac{Ml^2}{2EI}$ at B
(b)		$-\dfrac{Ml}{12EI}$ at C $+\dfrac{Ml}{24EI}$ at A, B	0 at C
(c)		$\pm\dfrac{Ml}{2EI}$ at A, B	$+\dfrac{Ml^2}{8EI}$ at C
(d)		$+\dfrac{Wl^2}{2EI}$ at B	$+\dfrac{Wl^3}{3EI}$ at B
(e)		$\pm\dfrac{Wl^2}{16EI}$ at A, B	$+\dfrac{Wl^3}{48EI}$ at C
(f)		0 at A, B, C	$+\dfrac{Wl^3}{192EI}$ at C
(g)		$+\dfrac{wl^3}{6EI}$ at B	$+\dfrac{wl^4}{8EI}$ at B
(h)		$\pm\dfrac{wl^3}{24EI}$ at A, B	$+\dfrac{5wl^4}{384EI}$ at C
(i)		0 at A, B, C	$+\dfrac{wl^4}{384EI}$ at C

following the simple rules associated with the use of step functions. The superposition and moment–area methods are valuable alternatives to the double integration method and choice depends on the nature of the problem. For example, it is obviously an advantage to use superposition if the case can be broken down into simple elements the solutions for which are readily available and can then be superposed. A final reminder may be made of the need to recognize the statically determinate nature of the problem; if not the case, then the treatment given in Chapter 8 is required. The principal values for slope and deflection for various basic support and loading conditions are given in Table 7.1.

PROBLEMS

7.1 A cantilevered deck is built in at the left end and is supported on a wall at 8 m from that end. The deck extends a further 4 m beyond this wall. The loading on each of the beams supporting the deck including self-weight is 20 kN/m and the flexural rigidity (EI) of each beam section is 200 MN-m^2. Determine the level of the top of the wall (assumed rigid) relative to the fixed support so that the bending moment is zero at that support.

7.2 A horizontal beam is subjected to the loading shown in Fig. 7.25. Calculate the beam deflection under the 8 kN force. The flexural rigidity (EI) of the beam is 1 MN-m^2.

Fig. 7.25

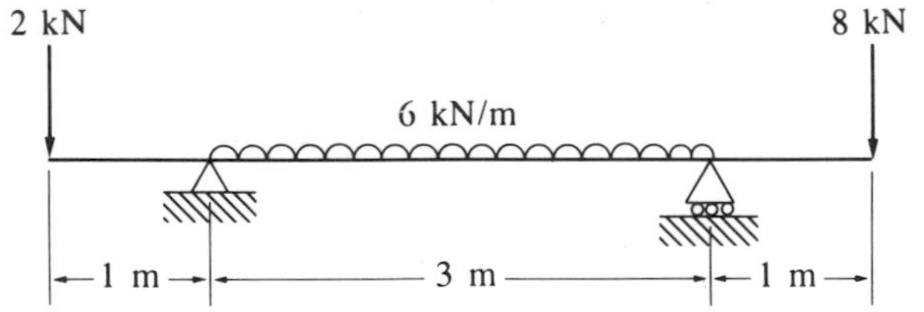

2 kN   8 kN

6 kN/m

$\leftarrow$1 m$\rightarrow$|$\leftarrow$ 3 m $\rightarrow$|$\leftarrow$1 m$\rightarrow$

7.3 A beam 10 m long is simply supported at each end and carries the loading shown in Fig. 7.26. Calculate the position and the magnitude of the maximum deflection. The flexural rigidity (EI) of the beam is 100 MN-m^2.

Fig. 7.26

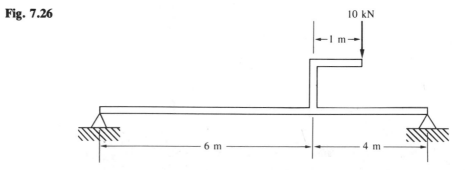

7.4 Calculate the deflection of the roof of the petrol station in Problem 6.9 at points A, B and C.

7.5 A flexible mounting pad 3 m in length and of flexural stiffness 0.55 MN-m^2 is simply-supported at one end and rests on a spring of stiffness 100 N/mm at the other end. Determine the overall displacement of a load of 10 kN placed on the pad at 2 m from the spring-supported end.

7.6 Part of a bridge is being assembled as shown in Fig. 7.27. The central section is hanging from a crane and is to be bolted to the left- and right-hand cantilever sections at A and B. In raising the section into position and before it is properly aligned it fouls at A and B, causing an upward force W exerted by the crane, on the beams. Show that the deflection of the beam at the crane hook, if E and I are the same for each of the beam sections, is

$$v = -\frac{15}{9}\frac{Wl^3}{EI}$$

Fig. 7.27

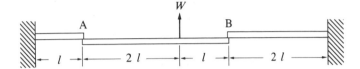

7.7 A sight-screen for a cricket field is illustrated in Fig. 7.28. The cross-beams AB and CD can be assumed to be rigid and act as simple supports to the screen itself. When the screen is subjected to uniform wind pressure the location of the cross-beams is to be such that the horizontal deflections of the top and bottom edges are the same as the horizontal deflection of the centre line. Hence determine the relationship between L and a.

Fig. 7.28

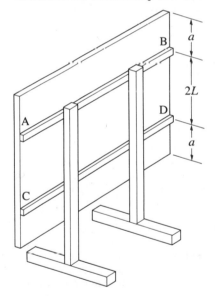

7.8 A horizontal shaft 1 m in length is simply-supported in bearings at 100 mm from each end. It carries loads of 0.3 kN at each end and 1.2 kN at the centre. Find the deflection under the loads and the slope at the bearings. Check the solution by the method of superposition. $EI = 7$ kN-m^2.

7.9 Calculate the vertical and horizontal components of deflection at the free end of the Z-section cantilever described in Problem 6.23. $E = 208$ GN/m^2.

7.10 A horizontal beam is simply-supported at each end of a 4-m span. It carries a distributed loading varying from 2 kN/m at the left end to 3 kN/m at the right-hand end. Find the position and magnitude of the maximum deflection. $E = 208$ GN/m^2; $I = 2 \times 10^{-4}$ m^4.

7.11 A floor beam 6 m in length is simply supported at each end and carries a varying distributed load of grain over the whole span. The loading distribution is represented by the equation $w = ax^2 + bx + c$, where w is the load intensity of the grain in kN/m at a distance x along the beam, and a, b and c are constants. The load intensity is zero at each end and has its maximum value of 4 kN/m at mid-span. Calculate the slope at each end of the beam if $E = 208$ GN/m² and $I = 10^8$ mm⁴.

7.12 A horizontal beam of rectangular section 25 mm wide by 50 mm deep is 1 m long and simply supported on (*a*) rigid rollers, (*b*) springs of stiffness 300 kN/m. A load of 15 kg falls 50 mm onto the beam at mid-span. Calculate the instantaneous maximum deflections and bending stresses. $E = 208$ GN/m².

7.13 A beam 3 m long is subjected to the loading shown in Fig. 7.29. Use the moment-area method to determine the deflection of the beam at the right-hand end. The flexural rigidity of the beam is 1 MN-m².

Fig. 7.29

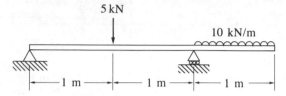

7.14 A vertical column 10 m in height is pinned at each end and is subjected to a horizontal load of 10 kN at 2 m above the lower end. Use the moment-area method to obtain the deflection of the column at the load. The flexural modulus of the column is 50 MN-m².

7.15 A beam 5 m long is subjected to the loading in Fig. 7.30. Use the moment-area method to find the position and magnitude of the maximum deflection. $EI = 10$ MN-m².

Fig. 7.30

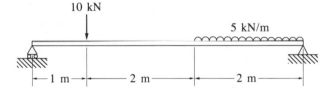

7.16 A stepped shaft consists of a 50 mm diameter section friction-welded to a 60 mm diameter section. Each section is 1 m long and the shaft is simply-supported at each end. If a moment of 10 kN-m is applied at the joint of the two shaft sections, use the moment-area method to determine the position and magnitude of the maximum deflection and the slope of the deflected shaft at the thin end. The modulus of the shaft material is 208 GN/m².

7.17 A shaft is simply-supported at each end of a length a having a second moment of area $2I$. There are also overhanging ends each of length a having a second moment of area I. The overhanging ends are subjected to uniform loading w per unit length. Calculate the deflection at either outer end.

Chapter 8

Statically-indeterminate beams

INTRODUCTION

The design of beams depends initially on the evaluation of shear-force and bending-moment distributions in order to calculate stresses and deflections. A prerequisite is the calculation of support reactions, and in the case of statically-determinate beam situations there are only two unknown reactions, which are found from the two equilibrium equations ($\Sigma M = 0$ and $\Sigma F = 0$). Thus a beam which is supported in such a way as to produce three or more reaction forces or moments is statically indeterminate. Some typical examples are shown in Fig. 8.1. The principal methods which are used for analysis are (i) double-integration with Macauley's method, (ii) superposition, (iii) moment–area. The application of these methods will be illustrated in a number of worked examples.

DOUBLE-INTEGRATION METHOD

This method was first developed in Chapter 7. To reiterate briefly, the curvature is expressed in terms of bending moment at any point along the beam, this equation is then integrated twice, and the constants of integration are found from known boundary conditions of slope and deflection at the supports, or elsewhere. Whereas, however, in the case of the statically-determinate beam the reactions could be found *prior* to the above procedure, this is not possible for the indeterminate beam and the reactions must be carried through the analysis as unknown quantities. Since there are *always* enough boundary conditions to determine all the unknowns in the equations, the reactions can be evaluated together with the constants of integration. A number of worked cases now follow.

Fig. 8.1

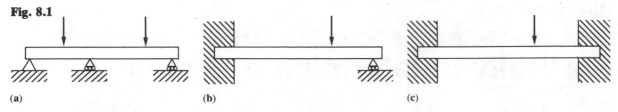

(a)   (b)   (c)

BEAM FIXED AT EACH END WITH UNIFORMLY DISTRIBUTED LOADING

The problem is illustrated in Fig. 8.2, and it is evident that, owing to symmetry, $M_1 = M_2$ and $R_1 = R_2 = wL/2$. However, we cannot determine M_1 and M_2 from a moment equilibrium equation. The next step is to write the equation for bending moment distribution as a function of x and equate this to $EI(d^2v/dx^2)$ (*see* eqn. (7.4)).

Fig. 8.2

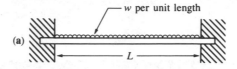

(a)

w per unit length

L

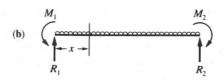

(b)

M_1 M_2

x

R_1 R_2

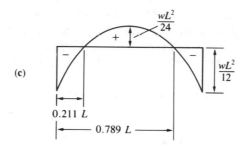

(c)

$\dfrac{wL^2}{24}$

$\dfrac{wL^2}{12}$

$0.211\,L$

$0.789\,L$

$$EI\frac{d^2v}{dx^2} = -M = -\frac{wLx}{2} + M_1 + \frac{wx^2}{2} \tag{8.1}$$

$$EI\frac{dv}{dx} = -\frac{wLx^2}{4} + M_1x + \frac{wx^3}{6} + A \tag{8.2}$$

$$EIv = -\frac{wLx^3}{12} + \frac{M_1x^2}{2} + \frac{wx^4}{24} + Ax + B \tag{8.3}$$

The boundary conditions are that when $x = 0$, $dv/dx = 0$ and $v = 0$ from which, respectively, $A = 0$ and $B = 0$; and when $x = L$, $dv/dx = 0$ and $v = 0$.

Either of these conditions can now be used to solve for M_1, which is found to be $M_1 = M_2 = wL^2/12$.

At mid-span

$$M = \frac{wL^2}{24}$$

The deflection is a maximum at mid-span and is

$$v_{max} = \frac{wL^4}{384EI}$$

The bending-moment diagram, Fig. 8.2, shows two points of contraflexure (where BM changes sign), occurring at $x = 0.211L$ and $0.789L$.

BEAM FIXED AT EACH END CARRYING A POINT LOAD

There are four unknown reactions illustrated in Fig. 8.3.

$$R_A + R_B - W = 0 \tag{8.4}$$

$$R_A L - W(L - a) - M_A + M_B = 0 \tag{8.5}$$

$$EI\frac{d^2v}{dx^2} = -R_A x + W[x - a] + M_A \tag{8.6}$$

$$EI\frac{dv}{dx} = -\frac{R_A x^2}{2} + \frac{W}{2}[x - a]^2 + M_A x + A \tag{8.7}$$

$$EIv = -\frac{R_A x^3}{6} + \frac{W}{6}[x - a]^3 + M_A \frac{x^2}{2} + Ax + B \tag{8.8}$$

Fig. 8.3

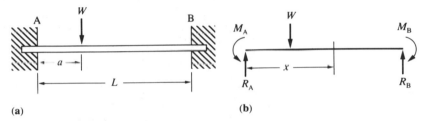

(a) (b)

The boundary conditions are,

(i) When $x = 0$, $v = 0$ and $dv/dx = 0$. (ii) When $x = L$, $v = 0$ and $dv/dx = 0$.

From (i)

$$A = 0 \quad \text{and} \quad B = 0$$

From (ii) and solving for R_A and M_A,

$$R_A = \frac{W}{L^3}(L - a)^2(L + 2a)$$

$$M_A = \frac{Wa}{L^2}(L - a)^2$$

Using eqn. (8.5),

$$M_B = \frac{Wa^2}{L^2}(L - a)$$

From eqn. (8.8) the deflection under the load is

$$v = \frac{Wa^3(L - a)^3}{3EIL^3} \tag{8.9}$$

For the particular case when $a = L/2$,

$$R_A = R_B = \frac{W}{2} \quad \text{and} \quad M_A = M_B = \frac{WL}{8}$$

$$v_{max} = \frac{WL^3}{192EI} \tag{8.10}$$

CANTILEVER WITH A PROP AT THE FREE END

This situation is illustrated in Fig. 8.4, in which the prop, considered as a simple support, is, for generality, assumed to be at a level Δ above the fixed end. The unknown reactions are P, R and M_0.

$$P + R - W = 0 \tag{8.11}$$

$$PL - \frac{WL}{2} + M_0 = 0 \tag{8.12}$$

Fig. 8.4

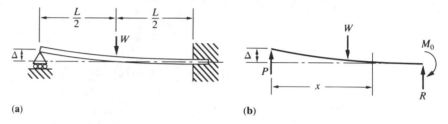

(a)   (b)

It is now convenient to apply the step-function or Macaulay method to deal with the discontinuous loading as in Chapter 7, so that

$$EI\frac{d^2v}{dx^2} = -Px + W\left[x - \frac{L}{2}\right] \tag{8.13}$$

$$EI\frac{dv}{dx} = -\frac{Px^2}{2} + \frac{W}{2}\left[x - \frac{L}{2}\right]^2 + A \tag{8.14}$$

$$EIv = -\frac{Px^3}{6} + \frac{W}{6}\left[x - \frac{L}{2}\right]^3 + Ax + B \tag{8.15}$$

The boundary conditions are,

(i) When $x = 0$, $v = -\Delta$. (ii) When $x = L$, $v = 0$ and $dv/dx = 0$.
For (i), from eqn. (8.8),

$$B = -EI\Delta$$

(The term in square brackets has to be omitted as it is negative.)
For (ii), from eqn. (8.14),

$$0 = -\frac{PL^2}{2} + \frac{WL^2}{8} + A \tag{8.16}$$

and from eqn. (8.15),

$$0 = -\frac{PL^3}{6} + \frac{WL^3}{48} + AL - EI\Delta \tag{8.17}$$

Solving eqns. (8.16) and (8.17),

$$P = \frac{5}{16}W + \frac{3EI\Delta}{L^3} \tag{8.18}$$

and

$$A = +\frac{WL^2}{32} + \frac{3EI\Delta}{2L}$$

Substituting for P in eqns. (8.11) and (8.12),

$$R = \frac{11}{16}W - \frac{3EI\Delta}{L^3}$$

and

$$M_0 = \frac{3}{16}WL - \frac{3EI\Delta}{L^2}$$

Any required S. F., B. M., slope or deflection can now be determined.

CONTINUOUS BEAM ON MULTIPLE SIMPLE SUPPORTS

A fairly general example is illustrated in Fig. 8.5 in which Δ_B and Δ_C are known displacements due to the supports not being at the same level.

$$R_A + R_B + R_C + R_D - W - wL = 0 \tag{8.19}$$

$$3R_AL + 2R_BL + R_CL - \frac{5}{2}WL - \frac{wL^2}{2} = 0 \tag{8.20}$$

$$EI\frac{d^2v}{dx^2} = -R_Ax + W\left[x - \frac{L}{2}\right] - R_B[x - L] - R_C[x - 2L]$$
$$+ \frac{w}{2}[x - 2L]^2 \tag{8.21}$$

$$EI\frac{dv}{dx} = -\frac{R_Ax^2}{2} + \frac{W}{2}\left[x - \frac{L}{2}\right]^2 - \frac{R_B}{2}[x - L]^2 - \frac{R_C}{2}[x - 2L]^2$$
$$+ \frac{w}{6}[x - 2L]^3 + A \tag{8.22}$$

$$EIv = -\frac{R_Ax^3}{6} + \frac{W}{6}\left[x - \frac{L}{2}\right]^3 - \frac{R_B}{6}[x - L]^3 - \frac{R_C}{6}[x - 2L]^3$$
$$+ \frac{w}{24}[x - 2L]^4 + Ax + B \tag{8.23}$$

Fig. 8.5

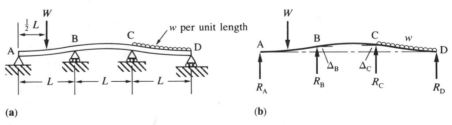

(a)

(b)

The boundary conditions are,

(i) $x = 0$, $v = 0$, (ii) $x = L$, $v = -\Delta_B$, (iii) $x = 2L$, $v = -\Delta_C$, (iv) $x = 3L$, $v = 0$.

From (i),

$$B = 0$$

From (ii),

$$A = -\frac{WL^2}{48} + \frac{R_AL^2}{6} - \frac{EI\Delta_B}{L}$$

From (iii),

$$EI\Delta_C = R_A L^3 - \frac{25WL^3}{48} + \frac{R_B L^3}{6} + 2EI\Delta_B$$

From (iv),

$$0 = -4R_A L^3 + \frac{122WL^3}{48} - \frac{8R_B L^3}{6} - \frac{R_C L^3}{6} + \frac{wL^4}{24} - 3EI\Delta_B$$

Although perhaps somewhat laborious, these latter two equations together with eqns. (8.19) and (8.20) can be solved to give values for R_A, R_B, R_C and R_D. From this stage any required aspect of this beam problem can be evaluated.

SUPERPOSITION METHOD

The principle of superposition can be very useful in finding redundant reactions, particularly if a problem can be split up into "standard" cases (*see* Table 7.1).

CANTILEVER WITH A PROP AT THE FREE END This problem is illustrated in Fig. 8.6 and can be represented by superposition of the two parts shown in Figs. 8.7(*a*) and (*b*). If there were no support at A there would be a downward deflection due to the distributed load. If there were no distributed load and the reaction, *P*, at the support was considered as a force which could cause an upward deflection of the beam, then the necessary boundary condition is that the sum of these deflections must be zero.

Fig. 8.6

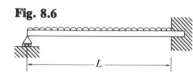

Fig. 8.7

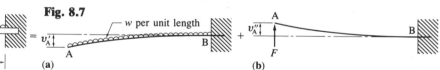

(a)   (b)

Due to loading, *w*,

$$v'_A = +\frac{wL^4}{8EI}$$

Due to reaction force, *P*,

$$v''_A = -\frac{PL^3}{3EI}$$

But

$$v'_A + v''_A = 0$$

Therefore

$$+\frac{wL^4}{8EI} - \frac{PL^3}{3EI} = 0 \tag{8.24}$$

Hence

$$P = \frac{3}{8}wL$$

From vertical and moment equilibrium,

$$R = \frac{5}{8}wL \quad \text{and} \quad M_0 = \frac{wL}{8}$$

ALTERNATIVE SUPERPOSITION

The case of Fig. 8.6 can be split up in the manner shown in Figs. 8.8 (b) and (c). The superposition must now satisfy the boundary condition of zero slope at B. Thus

$$\theta'_B + \theta''_B = 0$$

Now,

$$\theta'_B = -\frac{wL^3}{24EI}$$

and

$$\theta''_B = +\frac{M_0L}{3EI}$$

Therefore

$$-\frac{wL^3}{24EI} + \frac{M_0L}{3EI} - 0 \tag{8.25}$$

$$M_0 = \frac{wL^2}{8}$$

Knowing M_0 we can find P and R from the two equilibrium equations.

Fig. 8.8

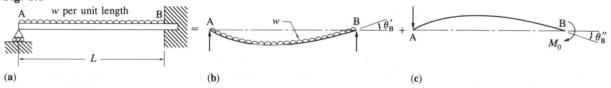

(a) (b) (c)

BEAM FIXED HORIZONTALLY AT EACH END

Before considering any particular form of applied loading, we will examine the effect of the fixing moments alone. Taking the general case of, say, $M_B > M_A$ as illustrated in Fig. 8.9(a), this may itself be put into the two parts in Figs. 8.9(b) and (c), giving slopes at each end of

$$\theta_A = -\frac{M_AL}{2EI} \quad \text{and} \quad \theta_B = +\frac{M_AL}{2EI}$$

$$\theta_A = -\frac{(M_B - M_A)L}{6EI} \quad \text{and} \quad \theta_B = +\frac{(M_B - M_A)L}{3EI}$$

215

Fig. 8.9

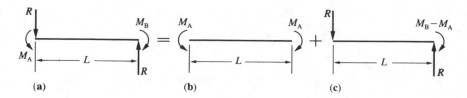

(a)   (b)   (c)

The resultant slopes are therefore

$$\theta_A = -\frac{(M_B + 2M_A)L}{6EI} \quad \text{and} \quad \theta_B = +\frac{(2M_B + M_A)L}{6EI} \tag{8.26}$$

FIXED BEAM WITH UNIFORMLY-DISTRIBUTED LOADING

This problem may be represented as in Fig. 8.10. The slopes at the ends for the simply-supported part are $\pm wL^3/24EI$. Using the condition of zero slope at each end and the results in eqn. (8.26).

$$+\frac{wL^3}{24EI} - \frac{(M_B + 2M_A)L}{6EI} = 0 \tag{8.27}$$

and

$$-\frac{wL^3}{24EI} + \frac{(2M_B + M_A)L}{6EI} = 0 \tag{8.28}$$

from which

$$M_A = M_B = \frac{wL^2}{12}$$

Fig. 8.10

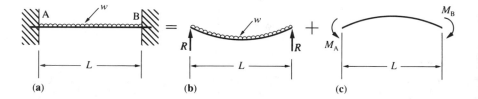

(a)   (b)   (c)

FIXED BEAM WITH ENDS NOT AT SAME LEVEL

This is an important structural situation since a considerable bending moment can be set up by the ends not being at the same level, even without any applied loading to the beam.

Let v be the difference in level between the ends of the beam, Fig. 8.11. A point of contraflexure occurs at mid-span, owing to symmetry, and each half of the beam may be taken as a cantilever, the free end of which is caused to deflect $\frac{1}{2}v$ by a force F at the free end. Then, using the basic solution for a cantilever carrying a load at the free end,

$$\frac{v}{2} = \frac{1}{3}\frac{F(\frac{1}{2}L)^3}{EI} \quad \text{so that} \quad F = \frac{12EIv}{L^3}$$

The bending-moment diagram on each half, due to F, is triangular, the

Fig. 8.11

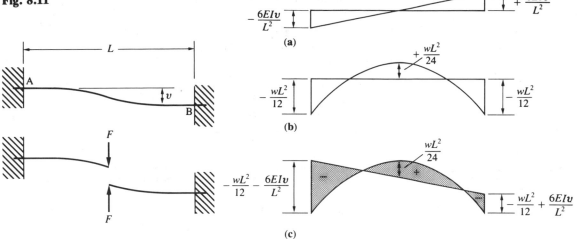

(a)

(b)

(c)

maximum ordinate being

$$M_F = \pm \frac{12EIv}{L^3} \frac{L}{2} = \pm \frac{6EIv}{L^2},$$

as shown in Fig. 8.11(*a*).

If, say, a distributed load *w* is now applied to the beam, then, if the ends were at the same level the bending moment diagram would be as shown in Fig. 8.11(*b*).

The fixing moments due to the distributed load *w* will be altered, owing to the differences in level of A and B, by the amount M_F, and hence this quantity must be added to, or subtracted from, M_A and M_B, depending on which side is the higher. When A is above B,

$$M_A = \frac{wL^2}{12} + \frac{6EIv}{L^2} \quad \text{and} \quad M_B = \frac{wL^2}{12} - \frac{6EIv}{L^2} \tag{8.29}$$

The combined diagram of bending moment, due to *F*, to the distributed load, and to the fixing of the ends, is shown in Fig. 8.11(*c*).

WILSON'S METHOD FOR CONTINUOUS BEAMS

In this method the reactions at the supports are found by equating the upward deflections caused by the supporting forces to the downward deflections which the loads would cause at the various supports, if in each case the beam were supported at the ends only.

In Fig. 8.12 assume the beam is supported at A and D only. Let

v_B = Sum of deflections at B caused by loads W_1, W_2 and W_3 separately
v_C = Sum of deflections at C caused by loads W_1, W_2 and W_3 separately

With the above loads removed, again assume the beam to be supported

Fig. 8.12

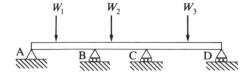

at A and D only and to carry loads R_B and R_C.
Let

$\delta_1{}^B$ = Deflection at B caused by unit load at B
$\delta_1{}^C$ = Deflection at B caused by unit load at C
$\delta_2{}^B$ = Deflection at C caused by unit load at B
$\delta_2{}^C$ = Deflection at C caused by unit load at C
$v_B = R_B\delta_1{}^B + R_C\delta_1{}^C$ $v_C = R_B\delta_2{}^B + R_C\delta_2{}^C$

From these equations R_B and R_C can be determined. R_A and R_D can then
be found by taking moments about A and D respectively.

MOMENT–AREA METHOD

For a beam of uniform section, the change in slope is given by $\theta = A/EI$
(*see* eqn. (7.50)). In the case of a horizontally-fixed beam, $\theta = 0$ for the
complete span, and since the product EI is not zero, it follows that A, the
resultant area of the moment diagram for the beam, must be zero. This
enables the deflection and slope at any point on the beam to be
evaluated, and various examples of its use will follow.

**FIXED BEAM
WITH IRREGULAR
LOADING**

Referring to Fig. 8.13, let A_1 be the area of the free bending moment
diagram AEFHB, and let $\bar{x}_1$ be the distance of its centroid from A; also
let A_2 be the area of the fixing moment diagram ACDB and $\bar{x}_2$ the
distance of its centroid from A. Then

$$A_2 = \frac{(M_A + M_B)L}{2}$$

Fig. 8.13 (a) Fixed beam
with irregular loading;
(b) free moment diagram;
(c) fixing moment
diagram; (d) combined
moment diagram

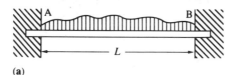

(a)

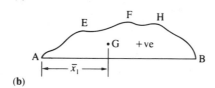

(b)

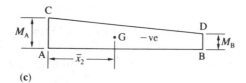

(c)

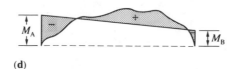

(d)

It is required that $A_1 + A_2 = 0$, thus

$$M_A + M_B = -\frac{2A_1}{L} \qquad (8.30)$$

since there is no relative deflection of each end of the beam, then

$$\frac{A_1 \bar{x}_1}{EI} + \frac{A_2 \bar{x}_2}{EI} = 0$$

Taking moments about A, for the areas under the fixing moment diagram

$$A_2 \bar{x}_2 = \frac{M_B L^2}{2} + \frac{(M_A - M_B)}{2} L \frac{L}{3}$$

$$= \frac{M_B L^2}{3} + \frac{M_A L^2}{6} \qquad (8.31)$$

therefore

$$-A_1 \bar{x}_1 = \frac{M_B L^2}{3} + \frac{M_A L^2}{6}$$

Hence

$$2M_B + M_A = -\frac{6A_1 \bar{x}_1}{L^2} \qquad (8.32)$$

From eqns. (8.30) and (8.32),

$$M_A = -\frac{4A_1}{L} + \frac{6A_1 \bar{x}_1}{L^2} \qquad (8.33)$$

and

$$M_B = -\frac{6A_1 \bar{x}_1}{L^2} + \frac{2A_1}{L} \qquad (8.34)$$

FIXED BEAM WITH CONCENTRATED CENTRAL LOAD

By combining the positive free bending moment and negative fixing moment diagram, Fig. 8.14, the resultant diagram AEFBGDC is obtained. Since area ABC + EACD must be zero for no change in slope at A compared with C

$$M_A L + \frac{WL^2}{8} = 0$$

Hence

$$M_A = -\frac{WL}{8} = M_C$$

and the points of contraflexure will be at G and F, distant x from each end, where $x = \frac{1}{4}L$.

The value of M at mid-span is numerically equal to that of M at each end, but is positive. Considering half the span, and taking a chosen line through one of the constraints, we have the deflection at mid-span given

Fig. 8.14 (a) Fixed beam
with central loading;
(b) moment diagram;
(c) deflected shape

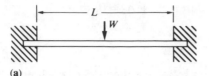

(a)

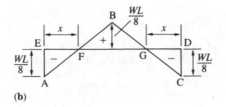

(b)

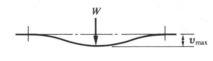

(c)

by eqn. (7.51)

$$v = \frac{A\bar{x}}{EI}$$

Therefore

$$v_{max} = \frac{1}{EI}\left[\left(\frac{WL}{4}\times\frac{L}{4}\times\frac{2}{3}\frac{L}{2}\right)+\left(-\frac{WL}{8}\times\frac{L}{2}\times\frac{L}{4}\right)\right]$$

$$= \frac{WL^3}{EI}\left[\frac{1}{48}-\frac{1}{64}\right]$$

$$= \frac{1}{192}\frac{WL^3}{EI} \tag{8.35}$$

CONTINUOUS BEAM ON MULTIPLE SUPPORTS

A beam resting on more than two supports is said to be *continuous*. Such a beam is represented by Fig. 8.15(a). Changes of curvature occur in each span, due to negative bending moments at the supports. In the case represented, the supports are assumed to be at different levels, being displaced v_0, v_1 and v_2 from a horizontal line AB. Suppose the loading to be such that the "free" (positive) and "fixing" (negative) bending-moment diagrams are as shown at (b), the resultant diagram being shown at (d). The area of the resultant diagram on the span l_1 is A_1, and the distance of its centroid G_1 from a vertical line through the left-hand support is $\bar{x}_1$; also the area of the resultant diagram on the span l_2 is A_2, and $\bar{x}_2$ is the distance of its centroid G_2 from a vertical line through the right-hand support.

(i) Bending moments at supports

Draw CD, a common tangent at the point of contact of the central support, and let α which is actually a small angle be its inclination to the horizontal; taking intercepts, between tangents to the deflected beam, on

Fig. 8.15 (a) Continuous beam; (b) free moment diagram; (c) fixing moment diagram; (d) resultant moment diagram

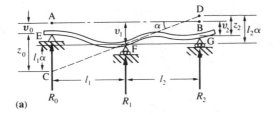

(a)

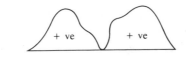

(b)

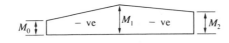

(c)

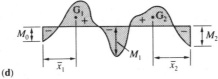

(d)

a vertical line as positive when measured downwards, and vice versa as negative upwards, the same convention as for deflections, then

$$z_0 = \frac{A_1 \bar{x}_1}{EI} = v_1 - v_0 + l_1 \alpha$$

$$-z_2 = \frac{A_2 \bar{x}_2}{EI} = -l_2 \alpha + (v_1 - v_2)$$

and

$$\frac{A_1 \bar{x}_1}{EIl_1} - \frac{v_1 - v_0}{l_1} = \alpha = -\frac{A_2 \bar{x}_2}{EIl_2} + \frac{v_1 - v_2}{l_2}$$

Therefore

$$\frac{A_1 \bar{x}_1}{EIl_1} + \frac{A_2 \bar{x}_2}{EIl_2} = \frac{v_1 - v_0}{l_1} + \frac{v_1 - v_2}{l_2}$$

and

$$\frac{A_1 \bar{x}_1}{l_1} + \frac{A_2 \bar{x}_2}{l_2} = \left[\frac{v_1 - v_0}{l_1} + \frac{v_1 - v_2}{l_2} \right] EI \tag{8.36}$$

When the supports are all at the same level, $v_0 = v_1 = v_2$, and

$$\frac{A_1 \bar{x}_1}{l_1} + \frac{A_2 \bar{x}_2}{l_2} = 0 \tag{8.37}$$

221

Let the areas of the "free" bending-moment diagrams be S_1 and S_2 and the distance of the centroid of S_1 from a vertical line through the left-hand support is x_1, the corresponding distance of the centroid of S_2 from a vertical through the right-hand support is x_2; then, the sum of the moments of the "free" and "fixing" moment diagrams are respectively

$$A_1 \bar{x}_1 = S_1 x_1 + \left\{ M_0 l_1 \frac{l_1}{2} + \frac{(M_1 - M_0)}{2} l_1 2 \frac{l_1}{3} \right\}$$

and

$$A_2 \bar{x}_2 = S_2 x_2 + \left\{ M_2 l_2 \frac{l_2}{2} + \frac{(M_1 - M_2)}{2} l_2 2 \frac{l_2}{3} \right\}$$

Substituting into eqn. (8.37) and rearranging gives

$$\frac{S_1 x_1}{l_1} + \frac{S_2 x_2}{l_2} + \frac{M_0 l_1}{6} + \frac{M_1}{3}(l_1 + l_2) + \frac{M_2 l_2}{6} = 0$$

and

$$M_0 l_1 + 2M_1(l_1 + l_2) + M_2 l_2 = -6 \left[\frac{S_1 x_1}{l_1} + \frac{S_2 x_2}{l_2} \right] \tag{8.38}$$

This is Clapeyron's *theorem of three moments,* and by taking the spans in pairs, sufficient equations are obtained to solve for the bending moments at the supports.

Fig. 8.16 (a) For a point load; (b) for a uniformly distributed load

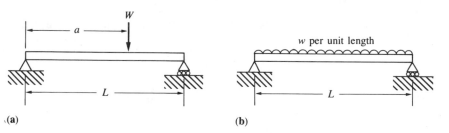

Useful forms of $6Sx/L$ are shown in Fig. 8.16:
(*a*) for a point load,

$$\frac{6Sx}{L} = \frac{Wa}{L}(L^2 - a^2)$$

(*b*) for a uniformly distributed load,

$$\frac{6Sx}{L} = \frac{wL^3}{4}$$

(ii) Reactions at supports

In order to calculate the support reactions, consider the beam in Fig. 8.15 and let the reactions at E, F and G be R_0, R_1, R_2 respectively. In Fig. 8.17 we have split the beam into two separate free bodies which have reactions R_0', R_1'' and R_1', R_2'' and bending moments at the supports, M_0, M_1 and M_2. Let the loading on each span be $f(W_1, w_1)$ and $f(W_2, w_2)$ and their centroidal distances from E and G be $\bar{x}_1$ and $\bar{x}_2$ respectively. Then

Fig. 8.17

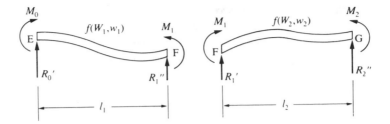

taking moments about E for span EF

$$R_1''l_1 + M_1 - M_0 - f(W_1, w_1)\bar{x}_1 = 0$$

$$R_1'' = \frac{1}{l_1} \{f(W_1, w_1)\bar{x}_1 - (M_1 - M_0)\} \tag{8.39}$$

Similarly for span FG take moments about G

$$R_1'l_2 + M_1 - M_2 - f(W_2, w_2)\bar{x}_2 = 0$$

$$R_1' = \frac{1}{l_2} \{f(W_2, w_2)\bar{x}_2 - (M_1 - M_2)\} \tag{8.40}$$

Now

$$R_1 = R_1'' + R_1' \tag{8.41}$$

therefore

$$R_1 = \frac{f(W_1, w_1)\bar{x}_1}{l_1} + \frac{f(W_2, w_2)\bar{x}_2}{l_2} - \frac{M_1 - M_0}{l_1} - \frac{(M_1 - M_2)}{l_2} \tag{8.42}$$

R_0' and R_2'' may be found from vertical equilibrium for each of the spans EF and FG and the above procedure can then be repeated for other adjacent spans to determine all the reactions. It then becomes a simple matter to plot the shear force diagram for the beam.

EXAMPLE 8.1

Draw the bending moment and shearing force diagrams for a continuous beam which is supported at three points at the same level, but free at its extremities. The spans are 15.2 m and 10.6 m; the 5.2 m span supports two loads of values 8900 N and 4450 N distant 6 m and 12 m respectively from a free end, and the 10.6 m span is loaded uniformly with 1459 N per metre run.

SOLUTION

The maximum ordinate of the free bending moment diagram (Fig. 8.18) for span BC is

$$\frac{wL^2}{8} = \frac{1459 \times 10.6 \times 10.6}{8} = 20\,500 \text{ Nm.}$$

Fig. 8.18

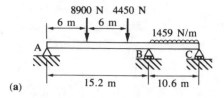

(a)

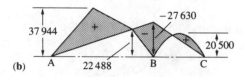

(b)

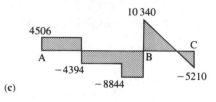

(c)

Taking moments about B for the free span AB

$$15.2\, R'_A - 8900 \times 9.2 - 4450 \times 3.2 = 0$$

$$R'_A = 6324$$

Bending moment at 8900 N load $= 6324 \times 6$
$$= 37\,944$$

Bending moment at 4450 N load $= (6324 \times 12) - (8900 \times 6)$
$$= 22488$$

Referring to Fig. 8.18 $M_A = 0$ and $M_C = 0$, since the ends are free. Then eqn. (8.38),

$$M_A l_1 + 2M_B(l_1 + l_2) + M_C l_2 = -6\left(\frac{S_1 x_1}{l_1} + \frac{S_2 x_2}{l_2}\right)$$

becomes

$$2M_B(l_1 + l_2) = -6\left(\frac{S_1 x_1}{l_1} + \frac{S_2 x_2}{l_2}\right)$$

$$\frac{S_1 x_1}{l_1} = \frac{1}{15.2}\left\{\left(\frac{37\,944 \times 6}{2} \times \frac{6 \times 2}{3}\right) + (22\,488 \times 6 \times 9)\right.$$

$$\left. + \left(\frac{15\,465 \times 6 \times 8}{2}\right) + \left(\frac{22\,488 \times 3.2}{2} \times 13.07\right)\right\}$$

$$= 165\,190 \text{ Nm}^2$$

$$\frac{S_2 x_2}{l_2} = \frac{1}{10.6}\left\{\frac{20\,500 \times 10.6 \times 2}{3} \times 5.3\right\} = 72\,433 \text{ Nm}^2$$

$$2 \times 25.8 M_B = -6(165\,190 + 72\,433)$$

$$M_B = -27\,630 \text{ Nm}.$$

The resultant bending moment diagram is shown in Fig. 8.18(b). Next we must find the reactions at A, B and C. Taking moments about B for span AB

$$15.2R_A - 8900 \times 9.2 - 4450 \times 3.2 + 27\,630 = 0$$

$$R_A = 4506\,\text{N}$$

Taking moments about B for span BC

$$-10.6R_C + 15\,465 \times 5.3 - 2\,7630 = 0$$

$$R_C = 5126\,\text{N}$$

Vertical equilibrium for the whole beam gives

$$-4506 - R_B - 5126 + 8900 + 4450 + 15\,465 = 0$$

$$R_B = 19183\,\text{N}$$

Note that this could also have been obtained by taking moments about A for span AB to get R_B'' and taking moments about C for BC to get R_B' and then $R_B = R_B'' + R_B'$.

The shearing force diagram can now be determined and is shown in Fig. 8.17.

EXAMPLE 8.2

A cantilever carries a uniformly distributed load w over its span L. Find the correct location for a prop which is to carry half the total load and to have the point of support at the same level as the fixed end as shown in Fig. 8.19.

Fig. 8.19

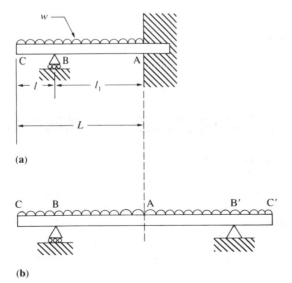

(a)

(b)

SOLUTION

Let the constraint be A, the prop B and the free end C. Then, imagining a similar span to the right of A, we can get the value of M_A.

$$M'_B = M_B = -\frac{wl^2}{2}$$

where $l = BC$ and w is the evenly distributed load. Then, if $AB = l_1$, $AC = l_1 + l = L$.

In eqn. (8.38),

$$S_1 = S_2 = \frac{wl_1^2}{8} \times \frac{2}{3} l_1 \quad \text{and} \quad x_1 = x_2 = \frac{l_1}{2}$$

Therefore

$$M_B l_1 + 2M_A(l_1 + l_1) + M'_B l_1 = -6\left[\frac{1}{l_1}\left(\frac{wl_1^2}{8} \times \frac{2}{3} l_1 \times \frac{l_1}{2}\right)\right]2$$

Substituting for M_B and M'_B,

$$-wl^2 + 4M_A = -\frac{wl_1^2}{2}$$

Therefore

$$M_A = -\left[\frac{wl_1^2}{2} - wl^2\right]\frac{1}{4}$$

Now R_B is equal to half the total load, $w(l_1 + l)$, and also, from eqn. 8.42,

$$R_B = \frac{wl_1}{2} - \frac{(M_B - M_A)}{l_1} + \frac{wl}{2} - \frac{(M_B - M_C)}{l} = \frac{w(l_1 + l)}{2}$$

So that $M_A - M_B(l + l_1) = 0$, since $M_C = 0$

Substituting for M_A and M_B and simplifying gives

$$l_1^2 - 4ll_1 - 6l^2 = 0$$

Therefore

$$l_1 = 5.16l = 0.840L$$

SUMMARY

The first consideration in relation to this chapter is the recognition of a statically-indeterminate beam situation, i.e. the appreciation that the support reactions cannot be determined from the two equilibrium equations for the beam. The next step is to choose one of the methods of solution that have been presented. This will depend on the nature of the problem, the availability of computation, etc. Double integration can be used for any case, but, for example, superposition is best suited to a convenient breakdown into simple separate elements. Moment-area can be quite a convenient method for some continuous beam situations as an alternative to double integration. Another important aspect is the correct application of the boundary conditions at support points. It is essential *in practice* to realize that supports which are supposed to be at the same level may not be so, and hence additional bending stresses can be induced. A so-called fixed or built-in support may not ensure zero deflection, or zero slope for that matter. It is also quite useful to remember that if points of contraflexure (zero B.M.) occur, then it may be more efficient to make a beam which has hinged connections at these points, for example in a multi-span bridge such as the Forth Bridge.

PROBLEMS

8.1 An I-section beam of length 5 m is built in horizontally and at the same level at each end. If the maximum allowable stress is $90\,\text{MN/m}^2$, determine what could be the maximum uniformly-distributed load. The depth of section is 400 mm, and the second moment of area is $3 \times 10^{-4}\,\text{m}^4$.

8.2 A horizontal beam of length L is fixed at one end and simply-supported at the other. A uniformly-distributed load w extends from the fixed end to mid-span. Determine all the reactions and the deflection at mid-span.

8.3 A bar of length L and flexural stiffness EI is built in horizontally and at the same level at each end. A clockwise couple M is applied at mid-span. Find the slope at this point and the deflection curve for the bar.

8.4 A beam of 20 m length is fixed at the left end and simply-supported at the right end, where a clockwise couple of 210 kN-m is applied. A uniformly distributed load of 4 kN/m extends from the fixed support to mid span, and a concentrated load of 60 kN is applied at 5 m from the right-hand end. Calculate the maximum deflection. $EI = 140\,\text{MNm}^2$.

8.5 A beam is fixed horizontally at the left-hand end, A, and is simply-supported at the same level at B and C, distant 4 m and 6 m from A. A uniformly-distributed load of 2.5 kN/m is carried between B and C. Determine the fixing moments and reactions by the double-integration method.

8.6 A beam in a small bridge deck has been damaged at a particular point and temporarily a prop is to be placed beneath that point to carry half the concentrated load occurring at that position. The beam is 4 m long and has both ends built in at the same level as shown in Fig. 8.20 and the concentrated load F occurs at 3 m from the left wall. The prop is to be a circular bar. Calculate its diameter so that, as stated, the beam and column each carry half the applied load. The second moment of area for the beam is $30 \times 10^7\,\text{mm}^4$ and the modulus of the beam material is three times the modulus of the column material.

Fig. 8.20

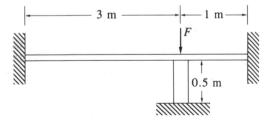

8.7 Use the principle of superposition to find the wall reactions and moments (a) in Problem 8.3. (b) in Problem 8.2.

8.8 Draw the bending-moment and shear-force diagrams for the beam loaded as shown in Fig. 8.21.

Fig. 8.21

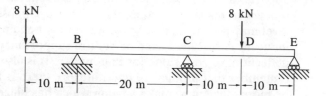

8.9 The beam AE is supported at four points as in Fig. 8.22. Draw the bending-moment and shear-force diagrams for the beam (*a*) if the beam is pinned at A and (*b*) if the beam just rests on A.

Fig. 8.22

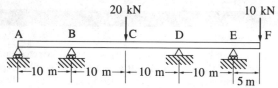

8.10 A continuous beam having three spans each of 10 m has the four simple supports A, B, C and D at the same level. Span AB carries a uniform load of 5 kN/m, and a concentrated load of 20 kN acts at 4 m to the left of D. Sketch the shear-force and bending-moment diagrams.

8.11 A horizontal beam is built in at one end and supported at three points as in Fig. 8.23. If it is subjected to point loads W at the middle of the three spans AB, BC and CD, calculate the length of each span so that $M_A = M_B = M_C = \dfrac{WL_1}{8}$.

Fig. 8.23

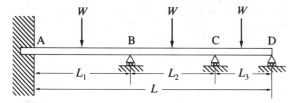

Chapter 9

Energy theorems

INTRODUCTION

The concept of stored elastic strain energy was introduced in Chapter 3, page 66; now in this chapter the topic will be developed on a broader front. For example, the performance of helical springs depends on the strain energy stored in wire that is bent and twisted. Then there are important energy theorems such as virtual work and strain and complementary energy functions which are valuable for the solution of beam deflections and structural displacements.

ELASTIC STRAIN ENERGY: NORMAL STRESS AND SHEAR STRESS

On pages 67 and 68 expressions were derived for the elastic strain energy stored in a material per unit volume under the action of normal stress or shear stress. It represented an area under the appropriate stress–strain curve, as shown in Figs. 9.1(a) and (b), and was given by

$$U = \frac{1}{2}\frac{\sigma^2}{E} \text{ per unit volume for normal stress}$$

and

$$U_s = \frac{1}{2}\frac{\tau^2}{G} \text{ per unit volume for shear stress}$$

$$(9.1)$$

STRAIN ENERGY IN TORSION

SOLID CIRCULAR SHAFT

If a solid shaft of radius R and length L is subjected to a torque which increases gradually from zero to a value T, and θ is the corresponding

Fig. 9.1

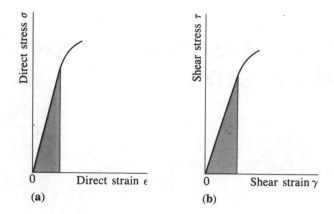

(a)   (b)

angle of twist, then the energy stored in the shaft is

$$U = \tfrac{1}{2}T\theta = \tfrac{1}{2}\frac{T^2 L}{JG}$$

Substituting for T and J in terms of $\hat{\tau}$ and R gives

$$U = \frac{\hat{\tau}^2}{4G}\pi R^2 L$$

Hence

$$U = \frac{\hat{\tau}^2}{4G} \times \text{volume of shaft} \tag{9.2}$$

HOLLOW CIRCULAR SHAFT

In the case of a hollow shaft of outer radius R and inner radius R', using the procedure above it is found that

$$U = \frac{\hat{\tau}^2}{4G}\frac{(R^2 + R'^2)}{R^2} \times \text{volume of shaft} \tag{9.3}$$

STRAIN ENERGY IN BENDING

The change in slope $\delta\theta$ between two cross-sections δs apart was shown in Chapter 7 to be related to bending moment by the relationship

$$\frac{\delta\theta}{\delta s} = \frac{M}{EI}$$

The work done by the moments acting on the two sections is $\tfrac{1}{2}M\,\delta\theta$; therefore

$$\text{Stored energy, } \delta U = \tfrac{1}{2}M\frac{M\,\delta s}{EI}$$

$$\text{Total strain energy, } U = \int \frac{M^2}{2EI}\,ds \tag{9.4}$$

between required limits of length along the beam.

BEAM SIMPLY SUPPORTED WITH LOAD AT MID-SPAN (FIG. 9.2)

Since the bending-moment relationship is discontinuous over the length of the beam, it is necessary to split the integral of eqn. (9.4) into two parts. Thus for $0 < x < \frac{1}{2}L$, $M = \frac{1}{2}Wx$, and

$$\text{Strain energy up to the load} = \int_0^{L/2} \frac{W^2 x^2}{8EI}\,\mathrm{d}x \tag{9.5}$$

Fig. 9.2

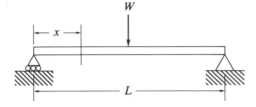

When $\frac{1}{2}L < x < L$, $M = \frac{1}{2}Wx - W(x - \frac{1}{2}L) = \frac{1}{2}W(L - x)$, and

$$\text{Strain energy stored in second portion of beam} = \int_{L/2}^{L} \frac{W^2}{8EI}(L - x)^2\,\mathrm{d}x$$

Therefore the total strain energy is

$$U = \int_0^{L/2} \frac{W^2 x^2}{8EI}\,\mathrm{d}x + \int_{L/2}^{L} \frac{W^2(L - x)^2}{8EI}\,\mathrm{d}x$$

$$U = \frac{W^2 L^3}{192EI} + \frac{W^2 L^3}{192EI}$$

$$U = \frac{W^2 L^3}{96EI} \tag{9.6}$$

Also

$$M_{max} = \frac{WL}{4} \quad \text{or} \quad W = \frac{4M_{max}}{L} = \frac{4\sigma_{max} I}{L y_{max}}$$

Therefore, substituting this expression in eqn. (9.6), the total strain energy in terms of the maximum bending stress is

$$U = \frac{1}{6}\frac{IL\sigma_{max}^2}{y_{max}^2 E}$$

In this particular problem, since the load is at mid-span, the total strain energy could have been obtained by doubling the first integral (*see* eqn. (9.5)). The total strain energy could also have been obtained by considering the deflection of the beam under the load; hence

$$\text{Work done} = \frac{1}{2}Wv_{max}$$

Therefore, with $v_{max} = WL^3/48EI$,

$$\text{Strain energy, } U = \frac{1}{2}W\frac{WL^3}{48EI} = \frac{W^2 L^3}{96EI}$$

CANTILEVER WITH UNIFORMLY-DISTRIBUTED LOADING, (FIG. 9.3)

Here

$$M = \frac{wx^2}{2}$$

Therefore eqn. (9.4) becomes

$$U = \frac{1}{EI}\int_0^L \frac{w^2x^4}{8}\,\mathrm{d}x$$

$$U = \frac{w^2L^5}{40EI}$$

(9.7)

Fig. 9.3

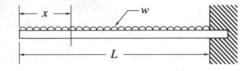

Since

$$M_{max} = \tfrac{1}{2}wL^2 = \frac{\sigma_{max}I}{y_{max}}$$

$$w = \frac{2\sigma_{max}I}{yL^2}$$

Total strain energy, $U = \dfrac{1}{10}\dfrac{IL}{y_{max}^2 E}\cdot\sigma_{max}^2$

BEAM IN PURE BENDING (FIG. 9.4)

If M is constant, eqn. (9.4) becomes

$$U = \frac{M^2}{2EI}\int_0^L \mathrm{d}x$$

Therefore

Total strain energy, $U = \dfrac{M^2L}{2EI}$

which in terms of stress is

$$U = \frac{1}{2}\frac{IL}{y_{max}^2 E}\sigma_{max}^2$$

Fig. 9.4

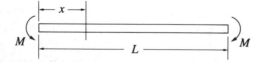

HELICAL SPRINGS

Springs are directly concerned with theories of torsion, bending and strain energy and form an excellent example of their application. There are comparatively few machines which do not incorporate a spring to assist in their operation. The principal function of a spring is to absorb energy, store it for a long or short period, and then return it to the surrounding material. Two extremes of operation are found in a watch and on an engine valve. In the former case energy is stored for a long period and in the latter case the process is very rapid. The energy stored per unit volume by a spring up to a certain limiting stress is termed the *resilience*. The force required to produce a unit deformation is called the *stiffness* of a spring.

The geometry of a helical spring is shown in Fig. 9.5. The centre-line of the wire forming the spring is a helix on a cylindrical surface such that the helix angle is α. Helical springs are designed and manufactured in two categories: close-coiled and open-coiled. In the former the helix angle α is very small and the coils almost touch each other. In the second case the helix angle is larger and the coils are spaced farther apart.

Fig. 9.5

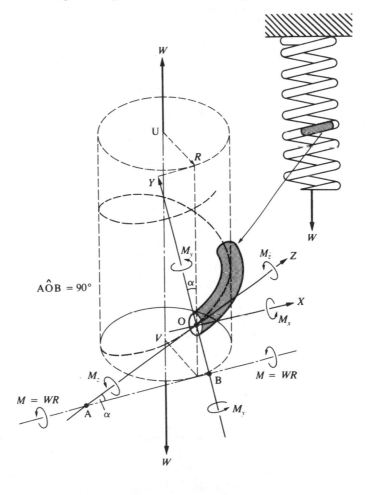

AXIAL LOAD ON OPEN-COILED SPRING

The most common form of loading on a spring is a force W acting along the central axis. Since this force acts at a distance R, the coil radius, from the axis of the wire, there will be torque and bending moment set up about the mutually perpendicular axes OX, OY and OZ on the cross-section at O in Fig. 9.5.

$M_z = WR \cos \alpha$ (causing torsion on the wire) $= T$

$M_y = WR \sin \alpha$ (causing bending about the y-axis and a change in R)

$M_x = 0$

In addition, cross-sections of the wire are subjected to a transverse shear force $W \cos \alpha$ and an axial force $W \sin \alpha$. The stresses due to these forces are considerably smaller than those due to torsion and bending and are generally neglected.

The work done in deflecting the spring δ by the axial load W is $\frac{1}{2}W\delta$, and the stored energies due to torsion and bending are $\frac{1}{2}T\theta$ and $\frac{1}{2}M\phi$, where θ is the angular twist of the wire and ϕ is the change in slope of the wire. Therefore

$$\tfrac{1}{2}W\delta = \tfrac{1}{2}T\theta + \tfrac{1}{2}M\phi$$
$$= \tfrac{1}{2}(WR \cos \alpha)\theta + \tfrac{1}{2}(WR \sin \alpha)\phi$$

Hence

$$\delta = R\left[\cos \alpha \frac{Tl}{JG} + \sin \alpha \frac{Ml}{EI}\right]$$

$$= WR^2 l\left[\frac{\cos^2 \alpha}{JG} + \frac{\sin^2 \alpha}{EI}\right] \qquad (9.8)$$

The length of wire in the spring is $2\pi n R \sec \alpha$, where n is the number of complete coils.

$$J = \frac{\pi d^4}{32} \quad \text{and} \quad I = \frac{\pi d^4}{64}$$

where d is the wire diameter: thus

$$\delta = \frac{64 WR^3 n \sec \alpha}{d^4}\left[\frac{\cos^2 \alpha}{G} + \frac{2\sin^2 \alpha}{E}\right] \qquad (9.9)$$

AXIAL LOAD ON CLOSE-COILED SPRING

The helix angle α is very small in the close-coiled spring so that $\sin \alpha \to 0$ and $\cos \alpha \to 1$, and eqn. (9.9) reduces to

$$\delta = \frac{64 WR^3 n}{Gd^4} \qquad (9.10)$$

EXAMPLE 9.1

A close-coiled helical spring is to have a stiffness of 1 kN/m compression, a maximum load of 50 N, and a maximum shearing stress of 120 MN/m². The solid length of the spring, i.e. when the coils are touching, is to be 45 mm. Find the diameter of the wire, the mean radius of the coils and the number of coils required. Shear modulus $G = 82$ GN/m².

SOLUTION

$$\text{Stiffness} = \frac{W}{\delta} = 1000 = \frac{Gd^4}{64R^3 n}$$

$$\text{Maximum torque } T = \frac{\tau \pi d^3}{16} = 120 \times 10^6 \frac{\pi d^3}{16} = 50R$$

Hence

$$R = \frac{120 \times 10^6}{800} \pi d^3$$

Closed length of spring $= nd = 0.045$

Substituting for n and R in the equation for stiffness above,

$$1000 = \frac{Gd^4 \times d}{64 \times (15 \times 10^4 \pi d^3)^3 \times 0.045}$$

$$d^4 = \frac{82 \times 10^9}{64 \times (15\pi)^3 \times 10^{12} \times 45} = 273 \times 10^{-12}$$

Thus $d = 0.004\,06$ m $= 4.06$ mm; $R = 31.6$ mm; and $n = 11$ coils.

EXAMPLE 9.2

Compare the stiffness of a close-coiled spring with that of an open-coiled spring of helix angle 30°. The two springs are made of the same steel and have the same coil radius, number of coils and wire diameter, and are subjected to axial loading. $E = 2.5G$.

SOLUTION

From eqn. (9.10), the stiffness is,

$$k_{cc} = \frac{W}{\delta} = \frac{Gd^4}{64R^3 n}$$

From eqn. (9.9) the stiffness is,

$$k_{oc} = \frac{W}{\delta} = \frac{d^4}{64R^3 n \sec \alpha \left[\dfrac{\cos^2 \alpha}{G} + \dfrac{2 \sin^2 \alpha}{E} \right]}$$

$$\frac{k_{cc}}{k_{oc}} = G \sec \alpha \left[\frac{\cos^2 \alpha}{G} + \frac{2 \sin^2 \alpha}{E} \right]$$

$$= \sec \alpha \left[\cos^2 \alpha + \frac{2 \sin^2 \alpha}{2.5} \right]$$

$$= \frac{2}{\sqrt{3}} \left[\frac{3}{4} + \frac{1}{5} \right] = 1.1$$

AXIAL TORQUE ON OPEN-COILED SPRING

The other type of loading on a spring which is of interest is that where the spring is subjected to a torque about the central axis. The resolved components of the axial torque $\bar{T}$ about any cross-section are $\bar{T} \sin \alpha$ about the axis of the wire and $\bar{T} \cos \alpha$ changing the curvature of the coils. If one end of the spring moves round the longitudinal axis an amount ψ relative to the other end due to the torque $\bar{T}$ then the work done is $\frac{1}{2}\bar{T}\psi$ and the stored energies are $\frac{1}{2}\bar{T}\theta \sin \alpha$ and $\frac{1}{2}\bar{T}\phi \cos \alpha$, so that

$$\psi = \theta \sin \alpha + \phi \cos \alpha$$

$$= \frac{Tl}{JG} \sin \alpha + \frac{Ml}{EI} \cos \alpha$$

$$= \bar{T}\pi n R \sec \alpha \left[\frac{\sin^2 \alpha}{JG} + \frac{\cos^2 \alpha}{EI} \right]$$

$$= \frac{128\bar{T}nR \sec \alpha}{d^4} \left[\frac{\sin^2 \alpha}{2G} + \frac{\cos^2 \alpha}{E} \right] \tag{9.11}$$

AXIAL TORQUE ON CLOSE-COILED SPRING

Putting $\sec \alpha = \cos \alpha = 1$ and $\sin \alpha = 0$ in eqn. (9.11) gives

$$\psi = \frac{128\bar{T}Rn}{Ed^4} \tag{9.12}$$

SHEAR DEFLECTION OF BEAMS

A deflection other than that due to bending moment occurs in beams owing to the shearing forces on transverse sections. This deflection may be found approximately from strain energy principles and by making use of the equation for shear stress at a point in the transverse section of a beam. For the majority of beams, where the span L is large compared to the cross-section of the beam, it is seen that the deflection due to shear is negligible in comparison with that due to bending.

CANTILEVER WITH LOAD AT FREE END (FIG. 9.6)

Assume the section to be rectangular, of breadth b and depth d, and the total length of the beam to be L. If v_s is the deflection, due to shear, at the free end, then

Work done by load $= \frac{1}{2}Wv_s$

The shear stress at a distance y from the neutral axis is obtained from

Fig. 9.6

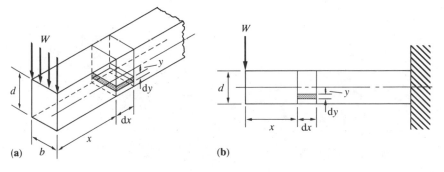

(a)   (b)

eqn. (6.35)

$$\tau = \frac{6W}{bd^3}\left(\frac{d^2}{4} - y^2\right)$$

where the shear force $Q = W$. Also, if dy is the height of the strip in the direction of the depth of the beam, and we consider a small portion of the beam of length dx and section bdy, we have

$$\text{Strain energy in strip} = \frac{1}{2}\frac{\tau^2}{G}\,dxb\,dy$$

and substituting for τ, from above

$$\text{Strain energy} = \frac{1}{2G}\,dx\,\frac{36W^2}{b^2d^6}\left(\frac{d^4}{16} - \frac{d^2y^2}{2} + y^4\right)b\,dy$$

Therefore the total strain energy for the piece of beam of length dx is

$$U = \frac{18W^2dx}{bd^6G}\int_{-d/2}^{+d/2}\left(\frac{d^4}{16} - \frac{d^2y^2}{2} + y^4\right)dy$$

$$= \frac{18W^2dx}{bd^6G}\left[\frac{d^4y}{16} - \frac{d^2y^3}{6} + \frac{y^5}{5}\right]_{-d/2}^{d/2}$$

$$= \frac{3}{5}\frac{W^2dx}{(bd)G} \tag{9.13}$$

The strain energy for the whole beam of length L is

$$U_t = \frac{3}{5}\frac{W^2}{bdG}\int_0^L dx = \frac{3}{5}\frac{W^2L}{bdG}$$

Equating the strain energy to the work done by W,

$$\tfrac{1}{2}Wv_s = \frac{3}{5}\frac{W^2L}{bdG}$$

Therefore

$$v_s = \frac{6}{5}\frac{WL}{bdG} \tag{9.14}$$

Thus the total deflection at the free end due to bending and shear is

$$v = v_b + v_s$$

$$= \frac{1}{3}\frac{WL^3}{EI} + \frac{6}{5}\frac{WL}{bdG}$$

SIMPLY-SUPPORTED BEAM WITH POINT LOAD AT MID-SPAN

If we treat each half of the span as a cantilever, L in eqn. (9.14) becomes $\tfrac{1}{2}L$, W becomes $\tfrac{1}{2}W$, and

$$v_s = \frac{3}{10}\frac{WL}{bdG}$$

Therefore

$$\text{Total deflection} = \frac{1}{48}\frac{WL^3}{EI} + \frac{3}{10}\frac{WL}{bdG} \tag{9.15}$$

CANTILEVER WITH UNIFORMLY DISTRIBUTED LOAD

On page 237 the shearing force Q is constant along the beam, but for a uniformly-loaded cantilever at distance x from the fixed end, the shearing force is $w(L-x)$; hence eqn. (9.13) becomes

$$U = \frac{3}{5}\frac{w^2(L-x)^2}{bdG}\,dx$$

The shearing force acting on a piece of length dx is $w(L-x)$, and the external work done by this force is $\frac{1}{2}w(L-x)\,dv_s$, where dv_s is the deflection of the piece due to shear. Equating the external work done to the strain energy,

$$\tfrac{1}{2}w(L-x)\,dv_s = \frac{3}{5}\frac{w^2(L-x)^2}{bdG}\,dx$$

$$dv_s = \frac{6}{5}\frac{w(L-x)}{bdG}\,dx$$

Therefore

$$v_s = \frac{6}{5}\frac{w}{bdG}\int_0^L (L-x)\,dx = \frac{3}{5}\frac{wL^2}{bdG} \tag{9.16}$$

$$\text{Total deflection} = \frac{1}{8}\frac{wL^4}{EI} + \frac{3}{5}\frac{wL^2}{bdG}$$

BEAM SIMPLY SUPPORTED WITH DISTRIBUTED LOADING

Treating each half of the span as a cantilever, L becomes $\frac{1}{2}L$ in eqn. (9.16). Therefore

$$v_s = \frac{3}{20}\frac{wL^2}{bdG}$$

$$\text{Total deflection} = \frac{5}{384}\frac{wL^4}{EI} + \frac{3}{20}\frac{wL^2}{bdG} \tag{9.17}$$

EXAMPLE 9.3

A beam of 3 m length is simply-supported at each end and is subjected to a couple of 9 kN-m at a point B, 2 m from the left end as shown in Fig. 9.7. Determine the slope at B. $EI = 30$ kN-m^2.

Fig. 9.7

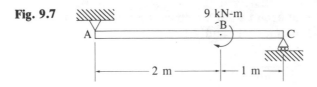

SOLUTION

The reactions at A and C are $\bar{M}/L = 9000/3 = 3$ kN.

When $0 < x < 2$ $\quad M = \dfrac{\bar{M}x}{L}$

When $2 < x < L$ $\quad M = \dfrac{\bar{M}}{L}(x - L)$

The strain energy stored is

$$U = \frac{1}{2EI}\int_0^2 \left(\frac{\bar{M}x}{L}\right)^2 dx + \frac{1}{2EI}\int_2^3 \left[\frac{\bar{M}}{L}(x - L)\right]^2 dx$$

$$= \frac{1}{60 \times 10^3}\left(\frac{9000}{3}\right)^2 \left\{\left[\frac{x^3}{3}\right]_0^2 + \left[\frac{(x - 3)^3}{3}\right]_2^3\right\}$$

$$= \frac{3}{60 \times 10^3}\left(\frac{9000}{3}\right)^2$$

The work done at B is

$$\tfrac{1}{2}\bar{M}\theta = \frac{9000}{2}\theta$$

Therefore

$$\frac{9000}{2}\theta = \frac{3}{60 \times 10^3}\left(\frac{9000}{3}\right)^2$$

and

$$\theta = 0.1 \text{ rad}$$

VIRTUAL WORK

The principle of *virtual work* is one of the most powerful tools in structural analysis. The principle may be stated as follows: if a system of forces acts on a particle which is in statical equilibrium and the particle is given any virtual displacement then the net work done by the forces is zero.

A virtual displacement is any arbitrary displacement which is mathematically conceived and does not actually have to take place, but must be geometrically possible. The forces must not only be in equilibrium but are assumed to remain constant and parallel to their original lines of action.

If a body is subjected to the system of forces shown in Fig. 9.8 and is

Fig. 9.8

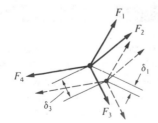

then given an arbitrary virtual displacement for which the corresponding displacements in the directions of the forces are δ_1, δ_2, δ_3 and δ_4, then static equilibrium exists for the system of forces if

$$F_1\delta_1 + F_2\delta_2 + F_3\delta_3 + F_4\delta_4 = 0$$

That is to say, if a body, subjected to a system of forces, is given any virtual displacement, the net work done by the forces must be zero for static equilibrium of the system.

Let the resultant force of the above system be P and assume a virtual displacement Δ in the direction of P; then for static equilibrium $P\Delta$ must be zero. Since Δ need not be zero, it follows that P must be. Thus the resultant of a system of forces in equilibrium is zero.

We can now make use of the above principle in deriving energy solutions for deflections. Consider the simple framework in Fig. 9.9 acted upon by forces F_1 and F_2. Let the displacements at the joints in the direction of the forces be δ_1 and δ_2. The internal reactions and deformations of the members of the frame are P_1, P_2, etc., and Δ_1, Δ_2, etc., respectively. Then, from the principle of virtual work,

$$F_1\delta_1 + F_2\delta_2 = P_1\Delta_1 + P_2\Delta_2 + \ldots P_n\Delta_n$$

or

$$\sum_j F\delta = \sum_n P\Delta \tag{9.18}$$

Thus the work done by external forces at the joints equals the internal work resulting from the tensions in the members. The summations in eqn. (9.18) cover all joints j, and all members n, for static equilibrium.

The use of the principle of virtual work to determine displacements and forces in frameworks is illustrated in Chapter 10.

Fig. 9.9

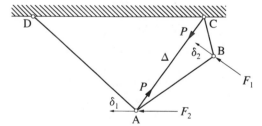

STRAIN AND COMPLEMENTARY ENERGY SOLUTIONS FOR DEFLECTIONS

Energy functions may be very usefully employed in the determination of deflections of frameworks, beams, shells, etc. The methods depend on the principle of virtual work as developed above. Now let us suppose that there is a small change in displacement at joint A of an amount $\delta\delta_1$, δ_2 remaining constant, which results in changes $\delta\Delta_1$, $\delta\Delta_2$, etc., in the members for compatibility. Then

$$F_1(\delta_1 + \delta\delta_1) + F_2\delta_2 = P_1(\Delta_1 + \delta\Delta_1) + P_2(\Delta_2 + \delta\Delta_2)$$
$$+ \ldots P_n(\Delta_n + \delta\Delta_n)$$
$$= \sum_n P(\Delta + \delta\Delta) \tag{9.19}$$

Subtracting eqn. (9.18) from eqn. (9.19)

$$F_1\delta\delta_1 = \sum_n P\,\delta\Delta \tag{9.20}$$

But $P\,\delta\Delta$ is the increment of strain energy stored in a member of the system due to the increments of deformation $\delta\Delta$ caused by the change in displacement $\delta\delta_1$. Therefore for the system

$$\sum_n P\,\delta\Delta = \delta U$$

or

$$F_1\delta\delta_1 = \delta U \tag{9.21}$$

Thus, for an infinitely small change in displacement,

$$F_1 = \frac{\partial U}{\partial \delta_1} \tag{9.22}$$

By a similar argument we have that

$$F_2 = \frac{\partial U}{\partial \delta_2}$$

Thus the external force on a member is given by the partial derivative of the strain energy with respect to the displacement at the point of application of and in the direction of the force.

We now return to the original proposition, and instead of changing the displacement δ_1, we change the force F_1 by an amount δF_1, keeping F_2 constant; then there will be a reaction in the system causing changes δP_1, δP_2, etc., in the internal forces in the members. Now, by the principle of virtual work, we have

$$(F_1 + \delta F_1)\delta_1 + F_2\delta_2 = (P_1 + \delta P_1)\Delta_1 + (P_2 + \delta P_2)\Delta_2$$
$$+ \ldots (P_n + \delta P_n)\Delta_n$$
$$= \sum_n (P + \delta P)\Delta \tag{9.23}$$

Subtracting eqn. (9.18) from eqn. (9.23)

$$\delta F_1\delta_1 = \sum_n \delta P\Delta \tag{9.24}$$

Now considering Fig. 9.10(a) or (b), the shaded area $P\,\delta\Delta$ is the increment of strain energy δU below the load-deformation curve used in eqn. (9.20). The shaded area above the load-deformation curve represents $\Delta\,\delta P$ in eqn. (9.24); this is termed the *complementary energy*

Fig. 9.10

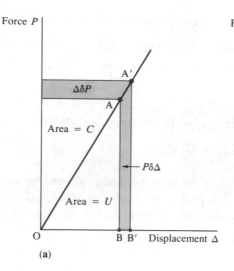

(a)

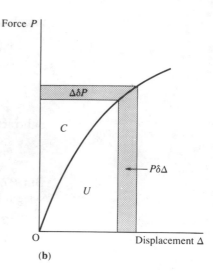

(b)

and is denoted by C; thus

$$\sum_n \delta P \Delta = \delta C$$

and therefore

$$\delta F_1 \delta_1 = \delta C \tag{9.25}$$

For an infinitely small change in the force,

$$\delta_1 = \frac{\partial C}{\partial F_1} \tag{9.26}$$

Thus the deflection at a point on a member in the direction of a force applied at that point is given by the partial derivative of the complementary energy with respect to the external force at the point.

The above energy theorems provide a most useful method of attack on many structural analysis problems, and examples of their use will be given in the next chapter.

A further point of interest is illustrated in the load-deformation characteristics of Figs. 9.10(a) and (b). The former illustrates linear elasticity for a member or system of members, while the latter represents non-linear elasticity, which can occur in certain frameworks and materials. In both cases the sum of the strain energy and complementary energy is given by the force times the deformation, i.e.

$$U + C = P\Delta \tag{9.27}$$

but in the particular case of linear elasticity,

$$\delta U = \delta C = P \, \delta \Delta = \Delta \, \delta P$$

and

$$U = C = \tfrac{1}{2} P \Delta \tag{9.28}$$

Because of this last relationship we can express displacements in terms of strain energy instead of complementary energy for linear elastic systems. One of the earliest theorems of this form was due to Castigliano (1875) in which it was stated that the partial derivative of the strain energy with respect to a force gives the displacement corresponding to that force, or

$$\frac{\partial U}{\partial P} = \Delta \tag{9.29}$$

This relationship may be proved in the following way. Consider a force P applied to a body giving a displacement Δ. Then the work done or the stored strain energy is equal to OAB which equals $\frac{1}{2}P\Delta$ in Fig. 9.10(a). If an additional force δP is applied giving an additional deformation $\delta\Delta$, then the extra strain energy is

$$\text{BAA}'\text{B}' = P\,\delta\Delta + \tfrac{1}{2}\delta P\,\delta\Delta = \delta U$$

or $\delta U/\delta\Delta = P$, neglecting second-order products. Therefore,

$$\text{Total energy, OA}'\text{B}' = \tfrac{1}{2}P\Delta + P\,\delta\Delta + \tfrac{1}{2}\delta P\,\delta\Delta$$

If both forces had acted simultaneously, the stored strain energy would have been OA$'$B$' = \frac{1}{2}(P + \delta P)(\Delta + \delta\Delta)$. Since work done is independent of the order of application of the forces, we have

$$\tfrac{1}{2}P\Delta + P\,\delta\Delta + \tfrac{1}{2}\delta P\,\delta\Delta = \tfrac{1}{2}(P + \delta P)(\Delta + \delta\Delta)$$

On simplifying, and neglecting small products, we find that

$$P\,\delta\Delta = \Delta\,\delta P \tag{9.30}$$

(i.e. $\delta U = \delta C$ for linear elasticity). Thus, substituting above,

$$\delta U = \Delta\,\delta P + \tfrac{1}{2}\delta P\,\delta\Delta$$

Therefore

$$\frac{\delta U}{\delta P} = \Delta$$

neglecting the second-order term on the right. Hence

$$\frac{\partial U}{\partial P} = \Delta$$

which proves *Castigliano's hypothesis*.

The simplest applications of eqn. (9.29) are related to the deformations in tension and in torsion.

(i) In the case of a bar under a simple tensile force F,

$$U = \frac{F^2 L}{2AE}$$

and

$$\frac{\partial U}{\partial F} = \frac{FL}{AE} = \Delta, \text{ the extension of the bar,}$$

(ii) and in torsion for a torque T,

$$U = \frac{T^2 L}{2GJ}$$

or

$$\frac{\partial U}{\partial T} = \frac{TL}{GJ} = \theta, \text{ the angle of twist.}$$

BENDING DEFLECTION OF BEAMS

The complementary energy function can be used very conveniently to solve for beam deflections, since

$$\frac{\partial C}{\partial F} = \delta$$

or, using the notation for beams,

$$\frac{\partial C}{\partial W} = v$$

where W is a concentrated load whose displacement (beam deflection) is v.

The complementary energy in bending of a small length of beam δx is shown in Fig. 9.11(a), which is the moment–slope relationship. The

Fig. 9.11

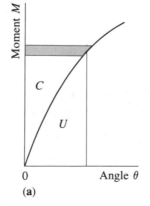

(a)

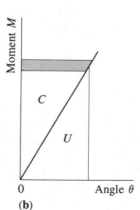

(b)

shaded area is

$$\delta C = \theta \, \delta M$$

or

$$C = \int_0^M \theta \, \mathrm{d}M \tag{9.31}$$

In the most frequently used case,

$$\theta = \frac{M}{EI} \delta x$$

for a linear elastic beam, Fig. 9.11(b). Therefore

$$C = \int_0^M \frac{M}{EI} \, \mathrm{d}x \, \mathrm{d}M = \frac{M^2}{2EI} \, \mathrm{d}x \tag{9.32}$$

For the whole length of beam, therefore,

$$C = \int_0^L \frac{M^2}{2EI} \, \mathrm{d}x \tag{9.33}$$

This result could also have been arrived at by the fact that $U = C$ in a linear elastic system, and it has already been shown, on page 230, that the strain energy is

$$U = \int_0^L \frac{M^2}{2EI} \, \mathrm{d}x$$

It is also apparent that

$$\frac{\partial C}{\partial W} = \frac{\partial U}{\partial W} = v$$

and the latter part of the expression shows that Castigliano's analysis can be applied to beam deflection.

We can solve for the deflection in one of two ways, either

$$v = \frac{\partial U}{\partial W} = \frac{\partial}{\partial W} \left[\int_0^L \frac{[Wf(x)]^2}{2EI} \, \mathrm{d}x \right] \tag{9.34}$$

where $Wf(x) = M$, or we can write

$$v = \frac{\partial U}{\partial W} = \int_0^L \frac{M}{EI} \frac{\partial M}{\partial W} \, \mathrm{d}x \tag{9.35}$$

It is merely a question of whether the bending moment expression in terms of W and x is substituted and the integral evaluated, followed by partial differentiation with respect to W, or the latter is carried out first, substituted in the integral and then evaluated.

CANTILEVER WITH CONCENTRATED LOAD AT FREE END

At any distance x from the free end, the bending moment is $M = Wx$; therefore

$$v = \frac{\partial}{\partial W} \int_0^L \frac{W^2 x^2}{2EI} \, dx = \frac{\partial}{\partial W} \frac{W^2 L^3}{6EI}$$

$$= \frac{WL^3}{3EI} \tag{9.36}$$

Alternatively, $\partial M / \partial W = x$; therefore

$$v = \int_0^L \frac{Wx}{EI} x \, dx = \int_0^L \frac{Wx^2}{EI} \, dx$$

$$= \frac{WL^3}{3EI}$$

BEAM SIMPLY SUPPORTED WITH UNIFORMLY-DISTRIBUTED LOAD

At any distance x from the left-hand end the bending moment is

$$M = \frac{wL}{2} x - \frac{wx^2}{2}$$

and $\partial / \partial W$ of the above is zero, indicating no deflection, which is obviously not true. To get round this difficulty, we introduce an imaginary concentrated load W at some point in the span, let us say mid-span for simplicity. Then, for $0 < x < L/2$,

$$M = \frac{W}{2} x + \frac{wL}{2} x - \frac{wx^2}{2} \tag{9.37}$$

and $\partial M / \partial W = \frac{1}{2} x$; therefore

$$v = 2 \int_0^{L/2} \frac{1}{EI} \left(\frac{W}{2} x + \frac{wL}{2} x - \frac{wx^2}{2} \right) \frac{x}{2} \, dx$$

$$= \frac{WL^3}{48EI} + \frac{5wL^4}{384EI}$$

Putting $W = 0$, we obtain

$$v_{max} = \frac{5}{384} \frac{wL^4}{EI} \tag{9.38}$$

If we require the deflection due to the point load only, we put $w = 0$; then

$$v = \frac{WL^3}{48EI} \tag{9.39}$$

EXAMPLE 9.4

A simply-supported beam (Fig. 9.12) carries a concentrated load at a distance a from the left-hand support and a distance b from the other support. Determine the deflection of the beam underneath the load.

Fig. 9.12

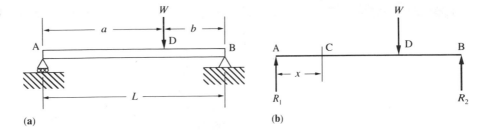

(a)   (b)

SOLUTION

$R_1 = Wb/L$ and the bending moment at C is $(Wb/L)x$. For the portion of beam AD, we have the complementary energy

$$C = \int_0^a \frac{(Wb/L)^2 x^2 \, dx}{2EI} = \int_0^a \frac{W^2 b^2}{L^2 2EI} x^2 \, dx = \frac{W^2 b^2 a^3}{6EIL^2} \qquad (9.40)$$

Similarly, we may write that C for the portion of beam DB is $W^2 a^2 b^3/6EIL^2$; therefore

$$\begin{aligned}\text{Total value of } C \\ \text{for the beam}\end{aligned} = \frac{W^2 b^2 a^3}{6EIL^2} + \frac{W^2 a^2 b^3}{6EIL^2} = \frac{W^2 a^2 b^2}{6EIL^2}(a+b)$$

$$= \frac{W^2 a^2 (L-a)^2}{6EIL} \qquad (9.41)$$

and

$$\text{Deflection underneath load} = \frac{dC}{dW} = \frac{W a^2 (L-a)^2}{3EIL} \qquad (9.42)$$

EXAMPLE 9.5

Determine the vertical and horizontal displacements of the end of the curved member shown in Fig. 9.13.

Fig. 9.13

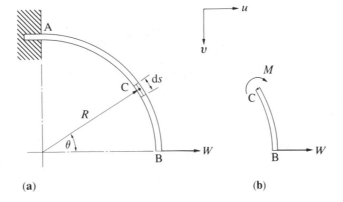

(a)   (b)

SOLUTION

Considering first the displacement u in the direction of the load,

$$U = \int_0^{\pi/2} \frac{M^2}{2EI} R \, d\theta$$

$$u = \frac{\partial U}{\partial W} = \int_0^{\pi/2} \frac{M}{EI} \frac{\partial M}{\partial W} R \, d\theta$$

$$M_c = -WR \sin \theta, \qquad \frac{\partial M_c}{\partial W} = -R \sin \theta$$

$$u = \int_0^{\pi/2} - \frac{WR \sin \theta}{EI} (-R \sin \theta) R \, d\theta$$

$$= \int_0^{\pi/2} \frac{WR^3}{EI} \sin^2 \theta \, d\theta$$

$$= \frac{\pi}{4} \frac{WR^3}{EI} \tag{9.43}$$

To find the vertical displacement v, an imaginary vertical force W_0 is applied at B. The bending moment on C will be

$$M_c = -WR \sin \theta + W_0 R (1 - \cos \theta)$$

and

$$\frac{\partial M_0}{\partial W_0} = R(1 - \cos \theta)$$

Putting $W_0 = 0$ in the expression for M_c,

$$v = \frac{\partial U}{\partial W_0} = \int_0^{\pi/2} - \frac{WR}{EI} \sin \theta R (1 - \cos \theta) R \, d\theta$$

$$= \int_0^{\pi/2} - \frac{WR^3}{EI} \sin \theta (1 - \cos \theta) \, d\theta$$

from which

$$v = -\frac{WR^3}{2EI} \tag{9.44}$$

EXAMPLE 9.6

Determine the horizontal deflection of the member shown in Fig. 9.14.

Fig. 9.14

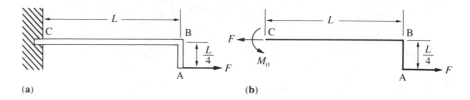

(a)   (b)

SOLUTION

The strain-energy function is scalar; therefore the separate strain-energy quantities for the two parts of the member can be added before proceeding to use the Castigliano theorem.

From A to B

$$M = +Fx \quad \text{and} \quad \frac{\partial M}{\partial F} = +x$$

From B to C

$$M = +\frac{FL}{4} \quad \text{and} \quad \frac{\partial M}{\partial F} = +\frac{L}{4}$$

$$\Delta = \frac{\partial U}{\partial F} = \frac{1}{EI} \int_0^{L/4} Fx\, x\, dx + \frac{1}{EI} \int_0^L \frac{FL}{4}\frac{L}{4}\, dx$$

$$= \frac{1}{EI}\left[\frac{Fx^3}{3}\right]_0^{L/4} + \frac{1}{EI}\left[\frac{FL^2 x}{16}\right]_0^L$$

$$= \frac{13FL^3}{192EI} \tag{9.45}$$

CASTIGLIANO'S SECOND THEOREM

This theorem is of value in finding redundant forces in members and frames. Let the force in a redundant member be R and also let it have an initial lack of fit λ. Then the theorem states that, if the total strain energy for the structure is partially differentiated with respect to the load in a redundant member, then any initial lack of fit of the member is obtained thus:

$$\frac{\partial U}{\partial R} = -\lambda \tag{9.46}$$

If there is no initial lack of fit,

$$\frac{\partial U}{\partial R} = 0 \tag{9.47}$$

which is a condition for a minimum value of the strain energy. This relationship also expresses what is termed the *principle of least work*.

From eqn. (9.47) the redundant force R can be evaluated. This is illustrated in the following example, and others will be found in Chapter 10.

EXAMPLE 9.7

Determine the bending moment for any cross-section of the slender ring shown in Fig. 9.15(a).

Fig. 9.15

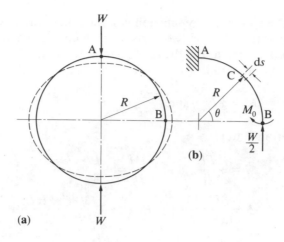

(a)

(b)

In view of the symmetry of the ring, only one quadrant need be considered as shown in Fig. 9.15(b). Cutting the ring at any section the bending moment is given by

$$M_c = M_0 - \frac{WR}{2}(1 - \cos\theta) \tag{9.48}$$

but M_0 is unknown and is a redundancy.
The strain energy is given by

$$U = \int_0^{\pi/2} \frac{1}{2EI}\left[M_0 - \frac{WR}{2}(1 - \cos\theta)\right]^2 R\,d\theta \tag{9.49}$$

and for $\partial U/\partial M_0 = 0$,

$$\frac{1}{EI}\int_0^{\pi/2}\left[M_0 - \frac{WR}{2}(1 - \cos\theta)\right]R\,d\theta = 0$$

from which

$$M_0 = WR\left(\tfrac{1}{2} - \frac{1}{\pi}\right) \tag{9.50}$$

and

$$M_c = WR\left(\tfrac{1}{2}\cos\theta - \frac{1}{\pi}\right) \tag{9.51}$$

THE RECIPROCAL THEOREM

In a linear structural system (Fig. 9.16), the deflection at point 1 due to forces F_1 at point 1 and F_2 at point 2 is, from the principle of superposition,

$$\Delta_1 = \Delta_{11} + \Delta_{12}$$

Fig. 9.16

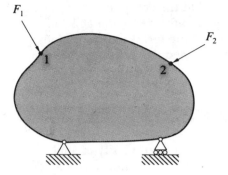

and the deflection at point 2 is

$$\Delta_2 = \Delta_{22} + \Delta_{21}$$

The deflections may be expressed in terms of *flexibility coefficients*, which are the displacements per unit force, as follows

$$\Delta_1 = f_{11}F_1 + f_{12}F_2 \tag{9.52}$$
$$\Delta_2 = f_{21}F_1 + f_{22}F_2 \tag{9.53}$$

If the strain energy of the system due to the application of these forces is U then we may write

$$\Delta_1 = \frac{\partial U}{\partial F_1} \tag{9.54}$$

and

$$\Delta_2 = \frac{\partial U}{\partial F_2} \tag{9.55}$$

Partially differentiating eqns. (9.52) and (9.54) with respect to F_2,

$$\frac{\partial \Delta_1}{\partial F_2} = f_{12}$$

and

$$\frac{\partial^2 U}{\partial F_1 \, \partial F_2} = f_{12} \tag{9.56}$$

Similarly, from eqns. (9.53) and (9.55),

$$\frac{\partial \Delta_2}{\partial F_1} = f_{21}$$

and

$$\frac{\partial^2 U}{\partial F_2 \, \partial F_1} = f_{21} \tag{9.57}$$

From eqns. (9.56) and (9.57),

$$f_{21} = f_{12} \tag{9.58}$$

This shows that the deflection at any point 1 due to a unit force at any point 2 is equal to the deflection at 2 due to a unit force at 1, providing the directions of the forces and deflections coincide in each of the two cases. This is termed the *reciprocal theorem*.

SUMMARY

It can be seen from the diversity of applications, e.g. from coil springs to structures, that energy theorems have a very important part to play in engineering design analysis. Some of the concepts may be a little difficult to grasp at first, especially in relation to a physical appreciation. However, it will be seen, for example, in the next chapter just how valuable these theorems can be in the solution of structural forces and displacements.

PROBLEMS

9.1 Two possible designs of hydraulic cylinder are being considered as shown in Fig. 9.17(*a*) and (*b*). One involves the use of long bolts and the other uses short bolts of the same diameter. On the basis that the bolts are each required to absorb the same amount of strain energy, decide which of the two designs would be the best.

Fig. 9.17

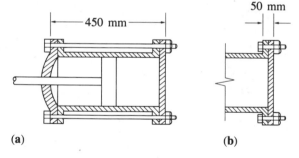

(a) (b)

9.2 Two identical rectangular section steel members are loaded in the following way:

(*a*) one is subjected to an axial tensile force W;
(*b*) the other is supported as a cantilever and subjected to a load W at its free end.

If each member is required to absorb the same strain energy, compare the maximum stresses which this would cause in each case.

9.3 Compare the strain energy stored in a beam simply-supported at each end and carrying a uniformly-distributed load, with that of the same beam carrying a concentrated load at mid-span and having the same value of maximum bending stress.

9.4 A solid shaft carries a flywheel of mass 150 kg and has a radius of gyration of 0.5 m. The shaft is rotating at a steady speed of 60 rev/min when a brake is applied at a point 4 m from the flywheel. Calculate the shaft diameter if the maximum instantaneous shear stress produced is 150 MN/m². Assume that the kinetic energy of the flywheel is taken up as torsional strain energy by the shaft. Neglect the inertia of the shaft. $G = 80$ GN/m².

9.5 A valve is controlled by two concentric close-coiled springs. The outer spring has twelve coils of 25 mm mean diameter, 3 mm wire diameter and 5 mm initial compression when the valve is closed. The free length of the inner spring is 6 mm longer than the outer. If the force required to open the valve 10 mm is 150 N, find the stiffness of the inner spring. If the diameter of the inner spring is 16 mm and the wire diameter is 2 mm, how many coils does it have? $G = 81$ GN/m²

9.6 When an open-coiled spring having ten coils is loaded axially the bending and torsional stresses are 140 MN/m² and 150 MN/m² respectively. Calculate the maximum permissible axial load and wire diameter for a maximum extension of 18 mm if the mean diameter of the coils is eight times the wire diameter. $G = 80$ GN/m², $E = 210$ GN/m²

9.7 A bar of rectangular section 1 m in length is simply-supported at each end and carries a uniformly distributed load. Determine the maximum depth of section so that the deflection due to shear shall not be greater than 2% of the total deflection. $E = 2.6\,G$

9.8 A beam of 4 m length carrying a concentrated load W at mid-span is simply-supported at the right-hand end and is pin-jointed at the same level at the left-hand end to the free end of a horizontal cantilever of lenth 2 m. Use Castigliano's theorem to find the deflection under the load. Both beams have a flexural stiffness EI.

9.9 A U-shaped pipe connecting two vessels has a radius R and a leg-length L. Show that the deflection caused by forces P applied at the free ends and perpendicular to the legs due to thermal expansion is

$$\Delta = \frac{P}{EI}\left\{\frac{2L^3}{3} + \pi L^2 R + \frac{\pi R^3}{2} + 4LR^2\right\}$$

9.10 A split ring of radius R is used as a retainer on a machine shaft, and, in order to install it, it is necessary to apply outward tangential forces F at the split to open up a gap δ. If the flexural stiffness of the cross-section is EI, determine the required value of the forces.

9.11 A curved beam is in the form of a semicircle as shown in Fig. 9.18. The free end is pinned but is restrained to move in the horizontal direction only. If a horizontal force F is applied at the free end show that the horizontal restraining force on the end of the beam is given by $4F/3\pi$.

Fig. 9.18

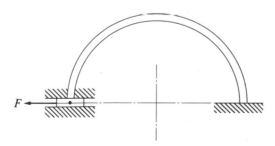

9.12 A curved beam is subjected to a load of 500 N at A, as shown in Fig. 9.19. If the product of Young's modulus and second moment of area for the beam section is given by $EI = 26$ kN-m², calculate the vertical and horizontal deflections at A.

Fig. 9.19

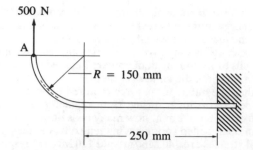

500 N

A

$R = 150$ mm

250 mm

9.13 The frame ABCD in Fig. 9.20 is fixed at A and a horizontal force F is applied at D. B and C are rigidly jointed and the lengths of the members are $AB = CD = h$ and $BC = L$. The relevant second moments of area are I_1 for AB and CD and I_2 for BC. Show that if D moves under load at an angle of 45° to the horizontal then

$$\frac{I_1}{I_2} = \frac{h}{3L}\left\{\frac{3L - 4h}{2h - L}\right\}$$

Ignore the effects of axial load and shear force on the frame.

Fig. 9.20

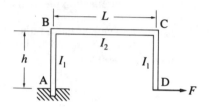

B $\vdash$———— L ————$\dashv$ C

I_2

h

I_1 I_1

A

D ➞ F

Chapter 10

Structural analysis

This chapter is intended to build on the elementary introduction of Chapter 1 in which a "structure" was defined as an assembly of members, which carry axial forces only, and may be either statically determinate or indeterminate in nature. Equilibrium of forces and free-body diagrams were used to establish the forces in the members of basic plane two-dimensional frameworks. The next stage of analysis for design purposes involves the determination of frame displacements to ensure that these are acceptable. Several methods will be explained in relation to statically-determinate structures.

Since in practice some structures are designed to contain "redundant" members and therefore are statically indeterminate, basic methods are developed for determining the forces in these types of structure. However, the chapter commences with a further equilibrium method, which is very useful for finding the forces in statically determinate plane and particularly space structures.

FORCES BY THE METHOD OF TENSION COEFFICIENTS

This method was proposed by Southwell in 1920 as a convenient way of solving for the forces in space frames, and it is of course also applicable to plane frames. It amounts to a shorthand notation for writing equilibrium equations of resolved force components at each joint.

The *tension coefficient* for a member is defined as the force, F, in the member divided by its length, L. Consider the member AB shown in Fig. 10.1: the resolved components of the force F_{AB} in the co-ordinate directions x and y will be $F_{AB} \cos \alpha$ and $F_{AB} \sin \alpha$. Now, $\cos \alpha$ and $\sin \alpha$ can be expressed as the ratio of the length of the member projected on to

Fig. 10.1

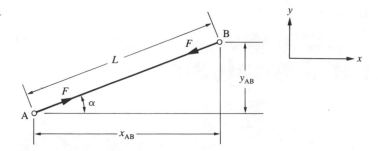

the x and y axes divided by the length, L_{AB}, of the member; thus

$$\cos \alpha = \frac{x_{AB}}{L_{AB}} \quad \text{and} \quad \sin \alpha = \frac{y_{AB}}{L_{AB}}$$

Therefore

$$F_{AB} \cos \alpha = F_{AB} \frac{x_{AB}}{L_{AB}} = t_{AB} x_{AB} \tag{10.1}$$

and

$$F_{AB} \sin \alpha = F_{AB} \frac{y_{AB}}{L_{AB}} = t_{AB} y_{AB} \tag{10.2}$$

where t_{AB} is the *tension coefficient* for the member.

Since the majority of configuration diagrams give dimensions which are related to the co-ordinate directions, it is simpler to express resolved components of force in terms of the tension coefficient and the projected length, rather than sines and cosines of angles. All members are assumed to be in tension initially and hence have a positive coefficient. If in the solution a tension coefficient turns out to be negative then this shows that the member is actually in compression. There is no need in this method to work out reactions at supports in advance as they can simply be included as unknowns in the equilibrium equations.

The first step is to choose a set of reference co-ordinate axes and directions and then to insert all the necessary reactions at the support points in the co-ordinate directions. The arrowhead direction is quite arbitrary. Now, commencing at any joint and considering one co-ordinate direction, write down the equilibrium equation. This will consist of the sum of the products of the tension coefficient t and the projected length, say x (for that co-ordinate direction), for each member at that joint plus any reaction or applied force at that point. It is important to remember that, taking the origin of the co-ordinate axes referenced at the joint, the projected lengths of the member *must* be given a sign appropriate to the positive and negative directions of the co-ordinate axes. Similarly, reactions and applied loads *must* also be given the appropriate sign relative to the axes. In the case of the space frame the above procedure must be repeated for each of the two other co-ordinate directions and then in turn for each joint.

Having obtained all the required equations these are solved for all the unknown tension coefficients and reactions. Each tension coefficient is then multiplied by the actual length of the member to give the force in the member. It will perhaps now be evident that, for clarity and ease of checking, a tabular system of solution is most desirable. The method will now be illustrated in the following example of a space frame.

EXAMPLE 10.1

Use the method of tension coefficients to determine the reactions and the forces in the space frame shown in Fig. 10.2.

Fig. 10.2

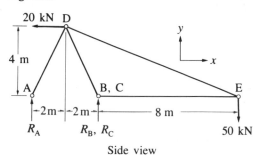

Side view

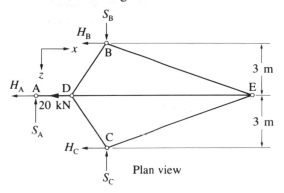

Plan view

SOLUTION

Eqn. no.	Joint/direction	Equilibrium equation
1	A − x	$-H_A + 2t_{AD} = 0$
2	A − y	$4t_{AD} + R_A = 0$
3	A − z	$-S_A = 0$
4	B − x	$-H_B - 2t_{BD} + 8t_{BE} = 0$
5	B − y	$R_B + 4t_{BD} = 0$
6	B − z	$S_B + 3t_{BD} + 3t_{BE} = 0$
7	C − x	$-H_C - 2t_{CD} + 8t_{CE} = 0$
8	C − y	$R_C + 4t_{CD} = 0$
9	C − z	$-S_C - 3t_{CD} - 3t_{CE} = 0$
10	D − x	$-20 - 2t_{AD} + 2t_{BD} + 2t_{CD} + 10t_{DE} = 0$
11	D − y	$-4t_{DA} - 4t_{DB} - 4t_{DC} - 4t_{DE} = 0$
12	D − z	$3t_{DC} - 3t_{DB} = 0$
13	E − x	$-8t_{EB} - 8t_{EC} - 10t_{ED} = 0$
14	E − y	$4t_{ED} - 50 = 0$
15	E − z	$-3t_{EB} + 3t_{EC} = 0$

Member	Tension coefficient	Length, m	Force, kN	Reaction, kN
AD	+20	4·48	+89·6	$H_A = 40$
BD	−16·25	5·39	−87·5	$R_A = -80$
CD	−16·25	5·39	−87·5	$S_A = 0$
BE	−7·8	8·55	−66·6	$H_B = -30$
CE	−7·8	8·55	−66·6	$R_B = 65$
DE	+12·5	10·78	+134·8	$S_B = 72·4$
				$H_C = -30$
				$R_C = 65$
				$S_C = 72·4$

DISPLACEMENTS BY THE CHANGE OF GEOMETRY METHOD

This method simply uses the changes in length of members under load to describe the change in geometry or configuration of the framework, thus giving the displacement of joints. The principle of the method will now be illustrated through the very simple situation of two pin-jointed members loaded as shown in Fig. 10.3. The forces in the members AB and AC can be calculated from two equilibrium equations, and thus, using the stress–strain relationship, the changes in length of the members are

$$\delta L_1 = \frac{P_1 L_2}{A_1 E} \quad \text{and} \quad \delta L_2 = \frac{P_2 L_2}{A_2 E} \tag{10.3}$$

where suffix 1 applies to AB and suffix 2 to AC.

Fig. 10.3

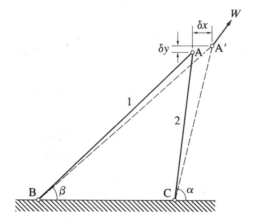

The joint A moves to A′ (exaggerated for clarity) under load where the displacements are δx and δy in the co-ordinate directions.

For small deformations the geometrical relationships are approximately

$$\left.\begin{array}{l} \delta L_1 = \delta x \cos \beta + \delta y \sin \beta \\ \delta L_2 = \delta x \cos \alpha + \delta y \sin \alpha \end{array}\right\} \tag{10.4}$$

The slight change in the angles α and β under load is ignored as a second-order quantity.

Using the values of δL_1 and δL_2 in eqns. (10.3) and solving eqns. (10.4) simultaneously, the displacement components, δx and δy, can be determined.

Although this method is quite satisfactory and is also applicable to three-dimensional frames, for any degree of complexity the arithmetic becomes rather tedious and other methods are therefore preferable.

DISPLACEMENTS BY THE WILLIOT AND WILLIOT–MOHR METHODS

Although perhaps graphical methods have been made somewhat obsolete with the development of the computer to solve numerical problems, a graphical solution can still on occasions provide a better physical understanding of what is happening.

WILLIOT METHOD

A very convenient graphical construction for determining displacements of frameworks was developed by Williot in 1877.

Referring to Fig. 10.3 again, the actual movement of the members may be regarded as a combination of change in axial length and a swing or rotation about one end. Since displacements are extremely small compared to the length of the members, the rotation component may be treated as a movement perpendicular to the direction of a member.

To draw the displacement diagram, shown in Fig. 10.4, the procedure is as follows. Mark a point b to represent B; then the same point also represents C, since there is no relative movement between B and C. Draw ba_1 to a suitable scale to represent δL_1 parallel to BA and in the correct sense for extension. Now draw a line through a_1 perpendicular to BA which is the line of the rotation component, although its length is not known. Next draw a line ca_2 parallel to CA to represent δL_2 and in the correct sense. Finally, draw a line from a_2 perpendicular to the direction of CA to intersect the previous rotation component. The point of intersection a defines the resultant position of A. The movement ot A relative to the "ground" points B and C can then be measured off the diagram.

Fig. 10.4

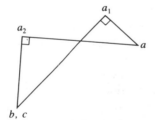

EXAMPLE 10.2

Determine the vertical and horizontal displacements of the joint D in the plane pin-jointed framework in Fig. 10.5(a). All members have a cross-sectional area of 1000 mm^2 and modulus of 200 GN/m^2.

SOLUTION

The first step is to calculate the forces in all the members by one of the methods described earlier. Next the change in length of each member is

Fig. 10.5

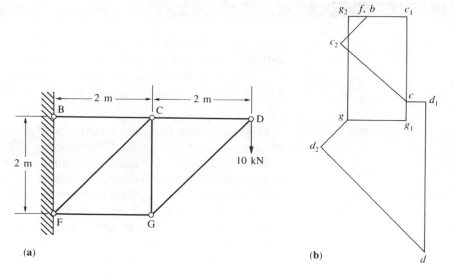

(a)

(b)

determined and it is convenient to tabulate the foregoing information:

Member	Force W (kN)	Length L (m)	Change in length, WL/EA (mm)
BC	+20	2	+0.2
CD	+10	2	+0.1
FG	−10	2	−0.1
FC	−10√2	2√2	−0.2
GC	+10	2	+0.1
GD	−10√2	2√2	−0.2

The displacement diagram (Fig. 10.5(b)) can now be drawn starting with the triangle BCF. BC extends to the right, so that c_1 is drawn to the right of b at a distance of 0.2 mm (to scale). C moves diagonally down to the left, so fc_2 is drawn in the same sense, 0.2 mm (to scale). The perpendiculars to the directions of BC and FC are c_1c and c_2c, respectively, meeting at c. We can now proceed to treat triangle FCG in the same manner, arriving at point g. Finally the displacement components of triangle CGD will give the position of D at d. From the completed diagram.

Vertical displacement of D = 1.266 mm

Horizontal displacement of D = 0.3 mm

WILLIOT–MOHR METHOD

The Williot diagram can only be drawn if there is a triangle of members with two of its apexes as "ground" points from which to commence. It is evident that there could be many framework configurations which did not satisfy this condition. For this reason Otto Mohr, in 1887, suggested an extension to the Williot method. It is best explained and understood by using a specific example.

Consider the plane frame in Fig. 10.6 which was solved for member forces in Chapter 1, Example 1.2. The first step is to determine the

Fig. 10.6

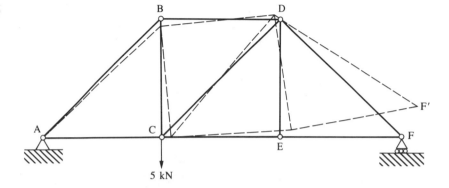

change in length of the members. Assuming a constant value of (area $A \times$ modulus E) for all members of 200 MN, the numerical values are given in the following table.

Member	Force (N)	Length (m)	Length change (mm)
AB	−4713	2.82	−0.0667
AC	+3333	2.00	+0.0333
BC	+3333	2.00	+0.0333
BD	−3333	2.00	−0.0333
CD	+2356	2.82	+0.0333
CE	+1667	2.00	+0.0167
DE	0	2.00	0
DF	−2356	2.82	−0.0333
EF	+1667	2.00	+0.0167

If now, for example, C is regarded temporarily as a position fixed joint, then a Williot diagram could be drawn starting from A and C. It would give the correct *distortion* of the framework and the *relative* displacements of joint to joint as in Fig. 10.7. However, as the assumption that AC is horizontal is not true, the Williot solution would be for the framework displaced relative to "ground" as shown rather exaggerated in Fig. 10.6. Mohr, then, proposed the required correction to the diagram to restore the real situation where F remains in the horizontal plane. This is achieved by giving the frame a rigid-body rotation until F′ coincides with F, which means that on the Williot diagram the apparent vertical distance between *a* and *f* must be eliminated. The other joints will move in a direction perpendicular (approximately) to lines connecting each joint to A by an amount equal to the distance of the joint from A multiplied by F′F/AF. This statement

Fig. 10.7

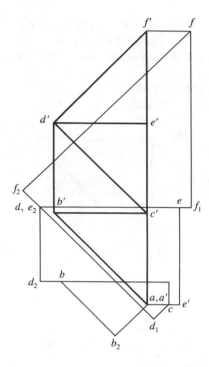

is represented on the Williot diagram by drawing the frame rotated through 90° (in the opposite sense of F′ moving to F), and to a suitable scale so that when a' is placed on a (since A does not move), f' is horizontally in line with f. The points a', b', c', d', e', f' now become the "ground" points for a, b, c, d, e, f, and the true displacements of the joints in the framework relative to the supports are given by the distances between the corresponding points $a'a$, $b'b$, $c'c$, etc.

DISPLACEMENTS BY THE VIRTUAL WORK METHOD

The principle of virtual work was introduced in Chapter 9, and will now be applied to structural analysis.

ACTUAL FORCES, EXTENSIONS AND DISPLACEMENTS

If a plane frame is subjected to *one* external load only and the displacement of the loaded joint is required *in the direction* of the load, then eqn. (9.18) can be used with all real values. The method will be illustrated in relation to part of Example 10.2. Let the vertical displacement of joint D, which is where the load is applied and in the same direction, be δ. Then, using numerical values from the table in Example 10.2,

$$\sum_{m} P\Delta = (20 \times 0.2) + (10 \times 0.1) + (-10 \times -0.1)$$

$$+ (-10\sqrt{2} \times -0.2) + (10 \times 0.1) + (-10\sqrt{2} \times -0.2)$$

$$= 12.66 \text{ kN-mm}$$

Therefore
$$10\delta = 12.66 \quad \text{and} \quad \delta = 1.266 \text{ mm}$$

The horizontal displacement of D cannot be found by the above method, since there is no force in the horizontal direction at D to give an external work term.

The above method of using the real forces and extensions cannot be used on a frame carrying several external loads, unless all the displacements at the loaded joints, except one, are known, since there would be several unknown δ's on the left-hand side of eqn. (9.18).

ACTUAL EXTENSIONS AND DISPLACEMENTS WITH FORCES HYPOTHETICAL

The previous section demonstrated the use of eqn. (9.18) in which the work terms were composed of the real forces and real displacements. However, the equation is equally valid if the displacements are those actually occurring due to the applied loading, but the load and force terms are completely fictitious, so long as they form an equilibrium system. This feature enables one to determine the displacement of any joint in any direction, whether loaded or not, by the use of a "dummy" load of unit magnitude.

Consider again Example 10.2 and the need to find the horizontal displacement at D. A dummy unit load is inserted horizontally at D. Then the equilibrium system of forces in the members *due to the unit load only* (the 10 kN is removed) can be found; the values are as given in the following table.

Member	Actual extensions, Δ, due to 10 kN load	Hypothetical forces, P', due to dummy load	$P'\Delta$
BC	+0.2	1	+0.2
FC	−0.2	0	0
FG	−0.1	0	0
CG	+0.1	0	0
CD	+0.1	1	+0.1
GD	−0.2	0	0
			$\Sigma = 0.3$

If the real horizontal displacement of D due to the actual load of 10 kN is δ, then the external virtual work is $1 \times \delta$. The internal virtual work is given by the sum of the products of the actual changes in length, Δ, of the members due to the actual load of 10 kN and the hypothetical equilibrium set of forces, P', resulting from the unit dummy load. The simple computation is shown in the right-hand column of the above table. Applying eqn. (9.18) gives
$$1\delta = \sum P'\Delta \quad \text{or} \quad \delta = 0.3 \text{ mm}$$

EXAMPLE 10.3

Determine the vertical displacement of joint G of the frame in Example 10.2.

SOLUTION

Following the same procedure as above, a unit dummy load is placed vertically at G; the corresponding force system *due to the unit load only* is shown in the table below.

Member	Actual extensions Δ	Forces due to unit load P'	P'Δ
BC	+0.2	+1	+0.2
FC	−0.2	−√2	+0.2√2
FG	−0.1	0	0
CG	+0.1	1	+0.1
CD	+0.1	0	0
GD	−0.2	0	0
			Σ = +0.5828

Thus

$$1\delta_G = 0.5828 \quad \text{or} \quad \delta_G = 0.5828 \text{ mm}$$

EXAMPLE 10.4

Determine the forces in the members of the plane framework illustrated in Fig. 10.4 and calculate the vertical deflection at the joint carrying the central load. All members are 3 m in length and 1500 mm² in cross-section. $E = 205$ GN/m².

SOLUTION

Firstly a force polygon is constructed for the actual loading on the frame (Fig. 10.8(b)). Although we are trying to find the displacement of a joint in the direction of the applied load, because there are three external loads having unknown displacements, eqn. (9.18) cannot be solved. Hence we resort to the method involving a dummy load. After removing the 10 kN loads insert a unit vertical load only at the central joint in the lower span. The appropriate force polygon is then drawn as in Fig. 10.8(c) (to a larger scale than at (b)).

Fig. 10.8

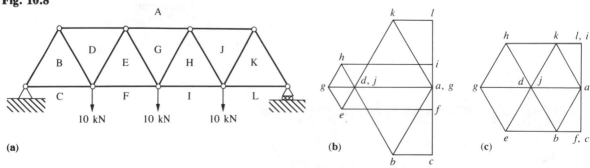

(a)   (b)   (c)

The forces are all tabulated below. Since A, L and E are the same for all members, the changes in length have one constant of proportionality to the forces and it is therefore wasted effort to include a specific column in the table.

Member	Actual forces P (kN)	Unit load forces, P' (kN)	PP' (kN²)
AD	−17.3	−0.575	9.950
AG	−23.1	−1.15	26.550
AJ	−17.3	−0.575	9.950
KL	+8.65	+0.289	2.475
IH	+20.2	+0.868	17.700
EF	+20.2	+0.868	17.700
BC	+8.65	+0.289	2.475
AB	−17.3	−0.575	9.950
BD	+17.3	+0.575	9.950
DE	−5.8	−0.575	3.325
EG	+5.8	+0.575	3.325
GH	+5.8	+0.575	3.325
HJ	−5.8	−0.575	3.325
JK	+17.3	+0.575	9.950
AK	−17.3	−0.575	9.950
			$\Sigma = 139.900$

At this point the factor L/AE required to give changes in length can be included. Therefore

$$1 \times \delta = \frac{L}{AE} \times 139.9$$

and

$$\delta = \frac{3 \times 10^3 \times 139.9}{1500 \times 10^{-6} \times 205 \times 10^6} = 1.363 \text{ mm}$$

STATICALLY-INDETERMINATE FRAMES

This is one of the main areas of study in the theory of structures, and it is almost presumptuous to attempt an introduction in the space available here. However, it is felt that a reasonable insight into the basic concepts can be got across now, and the reader who needs a greater depth of study may then proceed to the texts specializing in structural analysis.

There are three basic methods of analysis, namely (a) virtual work (b) compatibility (or flexibility) method due to Maxwell, (c) equilibrium (or stiffness) method due to Navier.

FORCES BY THE VIRTUAL WORK METHOD

**FORCES IN A
STATICALLY-
DETERMINATE
PLANE FRAME**

Before tackling the statically indeterminate structure it is useful to see how the virtual work method can be applied to find forces in a statically determinate plane frame. We now take the converse procedure to that above and use the true equilibrium system of forces in a framework in conjunction with a hypothetical set of compatible displacements to compose the virtual work eqn. (9.18).

Consider the plane frame in Fig. 10.9, which was analysed by equilibrium methods in Chapter 1. It is only possible to find one

Fig. 10.9

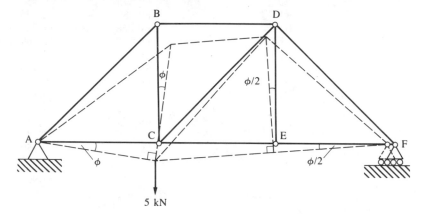

unknown force at a time using eqn. (9.18), so let this be, for example, that in member BD. It is also necessary to include the applied load of 5 kN at C in the left-hand side of eqn. (9.18). If no other forces are to appear in the equation, then the other members must not change in length and so rigid body rotations are given to the other parts of the frame. This results in the hypothetical displaced configuration of Fig. 10.9, from which eqn. (9.18) can be expressed as

$$5000\delta = (P\Delta)_{BD}$$

Let the rigid-body rotation of triangle ABC be ϕ; for small displacements $\sin \phi \approx \phi$ so that $\delta = \text{AC} \cdot \phi = 2\phi$.

For the system of displacements to be compatible, triangle CDF rotates through an angle $\phi/2$. The compression of BD can now be expressed as

$$\Delta_{BD} = \text{BC} \cdot \phi + \text{DE} \cdot \phi/2$$
$$= 2\phi + 2\phi/2$$

Substituting these values in the virtual work equation,

$$5000 \times 2\phi = P_{BD}(2\phi + 2\phi/2)$$

Hence

$$P_{BD} = 3333 \, \text{N}$$

The actual value of ϕ is now seen to be immaterial as it cancels through in the virtual work equation

Fig. 10.10

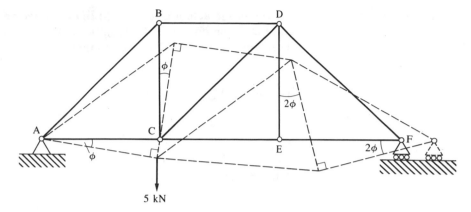

5 kN

To find the force in CE the displacement configuration in Fig. 10.10 is adopted giving

$$5000 \cdot AC \cdot \phi = P_{CE} \cdot BC \cdot \phi + P_{CE} \cdot DE \cdot 2\phi$$

from which

$$P_{CE} = 1667 \text{ N}$$

STATICALLY INDETERMINATE CASE

Now the method will be applied to a simple statically-indeterminate case. Consider the plane pin-jointed frame in Fig. 10.11, in which any one member may be regarded as redundant. Let the force in one member, say BD, be tensile of magnitude R. If this member is now removed and

Fig. 10.11

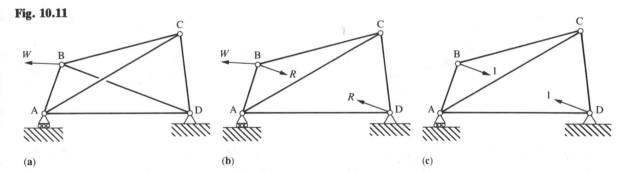

(a)   (b)   (c)

equal and opposite forces of magnitude R are applied at B and D, we still have the same status as in the original frame. However, the forces in all the other members can now be calculated in terms of the applied external load, W, and the force R. The deformations, Δ, of the members can then be found using the forces above in terms of W and R. The next step is to replace the forces R by unit loads and to remove all external loading. We now determine the forces, P, set up in the members by the unit loads and write the virtual work equation to express the real displacement δ_D of joint D in terms of the hypothetical forces, P, and the real deformations, Δ, of the members as follows:

$$1 \times \delta_D = \sum_m P\Delta \tag{10.5}$$

Now in the actual frame, since BD is subjected to tension R, then BD will *extend* an amount, say Δ_{BD}, which is in the opposite sense to δ_D. Therefore the compatibility condition is

$$\Delta_{BD} = -\delta_D$$

Thus eqn. (10.5) becomes

$$\Delta_{BD} + \sum_m P\Delta = 0 \tag{10.6}$$

and since all the quantities are in terms of W and R, the value of the latter redundant force can be found.

EXAMPLE 10.5

Determine the forces in the members of the plane frame shown in Fig. 10.12. The area and modulus are the same for all members.

Fig. 10.12

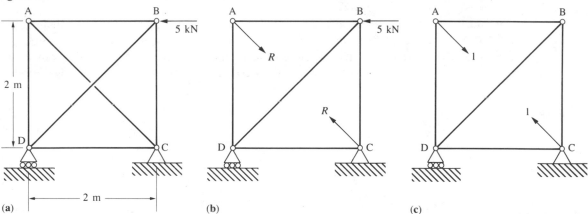

(a)   (b)   (c)

SOLUTION

This type of problem is most conveniently solved in tabular form (Table 10.1). Member AC will be considered as the redundancy.

Firstly, the forces are found by, say, resolution at joints due to the 5 kN load and the redundancy, R. These values are given in column (2) of the table. The next column (3) gives the deformations of the members. Now the 5 kN load and R are removed, and unit loads are placed at A and C from which forces in the members are found as shown in column (4). The virtual work is expressed in column (5) as the product of columns (3) and (4). Equation (10.6) may now be expressed as the summation of the terms in column (5) and R evaluated.

$$\left(R\frac{2\sqrt{2}}{EA} \times 1\right) + [-10(2 + \sqrt{2}) + 4R + 2\sqrt{2}R]\frac{1}{EA} = 0$$

from which $R = 3.53$ kN.

Table 10.1

(1) Member	(2) Force due to 5 kN and R	(3) Deformations due to 5 kN and R	(4) Forces due to unit loads at A and C	(5) (3) × (4)
AB	$-\dfrac{R}{\sqrt{2}}$	$-\dfrac{R}{\sqrt{2}}\dfrac{2}{EA}$	$-\dfrac{1}{\sqrt{2}}$	$+\dfrac{R}{EA}$
BC	$5-\dfrac{R}{\sqrt{2}}$	$\left(5-\dfrac{R}{\sqrt{2}}\right)\dfrac{2}{EA}$	$-\dfrac{1}{\sqrt{2}}$	$-\left(5-\dfrac{R}{\sqrt{2}}\right)\dfrac{\sqrt{2}}{EA}$
CD	$5-\dfrac{R}{\sqrt{2}}$	$\left(5-\dfrac{R}{\sqrt{2}}\right)\dfrac{2}{EA}$	$-\dfrac{1}{\sqrt{2}}$	$-\left(5-\dfrac{R}{\sqrt{2}}\right)\dfrac{\sqrt{2}}{EA}$
AD	$-\dfrac{R}{\sqrt{2}}$	$-\dfrac{R}{\sqrt{2}}\dfrac{2}{EA}$	$-\dfrac{1}{\sqrt{2}}$	$+\dfrac{R}{EA}$
AC	R	$R\dfrac{2\sqrt{2}}{EA}$	$+1$	$R\dfrac{2\sqrt{2}}{EA}\times 1$
BD	$-5\sqrt{2}+R$	$-(5\sqrt{2}-R)\dfrac{2\sqrt{2}}{EA}$	$+1$	$-(5\sqrt{2}-R)\dfrac{2\sqrt{2}}{EA}$

Substitution back into column (2) gives the values of all other member forces.

LACK OF FIT

In a statically determinate structure even if the length of one or more members is not correct to the design specification it would still be possible to assemble a pin-jointed framework since the overall geometry can adjust to accommodate the errors. Equally, the effect of change in temperature of one or more members causing change in length can be accommodated without inducing internal forces in a pin-jointed statically-determinate framework.

However, the situation is quite different in the case of statically-indeterminate or redundant structures, since only members of exact geometrical dimensions can be assembled without inducing internal forces. Hence *lack of fit* from the various possible causes will result in self-straining of the structure independently of external loading.

Denoting a positive lack of fit by λ, the member being longer than it should be, then eqn. (10.6) is written to include λ as follows:

$$(\lambda + \Delta_{BD}) + \sum_m P\Delta = 0 \qquad (10.7)$$

FORCES BY THE COMPATIBILITY METHOD

This method uses the principle that displacements and rotations at joints must be compatible with the deformation of the members. Forces and moments must be such as to satisfy the foregoing condition and thus a solution may be obtained.

EXAMPLE 10.6

Determine the forces in the members of the plane frame illustrated in Fig. 10.13(a).

Fig. 10.13

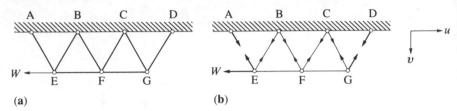

(a) (b)

SOLUTION

Since there are two redundancies in this framework the first step is to choose two members to remove, say AE and DG, and replace by the unknown forces, P and Q respectively, acting at the joints. The frame now appears as in Fig. 10.13(b). The next step is to determine the horizontal and vertical displacements of the joints E, F, G for the three situations, when W is acting alone, then P only and Q alone. Each of these cases is, of course, statically determinate, so forces and thus displacements are readily found using, say, a Williot diagram. The results are given in Table 10.2. The right-hand column gives the components of displacement due to the superposition of the W, P and Q cases.

Table 10.2

Displacement	Due to W	Due to P	Due to Q	Superposed displacements $\times \dfrac{L}{\mathrm{AE}}$
u_E	$-3\dfrac{WL}{\mathrm{AE}}$	$-3\dfrac{PL}{\mathrm{AE}}$	$+2\dfrac{QL}{\mathrm{AE}}$	$-3W - 3P + 2Q$
u_F	$-2\dfrac{WL}{\mathrm{AE}}$	$-2\dfrac{PL}{\mathrm{AE}}$	$+2\dfrac{QL}{\mathrm{AE}}$	$-2W - 2P + 2Q$
u_G	$-2\dfrac{WL}{\mathrm{AE}}$	$-2\dfrac{PL}{\mathrm{AE}}$	$+3\dfrac{QL}{\mathrm{AE}}$	$-2W - 2P + 3Q$
v_E	$-\dfrac{3}{\sqrt{3}}\dfrac{WL}{\mathrm{AE}}$	$-\dfrac{5}{\sqrt{3}}\dfrac{PL}{\mathrm{AE}}$	$+\dfrac{2}{\sqrt{3}}\dfrac{QL}{\mathrm{AE}}$	$-\dfrac{3W}{\sqrt{3}} - \dfrac{5P}{\sqrt{3}} + \dfrac{2Q}{\sqrt{3}}$
v_F	0	0	0	0
v_G	$+\dfrac{2}{\sqrt{3}}\dfrac{WL}{\mathrm{AE}}$	$+\dfrac{2}{\sqrt{3}}\dfrac{PL}{\mathrm{AE}}$	$-\dfrac{5}{\sqrt{3}}\dfrac{QL}{\mathrm{AE}}$	$\dfrac{2}{\sqrt{3}}W + \dfrac{2P}{\sqrt{3}} - \dfrac{5Q}{\sqrt{3}}$

The next stage is to ensure compatibility at joints E and G. Member AE will extend by an amount PL/AE, and DG, similarly, by QL/AE. These extensions can be expressed in terms of the joint displacements thus:

$$\frac{PL}{AE} = u_E \cos 60° + v_E \sin 60° = \frac{u_E}{2} + v_E \frac{\sqrt{3}}{2} \qquad (10.8)$$

$$\frac{QL}{AE} = -u_G \cos 60° + v_G \sin 60° = -\frac{u_G}{2} + v_G \frac{\sqrt{3}}{2} \qquad (10.9)$$

Substituting the appropriate values from the above table into eqns. (10.8) and (10.9) gives

$$\frac{2PL}{AE} = (-3W - 3P + 2Q)\frac{L}{AE} + (-3W - 5P + 2Q)\frac{L}{AE}$$

$$4Q - 10P = 6W \qquad (10.10)$$

and

$$\frac{2QL}{AE} = -(-2W - 2P + 3Q)\frac{L}{AE} + (2W + 2P - 5Q)\frac{L}{AE}$$

$$10Q - 4P = 4W \qquad (10.11)$$

Solving eqns. (10.10) and (10.11) gives

$$P = -\frac{11}{21}W \quad \text{and} \quad Q = \frac{4}{21}W$$

The forces in members AE and DG having been found, the remainder can be determined by simple statics. The complete solution is

AE	BE	BF	CF	CG	DG	EF	FG
$-\frac{11}{21}W$	$\frac{11}{21}W$	$-\frac{2}{7}W$	$\frac{2}{7}W$	$-\frac{4}{21}W$	$\frac{4}{21}W$	$\frac{10}{21}W$	$\frac{4}{21}W$

A somewhat shorter and simpler solution to the above, but by exactly the same method, would be obtained by taking EF and FG as the redundant members.

FORCES BY THE EQUILIBRIUM METHOD

In this method the forces in the members are expressed in terms of the components of joint displacement using the modulus, area and length of each member. The equilibrium equations for each joint are first expressed in terms of the forces in the members meeting at this joint. They are then transposed into displacement equations using the above force-displacement relations. After solving the equations for displacements, these are substituted back to give values for the forces.

EXAMPLE 10.7

Calculate the forces in the members of the plane frame illustrated in Fig. 10.14

Fig. 10.14

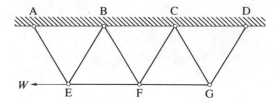

SOLUTION

Joints A, B, C and D are position fixed and joints E, F and G have displacement components u_E, u_F, u_G horizontally and v_E, v_F, v_G vertically.

The extensions of a member such as AE can be expressed in terms of the displacement components of E as

$$u_E \sin 60° + v_E \cos 60° = u_E \frac{\sqrt{3}}{2} + \frac{v_E}{2}$$

from which the force in the member can be expressed as

$$F_{AE} = \frac{AE}{L}\left(\frac{u_E}{2} + \frac{v_E}{2}\sqrt{3}\right)$$

The forces in all the other members, by a similar approach, are

$$F_{BE} = \frac{AE}{L}\left(-\frac{u_E}{2} + \frac{v_E}{2}\sqrt{3}\right); \qquad F_{BF} = \frac{AE}{L}\left(\frac{u_F}{2} + \frac{v_F}{2}\sqrt{3}\right)$$

$$F_{CF} = \frac{AE}{L}\left(-\frac{u_F}{2} + \frac{v_F}{2}\sqrt{3}\right); \qquad F_{CG} = \frac{AE}{L}\left(\frac{u_G}{2} + \frac{v_G}{2}\sqrt{3}\right)$$

$$F_{DG} = \frac{AE}{L}\left(-\frac{u_G}{2} + \frac{v_G}{2}\sqrt{3}\right); \qquad F_{EF} = \frac{AE}{L}(u_F - u_E)$$

$$F_{FG} = \frac{AE}{L}(u_G - u_F).$$

The above equations satisfy the geometry of deformation or compatibility condition. Next we must consider the equilibrium condition at each joint.

Joint E   Horizontal,

$$-W - F_{AE}\cos 60° + F_{EB}\cos 60° + F_{EF} = 0$$

Substituting for the forces from above,

$$-\frac{2WL}{AE} - (u_E + v_E\sqrt{3})\tfrac{1}{2} + (-u_E + v_E\sqrt{3})\tfrac{1}{2} + 2(u_F - u_E) = 0$$

$$-3u_E + 2u_F - \frac{2WL}{AE} = 0 \tag{10.12}$$

Vertical,

$$-F_{AE} \sin 60° - F_{BE} \sin 60° = 0$$

$$u_F + v_E \sqrt{3} - u_E + v_E \sqrt{3} = 0$$

Hence

$$v_E = 0$$

Joint F Horizontal,

$$-F_{EF} + F_{FG} - F_{BF} \cos 60° + F_{CF} \cos 60° = 0$$

$$2u_E - 5u_F + 2u_G = 0 \qquad\qquad (10.13)$$

Vertical,

$$-F_{BF} \sin 60° - F_{CF} \sin 60° = 0$$

$$u_F + v_F - u_F + v_F = 0$$

Hence

$$v_F = 0$$

Joint G Horizontal,

$$-F_{FG} - F_{CG} \cos 60° + F_{DG} \cos 60° = 0$$

$$-3u_G + 2u_F = 0 \qquad\qquad (10.14)$$

Vertical,

$$-F_{CG} \sin 60° - F_{DG} \sin 60° = 0$$

from which

$$v_G = 0$$

Eqns. (10.12), (10.13) and (10.14) can now be solved to give

$$u_E = -\frac{22}{21}\frac{WL}{AE}$$

$$u_F = -\frac{4}{7}\frac{WL}{AE}$$

$$u_G = -\frac{8}{21}\frac{WL}{AE}$$

These values are substituted back to find the above forces. For example,

$$F_{FG} = \frac{AE}{L}\left(-\frac{8}{21}\frac{WL}{AE} + \frac{12}{21}\frac{WL}{AE}\right) = \frac{4}{21}W$$

which is the same result as obtained by the compatibility method.

SUMMARY

In spite of the limited treatment that has been given to structural analysis, all the basic principles have been presented, albeit briefly. Firstly one must establish whether the framework will be statically determinate or indeterminate on the basis of the number of joints, members and support reactions, since this will determine what method of solution must be employed. The formulation of equilibrium equations relevant to free-body diagrams is the next important understanding one must have. Various methods can be adopted to determine structural displacements. Undoubtedly one of the most powerful is by the Principle of Virtual Work which is expressed in the equation, $\sum_j F\delta = \sum_m P\Delta$, and the only two necessary conditions are that the force system satisfies the requirements of equilibrium and that the deformation system satisfies the requirements of compatibility. This method together with other energy theorems introduced in Chapter 9 form the basis for most structural computations.

It should also be noted that in Chapter 18, there is an introduction to the analysis of structures using the stiffness matrix method.

PROBLEMS

10.1   Calculate the forces in each of the members of the pin-jointed space frame shown in Fig. 10.15.

Fig. 10.15

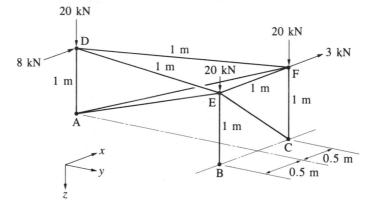

Fig. 10.16

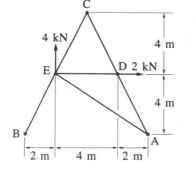

10.2 Use the method of tension coefficients to calculate the forces in the pin-jointed space frame shown in plan view in Fig. 10.16. The pinned supports A, B and C are at the same level, and DE is horizontal and at a height of 4 m above the supports.

10.3 Using the method of tension coefficients, determine the forces in each of the members of the framework illustrated in Fig. 10.17.

Fig. 10.17

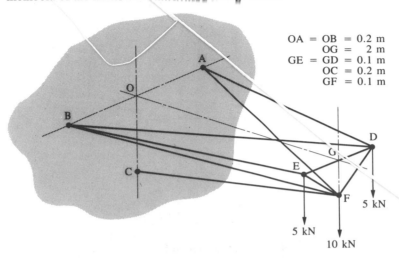

OA = OB = 0.2 m
OG = 2 m
GE = GD = 0.1 m
OC = 0.2 m
GF = 0.1 m

10.4 Determine the horizontal and vertical displacements of joint E in the plane pin-jointed truss illustrated in Fig. 10.18. Some of the member forces are given, and for all members. $EA = 100$ MN.

Fig. 10.18

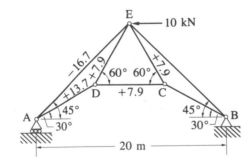

10.5 Determine the vertical and horizontal displacements of joint A of the plane pin-jointed framework illustrated in Fig. 10.19. For all members $AE = 45$ MN. Check the solution by a graphical construction.

Fig. 10.19

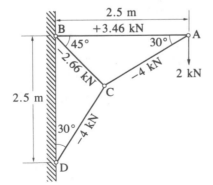

10.6 Obtain the vertical displacement of joint E in the framework shown in Fig. 10.20. Tension and compression members have cross-sectional areas of 500 and 1000 mm^2 respectively. $E = 208$ GN/m^2. Check the solution by graphical construction.

Fig. 10.20

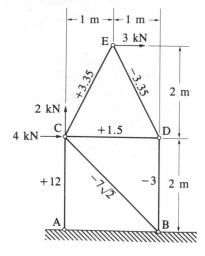

10.7 The plane pin-jointed framework shown in Fig. 10.21 has all members of area 2,700 mm^2. If the movement at the roller support on the inclined surface is limited to 5 mm, determine the maximum value for W and the vertical deflection at B for this value of W. $E = 200$ GN/m^2.

Fig. 10.21

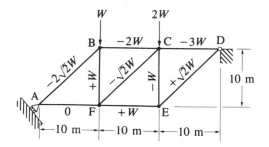

10.8 Using the principle of virtual work calculate the vertical deflection of joint E due to the loading shown in Fig. 10.22. For all members $AE = 260$ MN. Check the solution by constructing a Williot–Mohr diagram.

Fig. 10.22

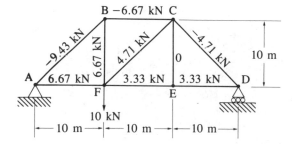

10.9 Determine the forces in the members of the plane pin-jointed redundant framework illustrated in Fig. 10.23.

Fig. 10.23

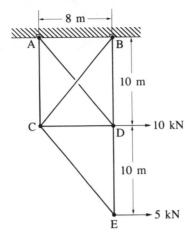

10.10   If the force in member AD of the framework in Problem 10.9 has to be limited to a value of 7 kN, determine the adjustment in initial length of AD that is required. For all members $EA = 18$ MN.

10.11   A plane pin-jointed framework is simply-supported and loaded as shown in Fig. 10.24. Calculate the load in the diagonal member EC. What will the load in EC become if its temperature is raised by 20 °C? $EA = 414$ MN, $\alpha = 12 \times 10^{-6}/$°C.

Fig. 10.24

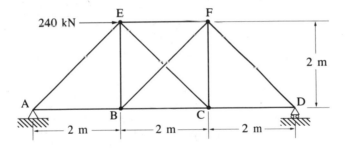

Chapter 11

![gray bar]

Buckling instability

INTRODUCTION

In earlier chapters basic analytical design procedures have been developed in which components and structural elements have been subjected to tension and compression forces, bending moment and torque. In this chapter we shall examine the specific effect of compressive forces in relation to the geometry and boundary conditions of members. Examples range from the compression force of combustion on the connecting-rod of an internal combustion engine to the vertical columns used in structural steelwork to support all the vertical mass and forces in a building. In addition to axial and eccentrically aligned compression forces, columns or struts may be subjected to transverse loading which contributes to buckling and these cases will also be examined.

STABILITY OF EQUILIBRIUM

In previous chapters a fundamental condition in all the problems was the equilibrium of internal and external forces. Now, if the system of forces is disturbed owing to a small displacement of a body, two principal situations are possible: either the body will return to its original configuration owing to restoring forces during displacement, or the body will accelerate farther away from its original state owing to displacing forces. The former situation is termed *stable equilibrium* and the latter is termed *unstable equilibrium*.

Consider the simple case in Fig. 11.1(*a*) of a vertical bar pinned at the lower end and carrying an axial tensile force at the upper end. If there is a slight displacement from the vertical, the force will tend to restore the bar to its original position. In Fig. 11.1(*b*), however, the same bar subjected to a compressive load when displaced slightly from the vertical will accelerate towards a horizontal position, illustrating unstable

Fig. 11.1

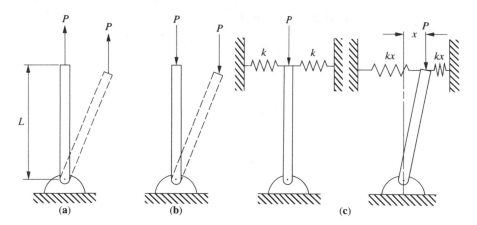

equilibrium. A slightly more sophisticated case is shown in Fig. 11.1(c), where the bar is assisted in remaining vertical by the action of the horizontal springs. When the bar is displaced by an amount x in either direction, there are a displacing moment Px and a restoring moment $2KxL$, where K is the stiffness of a spring; hence we have

$$Px < 2KxL \rightarrow \text{stable}$$

$$Px > 2KxL \rightarrow \text{unstable}$$

The critical condition is when

$$Px = 2KxL \quad \text{or} \quad P_c = 2KL$$

and P_c is termed the *critical load,* being the borderline between stable and unstable equilibrium.

The instability of structural members subjected to compressive loading may be regarded as a mode of failure, owing to excessive deformation and distortion of the structure, even though stress may remain elastic.

Columns or struts subjected to compression can be classified in three ways:

1. Up to about four diameters in length.
2. From four to about thirty diameters in length.
3. Above thirty diameters in length.

Each class can be subdivided into (a) those having axial loading and (b) those in which the loading is eccentric.

Columns under class 1(a) and (b) have been dealt with in earlier chapters. This chapter will be devoted to the investigation of columns in classes 2 and 3 for cases (a) and (b).

It will be assumed that the column is of uniform cross-section and is either ideally straight or has some initial curvature.

EULER THEORY OF BUCKLING FOR SLENDER COLUMNS

COLUMN WITH PINNED ENDS   The column shown in Fig. 11.2 is pin-jointed at each end and it will be assumed to be straight when unloaded. An axial compressive load is

Fig. 11.2

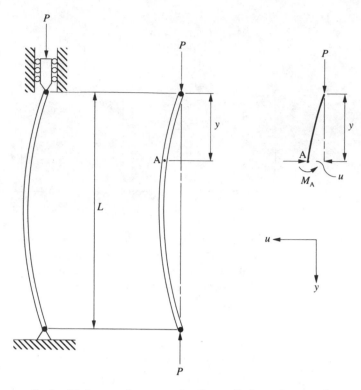

applied with increasing magnitude until the column takes up the deformed shape as shown. It will be noted that although bending is taking place so that displacements are in the horizontal direction the notation used is the same as for a horizontal beam by rotating the column anticlockwise through 90°.

At a distance y from the top joint the displacement is u and the bending moment M_A at section A is Pu in the positive sense. Therefore, using the form of the equation derived for beams (eqn. (7.4)), we have

$$EI\frac{d^2u}{dy^2} = -Pu \tag{11.1}$$

where I is the *least* second moment of area of the cross-section. Therefore

$$\frac{d^2u}{dy^2} = -\frac{P}{EI}u = -k^2u$$

where $k = \sqrt{(P/EI)}$. Hence

$$\frac{d^2u}{dy^2} + k^2u = 0 \tag{11.2}$$

The solution of this equation is

$$u = A\cos(ky) + B\sin(ky) \tag{11.3}$$

The boundary conditions are that $u = 0$ at $y = 0$ and L, therefore

$$A = 0 \quad \text{and} \quad 0 = B\sin(kL)$$

Now the condition $B = 0$ merely gives the trivial case of the undeflected strut but the condition $\sin(kL) = 0$ leads to the solution $kL = n\pi$. Buckling first occurs for $n = 1$, from which the critical load P_c is given by

$$\frac{P_c}{EI} = \frac{\pi^2}{L^2}$$

or

$$P_c = \frac{\pi^2 EI}{L^2} \tag{11.4}$$

If $u = u_{max}$ at $y = L/2$, from symmetry, then $B = u_{max}$ and the deflection curve is given by

$$u = u_{max} \sin(ky) \tag{11.5}$$

OTHER END CONDITIONS

Three other boundary conditions for the end restraint of columns are shown in Fig. 11.3.

Fig. 11.3

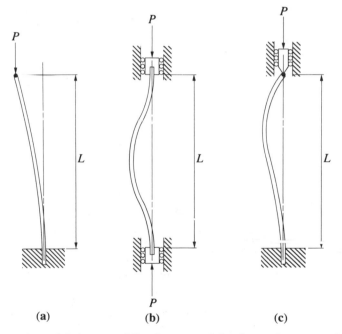

(a)   (b)   (c)

Case (a) (*one end fixed, one end free*) may be treated as a pin-ended strut of equivalent length $2L$, hence

$$P_c = \frac{\pi^2 EI}{(2L)^2} = \frac{\pi^2 EI}{4L^2} \tag{11.6}$$

Case (b) (*both ends fixed*) may be considered as represented by a pin-ended strut of equivalent length $L/2$, hence

$$P_c = \frac{\pi^2 EI}{(\frac{1}{2}L)^2} = \frac{4\pi^2 EI}{L^2} \tag{11.7}$$

Equations (11.6) and (11.7) contrast the difference in critical buckling load depending on the end conditions.

Case (c) (*one end fixed, the other end only free to rotate*) does not have a readily assessed "equivalent length" pin-ended strut and will be solved from first principles. The free-body diagram is illustrated in Fig. 11.4 and the bending moment resulting at point A is

$$M = Pu - Hy$$

Fig. 11.4

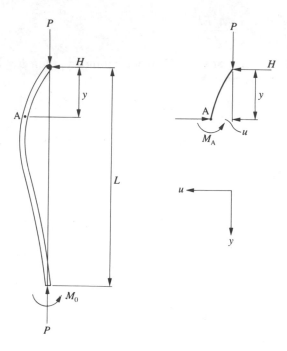

from which

$$\frac{d^2u}{dy^2} + \frac{P}{EI}u = +Hy \tag{11.8}$$

Let $\sqrt{(P/EI)} = k$. The solution of eqn. (11.8) is

$$u = A\cos(ky) + B\sin(ky) + \frac{H}{P}y \tag{11.9}$$

The boundary conditions are $u = 0$ at $y = 0$ and $du/dy = 0$ at $y = L$, from which

$$A = 0; \qquad B = -\frac{H}{Pk}\sec(kL)$$

The deflection curve is thus

$$u = -\frac{H}{Pk}\frac{\sin(ky)}{\cos(kL)} + \frac{H}{P}y \tag{11.10}$$

Finally $u = 0$ at $y = L$, and so

$$0 = -\frac{H}{Pk}\tan(kL) + \frac{H}{P}L$$

Therefore

$$\tan(kL) = kL. \tag{11.11}$$

The smallest value of P (other than $P = 0$) to satisfy eqn. (11.11) is

$$kL = 4.49$$

or

$$P_c = 20\frac{EI}{L^2} \approx \frac{2\pi^2 EI}{L^2} \tag{11.12}$$

This case is therefore approximately an "equivalent length" of 0.7 that for the pin-ended column.

In general the critical buckling load for columns can be expressed as $P_c = \beta(EI/L^2)$, where β has values as above or other values dependent on the end conditions. In practice these are rarely definable as "pinned" or "fixed" and the above formulae must only be applied with careful assessment.

BUCKLING CHARACTERISTICS FOR REAL STRUTS

The load–deflection behaviour for an ideal Euler strut is illustrated in Fig. 11.5(a). For applied loads up to the critical value P_c small transverse displacements u can be maintained under load in a stable-equilibrium state. However at P_c and above the smallest transverse displacement is unstable and will rapidly grow to "failure" of the strut. Obviously the strut can only accommodate a certain deflection up to its elastic limit; thereafter yielding and "failure" would occur by plasticity.

In the case of a real strut, which incorporates some deficiency such as eccentricity of loading, deflection will occur from the moment when load

Fig. 11.5

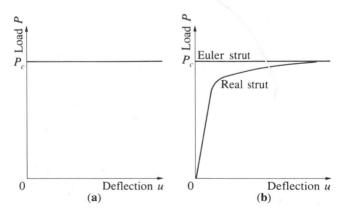

(a)   (b)

is applied as shown in Fig. 13.5(*b*). The curve becomes asymptotic to the Euler load at large deflection. Again, this situation will probably not be attained owing to yielding.

In order to appreciate the significance of stress during buckling behaviour we consider the Euler equation (11.4)

$$P_c = \frac{\pi^2 EI}{L^2} = \frac{\pi^2 EAr^2}{L^2}$$

where A is the cross-sectional area and r is the minimum radius of gyration of the section. Therefore

$$\sigma_c = \frac{P_c}{A} = \frac{\pi^2 E}{(L/r)^2} \qquad (11.13)$$

The ratio L/r is termed the *slenderness ratio*, and plotting this against σ_c gives a curve known as the *Euler hyperbola*, as shown in Fig. 11.6. If a comparison is now made with the compression stress–strain curve for the material, one sees that only struts of large L/r in the range B to C will buckle elastically.

Fig. 11.6

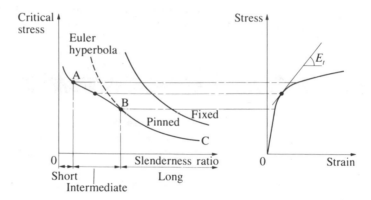

For intermediate values of L/r, from A to B, instability will be accompanied by yielding. At a particular point on the stress–strain curve the stiffness is given by E_t, known as the *tangent modulus*. It is found that, in this elastic–plastic range, buckling loads can be predicted from the original Euler expression if the ordinary modulus E is replaced by the tangent modulus E_t. For short columns, instability does not occur and the problem is one of simple compression.

Figure 11.6 also brings out the influence of end condition in relation to slenderness ratio.

EXAMPLE 11.1

A steel column is to be a fabricated I-section shown in Fig. 11.7. It is 6 m in height and is fixed at its lower end. It is to carry a design compressive load of 92 kN at the central axis O of the section. Determine the

Fig. 11.7

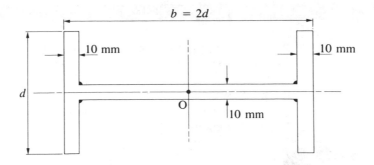

minimum dimensions of the column to resist buckling for (a) the upper end free, and (b) the upper end pinned but restrained in the horizontal direction by other structural members. What are the compressive stresses in each case? $E = 200 \, \text{GN/m}^2$.

SOLUTION

Case (a). The Euler buckling load is

$$P = \frac{\pi^2 EI}{4L^2}$$

$$I = \frac{4 \times 92 \times 10^3 \times 36}{\pi^2 \times 200 \times 10^9} = 6.72 \times 10^{-6} \, \text{m}^4$$

I for the section in Fig. 11.7 is given by

$$\frac{2 \times 10 \times d^3}{12} + \frac{(2d - 20) \times 10^3}{12}$$

Therefore $(1.67d^3 + 167d - 1670) \times 10^{-12} = 6.72 \times 10^{-6}$,

$$d^3 + 100d \approx 4.03 \times 10^6$$

from which $d = 159$ mm and $b = 318$ mm.

Cross-sectional area $= 6160 \times 10^{-6} \, \text{m}^2$

and

$$\text{Compressive stress} = \frac{92\,000}{6160} \times 10^6 = 14.9 \, \text{MN/m}^2$$

Case (b). The Euler buckling load is, from eqn. (11.12),

$$P = 2\pi^2 \frac{EI}{L^2}$$

Therefore $I = 0.84 \times 10^{-6} \, \text{m}^4$ and, from the cubic equation for d above, the solution is

$$d = 79 \, \text{mm} \quad \text{and} \quad b = 158 \, \text{mm}$$

Cross-sectional area $= 2960 \times 10^{-6} \, \text{m}^2$

and

$$\text{compressive stress} = \frac{92\,000}{2960} \times 10^6 = 31.1 \, \text{MN/m}^2$$

ECCENTRIC LOADING OF SLENDER COLUMNS

It is seldom in practice that a column or strut can be loaded exactly along its central axis as the Euler analysis implies. A general solution will now be developed for the case of a long column subjected to a load parallel to, but eccentric from, the central axis.

Consider the column illustrated in Fig. 11.8 displaced by the load P. Bending moment at D, $M_D = -P(e + a - u)$, where e is the eccentricity,

Fig. 11.8

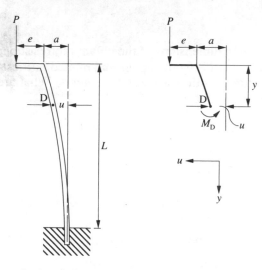

a is the deflection at the free end and u is the deflection at y from the fixed end. Then

$$EI\frac{d^2u}{dy^2} = P(e + a - u)$$

$$\frac{d^2u}{dy^2} + \frac{P}{EI}u = \frac{P}{EI}(e + a) \qquad (11.14)$$

Let $\sqrt{(P/EI)} = k$. Then

$$u = A\cos(ky) + B\sin(ky) + e + a \qquad (11.15)$$

The boundary conditions are that at $y = 0$, $du/dy = 0$; therefore $B = 0$. At $y = 0$, $u = 0$; therefore $A = -(a + e)$ and

$$u = (a + e)\{1 - \cos(ky)\} \qquad (11.16)$$

At $y = L$, $\hat{u} = a$, and substituting in eqn. (11.16) gives

$$a = e\{\sec(kL) - 1\}$$

The deflection curve is therefore

$$u = e\sec(kL)\{1 - \cos(ky)\} \qquad (11.17)$$

The maximum bending stress occurs at the fixed end and is given by

$$\sigma = \pm\frac{Pec\sec(kL)}{Ar^2}$$

where c is the half-depth of section in the plane of bending, r the radius of gyration and A the cross-sectional area. The maximum resultant compressive stress is

$$\sigma_c = -\frac{P}{A} - \frac{Pec \sec(kL)}{Ar^2} = -\frac{P}{A}\left\{1 + \frac{ec \sec(kL)}{r^2}\right\} \tag{11.18}$$

Hence

$$P = -\frac{\sigma_c A}{1 + \dfrac{ec \sec(kL)}{r^2}} \tag{11.19}$$

Since P is part of k it therefore exists in both sides of eqn. (11.19) a value for the critical buckling load can only be found by iteration.

The effect of eccentricity on maximum stress as a function of slenderness ratio is shown in Fig. 11.9.

Fig. 11.9

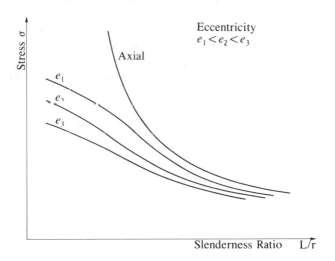

EXAMPLE 11.2

A vertical steel tube having 75 mm external and 62 mm internal diameters is 3 m long, fixed at the lower end and completely unrestrained at the upper end. The tube is subjected to a vertical compressive load parallel to, but eccentric by 6 mm from, the central axis. Determine the limiting value of the load so that there is no tensile stress at the base of the tube.

If the column had been loaded along its axis what would be the value of the Euler buckling load? $E = 208 \text{ GN/m}^2$.

SOLUTION

For zero tensile stress at the base

$$0 = -\frac{P}{A} + \frac{Pec \sec(kL)}{I}$$

Hence

$$\sec(kL) = \frac{I}{Aec} = \frac{0.815 \times 10^{-6}}{0.00138 \times 0.006 \times 0.0375} = 2.63$$

$$kL = 1.165$$

and

$$k = 0.388$$

$$P_c = 0.388^2 EI = 25.6 \text{ kN}$$

For the axially-loaded column

$$P_c = \frac{\pi^2 EI}{4L^2} = \frac{\pi^2 \times 208 \times 10^9 \times 0.815 \times 10^{-6}}{4 \times 9}$$

$$= 46.5 \text{ kN}$$

STRUTS HAVING INITIAL CURVATURE

After eccentricity the next practical departure from the Euler idealization is that in some cases a column or strut may not be perfectly straight before loading. This will influence the onset of instability. The following analysis was developed by Perry. The strut is illustrated in Fig. 11.10, in

Fig. 11.10

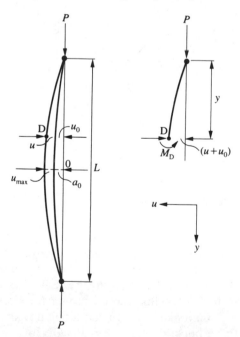

which the initial maximum deflection is a_0, the value of the deflection at D distant y from P is u_0, and

$$u_0 = a_0 \sin \frac{\pi y}{L}$$

When the buckling load P is applied, the deflection at y is increased by u and the bending moment at this point is M_D, where

$$M_D = P(u + u_0) = P\left(u + a_0 \sin \frac{\pi y}{L}\right)$$

Hence

$$EI\frac{\mathrm{d}^2 u}{\mathrm{d}y^2} = -P\left(u + a_0 \sin \frac{\pi y}{L}\right)$$

Therefore

$$\frac{\mathrm{d}^2 u}{\mathrm{d}y^2} + k^2\left(u + a_0 \sin \frac{\pi y}{L}\right) = 0 \tag{11.20}$$

where $k^2 = P/EI$; hence

$$u = A \cos (ky) + B \sin (ky) + \frac{k^2 a_0 \sin \dfrac{\pi y}{L}}{\dfrac{\pi^2}{L^2} - k^2}$$

The boundary conditions are that at $y = 0$ and $y = L$, $u = 0$; therefore $A = 0$ and $B \sin (kL) = 0$; and since k is not zero, it follows that $B = 0$; therefore

$$u = \frac{k^2 a_0 \sin \dfrac{\pi y}{L}}{\dfrac{\pi^2}{L^2} \quad k^2}$$

$$= \left(\frac{Pa_0}{\pi^2 \dfrac{EI}{L^2} - P}\right) \sin \frac{\pi y}{L}$$

$$= \left(\frac{Pa_0}{P_e - P}\right) \sin \frac{\pi y}{L} \tag{11.21}$$

where $P_e = \pi^2 EI/L^2$. Substituting for u_0,

$$u = \frac{u_0}{(P_e/P) - 1} \tag{11.22}$$

Thus the effect of the end thrust P is to increase the no-load maximum deflection a_0 by the multiplying factor $\{(P_e/P) - 1)\}^{-1}$.

From eqn. (11.22) it will be observed that, as the value of P approaches that of P_e, the basic Euler buckling load, the value of u increases, tending to become infinite. At $y = \frac{1}{2}L$, the increased deflection is given by

$$u' = \frac{a_0}{(P_e/P) - 1} \tag{11.23}$$

Plotting values of P and u, the curve shown in Fig. 11.11(a) is obtained. Failure of the strut would occur before P reached the theoretical value P_e.

Fig. 11.11

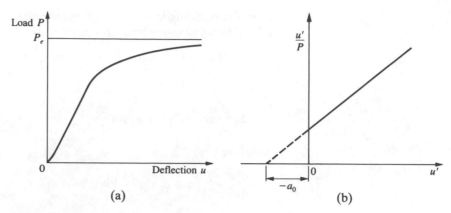

(a)　　　　　　　　(b)

Equation (11.23) may be written in the form

$$(P_e/P)u' - u' = a_0 \tag{11.24}$$

which shows that there is a linear relation between u' and u'/P, Fig. 11.11(b), the intercept on the axis of u' being equal to $-a_0$.

Let $\sigma = P/A$ and $\sigma_e = P_e/A$, where A is the cross-sectional area of the strut. Hence, from eqn. (11.21),

$$u = \left(\frac{\sigma}{\sigma_e - \sigma}\right)a_0 \sin\frac{\pi y}{L} \tag{11.21}$$

and total deflection at y is given by

$$u + u_0 = \left(\frac{\sigma}{\sigma_e - \sigma}\right)a_0 \sin\frac{\pi y}{L} + a_0 \sin\frac{\pi y}{L}$$

$$= \left(\frac{\sigma_e}{\sigma_e - \sigma}\right)a_0 \sin\frac{\pi y}{L} \tag{11.25}$$

The maximum deflection occurs at O, where $y = \frac{1}{2}L$; therefore

$$u_{max} = \left(\frac{\sigma_e}{\sigma_e - \sigma}\right)a_0 \tag{11.26}$$

Maximum bending moment is $a_0(\sigma_e/(\sigma_e - \sigma))P$ and the maximum compressive stress is given by

$$\sigma_c = -\frac{Pa_0\left(\dfrac{\sigma_e}{\sigma_e - \sigma}\right)c}{I} - \frac{P}{A} \tag{11.27}$$

where c is the distance from the neutral axis to the point of maximum compressive stress. If r is the least radius of gyration of the section then, substituting $P/A = \sigma$,

$$\sigma_c = \sigma\left(\frac{\eta\sigma_e}{\sigma_e - \sigma} + 1\right) \tag{11.28}$$

where $\eta = a_0c/r^2$. Taking σ_c equal to the yield stress in compression, σ_Y, we obtain a quadratic equation

$$\sigma^2 - \sigma\{\sigma_Y + \sigma_e(\eta + 1)\} + \sigma_Y\sigma_e = 0 \qquad (11.29)$$

from which the limiting value of σ and hence P can be determined.

EMPIRICAL FORMULAE FOR DESIGN

In practice many strut or column designs do not fall in the category of slenderness ratio relevant to Euler theory. Consequently a number of empirical formulae have been devised for different classes of materials (metal, timber, concrete) to cover the range A to B in Fig. 11.6. There is not space here to discuss these in detail, but some are given in summary and the reader is referred to British Standards and other design codes as required.

RANKINE–GORDON

$$P = \frac{\sigma A}{1 + a(L/r)^2} \qquad (11.30)$$

where symbols have their previous meaning and a is a constant dependent on end condition and material.

The formula applies for very short columns where buckling is not a factor, as well as for a range of larger slenderness ratios. Typical values for a for pin-ended struts are 0.000 1 for mild steel, 0.000 6 for cast iron, and 0.000 1 for timber.

For eccentric loading let the permissible load be P', then for the elastic limiting condition

$$\sigma_Y = -\frac{P'}{A} - \frac{P'ec}{Ar_b^2}$$

where r_b = radius of gyration in the plane of bending, and from eqn. (11.30) we can write

$$\sigma_Y = -\frac{P}{A} - \frac{aL^2P}{Ar^2}$$

Hence

$$P' = \frac{\left(1 + \dfrac{aL^2}{r^2}\right)P}{\left(1 + \dfrac{ec}{r_b^2}\right)} \qquad (11.31)$$

STRAIGHT LINE

This represents the region A to B in Fig. 11.6 by a straight line expressed as

$$P = \sigma A\{1 - c(L/r)\} \qquad (11.32)$$

where P = allowable load, σ = allowable compressive stress and c = a constant depending on the material and end restraint which is typically 0.005 for mild steel and 0.008 for cast iron.

PARABOLIC This formula, which is intended to agree with the Euler formula for long columns, is

$$P = \sigma A [1 - c(L/r)^2] \tag{11.33}$$

where c is a constant; the other symbols have the same meanings as in above. With pin ends and $L/r < 150$, then for mild steel c may be taken as 0.000 023.

EXAMPLE 11.3

A piece of timber of rectangular section 100×50 mm and 1 m in length can be regarded as having pinned ends. It is subjected to a compressive load P acting on one centre-line and eccentric from the other as shown in Fig. 11.12. Find the value of eccentricity which will result in an equal likelihood of reaching the limiting compressive stress for buckling in either of the principal planes. Use the Rankine–Gordon formula, with a constant of 1.33×10^{-3}. If the compressive stress is limited to 35 MN/m^2 calculate the allowable compressive load.

Fig. 11.12

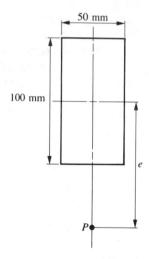

SOLUTION For buckling in either direction of the principal axes, P' must equal P in eqn. (11.31). Therefore

$$1 + \frac{ec}{r_b^2} = 1 + \frac{aL^2}{r^2}$$

$$e = \frac{aL^2 r_b^2}{cr^2}$$

$$I_l = \frac{0.1 \times 0.05^3}{12}, \qquad I_b = \frac{0.05 \times 0.1^3}{12}, \qquad A = 0.05 \times 0.1$$

$$r_l^2 = \frac{0.05^2}{12} \qquad r_b^2 = \frac{0.1^2}{12}$$

$$e = \frac{1.33 \times 10^{-3} \times 1 \times 0.1^2}{0.05 \times 0.05^2} = 106 \text{ mm}$$

From eqn. (11.30)

$$P = \frac{35 \times 10^6 \times 0.05 \times 0.1}{1 + \left[1.33 \times 10^{-3}\left(\dfrac{12}{0.05^2}\right)\right]} = \frac{175 \times 10^3}{7.4} \text{ N}$$
$$= 23.7 \text{ kN}$$

BUCKLING UNDER COMBINED COMPRESSION AND TRANSVERSE LOADING

Since transverse loading of a slender member causes bending, the effect on instability under axial compression is similar to that of initial curvature of the strut, and consequently a more rapid rate of deflection occurs.

This situation is also described as a beam-column and one example has already been studied. The eccentrically-loaded pin-ended column can be regarded as a beam-column carrying axial load P together with end moments P_e. Before proceeding to specific transverse loading cases it is appropriate to derive the differential relationships as was done in Chapter 6 for bending without end load.

Fig. 11.13

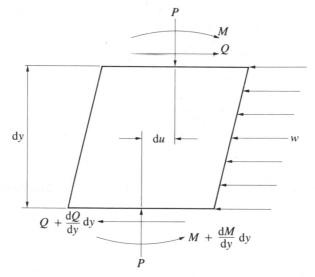

The equilibrium equations for the element shown in Fig. 11.13 are

$$w \, dy + \left(Q + \frac{dQ}{dy} \, dy\right) - Q = 0$$

and

$$-M - Q \, dy + w \, dy \, \frac{dy}{2} + \left(M + \frac{dM}{dy} \, dy\right) - P \, du = 0$$

These reduce to

$$\frac{dQ}{dy} = -w \tag{11.34}$$

and

$$\frac{dM}{dy} - Q - \frac{P\,du}{dy} = 0 \qquad (11.35)$$

When $P = 0$ these equations are the same as (6.1) and (6.2). Eliminating the shear force Q between these equations

$$\frac{d^2M}{dy^2} - \frac{P\,d^2u}{dy^2} = -w$$

Substituting for M, the curvature $-EI(d^2u/dy^2)$ gives

$$EI\frac{d^4u}{dy^4} + \frac{P\,d^2u}{dy^2} = w$$

or

$$\frac{d^4u}{dy^4} + \frac{P}{EI}\frac{d^2u}{dy^2} = \frac{w}{EI} \qquad (11.36)$$

Equation (11.36) is the general differential equation for beam-column-type situations. If the lateral loading is zero then the equation is of the form that was used for pure column buckling cases.

The standard form of solution for eqn. (11.36) is

$$u = A\sin(ky) + B\cos(ky) + C_1 wy^2 + C_2 wy + C_3$$

where $k = \sqrt{(P/EI)}$ and A, B, C_1, C_2, and C_3 are constants related to the boundary conditions.

PIN-ENDED STRUT CARRYING A UNIFORMLY-DISTRIBUTED LATERAL LOAD

The bending-moment at an arbitrary section A along the beam in Fig. 11.14 is

$$M_A = Pu + \frac{wL}{2}y - \frac{wy^2}{2} \qquad (11.37)$$

$$EI\frac{d^2u}{dy^2} = -Pu - \frac{wL}{2}y + \frac{wy^2}{2}$$

$$\frac{d^2u}{dy^2} + k^2u = \frac{wk^2y^2}{2P} - \frac{wLk^2y}{2P} \qquad (11.38)$$

The standard solution for this type of equation is

$$u = A\sin(ky) + B\cos(ky) + \frac{wy^2}{2P} - \frac{wL}{2P}y - \frac{w}{Pk^2} \qquad (11.39)$$

Fig. 11.14

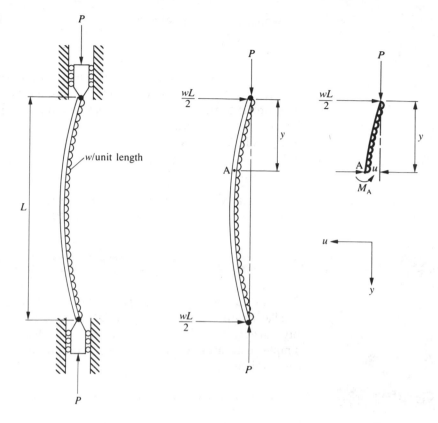

The boundary conditions are that at $y = 0$, $u = 0$; hence $B = w/Pk^2$; and at $y = L$, $u = 0$.

$$0 = A \sin (kL) + \frac{w}{Pk^2} \cos (kL) - \frac{w}{Pk^2}$$

$$A = \frac{w}{Pk^2} \{\operatorname{cosec} (kL) - \cot(kL)\}$$

$$= \frac{w}{Pk^2} \tan \frac{kL}{2}$$

The deflection equation (11.39) becomes

$$u = \frac{w}{Pk^2} \left\{ \tan \frac{kL}{2} \sin (ky) + \cos (ky) \right\} + \frac{wy^2}{2P} - \frac{wLy}{2P} - \frac{w}{Pk^2}$$

At $y = L/2$, $u = u_{max}$, therefore

$$u_{max} = \frac{w}{Pk^2} \left(\sec \frac{kL}{2} - 1 \right) - \frac{wL^2}{8P} \tag{11.40}$$

Now, $u_{max} \to \infty$ when $k(L/2) \to (n\pi/2)$. Hence

$$P_c = \frac{n^2\pi^2}{L^2} EI$$

for which the lowest value is

$$P_c = \frac{\pi^2 EI}{L^2}$$

which is the Euler load for a simple strut without transverse loading.

Although theoretically the load–deflection relationship would become asymptotic to the Euler load, in fact "failure" is governed by yielding at a lower load. Substituting the value of u_{max} from eqn. (11.40) and $y = L/2$ in eqn. (11.37) gives

$$M_{max} = \frac{w}{k^2} \left(\sec \frac{kL}{2} - 1 \right) \tag{11.41}$$

from which, with a limiting value of σ, either an allowable value of w can be determined for a particular end load P, or vice versa.

EXAMPLE 11.4

A part of a machine mechanism is illustrated in Fig. 11.15(a). The oscillating end portions AB and DE may be regarded as rigid, but can pivot freely at A, B, D and E. The central slender strut BCD has a simple restraint at C. What is the maximum load necessary to drive the mechanism?

SOLUTION

The deflected position of the members is shown in Fig. 11.15(b) with the

Fig. 11.15

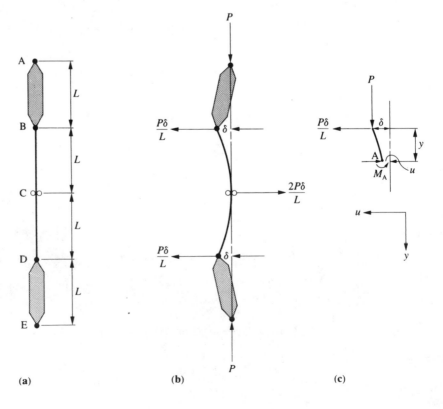

(a) (b) (c)

appropriate forces acting on BCD. The basic differential equation is

$$EI\frac{d^2u}{dy^2} = P(\delta - u) + \frac{P\delta}{L}y$$

$$\frac{d^2u}{dy^2} + k^2u = k^2\delta\left(\frac{y}{L} + 1\right)$$

Differentiating,

$$\frac{d^3u}{dy^3} + k^2\frac{du}{dy} = k^2\frac{\delta}{L}$$

The solution of this equation is

$$u = A + B\sin(ky) + C\cos(ky) + \frac{\delta}{L}y$$

The boundary conditions are (i) $y = 0$, $M = 0$; (ii) $y = L$, $du/dy = 0$; (iii) $y = 0$, $u = \delta$; (iv) $y = L$, $u = 0$. From (iii) and (iv),

$$\delta = A + C \quad \text{and} \quad 0 = A + B\sin(kL) + C\cos(kL) + \delta$$

From (ii),

$$\frac{du}{dy} = Bk\cos(ky) - Ck\sin(ky) + \frac{\delta}{L}$$

$$0 = Bk\cos(kL) - Ck\sin(kL) + \frac{\delta}{L}$$

From (i),

$$\frac{d^2u}{dy^2} = -Bk^2\sin(ky) - Ck^2\cos(ky)$$

$$0 = -Ck^2$$

Hence $C = 0$, $A = \delta$, and

$$B = \frac{\delta}{kL\cos(kL)} = -\frac{2\delta}{\sin(kL)}$$

from which

$$\tan kL = 2kL$$

and

$$kL = 1.166$$

Hence the load required to drive the mechanism is

$$P_c = \left(\frac{1.166}{L}\right)^2 EI = 1.36\frac{EI}{L^2}$$

OTHER EXAMPLES OF INSTABILITY

Instability is primarily a function of geometrical proportions, and in the case of columns this was expressed in terms of slenderness ratio in relation to compressive forces. Two other examples of members which become unstable with increasing load are shown in Fig. 11.16(a) and (b).

Fig. 11.16

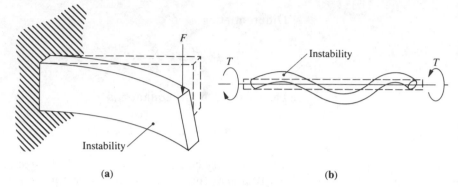

(a)   (b)

In the first case, because of the small width-to-depth ratio of the beam it is not possible to maintain the initial plane of bending, and the cross-section twists out of plane as shown. In the second case we again have a slenderness ratio influence in that the length of the bar under torsion is large compared to the diameter and it is not possible to maintain a straight axis of twist. Instability occurs through the shaft adopting a spiral axis when the torque reaches a critical level. There is not space here to analyse these cases but it is important for the designer to be aware of the possibility of these modes of deformation.

The final example relates to sheet metal structures of which perhaps the aeroplane is the most general case.

Fig. 11.17

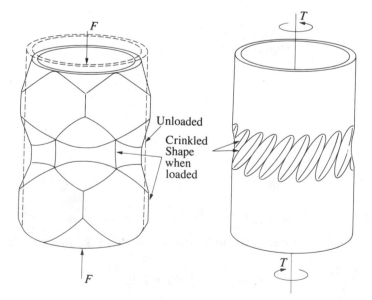

The surface skins will be a millimetre or less in thickness, and in areas subjected to tensile loading there is no problem. However, all parts of the aircraft are subjected to some kind of bending action in flight and so some skins will be subject to compressive loading. The buckling tendency of the skin has to be limited by frequent stiffeners, e.g. ribs, stringers, etc. Even the web of a spar has to have a series of stiffeners in order to carry shear, since this gives rise to diagonal tension and compression, the latter causing wrinkling.

Thin sheet tubular members subjected to compression or torsion may be stable in respect of the overall geometry, but will reveal local buckling instability characteristics as shown in Fig. 11.17.

SUMMARY

Instability in structural elements has been shown to be an important factor in design as it is a mode of "failure" dependent largely on compressive loading and geometrical proportions. The critical compressive load for buckling of columns or struts can be expressed as

$$P_c = \frac{\beta EA}{(L/r)^2}$$

where β is a constant depending on the material and end conditions. The idealized Euler theory is only applicable at large slenderness ratios and "real" struts will "fail" due to exceeding the yield stress of the material long before attaining the Euler load. Eccentricity of loading, initial curvature and transverse loading can each contribute to a lowering of the allowable "buckling" load as a function of the yield stress. Empirical formulae and design codes are now established for the design of columns for structural situations.

Local buckling of thin sheet material is also an important design consideration for which specialized texts such as *Theory of Elastic Stability* by Timoshenko and Gere, and design data sheets (R.Ae.S.) should be consulted.

PROBLEMS

11.1   A machine mechanism consists of two rigid members each of length 400 mm connected by a frictionless hinge at B and pinned at A and D as illustrated in Fig. 11.18. A spring of stiffness 30 N/mm is attached to the lower member at C as shown. Determine the critical load, *P,* for the system.

11.2   A vertical mechanism linkage consists of a slender member of length 500 mm and stiffness 500 N-m^2, built in at the lower end and pinned at the upper end to a rigid member of length 250 mm. The upper end of the latter is pinned between rollers which are axially aligned with the whole strut. Determine the critical compressive load, when applied at the roller bearing, which will cause buckling.

11.3   A straight slender column of height 2.77 m is fixed at the lower end and is entirely free at the upper end. The design criterion is to limit the maximum compressive

Fig. 11.18

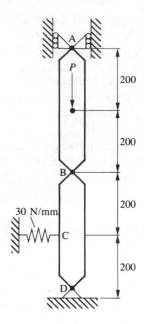

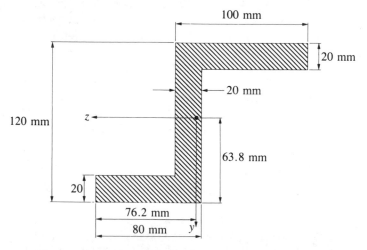

strain prior to buckling to 0.0008. Determine the required least radius of gyration.

11.4 A 4 m long strut has the cross-section shown in Fig. 11.19. Calculate the Euler buckling load for the strut if it has fixed ends. The Young's modulus for the strut material is 208 GN/m²; $r_z = 43.6$ mm, $r_y = 36.7$ mm, where r is the radius of gyration.

Fig. 11.19

11.5 The column shown in Fig. 11.20 has pinned ends. The upper part, of length $(1 - n)L$, which is slender and of constant stiffness EI, is fixed to the lower part, of length nL, which is rigid. Show that, at instability, $\tan \{k(1 - n)L\} = -knL$. What is the critical load if the upper and lower parts are of equal length?

11.6 In a temperature-control device a copper strip measuring 8 mm × 4 mm and 100 mm long is pinned at each end. How much axial pre-compression is required so that buckling will occur after a temperature rise of 50 °C? $E = 100$ GN/m², $\alpha = 18 \times 10^{-6}$ per deg C.

Fig. 11.20

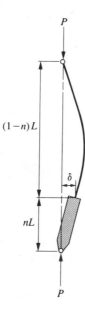

11.7   A strut 2 m long and pinned at both ends is subjected to an axial compressive force. If the strut cross-section is that shown in Fig. 11.21, calculate the Euler buckling load. $E = 207 \text{ GN/m}^2$.

Fig. 11.21

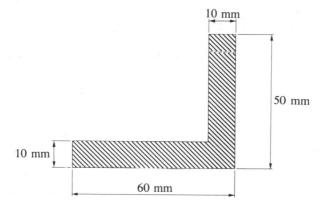

11.8   A circular steel column has a length of 2.44 m, an external diameter 101 mm and an internal diameter of 89 mm with its ends position fixed. Assuming that the centre-line is sinusoidal in shape with a maximum displacement at mid-length of 4.5 mm, determine the maximum stress due to an axial compressive load of 10 kN. $E = 205 \text{ GN/m}^2$.

11.9   A column is made up of two identical steel angle sections, as shown in Fig. 11.22 which are fixed at the lower end and free at the upper end. If the length of the column is 2.5 m and it is subjected to a compressive load of 8 kN calculate the safety factor in relation to buckling if:

(a) the two angle sections are touching along AA but not connected in any way;
(b) the two angle sections are fastened together along AA over the full length of the column.

The Young's modulus for steel is 207 GN/m^2 and the radii of gyration about xx or yy is 9.24 mm and about zz is 5.92 mm.

Fig. 11.22

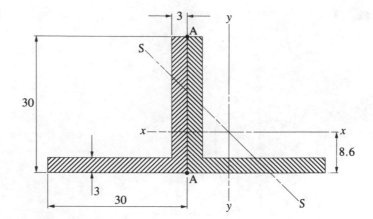

11.10 A tubular cast-iron column 5 m long has fixed ends and an external diameter of 250 mm. Calculate a suitable tube thickness if the column supports a load of 1 MN. Assume a constant of 1/6400 in the Rankine formula and a stress of 80 MN/m².

11.11 A column of length L is pinned at each end and is subjected to an axial compressive load P. A horizontal force F is now applied at mid-height to the column. Show that the maximum bending-moment is obtained in the form

$$M_{max} = \frac{F}{2k} \tan\left(\frac{kL}{2}\right) \quad \text{where} \quad k = \sqrt{(P/EI)}$$

11.12 A horizontal beam of length 3.6 m is simply-supported at each end and is tubular, having internal and external diameters of 46 and 50 mm respectively. It is subjected to axial compression of 5 kN and uniformly-distributed transverse loading of 50 N/m. Determine the maximum surface stress. $E = 200$ GN/m².

Chapter 12

Stress and strain transformations

INTRODUCTION

The preceding chapters have been concerned with "one-dimensional" problems of stress and strain, i.e. in any particular example consideration has only been given to one type of stress acting in one direction. However, the majority of engineering components and structures are subjected to loading conditions, or are of such a shape that, at any point in the material, a complex state of stress and strain exists involving normal (tension, compression) and shear components in various directions. A simple example of this is a shaft which transmits power through a pulley and belt drive. An element of material in the shaft would be subjected to normal stress and shear stress due to bending action and additional shear stress from the torque required to transmit power. It is not necessarily sufficient to be able to determine the individual values of these stresses in order to select a suitable material from which to make the shaft because, as will be seen later, on certain planes within the element more severe conditions of stress and strain exist. The identification of these maximum stresses is an essential step in the design process.

In practice, however, it should be noted that stress cannot be measured directly. It can only be calculated from a knowledge of the strains in the material. These may be measured easily using electrical resistance strain gauges and generally a rosette of such gauges is used to measure strains in precise directions on the surface of the component. This chapter describes how the information from strain gauges may be analysed.

Finally the chapter introduces the way in which stresses and strains in fibre-composite materials may be analysed. These materials are different to anything considered so far in the sense that they are anisotropic, i.e. they have different properties in different directions. However, it will be seen that the procedures developed in the chapter for the transformation of stresses and strains to different planes is directly applicable to fibre composites.

SYMBOLS, SIGNS AND ELEMENTS

Since conditions will be studied in which several different stresses occur simultaneously, it is essential to be consistent in the use of distinctive symbols, and a sign convention must be established and adhered to. A subscript notation will be used as follows:

Direct stress $\quad \sigma_x, \sigma_y, \sigma_z$

where the subscript denotes the direction of the stress.

Shear stress $\quad \tau_{xy}, \tau_{yx}, \tau_{yz}, \tau_{zy}, \tau_{xz}, \tau_{zx}$

where the first subscript denotes the direction of the normal to the plane on which the shear stress acts, and the second subscript the direction of the shear stress.

As in the previous work, tensile stress will be taken as positive and compressive stress negative. A shear stress is defined as positive when the direction of the stress vector and the direction of the normal to the plane are both in the positive sense or both in the negative sense in relation to the co-ordinate axes. If the directions of the shear stress and the normal to the plane are opposed in sign, then the shear stress is negative. Pairs of complementary shear stress components are therefore either both positive or both negative.

The angle between an inclined plane and a co-ordinate axis is positive when measured in the anticlockwise sense from the co-ordinate axis.

A general three-dimensional stress system is shown in Fig. 12.1.

Fig. 12.1

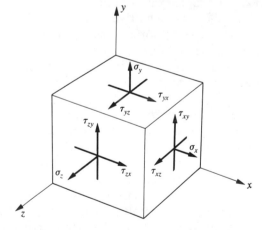

PLANE STRESS
If there are no normal and shear stresses on the two planes perpendicular to one of the co-ordinate directions, which implies that the complementary shear stresses τ_{yz} and τ_{xz} are also zero, then this system is known as *plane stress* (*see* also page 64). It is a situation found, or approximately so, in a number of important engineering problems. The analysis of complex stresses which follows is only concerned with plane stress conditions.

STRESSES ON A PLANE INCLINED TO THE DIRECTION OF LOADING

In preceding chapters the analysis has dealt with stress set up on a plane perpendicular (normal stress) or parallel (shear stress) to the direction of loading. However, if a piece of material is cut along a plane inclined at some angle θ to the direction of loading, then, in order to maintain equilibrium, a system of forces would have to be applied to the plane. This implies that there must be a stress system acting on that plane.

In Fig. 12.2 a bar is shown subjected to an axial tensile force F. The area of cross-section normal to the axis of the bar is denoted by A. If the bar is cut along the plane BC, inclined at an angle θ to the y-direction, then the force F applied to this portion of the bar can be reacted by component forces F_n normal and F_s tangential to the plane BC. Thus for equilibrium,

$$F_n = F \cos \theta \quad \text{and} \quad F_s = F \sin \theta$$

The area of the plane BC will be the area of the bar normal to its axis multiplied by $\sec \theta$.

Fig. 12.2

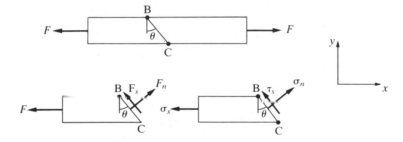

Denoting on the plane BC the positive direct stress by σ_n and the positive shear stress by τ_s, then

$$\sigma_n = \frac{F_n}{A \sec \theta} = \frac{F \cos \theta}{A \sec \theta} = \frac{F}{A} \cos^2 \theta$$

and

$$\tau_s = \frac{F_s}{A \sec \theta} = \frac{-F \sin \theta}{A \sec \theta} = \frac{-F}{A} \sin \theta \cos \theta$$

F/A is the direct stress normal to the axis of the bar and is equal to σ_x. Therefore

$$\sigma_n = \sigma_x \cos^2 \theta \tag{12.1}$$

and

$$-\tau_s = \sigma_x \sin \theta \cos \theta = \tfrac{1}{2}\sigma_x \sin 2\theta \tag{12.2}$$

Note that, in eqn. (12.1), when $\theta = 0$, σ_n is a maximum and equal to σ_x, and for $\theta = 90°$, σ_n is zero, indicating that there is no transverse stress in the bar.

Again, in eqn. (12.2), τ_s will be a maximum when $\sin 2\theta$ is a maximum, i.e. when $2\theta = 90°$ and $270°$, or $\theta = 45°$ and $135°$; the value of τ_s is then $\frac{1}{2}\sigma_x$ on the planes prescribed by $\theta = 45°$ and $135°$. This result is borne out in practice, and for materials whose shear strength is less than half the tensile strength, direct tensile loading results in failure along planes of maximum shear stress.

ELEMENT SUBJECTED TO NORMAL STRESSES

The rectangular element of material of unit thickness, Fig. 12.3, is subjected to tensile stresses in the x- and y-directions as shown.

Fig. 12.3

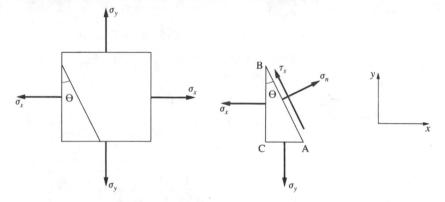

Considering a corner cut off the element by the plane AB inclined at θ to the y-axis, then positive normal and shear stresses will act on the plane as shown. For equilibrium of ABC, the forces on AB, BC and CA must be in balance. As the element is of constant unit thickness, the areas of the faces are proportional to the lengths of the sides of the triangle. Resolving *forces* normal to the plane AB,

$$\sigma_n AB - \sigma_x BC \cos\theta - \sigma_y AC \sin\theta = 0 \tag{12.3}$$

Dividing through by AB,

$$\sigma_n - \sigma_x \frac{BC}{AB} \cos\theta - \sigma_y \frac{AC}{AB} \sin\theta = 0$$

Therefore

$$\sigma_n = \sigma_x \cos^2\theta + \sigma_y \sin^2\theta = \tfrac{1}{2}(\sigma_x + \sigma_y) + \tfrac{1}{2}(\sigma_x - \sigma_y)\cos 2\theta \tag{12.4}$$

Resolving *forces* parallel to AB,

$$\tau_s AB + \sigma_x BC \sin\theta - \sigma_y AC \cos\theta = 0 \tag{12.5}$$

Dividing by AB,

$$\tau_s + \sigma_x \frac{BC}{AB} \sin\theta - \sigma_y \frac{AC}{AB} \cos\theta = 0$$

$$-\tau_s = \sigma_x \cos\theta \sin\theta - \sigma_y \sin\theta \cos\theta = \tfrac{1}{2}(\sigma_x - \sigma_y)\sin 2\theta \tag{12.6}$$

If σ_y is made zero, eqns. (12.4) and (12.6) reduce to equations (12.1) and (12.2) for the normal and shear stresses on a plane inclined to the direction of loading.

ELEMENT SUBJECTED TO SHEAR STRESSES

The rectangular element of the previous section is now considered with shear stresses on the faces instead of normal stresses. The plane AB inclined at θ to the y-axis, Fig. 12.4, is subjected to positive normal and

Fig. 12.4

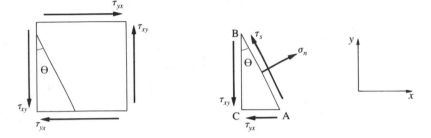

shear stresses as previously. Consider the equilibrium of the triangular portion ABC. Resolving *forces* normal to the plane AB,

$$\sigma_n AB - \tau_{xy} BC \sin \theta - \tau_{yx} AC \cos \theta = 0 \qquad (12.7)$$

Dividing through by AB as before,

$$\sigma_n - \tau_{xy} \frac{BC}{AB} \sin \theta - \tau_{yx} \frac{AC}{AB} \cos \theta = 0$$

$$\sigma_n = \tau_{xy} \cos \theta \sin \theta + \tau_{yx} \sin \theta \cos \theta \qquad (12.8)$$

But, from a consideration of complementary shear stresses,

$$\tau_{xy} = \tau_{yx} \qquad (12.9)$$

Therefore

$$\sigma_n = 2\tau_{xy} \sin \theta \cos \theta = \tau_{xy} \sin 2\theta \qquad (12.10)$$

Resolving forces parallel to the plane AB,

$$\tau_s AB - \tau_{xy} BC \cos \theta + \tau_{yx} AC \sin \theta = 0 \qquad (12.11)$$

Dividing by AB,

$$\tau_s - \tau_{xy} \frac{BC}{AB} \cos \theta + \tau_{yx} \frac{AC}{AB} \sin \theta = 0$$

Therefore

$$-\tau_s = \tau_{yx} \sin^2 \theta - \tau_{xy} \cos^2 \theta = -\tau_{xy} \cos 2\theta \qquad (12.12)$$

ELEMENT SUBJECTED TO GENERAL TWO-DIMENSIONAL STRESS SYSTEM

The general two-dimensional stress system shown in Fig. 12.5 may be obtained by a summation of the conditions of stress in Figs. 12.3 and 12.4. Hence the equations obtained for σ_n and τ_s under normal stresses and shear stresses separately may be added together to give values for the normal and shear stress on the inclined plane AB, Fig. 12.5, in the general stress system. Therefore, from eqns. (12.4) and (12.10)

$$\sigma_n = \tfrac{1}{2}(\sigma_x + \sigma_y) + \tfrac{1}{2}(\sigma_x - \sigma_y)\cos 2\theta + \tau_{xy}\sin 2\theta \tag{12.13}$$

and from eqns. (12.6) and (12.12)

$$-\tau_s = \tfrac{1}{2}(\sigma_x - \sigma_y)\sin 2\theta - \tau_{xy}\cos 2\theta \tag{12.14}$$

Fig. 12.5

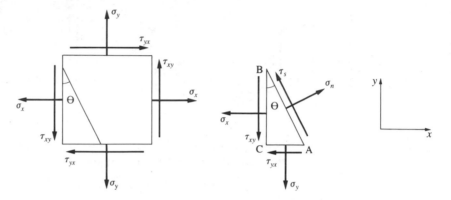

The validity of this method of superposition may be checked by considering the equilibrium of the triangular element of Fig. 12.5 and resolving forces perpendicular and parallel to the plane AB as was done in the previous sections.

It must be remembered that the signs in eqns. (12.13) and (12.14) are dependent on the directions chosen for the arrows (indicating stress) on the element in Fig. 12.5. If for any reason the directions of the stresses are different, e.g. compression instead of tension, then the appropriate signs in eqns. (12.13) and (12.14) must be changed.

EXAMPLE 12.1

A marine propellor shaft of 200 mm diameter is subjected to a torque of 126 kN-m and a pure bending moment of 157 kN-m. For each of the points A, B and C on the surface of the shaft (shown in Fig. 12.6) determine the normal and shear stresses on a plane at 60° clockwise to the longitudinal axis of the shaft.

SOLUTION

This is an example to illustrate the common engineering situation of shafts subjected to combined bending and torsion.

Fig. 12.6

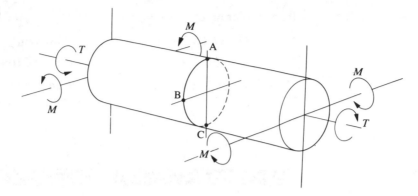

The second moments of area are as follows:

$$I = \frac{\pi d^4}{64} = \frac{\pi (0.2)^4}{64} = 78.54 \times 10^{-6} \, \text{m}^4$$

$$J = \frac{\pi d^4}{32} = 157.1 \times 10^{-6} \, \text{m}^4$$

Using the appropriate equations for bending and torsion,

$$\sigma = \frac{My}{I} = \pm \frac{157 \times 10^3 \times 0.1}{78.54 \times 10^{-6}} = \pm 200 \, \text{MN/m}^2$$

$$\tau = \frac{Tr}{J} = \frac{126 \times 10^3 \times 0.1}{157.1 \times 10^{-6}} = 80 \, \text{MN/m}^2$$

The stress components at A, B and C are

$$\sigma_x^A = +200 \, \text{MN/m}^2 \qquad \tau_{xz}^A = 80 \, \text{MN/m}^2$$
$$\sigma_x^B = 0 \qquad \tau_{xz}^B = 80 \, \text{MN/m}^2$$
$$\sigma_x^C = -200 \, \text{MN/m}^2 \qquad \tau_{xz}^C = 80 \, \text{MN/m}^2$$

Using eqns. (12.13) and (12.14) and noting that the correct value of θ on the element is 30° as shown in Fig. 12.7,

$$\sigma_n^A = 100 + 100 \cos 60 + 80 \sin 60$$
$$= +219.3 \, \text{MN/m}^2$$
$$- \tau_s^A = +100 \sin 60 - 80 \cos 60$$
$$\tau_s^A = -46.6 \, \text{MN/m}^2$$

Fig. 12.7

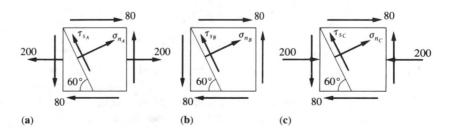

(a)   (b)   (c)

By a similar analysis we find

$$\sigma_n{}^B = +69.3 \text{ MN/m}^2 \qquad \tau_s{}^B = +40 \text{ MN/m}^2$$
$$\sigma_n{}^C = -80.7 \text{ MN/m}^2 \qquad \tau_s{}^C = +126.6 \text{ MN/m}^2$$

The stress components above are illustrated acting on the planes in Fig. 12.7. This example is extended in relation to principal stresses in Example 12.5.

MOHR'S STRESS CIRCLE

Considering eqns. (12.13) and (12.14) once more and rewriting,

$$\sigma_n - \tfrac{1}{2}(\sigma_x + \sigma_y) = \tfrac{1}{2}(\sigma_x - \sigma_y)\cos 2\theta + \tau_{xy}\sin 2\theta \qquad (12.15)$$

$$-\tau_s = \tfrac{1}{2}(\sigma_x - \sigma_y)\sin 2\theta - \tau_{xy}\cos 2\theta \qquad (12.16)$$

Squaring both sides and adding the equations,

$$[\sigma_n - \tfrac{1}{2}(\sigma_x + \sigma_y)]^2 + \tau_s^2 = \tfrac{1}{4}(\sigma_x - \sigma_y)^2 + \tau_{xy}^2 \qquad (12.17)$$

This is the equation of a circle of radius

$$\sqrt{[\tfrac{1}{4}(\sigma_x - \sigma_y)^2 + \tau_{xy}^2]}$$

and whose centre has the co-ordinates

$$[\tfrac{1}{2}(\sigma_x + \sigma_y), \, 0]$$

The circle represents all possible states of normal and shear stress on any plane through a stressed point in a material, and was developed by the German engineer Otto Mohr. The element of Fig. 12.5 and the corresponding Mohr diagram are shown in Fig. 12.8.

Fig. 12.8

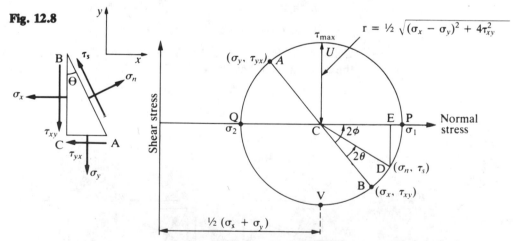

The sign convention used on the circle will be, for normal stress, positive to the right and negative to the left of the origin. Shear stresses which might be described as trying to cause a clockwise rotation of an element are plotted above the abscissa axis, i.e. "positive", and shear stresses appearing as anticlockwise rotation are plotted below the axis, i.e. "negative".

It is important to remember that shear stress plotted, say, below the σ-axis, although being regarded as negative in the circle construction, may be either positive or negative on the physical element according to the shear stress convention previously defined. Likewise, positive shear stress on the circle may be either positive or negative on the element.

The diagram is constructed as follows: using co-ordinate axes of normal stress and shear stress, both to the same scale, point B (σ_x, τ_{xy}) is plotted representing the direct and shear stress acting on the plane BC of the element. Assuming in this case $\sigma_y < \sigma_x$, the point A (σ_y, τ_{yx}) is plotted to represent the stresses on the plane AC of the element. The normal stress axis bisects the line joining AB at C, and with centre C and radius AC a circle is drawn.

An angle equal to twice that in the element, i.e. 2θ, is set off from BC in the anticlockwise direction (the same sense as in the element), and the line CD then cuts the circle at the point whose co-ordinates are (σ_n, τ_s). These are then the normal and shearing stresses on the plane AB in the element.

The validity of the diagram is demonstrated thus:

$$\sigma_n = OC + CE = \tfrac{1}{2}(\sigma_x + \sigma_y) + r \cos (2\phi - 2\theta)$$
$$= \tfrac{1}{2}(\sigma_x + \sigma_y) + r \cos 2\theta \cos 2\phi + r \sin 2\theta \sin 2\phi$$

But

$$\cos 2\phi = \frac{\tfrac{1}{2}(\sigma_x - \sigma_y)}{r} \quad \text{and} \quad \sin 2\phi = \frac{\tau_{xy}}{r}$$

Therefore

$$\sigma_n = \tfrac{1}{2}(\sigma_x + \sigma_y) + \tfrac{1}{2}(\sigma_x - \sigma_y) \cos 2\theta + \tau_{yx} \sin 2\theta$$
$$\tau_s = DE = r \sin (2\phi - 2\theta)$$
$$= r(\sin 2\phi \cos 2\theta - \cos 2\phi \sin 2\theta)$$

Substituting for $\cos 2\phi$ and $\sin 2\phi$,

$$-\tau_s = \tfrac{1}{2}(\sigma_x - \sigma_y) \sin 2\theta - \tau_{xy} \cos 2\theta$$

These expressions for σ_n and τ_s are seen to be the same as those derived from equilibrium of the element. Thus, if at a point in a material the stress conditions are known on two planes, then the normal and shear stresses on any other plane through the point can be found using Mohr's circle.

Certain features of the diagram are worthy of note. The sides of the element AC and CB, which are 90° apart, are represented on the circle by AC and CB, 180° apart. A compressive direct stress would be plotted to the left of the shear-stress axis. The maximum shear stress in an element is given by the top and bottom points of the circle, i.e.

$$\tau_{s\,max} = \pm\sqrt{[\tfrac{1}{4}(\sigma_x - \sigma_y)^2 + \tau_{xy}^{\,2}]} \tag{12.18}$$

and the corresponding normal stress is $\tfrac{1}{2}(\sigma_x + \sigma_y)$. The angle θ to the plane on which a maximum shear stress acts is obtained from the circle as

$$\tan 2\theta = \tan (90° + 2\phi) = -\cot 2\phi$$

Therefore

$$\tan 2\theta = -\left(\frac{\sigma_x - \sigma_y}{2\tau_{xy}}\right) \tag{12.19}$$

The second plane of maximum shear stress is displaced by 90° from that above.

EXAMPLE 12.2

At a point in a complex stress field $\sigma_x = 40 \text{ MN/m}^2$, $\sigma_y = 80 \text{ MN/m}^2$, and $\tau_{xy} = -20 \text{ MN/m}^2$. Use Mohr's circle solution to find the normal and shear stresses on a plane at 45° to the y-axis.

SOLUTION

The stresses on the element are as shown in Fig. 12.9(b) and the corresponding Mohr's circle is shown in Fig. 12.9(a). From Fig. 12.9(a)

Fig. 12.9

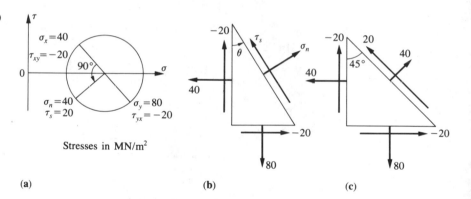

Stresses in MN/m²

(a)   (b)   (c)

for $\theta = 45°$ on the element which is 90° on the circle.

$$\sigma_n = 40 \text{ MN/m}^2 \quad \text{and} \quad \tau_s = 20 \text{ MN/m}^2$$

Note that τ_s will be upwards to the left as shown in Fig. 12.9(c) because it is "negative" on the circle (i.e. would turn element anticlockwise). The values of σ_n and τ_s may be checked by calculation using eqns. (12.13) and (12.14).

EXAMPLE 12.3

Construct Mohr's circle for the following point stresses: $\sigma_x = 60 \text{ MN/m}^2$, $\sigma_y = 10 \text{ MN/m}^2$ and $\tau_{xy} = +20 \text{ MN/m}^2$, and hence determine the stress components and planes on which the shear stress is a maximum.

SOLUTION

The stresses on the element are as shown in Fig. 12.10(*b*) and the corresponding Mohr's circle is shown in Fig. 12.10(*a*).

Fig. 12.10

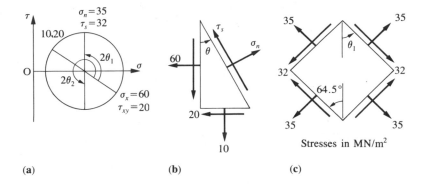

(a)   (b)   (c)

Stresses in MN/m²

From Fig. 12.10, the normal and maximum shear stress components are

$$\sigma_n = 35 \text{ MN/m}^2 \quad \text{and} \quad \tau_s = 32 \text{ MN/m}^2$$

acting on the planes $\theta_1 = 64.5°$, $\theta_2 = 154.5°$. The orientation of the element experiencing these stresses is shown in Fig. 12.10(*c*).

PRINCIPAL STRESSES AND PLANES

It has been shown that Mohr's circle represents all possible states of normal and shear stress at a point. From Fig. 12.8 it can be seen that there are two planes, QC and CP, 180° apart on the diagram and therefore 90° apart in the material, on which the shear stress τ_s is zero. These planes are termed *principal planes* and the normal stresses acting on them are termed *principal stresses*. The latter are denoted by σ_1 and σ_2 at P and Q respectively, and are the maximum and minimum values of normal stress that can be obtained at a point in a material. The values of the principal stresses can be found either from eqn. (12.17) by putting $\tau_s = 0$, or directly from the Mohr diagram; hence

$$\sigma_1 = \frac{\sigma_x + \sigma_y}{2} + \tfrac{1}{2}\sqrt{[(\sigma_x - \sigma_y)^2 + 4\tau_{xy}^2]} \tag{12.20}$$

$$\sigma_2 = \frac{\sigma_x + \sigma_y}{2} - \tfrac{1}{2}\sqrt{[(\sigma_x - \sigma_y)^2 + 4\tau_{xy}^2]} \tag{12.21}$$

where σ_1 is the maximum and σ_2 the minimum principal stress. The planes are specified by

$$2\theta = 2\phi \quad \text{and} \quad 180° + 2\phi$$

or

$$\theta = \phi \quad \text{and} \quad 90° + \phi$$

But

$$\tan 2\phi = \frac{\tau_{xy}}{\frac{1}{2}(\sigma_x - \sigma_y)} \qquad \left(\sin 2\phi = \frac{\tau_{xy}}{\sqrt{[\frac{1}{2}(\sigma_x - \sigma_y)^2 + \tau_{xy}^2]}}\right)$$

Therefore

$$\theta = \frac{1}{2}\tan^{-1}\left(\frac{2\tau_{xy}}{\sigma_x - \sigma_y}\right) \quad \text{and} \quad 90° + \frac{1}{2}\tan^{-1}\left(\frac{2\tau_{xy}}{\sigma_x - \sigma_y}\right) \qquad (12.22)$$

Thus the magnitude and direction of the principal stresses at any point in a material depend on σ_x, σ_y and τ_{xy} at that point, Fig. 12.11.

Fig. 12.11

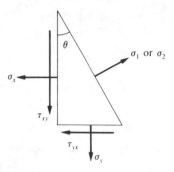

MAXIMUM SHEAR STRESS IN TERMS OF PRINCIPAL STRESSES

It was shown earlier, eqn. (12.18), that the maximum shear stress at a point is given by

$$\tau_{s\,max} = \pm\frac{1}{2}\sqrt{[(\sigma_x - \sigma_y)^2 + 4\tau_{xy}^2]}$$

If expressions (12.20) and (12.21) for σ_1 and σ_2 are subtracted, then

$$\sigma_1 - \sigma_2 = \pm\sqrt{[(\sigma_x - \sigma_y)^2 + 4\tau_{xy}^2]} \qquad (12.23)$$

and therefore

$$\tau_{s\,max} = \frac{1}{2}(\sigma_1 - \sigma_2) \qquad (12.24)$$

It should be noted that principal stresses are considered a maximum or minimum mathematically, e.g. a compressive or negative stress is less than a positive stress, irrespective of numerical value.

In Mohr's circle the principal planes PC and QC are at 90° to those of maximum shear stress, UC and VC, and therefore in the material the angles between these two sets of planes become 45°, or the maximum shear-stress planes bisect the principal planes.

EXAMPLE 12.4

Determine the principal stresses and maximum shear stresses at points A and B by calculation and at point C by a Mohr's circle construction for the I-section beam shown in Fig. 12.12.

Fig. 12.12

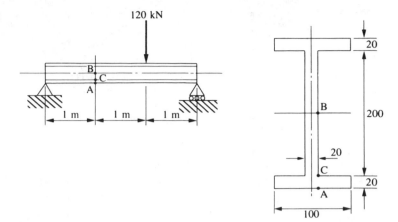

SOLUTION

The reaction at the left-hand end is 40 kN and hence at the required cross-section of the beam

$$M = 40 \times 1 = 40 \text{ kN-m}; \qquad Q = 40 \text{ kN}; \qquad I = 61\,867 \times 10^3 \text{ mm}^4$$

$$y_A = +120 \text{ mm}; \ y_B = 0; \ y_C = +100 \text{ mm}.$$

The bending stresses are:

$$\sigma_x{}^A = \frac{40 \times 10^3 \times (+0.12)}{61\,867 \times 10^{-9}} = 77.5 \text{ MN/m}^2$$

$$\sigma_x{}^B = 0$$

$$\sigma_x{}^C = +77.5 \times \frac{0.1}{0.12} = +64.5 \text{ MN/m}^2$$

The shear stresses are

$$\tau_{xy}{}^B = \frac{QA\bar{y}}{bI} = \frac{40 \times 10^3}{61\,867 \times 10^{-9}} \left\{ \frac{(0.2)^2}{8} + \frac{0.1}{8 \times 0.02} (0.24^2 - 0.2^2) \right\}$$

$$= 10.34 \text{ MN/m}^2$$

$$\tau_{xy}{}^C = \frac{40 \times 10^3 \times 2000 \times 10^{-6} \times 0.11}{0.02 \times 61\,867 \times 10^{-9}} = 7.11 \text{ MN/m}^2$$

$$\tau_{xy}{}^A = \frac{40 \times 10^3 \times (0.24 - 0.02)(0.1 - 0.02)}{4 \times 61\,867 \times 10^{-9}} = 2.84 \text{ MN/m}^2$$

Using eqns. (12.20) and (12.21),

$$\sigma_1{}^A = \frac{77.5}{2} + \tfrac{1}{2}(77.5^2 + 4 \times 2.84^2)^{1/2} = +77.6 \text{ MN/m}^2$$

$$\sigma_2{}^A = \frac{77.5}{2} - \tfrac{1}{2}(77.5^2 + 4 \times 2.84^2)^{1/2} = -0.1 \text{ MN/m}^2$$

$$\sigma_1{}^B = +\tau_{xy}{}^B = +10.34 \text{ MN/m}^2; \qquad \sigma_2{}^B = -\tau_{xy}{}^B = -10.34 \text{ MN/m}^2$$

$$\tau_{max}^A = \frac{77.6 - (-0.1)}{2} = 38.8 \text{ MN/m}^2;$$

$$\tau_{max}^B = \frac{10.34 - (-10.34)}{2} = 10.34 \text{ MN/m}^2$$

Fig. 12.13

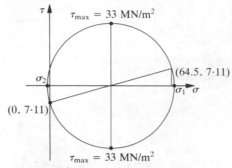

The Mohr's circle construction is shown in Fig. 12.13, from which the principal stresses and maximum shear stress are:

$$\sigma_1{}^C = +65.3 \text{ MN/m}^2; \quad \sigma_2{}^C = -0.77 \text{ MN/m}^2; \quad \tau_{max}^C = 33 \text{ MN/m}^2$$

EXAMPLE 12.5

Derive expressions, in terms of M and T, for the magnitude and direction of the principal stresses at points A, B and C on the shaft in Example 12.1.

SOLUTION

The maximum bending and shear stresses occur at the outer surface of the shaft and are given by

$$\sigma_x = \frac{32M}{\pi d^3} \quad \text{and} \quad \tau_{xz} = \frac{16T}{\pi d^3}$$

where d is the diameter of the shaft.

The stress components on an element of material at the surface points A, B and C are shown in Fig. 12.7. The shear stresses are the same in each; however, the bending stress is maximum tension at A, zero at B (the neutral plane) and maximum compression at C. The principal stresses at these three points are therefore

At A $\quad \sigma_1, \sigma_2 = \tfrac{1}{2}\sigma_x \pm \tfrac{1}{2}\sqrt{(\sigma_x^2 + 4\tau_{xz}^2)}$

At B $\quad \sigma_1 = -\sigma_2 = \tau_{xz}$

At C $\quad \sigma_1, \sigma_2 = -\tfrac{1}{2}\sigma_x \pm \tfrac{1}{2}\sqrt{(\sigma_x^2 + 4\tau_{xz}^2)}$

At B, therefore, there is a state of pure shear.

The principal stresses can be expressed in terms of the bending moment and torque by substituting for σ_x and τ_{xz}:

At A $\quad \sigma_1, \sigma_2 = \dfrac{16M}{\pi d^3} \pm \sqrt{\left[\left(\dfrac{16M}{\pi d^3}\right)^2 + \left(\dfrac{16T}{\pi d^3}\right)^2\right]}$

$$= \frac{16}{\pi d^3}[M \pm \sqrt{(M^2 + T^2)}] \tag{12.25}$$

At B $\quad \sigma_1 = -\sigma_2 = \dfrac{16T}{\pi d^3}$

At C the values are the same as at A but are negative.

The inclinations of the principal planes to the z-axis are

$$\theta = \tfrac{1}{2}\tan^{-1}\frac{T}{M} \quad \text{and} \quad 90° + \tfrac{1}{2}\tan^{-1}\frac{T}{M}$$

The maximum shear stress is given by

$$\tau = \frac{16}{\pi d^3}\sqrt{(M^2 + T^2)} \tag{12.26}$$

It is sometimes useful to express the above principal stresses in terms of a shaft subjected to bending only. If the equivalent bending moment is M_e, then

$$\frac{32M_e}{\pi d^3} = \frac{16}{\pi d^3}[M \pm \sqrt{(M^2 + T^2)}]$$

and

$$M_e = \tfrac{1}{2}[M \pm \sqrt{(M^2 + T^2)}] \tag{12.27}$$

If expressed in terms of an equivalent torque only,

$$T_e = M \pm \sqrt{(M^2 + T^2)} \tag{12.28}$$

MAXIMUM SHEAR STRESS IN THREE DIMENSIONS

So far the analysis of complex stress situations has been restricted to the two-dimensional situation. Therefore the value of the maximum shear stress obtained from the equations or the Mohr diagram represent the maximum shear stress in the (x, y) plane. However, even for a state of plane stress this may not be the maximum shear stress in the material. To obtain the true maximum shear stress for use in design calculations (*see* Chapter 13) it is necessary to consider all three principal planes. The three-dimensional element subjected to the principal stresses is shown in Fig. 12.14. The principal stress σ_3 is zero in this particular case since we

Fig. 12.14

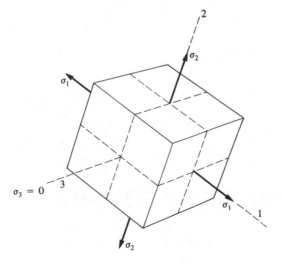

Fig. 12.15

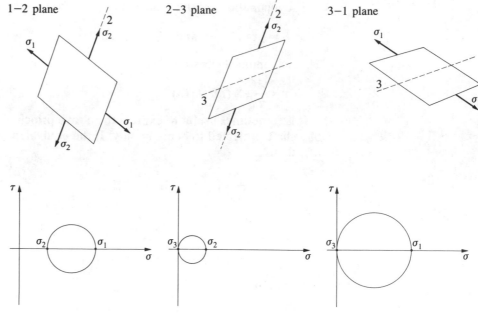

have been concerned only with plane stress situations. For convenience, consider the axis of the three principal stresses to be labelled 1, 2 and 3. Considering each of the three planes (1, 2), (2, 3) and (3, 1) in turn it is possible to construct a Mohr diagram for each, as shown in Fig. 12.15. It is then apparent that a composite Mohr diagram could be constructed, as shown in Fig. 12.16, by superimposing these three diagrams. This Mohr diagram then enables the true maximum shear stress in the material to be determined.

Fig. 12.16 Composite Mohr diagram for three-dimensional system

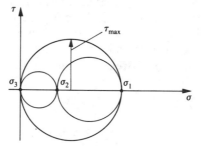

It is evident from the composite Mohr diagram that if σ_1 and σ_2 are either both positive or both negative then τ_{max} in the (1, 2) plane will not be the maximum shear stress in the material. It should also be noted that in the above superposition of the Mohr diagrams for the three planes, it is not essential that $\sigma_3 = 0$, so that in fact Mohr diagrams may be used generally for three-dimensional stress systems.

GENERAL TWO-DIMENSIONAL STATE OF STRESS AT A POINT

In the preceding paragraphs certain specific stress conditions which exist on planes in an element subjected to normal and shear stress have been

Fig. 12.17

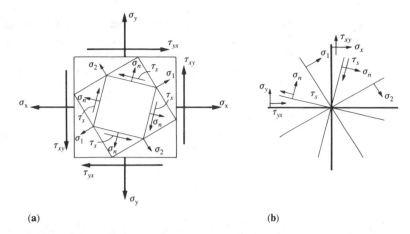

(a)　　　　　　　　　　　　(b)

derived, and these are shown on elements of the material in Fig. 12.17(*a*). Although it has been convenient in the foregoing analyses to draw an element of material it must be remembered that the stresses actually act at a *point* in the material. Figure 12.17(*b*) therefore illustrates the general two-dimensional state of stress at a point.

STATES OF STRAIN

In the following analysis all strains are considered to be small in magnitude. *Normal strains* are defined as the ratio of change in length to original length in a particular direction, and a subscript notation similar to that for stresses will be adopted, ε_x and ε_y being the direct strains in the *x*- and *y*-directions respectively, positive for tension and negative for compression. *Shear strain* is defined as the change in angle between two planes initially at right angles, and the symbol and subscripts γ_{xy} will be used for shear referred to the *x*- and *y*-planes.

PLANE STRAIN　　*Plane strain* is the term used to describe the strain system in which the normal strain in, say, the *z*-direction, along with the shear strains γ_{zx} and

Fig. 12.18

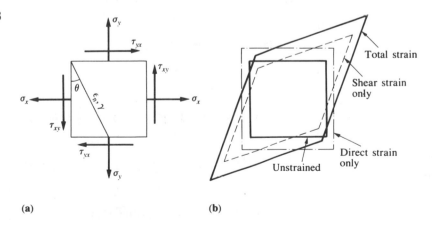

(a)　　　　　　　　　　　　(b)

γ_{zy}, are zero. It should be noted that plane stress is not the stress system associated with plane strain. It will be evident from page 64 that plane strain, i.e. $\varepsilon_z = 0$, is associated with a three-dimensional stress system, and likewise plane stress is related to a three-dimensional strain system.

The stress system in Fig. 12.18(a) will give rise to a strain system combining direct and shear strains as shown in an exaggerated manner at (b). The object now is to determine the direct, ε_n, and shear, γ_s, strains for directions normal and tangential to a plane, inclined at θ to a co-ordinate direction, in terms of the direct, ε_x, ε_y, and shear, γ_{xy}, γ_{yx}, strains referred to the co-ordinate planes.

NORMAL STRAIN IN TERMS OF CO-ORDINATE STRAINS

Referring to Fig. 12.19 $(A'C' - AC)/AC$ gives the strain normal to the plane FB related to the normal stress σ_n. Considering the triangles ACD and A'C'D' and with δx as the increase in length from AD to A'D', and δy the increase in length from CD to C'D', then

$$A'D' = AD + \delta x = AD\left(1 + \frac{\delta x}{AD}\right) = AD(1 + \varepsilon_x)$$

$$C'D' = CD + \delta y = CD\left(1 + \frac{\delta y}{CD}\right) = CD(1 + \varepsilon_y)$$

Fig. 12.19
(a) Unstrained; (b) strained

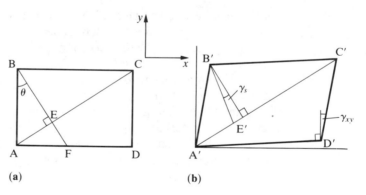

(a)　　　　　　　　　　(b)

Similarly

$$A'C' = AC(1 + \varepsilon_n)$$

Now,

$$(A'C')^2 = (A'D')^2 + (C'D')^2 - 2A'D' \cdot C'D' \cos(90° + \gamma_{xy})$$

or

$$(AC)^2(1 + \varepsilon_n)^2 = (AD)^2(1 + \varepsilon_x)^2 + (CD)^2(1 + \varepsilon_y)^2$$
$$+ 2AD(1 + \varepsilon_x)CD(1 + \varepsilon_y)\sin\gamma_{xy}$$

Since strains are assumed small, then $\sin\gamma_{xy} \approx \gamma_{xy}$ and second-order powers may be neglected. We have

$$(AC)^2(1 + 2\varepsilon_n) = (AD)^2(1 + 2\varepsilon_x) + (CD)^2(1 + 2\varepsilon_y)$$
$$+ 2AD \,.\, CD\gamma_{xy}$$

which, with $(AC)^2 = (AD)^2 + (CD)^2$, reduces to

$$2\varepsilon_n(AC)^2 = 2\varepsilon_x(AD)^2 + 2\varepsilon_y(CD)^2 + 2AD \,.\, CD\gamma_{xy}$$

Dividing through by $2(AC)^2$ and introducing $\sin\theta$ and $\cos\theta$,

$$\varepsilon_n = \varepsilon_x \cos^2\theta + \varepsilon_y \sin^2\theta + \gamma_{xy}\sin\theta\cos\theta \qquad (12.29)$$

Therefore

$$\varepsilon_n = \frac{\varepsilon_x + \varepsilon_y}{2} + \frac{\varepsilon_x - \varepsilon_y}{2}\cos 2\theta + \tfrac{1}{2}\gamma_{xy}\sin 2\theta \qquad (12.30)$$

SHEAR STRAIN IN TERMS OF CO-ORDINATE STRAINS

Referring to Fig. 12.19, the shear strain γ_s related to the shear stress τ_s, Fig. 12.5, is given by the change in angle between EB and AE, or $\angle AEB - \angle A'E'B'$. Considering the triangles AEB and A'E'B', then, as before,

$$A'B' = AB(1 + \varepsilon_y)$$

$$A'E' = AE(1 + \varepsilon_n)$$

$$E'B' = EB(1 + \varepsilon_{n+90})$$

ε_{n+90} is the normal strain in a direction at $90°$ to ε_n. Now

$$(A'B')^2 = (A'E')^2 + (E'B')^2 - 2A'E' \,.\, E'B' \cos(90° + \gamma_s) \qquad (12.31)$$

or

$$(AB)^2(1 + \varepsilon_y)^2 = (AE)^2(1 + \varepsilon_n)^2 + (EB)^2(1 + \varepsilon_{n+90})^2$$
$$-2AE(1 + \varepsilon_n)EB(1 + \varepsilon_{n+90})\cos(90° + \gamma_s)$$

But $\cos(90° + \gamma_s) = \sin\gamma_s \approx -\gamma_s$, and, neglecting the second order of small quantities,

$$(AB)^2(1 + 2\varepsilon_y) = (AE)^2(1 + 2\varepsilon_n) + (EB)^2(1 + 2\varepsilon_{n+90}) + 2AE \,.\, EB\gamma_s$$

But $(AB)^2 = (AE)^2 + (EB)^2$, and dividing by $2(AB)^2$ gives

$$\varepsilon_y = \varepsilon_n \sin^2\theta + \varepsilon_{n+90}\cos^2\theta + \gamma_s\sin\theta\cos\theta$$

Rewriting in terms of 2θ,

$$-\tfrac{1}{2}\gamma_s\sin 2\theta = \frac{\varepsilon_{n+90} + \varepsilon_n}{2} + \frac{\varepsilon_{n+90} - \varepsilon_n}{2}\cos 2\theta - \varepsilon_y \qquad (12.32)$$

Now,

$$\varepsilon_{n+90} = \frac{\varepsilon_x + \varepsilon_y}{2} + \frac{\varepsilon_x - \varepsilon_y}{2} \cos 2(\theta + 90°) + \tfrac{1}{2}\gamma_{xy} \sin 2(\theta + 90°)$$

$$= \frac{\varepsilon_x + \varepsilon_y}{2} - \frac{\varepsilon_x - \varepsilon_y}{2} \cos 2\theta - \tfrac{1}{2}\gamma_{xy} \sin 2\theta$$

and

$$\varepsilon_n = \frac{\varepsilon_x + \varepsilon_y}{2} + \frac{\varepsilon_x - \varepsilon_y}{2} \cos 2\theta + \tfrac{1}{2}\gamma_{xy} \sin 2\theta$$

Therefore

$$\frac{\varepsilon_{n+90} + \varepsilon_n}{2} = \frac{\varepsilon_x + \varepsilon_y}{2}$$

and

$$\frac{\varepsilon_{n+90} - \varepsilon_n}{2} = -\frac{\varepsilon_x - \varepsilon_y}{2} \cos 2\theta - \tfrac{1}{2}\gamma_{xy} \sin 2\theta$$

Substituting the above expressions in eqn. (12.32),

$$-\tfrac{1}{2}\gamma_s \sin 2\theta = \frac{\varepsilon_x + \varepsilon_y}{2} - \frac{\varepsilon_x - \varepsilon_y}{2} \cos^2 2\theta - \tfrac{1}{2}\gamma_{xy} \sin 2\theta \cos 2\theta - \varepsilon_y$$

$$= \frac{\varepsilon_x - \varepsilon_y}{2}(1 - \cos^2 2\theta) - \tfrac{1}{2}\gamma_{xy} \sin 2\theta \cos 2\theta$$

Dividing through by $\sin 2\theta$,

$$-\tfrac{1}{2}\gamma_s = \frac{\varepsilon_x - \varepsilon_y}{2} \sin 2\theta - \tfrac{1}{2}\gamma_{xy} \cos 2\theta \tag{12.33}$$

MOHR'S STRAIN CIRCLE

The graphical method of obtaining normal and shear stress on any plane by Mohr's circle can also be employed to determine normal and shear strains at a point. Considering eqns. (12.30) and (12.33) and rewriting, we have

$$\varepsilon_n - \frac{\varepsilon_x + \varepsilon_y}{2} = \frac{\varepsilon_x - \varepsilon_y}{2} \cos 2\theta + \tfrac{1}{2}\gamma_{xy} \sin 2\theta$$

and

$$-\tfrac{1}{2}\gamma_s = \frac{\varepsilon_x - \varepsilon_y}{2} \sin 2\theta - \tfrac{1}{2}\gamma_{xy} \cos 2\theta$$

Squaring each and adding produces the expression

$$\left(\varepsilon_n - \frac{\varepsilon_x + \varepsilon_y}{2}\right)^2 + (\tfrac{1}{2}\gamma_s)^2 = \left(\frac{\varepsilon_x - \varepsilon_y}{2}\right)^2 + (\tfrac{1}{2}\gamma_{xy})^2 \tag{12.34}$$

Fig. 12.20

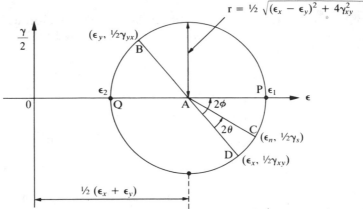

which is the equation of a circle of radius $\frac{1}{2}\sqrt{[(\varepsilon_x - \varepsilon_y)^2 + \gamma_{xy}^2]}$, and with centre at $[\frac{1}{2}(\varepsilon_x + \varepsilon_y), 0]$ relating ε_n and γ_s.

The Mohr's circle as shown in Fig. 12.20 is constructed in the same manner as for stresses. The correct position for plotting shear strain on the circle, i.e. above or below the ε-axis, may be found either by relating the deformation of the element to the corresponding shear stress system, or by the convention that a positive shear strain in the element corresponds to the sides of the deformed element having positive slope in relation to the co-ordinate axes. On co-ordinate axes of normal strain ε and semi-shear strain $\frac{1}{2}\gamma$, each to the same scale, the points $(\varepsilon_x, -\frac{1}{2}\gamma_{xy})$ and $(\varepsilon_y, +\frac{1}{2}\gamma_{xy})$ are set up, and a circle is drawn with the line joining these two points as diameter. The normal and semi-shear strain ε_n, $\frac{1}{2}\gamma_s$ in a direction at θ to the x-direction are obtained from the intersection of a radius with the circle, at 2θ (anticlockwise) from AD.

EXAMPLE 12.6

Two electrical resistance strain gauges are fitted at $\pm45°$ to the axis of a 75 mm diameter shaft. The shaft is rotating, and in addition to transmitting power, it is subjected to an unknown bending moment and a direct thrust. The readings of the gauges are recorded, and it is found that the maximum or minimum values for each gauge occur at 180° intervals of shaft rotation and are -0.0006 and $+0.0003$ for the two gauges at one instant and -0.0005 and $+0.0004$ for the same gauges 180° of rotation later. Determine the transmitted torque, the applied bending moments and the end thrust. Assume all the forces and moments are steady, i.e. do not vary during each rotation of the shaft.
$E = 208 \text{ GN/m}^2$, $\nu = 0.29$, $G = 80 \text{ GN/m}^2$. The shaft and strain gauges are illustrated diagrammatically in Fig. 12.21.

Fig. 12.21

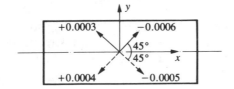

SOLUTION

The simplest starting point is to find the torque. This may be found from the shear stress which is in turn related to shear strain. From eqn. (12.33) we have

$$-\gamma_{xy} = (\varepsilon_p - \varepsilon_q) \sin 2\theta - \gamma_{pq} \cos 2\theta$$
$$= (0.0003 + 0.0006) \sin 90 - \gamma_{pq} \cos 90°$$
$$\gamma_{xy} = -0.0009$$
$$\tau_{xy} = -0.0009 \times 80 = -72 \text{ MN/m}^2$$

Hence the torque is

$$T = \frac{\pi}{16} \times 0.075^3 \times 72 \times 10^6 = 5.97 \text{ kN-m}$$

In order to find the bending moment and end thrust it is necessary to know the strains in the x-direction for the two specific rotational positions of the shaft. From eqn. (12.30) we have

$$Top \quad -0.0006 = \frac{\varepsilon_x^T + \varepsilon_y^T}{2} - 0.00045$$

and

$$+0.0003 = \frac{\varepsilon_x^T + \varepsilon_y^T}{2} + 0.00045$$

Hence

$$\varepsilon_x^T + \varepsilon_y^T = -0.0003$$
$$Bottom \quad -0.0005 = \frac{\varepsilon_x^B + \varepsilon_y^B}{2} - 0.00045$$

or

$$+0.0004 = \frac{\varepsilon_x^B + \varepsilon_y^B}{2} + 0.00045$$

Hence

$$\varepsilon_x^B + \varepsilon_y^B = -0.0001$$

To eliminate ε_y from the above equations we use the relationship:

$$\varepsilon_y = -\nu \varepsilon_x$$
$$\varepsilon_x^T = -\frac{0.0003}{1 - \nu} = -0.000423$$
$$\varepsilon_x^B = -\frac{0.0001}{1 - \nu} = -0.000141$$

These strains are the sum of the bending strain which reverses in sign for 180° rotation and the steady compressive strain (due to end thrust).

Bending strain $= \pm\frac{1}{2}(\varepsilon_x^T - \varepsilon_x^B) = \pm0.000141$

Compressive strain $= -\frac{1}{2}(\varepsilon_x^T + \varepsilon_x^B) = -0.000282$

Hence

$$\sigma_{xb} = \pm 0.000\,141 \times 208 \times 10^9 = \pm 29.3 \, \text{MN/m}^2$$

from which

$$M = \frac{\pi}{32} \times 0.075^3 \times 29.3 \times 10^6 = 1.21 \, \text{kN-m}$$

$$\sigma_{xc} = -0.000\,282 \times 208 \times 10^9 = -58.6 \, \text{MN/m}^2$$

and so the end thrust F is given by

$$F = \frac{\pi}{4} \times 0.075^2 \times (-58.6) \times 10^6 = -259 \, \text{kN}$$

PRINCIPAL STRAIN AND MAXIMUM SHEAR STRAIN

The maximum and minimum values of the normal strain at a point are given by P and Q in Fig. 12.20, whence

$$\varepsilon_1 = \frac{\varepsilon_x + \varepsilon_y}{2} + \tfrac{1}{2}\sqrt{[(\varepsilon_x - \varepsilon_y)^2 + \gamma_{xy}^2]} \quad \text{(maximum)} \tag{12.35}$$

and

$$\varepsilon_2 = \frac{\varepsilon_x + \varepsilon_y}{2} - \tfrac{1}{2}\sqrt{[(\varepsilon_x - \varepsilon_y)^2 + \gamma_{xy}^2]} \quad \text{(minimum)} \tag{12.36}$$

These are termed the *principal strains* and may be compared for similarity with the expressions for principal stresses. The former occur on mutually perpendicular planes making angles $2\theta = 2\phi$ and $180° + 2\phi$, or $\theta = \phi$ and $90° + \phi$ with the x-direction. From the diagram, when $2\theta = 2\phi$,

$$\theta = \tfrac{1}{2}\tan^{-1}\frac{\gamma_{xy}}{\varepsilon_x - \varepsilon_y} \quad \text{and} \quad 90° + \tfrac{1}{2}\tan^{-1}\frac{\gamma_{xy}}{\varepsilon_x - \varepsilon_y} \tag{12.37}$$

Either by substituting for θ in eqn. (12.33) or by reference to the circle, it is found that the shear strain is zero for the planes of principal strain.

Since $\tau_{xy} = G\gamma_{xy}$ when the shear strain is zero, so also must be the shear stress. But it was previously shown that the shear stress was zero on the principal stress planes, and therefore the planes of principal stress and principal strain must coincide.

The maximum shear strain occurs at the top and bottom points of the circle when $\varepsilon_n = \tfrac{1}{2}(\varepsilon_x + \varepsilon_y)$ and $\tfrac{1}{2}\gamma_s = \tfrac{1}{2}\sqrt{[(\varepsilon_x - \varepsilon_y)^2 + \gamma_{xy}^2]}$; therefore

$$\hat{\gamma}_s = \sqrt{[(\varepsilon_x - \varepsilon_y)^2 + \gamma_{xy}^2]} \tag{12.38}$$

Alternatively, this may be written in terms of the principal strains as

$$\tfrac{1}{2}\hat{\gamma}_s = \frac{\varepsilon_1 - \varepsilon_2}{2}$$

or

$$\hat{\gamma}_s = \varepsilon_1 - \varepsilon_2 \tag{12.39}$$

ROSETTE STRAIN COMPUTATION AND CIRCLE CONSTRUCTION

For the complete determination of strain at a point on the surface of a component, it is necessary to measure the strain in three directions at the point. This is achieved by cementing an electrical resistance rosette strain gauge to the surface.

Let the three measured strains be ε_l, ε_m and ε_n and the angle between the l and m, and m and n, directions be 45° in each case. Then this arrangement is known as a 45° rosette as shown in Fig. 12.22. If the angle

Fig. 12.22

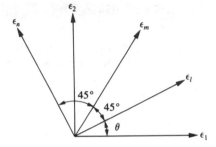

between ε_l and the principal strain ε_1 is θ, then from eqn. (12.29), the principal strains are related to the measured strains as follows:

$$\left. \begin{array}{l} \varepsilon_l = \varepsilon_1 \cos^2 \theta + \varepsilon_2 \sin^2 \theta \\ \varepsilon_m = \varepsilon_1 \cos^2 (\theta + 45°) + \varepsilon_2 \sin^2 (\theta + 45°) \\ \varepsilon_n = \varepsilon_1 \cos^2 (\theta + 90°) + \varepsilon_2 \sin^2 (\theta + 90°) \end{array} \right\} \tag{12.40}$$

These equations may be rewritten as

$$\left. \begin{array}{l} \varepsilon_l = \tfrac{1}{2}(\varepsilon_1 + \varepsilon_2) + \tfrac{1}{2}(\varepsilon_1 - \varepsilon_2) \cos 2\theta \\ \varepsilon_m = \tfrac{1}{2}(\varepsilon_1 + \varepsilon_2) - \tfrac{1}{2}(\varepsilon_1 - \varepsilon_2) \sin 2\theta \\ \varepsilon_n = \tfrac{1}{2}(\varepsilon_1 + \varepsilon_2) - \tfrac{1}{2}(\varepsilon_1 - \varepsilon_2) \cos 2\theta \end{array} \right\} \tag{12.41}$$

Solving the above equations simultaneously gives

$$\left. \begin{array}{l} \varepsilon_1 = \tfrac{1}{2}(\varepsilon_l + \varepsilon_n) + \dfrac{\sqrt{2}}{2} \sqrt{[(\varepsilon_l - \varepsilon_m)^2 + (\varepsilon_m - \varepsilon_n)^2]} \\[2mm] \varepsilon_2 = \tfrac{1}{2}(\varepsilon_l + \varepsilon_n) - \dfrac{\sqrt{2}}{2} \sqrt{[(\varepsilon_l - \varepsilon_m)^2 + (\varepsilon_m - \varepsilon_n)^2]} \end{array} \right\} \tag{12.42}$$

$$\tan 2\theta = \frac{2\varepsilon_m - \varepsilon_l - \varepsilon_n}{\varepsilon_l - \varepsilon_n} \tag{12.43}$$

To obtain principal stresses from principal strains we use the stress–strain relationships,

$$\left.\begin{aligned} \varepsilon_1 &= \frac{\sigma_1}{E} - \frac{v\sigma_2}{E} \\ \varepsilon_2 &= \frac{\sigma_2}{E} - \frac{v\sigma_1}{E} \end{aligned}\right\} \tag{12.44}$$

from which

$$\left.\begin{aligned} \sigma_1 &= \frac{E}{1-v^2}(\varepsilon_1 + v\varepsilon_2) \\ \sigma_2 &= \frac{E}{1-v^2}(\varepsilon_2 + v\varepsilon_1) \end{aligned}\right\} \tag{12.45}$$

An alternative approach to the analysis of the strain rosette is to use the Mohr's circle construction.

Consider the general case of three strain gauges, a, b and c, having arbitrary orientation as shown in Fig. 12.23(a) and strain readings, ε_a, ε_b

Fig. 12.23

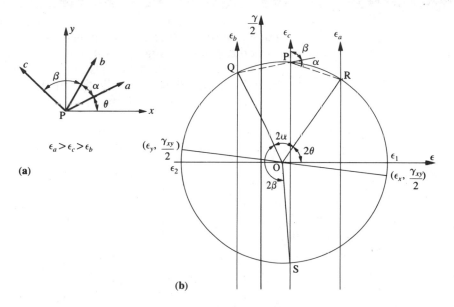

(a)

(b)

and ε_c in the given directions. As an example to illustrate the method assume $\varepsilon_a > \varepsilon_c > \varepsilon_b$. The procedure is as follows:

1. Set up a vertical axis to represent $\varepsilon = 0$ (which will subsequently be the semi-shear-strain axis).
2. Draw three lines parallel to the above axis at the appropriate distances representing the values (positive or negative) of ε_a, ε_b and ε_c.
3. On the middle line of these three (representing the middle value of the three strains ε_c) mark a point P representing the origin of the rosette.
4. Draw the rosette configuration at the point P *but lining up gauge c* (in this particular example) *along its vertical ordinate.*

5. Project the directions of gauges *a* and *b* to cut their respective vertical ordinates at Q and R.
6. Construct perpendicular bisectors of PQ and PR; where these intersect is the centre of the strain circle, O.
7. Draw the circle on this centre, which of course should pass through the points P, Q and R. Insert the horizontal strain abscissa through O.
8. Join O to Q, R and S, where S is the other intersection of the circle with the middle vertical line.
9. The lines OQ, OR and OS represent the three gauges *on the circle* where 2α and 2β are the angles between OR and OQ, and OQ and OS, respectively.
10. From the circle read off as required the principal strains ε_1, ε_2 or the chosen co-ordinate direction strains ε_x, ε_y, γ_{xy}.

EXAMPLE 12.7

At a point on the surface of a component, a 60° rosette strain gauge positioned as shown in Fig. 12.24(*a*) measures strains of $\varepsilon_l = 0.000\,46$, $\varepsilon_m = 0.0002$ and $\varepsilon_n = -0.000\,16$. Use Mohr's strain circle to determine the magnitude and direction of the principal strains and hence the principal stresses. $E = 208 \text{ GN/m}^2$, $v = 0.29$.

Fig. 12.24

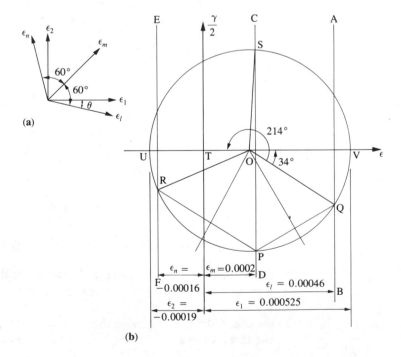

(a)

(b)

SOLUTION

To construct the strain circle, Fig. 12.24(*b*), we use the procedure described above. The principal strain values are represented by TV and

TU, and therefore

$$\varepsilon_1 = 0.000525 \quad \text{and} \quad \varepsilon_2 = -0.00019$$

The angle between ε_l and ε_1 on the circle is $34°$ and between ε_l and ε_2, $214°$. Therefore

$$\theta_1 = 17° \quad \text{and} \quad \theta_2 = 107°$$

The principal stresses are given by eqns. (12.45); thus

$$\sigma_1 = \frac{208 \times 10^9}{1 - 0.29^2} (0.000525 + 0.29 \times (-0.00019))$$

$$\sigma_2 = \frac{208 \times 10^9}{1 - 0.29^2} (-0.00019 + 0.29 \times 0.000525)$$

from which

$$\sigma_1 = 106.3 \, \text{MN/m}^2$$
$$\sigma_2 = -86.0 \, \text{MN/m}^2$$

RELATIONSHIPS BETWEEN THE ELASTIC CONSTANTS

In Chapter 3 four constants of elasticity were defined relating various conditions of stress and strain. These are Young's modulus, E, shear modulus, G, bulk modulus, K, and Poisson's ratio, v. It will now be shown that these constants are not independent of one another.

RELATIONSHIP BETWEEN K, E AND v

It was shown in eqn. (3.1) that volumetric strain is given by the sum of the three linear strains along the axes of the element. Hence

$$\varepsilon_v = \varepsilon_x + \varepsilon_y + \varepsilon_z$$

But from eqn. (3.2)

$$\varepsilon_x = \frac{\sigma_x}{E} - \frac{v}{E}(\sigma_y + \sigma_z)$$

Also for hydrostatic stress, $\sigma_x = \sigma_y = \sigma_z = \sigma$; therefore

$$\varepsilon_x = \frac{\sigma}{E}(1 - 2v)$$

Similarly

$$\varepsilon_y = \frac{\sigma}{E}(1 - 2v) \quad \text{and} \quad \varepsilon_z = \frac{\sigma}{E}(1 - 2v)$$

Therefore, summing the above three strains, we have

$$\varepsilon_v = \frac{3\sigma}{E}(1 - 2v)$$

329

or

$$E = 3\frac{\sigma}{\varepsilon_v}(1 - 2v)$$

But $\sigma/\varepsilon_v = K$, the bulk modulus; therefore

$$E = 3K(1 - 2v) \tag{12.46}$$

RELATIONSHIP BETWEEN E, G AND v

The square element of unit thickness shown in Fig. 12.25(a) is acted on by pure shearing stresses. This system is equivalent to the system of direct stresses on the element of Fig. 12.25(b), and from equilibrium $\sigma_{n_1} = \sigma_{n_2} = \tau_s$. The strain along the diagonal AB in terms of the stresses is given by

$$\varepsilon_{n_1} = \frac{\sigma_{n_1}}{E} - \frac{v}{E}(-\sigma_{n_2})$$

Fig. 12.25 (a) Pure shear stress; (b) equivalent direct stress; (c) strain system

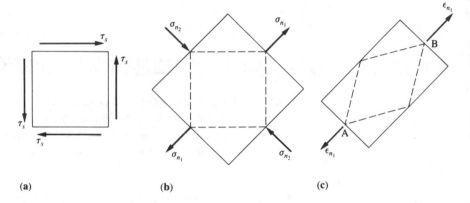

(a)　　　　　　　　(b)　　　　　　　　(c)

that is, the extension due to σ_{n_1} plus the lateral expansion in the direction AB due to the compression σ_{n_2}. But

$$\sigma_{n_1} = \sigma_{n_2} = \tau_s$$

Therefore

$$\varepsilon_{n_1} = \frac{\tau_s}{E}(1 + v)$$

Since for pure shear ε_x and ε_y are zero, then from eqn. (12.30),

$$\varepsilon_{n_1} = \tfrac{1}{2}\gamma_s \sin 2\theta$$

In this case $\theta = 45°$; therefore

$$\varepsilon_{n_1} = \tfrac{1}{2}\gamma_s$$

Equating the above expressions,

$$\tfrac{1}{2}\gamma_s = \frac{\tau_s}{E}(1 + v)$$

or

$$E = \frac{2\tau_s}{\gamma_s}(1 + v)$$

But $\tau_s/\gamma_s = G$, the shear modulus; therefore

$$E = 2G(1 + v) \qquad (12.47)$$

RELATIONSHIP BETWEEN K, G AND v

It follows from eqns. (12.46) and (12.47) that

$$3K(1 - 2v) = 2G(1 + v)$$

or

$$K = \frac{2G(1 + v)}{3(1 - 2v)} \qquad (12.48)$$

Thus if any two of the four constants are known, or can be measured, then the other two can be determined.

STRESS AND STRAIN TRANSFORMATIONS IN COMPOSITES

In recent years there has been a rapid growth in the use of fibre-reinforced composites. The major advantage of such materials is that high strength and stiffness can be achieved at low weight. Products that have benefited from the use of composites include aircraft, ships, automobiles, chemical vessels and sporting goods. In these industries the base material is usually metal or plastic and the fibres used include glass, carbon, aramid ("Kevlar"), boron and asbestos. In some cases short ("chopped") fibres are used and this provides a significant property enhancement over the base resin. However, by far the greatest improvement in properties is observed if the fibres are continuous. For example, if undirectional carbon fibres are added to an epoxy resin, the modulus of the resulting composite is improved by a factor of about 60 (*see* Table 12.1) and the strength by a factor of about 30 compared with the unreinforced base resin. However, the composite is markedly anisotropic in that in the direction perpendicular to the fibre axis the modulus is only improved by a factor of about 2 and the strength is likely

Table 12.1
Typical properties of some materials

Material	E_x* (GN/m^2)	E_y† (GN/m^2)	G_{xy} (GN/m^2)	v_x	Density (kg/m^3)
Carbon fibre/epoxy	180	10	7.2	0.28	1600
Kevlar/epoxy	76	5.5	2.3	0.34	1460
Glass/epoxy	39	8.4	4.2	0.26	1800
Spruce	8.9		2.5		400
Aluminium alloy	70	70	27		2770

* Parallel to fibres or grain.
† Perpendicular to fibres.

to be reduced. Therefore, in the aircraft industry, for example, in order to get property enhancement in all the required directions within the component, it is normal practice to build up a laminate structure where each layer has fibres arranged in the desired direction, as shown in Fig. 12.26.

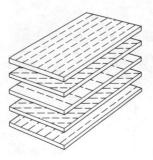

Fig. 12.26 Arrangement of fibre orientation in laminate

As it is becoming increasingly likely that engineers and designers will at some stage have to become involved in the design of components made from fibre composites it is important that they should have an appreciation of the stages in the design process. In the following sections a brief introduction is given to the laminate theory involved in fibre-reinforced composites.

ANALYSIS OF A LAMINA

In this analysis it is necessary to consider both the local co-ordinates (x, y) for the lamina and the global co-ordinates (X, Y) for the applied stress system.

ON-AXIS PROPERTIES

Consider first of all a single lamina in which the fibres are all aligned in the global X-direction as shown in Fig. 12.27(a). The lamina is thin in relation to its transverse dimensions and therefore it will be in a state of plane stress when forces are applied to it.

Fig. 12.27 (a) Fibres aligned in global X-direction; (b) fibres aligned at an angle to global X-direction

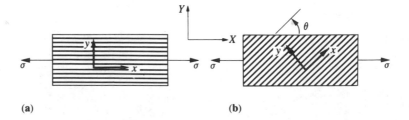

(a)   (b)

Recognizing that the lamina is anisotropic with modulus values of E_x and E_y in its x- and y-directions and Poisson's ratio terms related by

$v_x E_y = v_y E_x$ then the strains in the lamina may be written as

$$
\left.\begin{aligned}
\varepsilon_X &= \frac{\sigma_X}{E_x} - \frac{v_y \sigma_Y}{E_y} \\[2ex]
\varepsilon_Y &= \frac{\sigma_Y}{E_x} - \frac{v_x \sigma_X}{E_y} \\[2ex]
\gamma_{XY} &= \frac{\tau_{XY}}{G_{xy}}
\end{aligned}\right\}
\tag{12.49}
$$

It is convenient for subsequent analysis to write these three equations in matrix form (a brief introduction to the use of matrices is given in Appendix B).

$$
\left\{\begin{array}{c} \varepsilon_X \\[2ex] \varepsilon_Y \\[2ex] \gamma_{XY} \end{array}\right\} =
\left[\begin{array}{ccc}
\dfrac{1}{E_x} & -\dfrac{v_y}{E_y} & 0 \\[2ex]
-\dfrac{v_x}{E_x} & \dfrac{1}{E_y} & 0 \\[2ex]
0 & 0 & \dfrac{1}{G_{xy}}
\end{array}\right]
\left\{\begin{array}{c} \sigma_X \\[2ex] \sigma_Y \\[2ex] \tau_{XY} \end{array}\right\}
\tag{12.50}
$$

or in abbreviated form

$$
\{\varepsilon\} = |S|\,\{\sigma\}
\tag{12.51}
$$

where $|S|$ is referred to as the *compliance* matrix for the lamina.

At this point it is worth noting that in order to describe completely an anisotropic material subject to a triaxial stress system, the compliance matrix will have thirty-six terms.

$$
\left\{\begin{array}{c} \varepsilon_X \\ \varepsilon_Y \\ \varepsilon_Z \\ \gamma_{YZ} \\ \gamma_{ZX} \\ \gamma_{XY} \end{array}\right\} =
\left[\begin{array}{cccccc}
S_{11} & S_{12} & S_{13} & S_{14} & S_{15} & S_{16} \\
S_{21} & S_{22} & S_{23} & S_{24} & S_{25} & S_{26} \\
S_{31} & S_{32} & S_{33} & S_{34} & S_{35} & S_{36} \\
S_{41} & S_{42} & S_{43} & S_{44} & S_{45} & S_{46} \\
S_{51} & S_{52} & S_{53} & S_{54} & S_{55} & S_{56} \\
S_{61} & S_{62} & S_{63} & S_{64} & S_{65} & S_{66}
\end{array}\right]
\left\{\begin{array}{c} \sigma_X \\ \sigma_Y \\ \sigma_Z \\ \tau_{YZ} \\ \tau_{ZX} \\ \tau_{XY} \end{array}\right\}
\tag{12.52}
$$

In practice for isotropic materials the number of constants may be reduced if we assume the following:

1. Shear stresses do not affect normal strains and normal stresses do not affect shear strains. Hence

$$
S_{14} = S_{15} = S_{16} = S_{24} = S_{25} = S_{26} = S_{34} = S_{35} = S_{36} = 0
$$

2. Shear strains are only affected by shear stresses in the same plane. Hence

$$
S_{45} = S_{46} = S_{54} = S_{56} = S_{65} = S_{64} = 0
$$

3. The effect of σ_X on ε_X is the same as the effect of σ_Y on ε_Y, etc.

Hence

$$S_{11} = S_{22} = S_{33}$$

4. The effect of σ_Y on ε_X is the same as the effect of σ_Z on ε_X, etc. Hence

$$S_{12} = S_{13} = S_{21} = S_{23} = S_{31} = S_{32}$$

5. The effect of τ_{XY} on γ_{XY} is the same as the effect of τ_{YZ} on γ_{YZ}, etc. Hence

$$S_{44} = S_{55} = S_{66}$$

Hence for isotropic materials the matrix in eqn. (12.52) reduces to one in which there are only three constants S_{11}, S_{12} and S_{66}, the values of which are

$$S_{11} = \frac{1}{E}, \qquad S_{12} = -\frac{v}{E}, \qquad S_{66} = \frac{1}{G}$$

It may be seen that eqn (12.52) then reduces to eqns. (3.2) and (3.4). When introducing the stress analysis of isotropic materials it is generally more convenient to use these simple equations, but it should be remembered that they are derived from a much more general situation.

However, in the analysis of laminates, which by their nature are anisotropic, the basic equations are a little more complex and it is generally found that their manipulation is simplified by the use of matrix algebra. For a lamina it may be shown (*see* Jones,[1] for example) that eqn. (12.52) reduces to the form

$$|S| = \begin{bmatrix} S_{11} & S_{12} & S_{16} \\ S_{21} & S_{22} & S_{26} \\ S_{61} & S_{62} & S_{66} \end{bmatrix}$$

where, from eqn. (12.50)

$$S_{11} = \frac{1}{E_x}; \qquad S_{22} = \frac{1}{E_y}; \qquad S_{66} = \frac{1}{G_{xy}}$$

$$S_{12} = S_{21} = -\frac{v_x}{E_x} = -\frac{v_y}{E_y}$$

$$S_{16} = S_{61} = S_{26} = S_{62} = 0$$

If eqn. (12.49) is rearranged to give stresses in terms of strains, then

$$\sigma_X = (\varepsilon_X + v_y\varepsilon_Y)E_x/(1 - v_xv_y)$$
$$\sigma_Y = (\varepsilon_Y + v_y\varepsilon_X)E_y/(1 - v_xv_y) \tag{12.53}$$
$$\tau_{XY} = G\gamma_{XY}$$

In matrix form this may be written as

$$\begin{Bmatrix} \sigma_X \\ \sigma_Y \\ \tau_{XY} \end{Bmatrix} = \begin{bmatrix} Q_{11} & Q_{12} & Q_{16} \\ Q_{21} & Q_{22} & Q_{26} \\ Q_{61} & Q_{62} & Q_{66} \end{bmatrix} \begin{Bmatrix} \varepsilon_X \\ \varepsilon_Y \\ \gamma_{XY} \end{Bmatrix} \tag{12.54}$$

where $|Q|$ is called the *stiffness* matrix. It is symmetrical and the individual terms are

$Q_{11} = E_x/(1 - v_y v_x)$

$Q_{22} = E_y/(1 - v_y v_x)$

$Q_{66} = G_{xy}$

$Q_{12} = Q_{21} = v_x E_y/(1 - v_x v_y) = v_y E_x/(1 - v_y v_x)$

$Q_{16} = Q_{61} = Q_{26} = Q_{62} = 0$

Note that eqn. (12.51) may be rewritten as

$$\{\sigma\} = |S|^{-1}\{\varepsilon\}$$

and comparing this with eqn. (12.54) it is apparent that the stiffness matrix is the inverse of the compliance matrix, so that

$$|Q| = |S|^{-1} \tag{12.55}$$

OFF-AXIS PROPERTIES

Consider now a situation where the x–y co-ordinates of the lamina do not coincide with the global X–Y co-ordinates. This is shown in Fig. 12.28.

Fig. 12.28 Fibres aligned at an angle to global X-direction

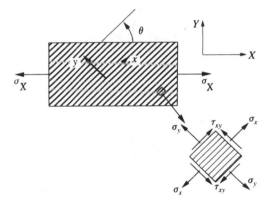

When stresses are applied to the lamina in the global (X–Y) co-ordinates these must be transformed to the local (x–y) axes for the lamina since it is in these directions that its properties are known. The transformation necessary has already been performed on pages 306 to 308. Therefore using eqns. (12.13) and (12.14) we may write:

$$\sigma_x = \sigma_X \cos^2 \theta + \sigma_Y \sin^2 \theta + 2\tau_{XY} \sin \theta \cos \theta \tag{12.56}$$

$$\sigma_y = \sigma_X \sin^2 \theta + \sigma_Y \cos^2 \theta - 2\tau_{XY} \sin \theta \cos \theta \tag{12.57}$$

$$\tau_{xy} = -\sigma_X \sin \theta \cos \theta + \sigma_Y \sin \theta \cos \theta + \tau_{XY}(\cos^2 \theta - \sin^2 \theta) \tag{12.58}$$

Note that σ_Y is obtained by letting $\theta = (\theta + 90)$ in eqn. (12.13).
Again using matrix notation and letting $s = \sin \theta$, $c = \cos \theta$, eqns.

(12.56), (12.57) and (12.58) may be written as

$$\left\{ \begin{array}{c} \sigma_x \\ \sigma_y \\ \tau_{xy} \end{array} \right\} = \left[\begin{array}{ccc} c^2 & s^2 & 2sc \\ s^2 & c^2 & -2sc \\ -sc & sc & (c^2 - s^2) \end{array} \right] \left\{ \begin{array}{c} \sigma_X \\ \sigma_Y \\ \tau_{XY} \end{array} \right\} \tag{12.59}$$

or, in shorthand form,

$$\{\sigma\}_{xy} = |T| \, \{\sigma\}_{XY} \tag{12.60}$$

where $|T|$ is the transformation matrix which may also be inverted to give global stress components in terms of local stress components.

$$\{\sigma\}_{XY} = |T|^{-1} \{\sigma\}_{xy} \tag{12.61}$$

The inversion of $|T|$ gives $|T|^{-1}$ as

$$|T|^{-1} = \left[\begin{array}{ccc} c^2 & s^2 & -2sc \\ s^2 & c^2 & 2sc \\ sc & -sc & (c^2 - s^2) \end{array} \right] \tag{12.62}$$

From the analysis on pages 320 to 322 it may be seen that the transformation of strain is similar to that of stress, so we may write

$$\left\{ \begin{array}{c} \varepsilon_x \\ \varepsilon_y \\ \frac{1}{2}\gamma_{xy} \end{array} \right\} = |T| \left\{ \begin{array}{c} \varepsilon_X \\ \varepsilon_Y \\ \frac{1}{2}\gamma_{XY} \end{array} \right\} \tag{12.63}$$

At this stage we have developed methods of transforming the values of stresses and strains independently from the loading axes to the fibre axis or vice versa. In practice it is much more important to be able to establish the contribution to overall stiffness which is made by a lamina in which the fibres are at an angle to the loading axis. This is because in the construction of laminates (page 332) a number of lamina will be arranged at different orientation and bonded together. The stiffness of the laminate will be the sum of the contribution made by each of the individual lamina.

In order to determine the stiffness in the global directions for a lamina in which the fibres are at an angle θ to the global X-direction then the following three steps are involved:

1. Determine the strains in the local (x, y) directions by transforming the applied strains through $\theta°$ from the global (X, Y) directions.
2. Calculate stresses in local directions using on-axis stiffness matrix $|Q|$.
3. Transform stresses back to the global directions through an angle of $-\theta°$.

Using matrix notation these steps are shown below.

Step 1
$$\left\{ \begin{array}{c} \varepsilon_x \\ \varepsilon_y \\ \gamma_{xy} \end{array} \right\} = \left[\begin{array}{ccc} c^2 & s^2 & sc \\ s^2 & c^2 & -sc \\ -2sc & 2sc & (c^2 - s^2) \end{array} \right] \left\{ \begin{array}{c} \varepsilon_X \\ \varepsilon_Y \\ \gamma_{XY} \end{array} \right\}$$

Note the modification to the transformation matrix $|T|$ so that we may write γ_{xy} instead of $\frac{1}{2}\gamma_{xy}$.

Step 2

$$\left\{\begin{array}{c} \sigma_x \\ \sigma_y \\ \tau_{xy} \end{array}\right\} = \left[\begin{array}{ccc} Q_{11} & Q_{12} & 0 \\ Q_{21} & Q_{22} & 0 \\ 0 & 0 & Q_{66} \end{array}\right] \left\{\begin{array}{c} \varepsilon_x \\ \varepsilon_y \\ \gamma_{xy} \end{array}\right\}$$

Step 3

$$\left\{\begin{array}{c} \sigma_X \\ \sigma_Y \\ \tau_{XY} \end{array}\right\} = \left[\begin{array}{ccc} c^2 & s^2 & -2sc \\ s^2 & c^2 & 2sc \\ sc & -sc & (c^2-s^2) \end{array}\right] \left\{\begin{array}{c} \sigma_x \\ \sigma_y \\ \tau_{xy} \end{array}\right\}$$

Hence to express global stresses in terms of global strains we perform the following matrix multiplication:

$$\left\{\begin{array}{c} \sigma_X \\ \sigma_Y \\ \tau_{XY} \end{array}\right\} =$$

$$\left[\begin{array}{ccc} c^2 & s^2 & -2sc \\ s^2 & c^2 & 2sc \\ sc & -sc & (c^2-s^2) \end{array}\right] \left[\begin{array}{ccc} Q_{11} & Q_{12} & 0 \\ Q_{21} & Q_{22} & 0 \\ 0 & 0 & Q_{66} \end{array}\right] \left[\begin{array}{ccc} c^2 & s^2 & sc \\ s^2 & c^2 & -sc \\ -2sc & 2sc & (c^2-s^2) \end{array}\right] \left\{\begin{array}{c} \varepsilon_X \\ \varepsilon_Y \\ \gamma_{XY} \end{array}\right\}$$

which provides an overall stiffness matrix $|\bar{Q}|$ where

$$\left\{\begin{array}{c} \sigma_{X'} \\ \sigma_Y \\ \tau_{XY} \end{array}\right\} = \left[\begin{array}{ccc} \bar{Q}_{11} & \bar{Q}_{12} & \bar{Q}_{16} \\ \bar{Q}_{21} & \bar{Q}_{22} & \bar{Q}_{26} \\ \bar{Q}_{61} & \bar{Q}_{62} & \bar{Q}_{66} \end{array}\right] \left\{\begin{array}{c} \varepsilon_X \\ \varepsilon_Y \\ \gamma_{XY} \end{array}\right\} \tag{12.64}$$

$$\bar{Q}_{11} = \frac{1}{\lambda}\{E_x \cos^4\theta + E_y \sin^4\theta + (2\nu_x E_y + 4\lambda G)\cos^2\theta \sin^2\theta\}$$

$$\bar{Q}_{21} = \bar{Q}_{12} = \frac{1}{\lambda}\{\nu_x E_y(\cos^4\theta + \sin^4\theta) + (E_x + E_y - 4\lambda G)\cos^2\theta \sin^2\theta\}$$

$$\bar{Q}_{61} = \bar{Q}_{16}$$

$$= \frac{1}{\lambda}\{\cos^3\theta \sin\theta(E_x - \nu_x E_y - 2\lambda G) - \cos\theta \sin^3\theta(E_y - \nu_x E_y - 2\lambda G)\}$$

$$\bar{Q}_{22} = \frac{1}{\lambda}\{E_y \cos^4\theta + E_x \sin^4\theta + \sin^2\theta \cos^2\theta(2\nu_x E_y + 4\lambda G)\}$$

$$\bar{Q}_{62} = \bar{Q}_{26}$$

$$= \frac{1}{\lambda}\{\cos\theta \sin^3\theta(E_x - \nu_x E_y - 2\lambda G) - \cos^3\theta \sin\theta(E_y - \nu_x E_y - 2\lambda G)\}$$

$$\bar{Q}_{66} = \frac{1}{\lambda}\{\sin^2\theta \cos^2\theta(E_x + E_y - 2\nu_x E_y - 2\lambda G) + \lambda G(\cos^4\theta + \sin^4\theta)\}$$

in which $\lambda = (1 - \nu_x \nu_y)$.

By a similar analysis it may be shown that, for applied stresses (rather than applied strains) in the global directions, the overall compliance matrix $|\bar{S}|$ has the form

$$
\left\{\begin{array}{c} \varepsilon_X \\ \varepsilon_Y \\ \gamma_{XY} \end{array}\right\} = \left[\begin{array}{ccc} \bar{S}_{11} & \bar{S}_{12} & \bar{S}_{16} \\ \bar{S}_{21} & \bar{S}_{22} & \bar{S}_{26} \\ \bar{S}_{61} & \bar{S}_{62} & \bar{S}_{66} \end{array}\right] \left\{\begin{array}{c} \sigma_X \\ \sigma_Y \\ \tau_{XY} \end{array}\right\}
\tag{12.65}
$$

where

$$\bar{S}_{11} = S_{11} \cos^4 \theta + S_{22} \sin^4 \theta + (2S_{12} + S_{66}) \cos^2 \theta \sin^2 \theta$$

$$\bar{S}_{21} = \bar{S}_{12} = (S_{11} + S_{22} - S_{66}) \cos^2 \theta \sin^2 \theta + S_{12}(\cos^4 \theta \sin^4 \theta)$$

$$\bar{S}_{61} = \bar{S}_{16} = (2S_{22} - 2S_{12} - S_{66}) \cos^3 \theta \sin \theta - (2S_{22} - 2S_{12} - S_{66}) \sin^3 \theta \cos \theta$$

$$\bar{S}_{22} = S_{11} \sin^4 \theta + S_{22} \cos^4 \theta + (2S_{12} + S_{66}) \cos^2 \theta \sin^2 \theta$$

$$\bar{S}_{62} = \bar{S}_{26} = (2S_{11} - 2S_{12} - S_{66}) \cos \theta \sin^3 \theta - (2S_{22} - 2S_{12} - S_{66}) \sin \theta \cos^3 \theta$$

$$\bar{S}_{66} = 4(S_{11} - 2S_{12} + S_{66}) \cos^2 \theta \sin^2 \theta + S_{66}(\cos^2 \theta - \sin^2 \theta)^2$$

ANALYSIS OF A LAMINATE

At this stage we are in a position to describe the stress–strain behaviour in any co-ordinate direction for a lamina consisting of unidirectional fibres. Such a lamina is used in beams and tension/compression members where the excellent longitudinal properties can be used to good advantage. However, in many cases the low transverse properties could not be tolerated. For such applications it is usual to build up a laminate in which laminae are arranged at different orientations in order to achieve the desired overall properties in the laminate. The orientations of the individual lamina can be in any desired combination. Generally there are two broad categories of laminates – those which are symmetric about the mid-plane and those which are unsymmetric, as shown in Fig. 12.29.

We will consider only those laminates which have mid-plane symmetry and the reader should refer to textbooks on laminate theory[1-3] for the unsymmetric cases. Laminates which are symmetric can be analysed by a simple extension of the theory developed in the previous section for unidirectional composites. In both cases their mechanical behaviour is described by three sets of elastic constants. Thus a symmetric laminate

Fig. 12.29 Different types of laminate construction: (a) symmetric angle ply; (b) non-symmetric angle ply

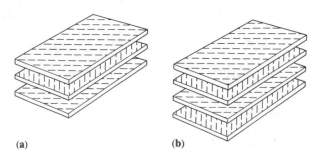

(a)   (b)

behaves like a homogeneous anisotropic plate and under uniaxial loading its effective modulus is simply the arithmetic average of the constituent laminae.

IN-PLANE BEHAVIOUR OF A SYMMETRIC LAMINATE

During the manufacture of a laminate the individual plies or laminae are bonded securely together so that when loaded they all experience the same strain. However, because the stiffnesses of each ply are all different, the stresses will not be the same in each case. This is illustrated in Fig. 12.30.

Fig. 12.30 Stresses and strains in uniaxially loaded symmetrical laminate

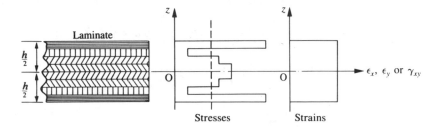

When defining the overall stress–strain behaviour of a laminate it is necessary to use average stresses. These are defined as

$$\bar{\sigma}_X = \frac{1}{h} \int_{-h/2}^{h/2} \sigma_X dZ \tag{12.66}$$

$$\bar{\sigma}_Y = \frac{1}{h} \int_{-h/2}^{h/2} \sigma_Y dZ \tag{12.67}$$

$$\bar{\tau}_{XY} = \frac{1}{h} \int_{-h/2}^{h/2} \tau_{XY} dZ \tag{12.68}$$

or in matrix form

$$\begin{Bmatrix} \bar{\sigma}_X \\ \bar{\sigma}_Y \\ \bar{\tau}_{XY} \end{Bmatrix} = \frac{1}{h} \int_{-h/2}^{h/2} \begin{Bmatrix} \sigma_X \\ \sigma_Y \\ \tau_{XY} \end{Bmatrix} dZ \tag{12.69}$$

$$= \frac{1}{h} \int_{-h/2}^{h/2} \begin{bmatrix} \bar{Q}_{11} & \bar{Q}_{12} & \bar{Q}_{16} \\ & \bar{Q}_{22} & \bar{Q}_{26} \\ \text{sym.} & & \bar{Q}_{66} \end{bmatrix} \begin{Bmatrix} \varepsilon_X \\ \varepsilon_Y \\ \gamma_{XY} \end{Bmatrix} dZ$$

As the strains are independent of Z they can be taken outside the integral:

$$\left\{\begin{array}{c} \bar{\sigma}_X \\ \bar{\sigma}_Y \\ \bar{\tau}_{XY} \end{array}\right\} = \frac{1}{h}\int_{-h/2}^{h/2} \begin{bmatrix} \bar{Q}_{11} & \bar{Q}_{12} & \bar{Q}_{16} \\ & \bar{Q}_{22} & \bar{Q}_{26} \\ \text{sym.} & & \bar{Q}_{66} \end{bmatrix} dZ \left\{\begin{array}{c} \varepsilon_X \\ \varepsilon_Y \\ \gamma_{XY} \end{array}\right\}$$

$$\left\{\begin{array}{c} \bar{\sigma}_X \\ \bar{\sigma}_Y \\ \bar{\tau}_{XY} \end{array}\right\} = \frac{1}{h}|A| \left\{\begin{array}{c} \varepsilon_X \\ \varepsilon_Y \\ \gamma_{XY} \end{array}\right\} \tag{12.70}$$

where, for example,

$$A_{11} = \frac{1}{h}\int_{-h/2}^{h/2} \bar{Q}_{11} dZ = \frac{2}{h}\int_0^{h/2} \bar{Q}_{11} dZ$$

Within a single lamina, such as the ith, the $\bar{Q}$ terms are constant so, referring to Fig. 12.31, the integral may be replaced by a summation.

$$A_{11} = \frac{2}{h}\sum \bar{Q}_{11}{}^i \int_{Z_{i-1}}^{Z_i} dZ = \frac{2}{h}\sum \bar{Q}_{11}{}^i h_i \tag{12.71}$$

$$A_{11} = \sum \bar{Q}_{11}{}^i \left(\frac{2h_i}{h}\right) \tag{12.72}$$

Fig. 12.31

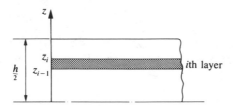

Thus the stiffness matrix for a symmetric laminate may be obtained by adding, in proportion to the lamina thickness, the corresponding terms in the stiffness matrix for each of the laminae. Note also that the volume fraction of fibres in the ith lamina is given by

$$v_i = \left(\frac{2h_i}{h}\right)$$

and this is a convenient substitution to use when performing the summations indicated by eqn. (12.72). The sum of all the v_i terms for the different laminae will of course be 1.

Having obtained all the terms for the stiffness matrix $|A|$ as indicated above, this may then be inverted to give the compliance matrix $|a|$ and hence the modulus values for the laminate. The inversion of the (3×3) stiffness matrix will give

$$a_{11} = (A_{11}A_{66} - A_{26}{}^2)/\Delta; \qquad a_{22} = (A_{11}A_{66} - A_{16}{}^2)/\Delta$$
$$a_{21} = a_{12} = (A_{16}A_{26} - A_{12}A_{66})/\Delta; \qquad a_{66} = (A_{11}A_{22} - A_{12}{}^2)/\Delta$$
$$a_{61} = a_{16} = (A_{12}A_{26} - A_{16}A_{22})/\Delta; \qquad a_{26} = (A_{16}A_{12} - A_{11}A_{26})/\Delta$$

where

$$\Delta = A_{11}A_{22}A_{66} - A_{11}A_{26}{}^2 + 2A_{12}A_{26}A_{16} - A_{12}{}^2A_{66} - A_{16}{}^2A_{22}.$$

For cross-ply laminates and balanced angle-ply laminates these equations will be simplified by the fact that $A_{16} = A_{61} = 0$; $A_{26} = A_{62} = 0$.

The following laminate properties may then be obtained from inspection of the compliance matrix (see eqn. (12.50)).

$$E_X = \frac{1}{a_{11}}; \qquad E_Y = \frac{1}{a_{22}}; \qquad G = \frac{1}{a_{66}}$$

$$v_X = \frac{-a_{12}}{a_{11}}; \qquad v_Y = \frac{-a_{12}}{a_{22}}$$

EXAMPLE 12.8

A filament-wound composite cylindrical pressure vessel is made up from ten plies of continuous carbon fibres in an epoxy resin. The arrangement of the plies is as shown in Fig. 12.32. There are two plies at $60°$, two plies at $-60°$ and the remainder are in the hoop direction. Calculate the maximum permissible pressure in the cylinder if the hoop strain is not to exceed 1%. At this pressure calculate the axial strain in the cylinder. The properties of the individual plies are $E_x = 180 \text{ GN/m}^2$, $E_y = 10 \text{ GN/m}^2$, $G_{xy} = 7 \text{ GN/m}^2$, $v_x = 0.28$.

Fig. 12.32 Composite cylindrical pressure vessel

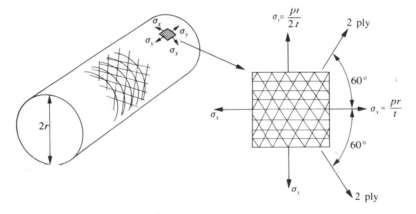

SOLUTION

The first step in the solution is to get the stiffness matrix terms for each ply in the global co-ordinate directions. Thus from eqn. (12.64):

$$(\bar{Q}_{11})_{60} = \frac{1}{\lambda} \{ E_x \cos^4 60 + E_y \sin^4 60$$

$$+ (2v_x E_y + 4\lambda G) \cos^2 60 \sin^2 60 \} = 23.24 \text{ GN/m}^2$$

Similarly,

$$(\bar{Q}_{11})_{-60} = 23.24 \text{ GN/m}^2 \quad \text{and} \quad (\bar{Q}_{11})_0 = 180.78 \text{ GN/m}^2$$

Then, from eqn. (12.72),

$$A_{11} = 0.2(\bar{Q}_{11})_{60} + 0.2(\bar{Q}_{11})_{-60} + 0.6(\bar{Q}_{11})_0$$
$$= 117.8 \text{ GN/m}^2$$

In a similar way the other terms in the stiffness matrix for the laminate may be calculated to give

$$|A| = \begin{bmatrix} 117.8 & 14.6 & 0 \\ 14.6 & 49.5 & 0 \\ 0 & 0 & 18.8 \end{bmatrix}$$

This may then be inverted to give

$$|a| = \begin{bmatrix} 8.81 & -2.6 & 0 \\ -2.6 & 20.97 & 0 \\ 0 & 0 & 53.2 \end{bmatrix} \times 10^{-3}$$

From which

$$E_X = \frac{1}{a_{11}} = \frac{1}{8.81 \times 10^{-3}} = 113.5 \text{ GN/m}^2$$

$$E_Y = \frac{1}{a_{22}} = \frac{1}{20.97 \times 10^{-3}} = 47.7 \text{ GN/m}^2$$

$$G_{XY} = \frac{1}{a_{66}} = \frac{1}{53.2 \times 10^{-3}} = 18.8 \text{ GN/m}^2$$

$$v_X = \frac{-a_{12}}{a_{11}} = \frac{2.6}{8.81} = 0.295$$

$$v_Y = \frac{-a_{12}}{a_{22}} = \frac{2.6}{20.97} = 0.124$$

Then, expressing the axial and hoop strains in terms of σ_Y and σ_X,

$$\begin{Bmatrix} \varepsilon_X \\ \varepsilon_Y \\ \gamma_{XY} \end{Bmatrix} = \begin{bmatrix} 8.81 & -2.6 & 0 \\ -2.6 & 20.97 & 0 \\ 0 & 0 & 53.2 \end{bmatrix} 10^{-3} \begin{Bmatrix} \sigma_X \\ \sigma_Y \\ \tau_{XY} \end{Bmatrix}$$

$$= \begin{bmatrix} 8.81 & -2.6 & 0 \\ -2.6 & 20.97 & 0 \\ 0 & 0 & 53.2 \end{bmatrix} 10^{-3} \begin{Bmatrix} 350 \\ 175 \\ 0 \end{Bmatrix} p$$

where p is the internal pressure

Hoop strain $\varepsilon_X = (8.81 \times 350 - 2.6 \times 175)10^{-3}p = 0.01$

$$p = 3.8 \text{ MN/m}^2$$

Also, at this pressure the axial strain

$$\varepsilon_Y = (-2.6 \times 350 + 20.97 \times 175)10^{-3} \times 3.8 \times 10^{-3} = 1.05\%$$

FLEXURAL BEHAVIOUR OF A SYMMETRIC LAMINATE

When a laminate is subjected to flexure there will be a uniform strain gradient across the section but, as shown in Fig. 12.33, the stress variation will be non-linear due to the different stiffnesses of the individual laminae.

Fig. 12.33 Stresses and strains in laminate subjected to flexure

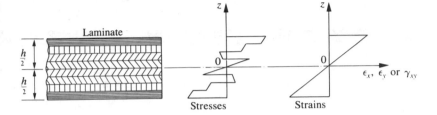

Using an analysis similar to that in the previous section it may be shown that the terms in the stiffness matrix for the laminate are given by

$$D_{11} = \sum \bar{Q}_{11}{}^{i} \left(\frac{I_i}{I_{COMP.}} \right) \tag{12.73}$$

where I_i and $I_{COMP.}$ are the second moments of area for the ith lamina and the complete laminate respectively. Inversion of the $|D|$ matrix will then enable moduli values to be obtained as before.

SUMMARY

The design of engineering components depends on the assessment of the critical stress levels occurring. In general these are the principal stresses, the importance of which will be demonstrated in the next chapter. It is therefore most important either through the analytical derivations or the Mohr's circle construction to be able to investigate the state of stress at a point in the material. It will be appreciated that for a component that already exists we cannot directly measure stress due to applied forces.

However, we can measure *in situ* displacements and strains, from which the associated stress system can be derived, using the stress–strain relationships. Therefore it is equally important to understand two-dimensional strain analysis for which the Mohr strain circle is most useful. Finally, although we seldom have to derive one elastic constant from two of the remainder, it is fundamental to the elastic behaviour of materials that the four constants are interrelated.

This chapter has also introduced the theory of fibre composite materials which by their nature are markedly anisotropic. It will be seen that the stresses and strains in a laminate, although apparently complex in nature, may be determined using the stress and strain transformations developed at the beginning of the chapter.

REFERENCES

1. Jones, R. M. *Mechanics of Composite Materials,* Scripta Book Co., Washington, 1975.
2. Tsai, S. W. and Hahn, H. T. *Introduction to Composite Materials,* Technomic Publ. Co., Westport, Conn., 1980.
3. Agarwal, B. D. and Broutman, L. J. *Analysis and Performance of Fiber Composites,* John Wiley & Sons, New York, 1980.

PROBLEMS

12.1 For the engineering components illustrated in Fig. 12.34 it is necessary to determine the normal and shear stresses at the points shown. Sketch an element in each case showing the magnitude and sense of the stresses on each face.

Fig. 12.34

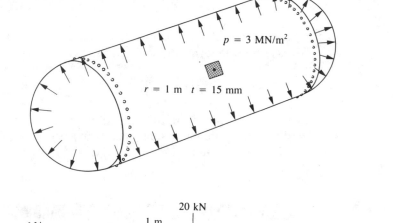

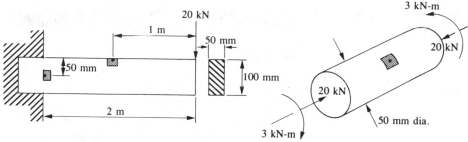

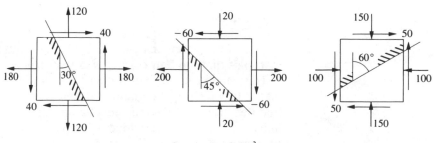

12.2 For the elements illustrated in Fig. 12.35 calculate the stress components on the inclined planes shown.

Fig. 12.35

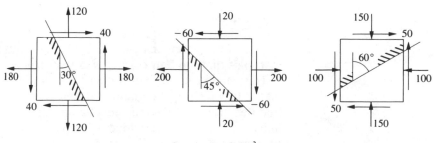

Stresses in MN/m²

344

12.3 At a point in a boiler rivet the material of the rivet is undergoing the action of a shear stress of 50 MN/m^2 while resisting movement between the boiler plates and a tensile stress of 40 MN/m^2 due to the extension of the rivet. Find the magnitude of the tensile stresses at the same point acting on two planes making an angle of 80° to the axis of the rivet.

12.4 Construct Mohr's circles for the stress systems given in Problem 12.2 and check the solutions for the stresses on the inclined planes.

12.5 At a point in the cross-section of a girder there is a tensile stress of 50 MN/m^2 and a positive shearing stress of 25 MN/m^2. Find the principal planes and stresses, and sketch a diagram showing how they act.

12.6 Draw the Mohr stress circles for the states of stress at a point given in Figs. 12.36(a) and (b). For (a) determine and show the magnitude and orientation of the principal stresses. For (b) show the stress components on the inclined plane.

Fig. 12.36

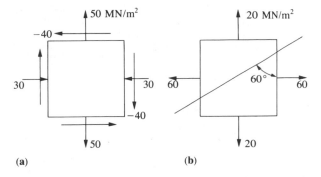

(a) (b)

12.7 Determine the principal stresses, maximum shear stresses and their orientations for locations A, B and C of Example 12.1 (p. 308).

12.8 The loads applied to a piece of material cause a shear stress of 40 MN/m^2 together with a normal tensile stress on a certain plane. Find the value of this tensile stress if it makes an angle of 30° with the major principal stress. What are the values of the principal stresses?

12.9 The I-section beam shown in Fig. 12.37 is simply-supported over a length of 6 m and subjected to a point load of 12 kN at mid-span. Calculate the values of the principal stresses at a point on the cross-section 54.5 mm above the neutral axis.

Fig. 12.37

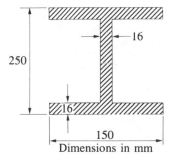

Dimensions in mm

12.10 A pulley of 250 mm diameter is keyed to the unsupported end of a 50 mm diameter shaft which overhangs 200 mm from a bearing. The pulley belt tension on the tight side is three times that on the slack side. Determine the largest values for these tensions if the maximum principal stress in the shaft is not to exceed 150 MN/m^2.

12.11 For a curved bar loaded as in Fig. 12.38 determine at what position the maximum principal stress has its greatest value. Calculate the latter and also the maximum

Fig. 12.38

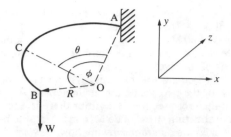

shear stress for a split ring of radius 250 mm, bar diameter 50 mm and loaded with 5 kN at one end perpendicular to the plane of the ring.

12.12 A 45° rosette is fixed to a short rectangular section pillar as illustrated in Fig. 12.39. If the gauges read $\varepsilon_a = 72 \times 10^{-6}$, $\varepsilon_b = 100 \times 10^{-6}$ and $\varepsilon_c = -240 \times 10^{-6}$ determine the values of the forces F and W. The cross-sectional area of the pillar is 600 mm^2 and it is made from steel for which $E = 207$ GN/m^2 and $v = 0.3$.

Fig. 12.39

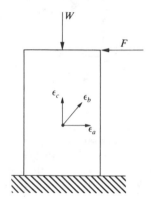

12.13 A state of two-dimensional strain is $\varepsilon_x = 0.0007$, $\varepsilon_y = -0.0006$; $\gamma_{xy} = \gamma_{yx} = 0.0003$. Calculate the principal strains in magnitude and direction and check the results using Mohr's circle construction.

12.14 At a particular point on the surface of a component the principal strain directions are known, but it is not convenient to attach electrical resistance strain gauges in these directions. However, it is possible to cement gauges at 30° and 60° anticlockwise from the major principal strain direction, and the readings from these gauges are +0.0009 and −0.0006 respectively. Construct the Mohr strain circle and find the value of the principal stresses and maximum shear stress. $E = 208$ GN/m^2, $v = 0.29$.

12.15 The principal strains, ε_1 and ε_2, are measured at a point on the surface of a shaft which is subjected to bending and torsion. The values are $\varepsilon_1 = 0.0011$ and $\varepsilon_2 = -0.0006$, and ε_1 is inclined at 20° to the axis of the shaft. If the diameter of the shaft is 51 mm and the rigidity modulus for the material is 83 GN/m^2, determine analytically the applied torque and the maximum shear stress in the material at the point concerned and check graphically.

12.16 A 60 mm diameter solid shaft has a strain gauge mounted at 65° to the axis of the shaft. In service a torque is applied to the shaft and the strain gauge reads 200×10^{-6}. Calculate the value of the torque if the shaft is made from steel with $E = 207$ GN/m^2 and $v = 0.3$.

12.17 Three strain gauges A, B and C are fixed to a point on the surface of a test plate at 120° intervals, and the strains recorded are $\varepsilon_A = +0.00108$, $\varepsilon_B = +0.00064$, $\varepsilon_C = +0.00090$. Draw Mohr's strain circle for this problem and determine the principal strains and the inclination of gauge A to the direction of the greater principal strain.

12.18 At a certain point in a steel structural element the directions of the principal stresses σ_1 and σ_2 are known. Measurements by strain gauges show that there is a tensile strain of 0.00083 in the direction of σ_1 and a compressive strain of 0.00052 in the direction of σ_2. Find the magnitudes of σ_1 and σ_2, stating whether tensile or compressive, and the maximum shear stress. $v = 0.28$, $E = 207\ \text{GN/m}^2$.

12.19 A thin-walled aluminium alloy pressure vessel of 200 mm diameter and 3 mm wall thickness is subjected to an internal pressure of $6\ \text{MN/m}^2$. Strain gauges which are bonded to the outer surface in the hoop and axial directions give readings of 0.00243 and 0.00057 at full pressure respectively. Determine the four elastic constants for the material.

12.20 A chemical pressure vessel is to be manufactured from glass fibres in an epoxy matrix as illustrated in Fig. 12.40. If the optimum fibre orientation is that in which the fibres are subjected to tensile stresses with no transverse or shear stresses, determine the optimum value of α.

Fig. 12.40

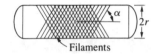

Filaments

12.21 A cylindrical pressure vessel is made up of continuous carbon fibres in an epoxy matrix. The fibres are wound at $\pm45°$ from the cylinder axis. Calculate the axial and hoop strains in the cylinder when the internal pressure is $5\ \text{MN/m}^2$. The cylinder diameter is 1 m and the wall thickness is 12 mm. Unidirectional carbon fibres in an epoxy matrix have the following properties: $E_x = 130\ \text{GN/m}^2$, $E_y = 7\ \text{GN/m}^2$; $G_{xy} = 5.6\ \text{GN/m}^2$, $v_{xy} = 0.3$.

12.22 Part of the hull of a speedboat is in the form of a flat sheet of fibre-reinforced polyester. The fibres are continuous glass strands and the lay-up is such that there are four plies at 45°, four plies at −45° and two plies at 0°. Calculate the in-plane stiffness of the sheet in the 0° and 90° directions. A unidirectional glass-fibre composite has the following properties: $v_{xy} = 0.3$, $E_x = 40\ \text{GN/m}^2$, $E_y = 9.8\ \text{GN/m}^2$, $G_{xy} = 2.8\ \text{GN/m}^2$.

Chapter 13

Yield criteria and stress concentration

INTRODUCTION

All the theoretical analysis of the previous chapters has made use of a proportional stress–strain relationship. This is because Hooke's law established that metals have a linear elastic stress–strain range. However, if a ductile metal is subjected to simple axial loading, it is found that beyond a certain point, stress is no longer proportional to strain and the material is said to be *yielding*. Knowing the stress at which this latter behaviour commenced, it would then be a simple matter to design a second specimen of the same material to withstand a particular axial load without any yielding occurring. This example is simple as there is only one principal stress to consider.

The problem of designing a pressure vessel, rotating disc, or some component containing a complex principal stress system so that the material remains elastic, i.e. no yielding, when under full load is rather more complex. One could adopt a trial-and-error method of building a component and testing it to find when the deformations were no longer proportional to the applied load, but this would obviously be very uneconomical. It is therefore essential to find some criterion based on stresses, or strains, or perhaps strain energy in the complex system which can be related to the simple axial conditions mentioned above. If a theoretical criterion can be established which predicts complex material behaviour, it is then only necessary to establish experimentally the yield point in a simple tension or compression test.

A number of theoretical criteria for yielding have been proposed over the past century but only those now currently accepted and used for ductile and brittle materials will be discussed.

YIELD CRITERIA: DUCTILE MATERIALS

Materials which exhibit "yielding" followed by some plastic deformation prior to fracture as measured under simple tensile or compressive stress are termed *ductile*. This is a very important property as it provides a design reserve for materials if they should exceed the elastic range during service. It also has significance in relation to stress concentration as explained later in the chapter. It is therefore essential to have a method of designing to avoid the possibility of yielding under complex stress situations.

Plastic deformation is related to "slip", as it is termed, in the grain structure which is dependent on shearing action. There are two yield criteria based on concepts of shear which are now accepted and used for ductile materials. These are known as the maximum shear stress (Tresca) criterion and the shear strain energy (von Mises) criterion. These will now be developed.

MAXIMUM SHEAR STRESS (TRESCA) CRITERION

Theories of yielding are generally expressed in terms of principal stresses, since these completely determine a general state of stress. The element of material shown in Fig. 13.1 is subjected to three principal stresses and it will be taken that $\sigma_1 > \sigma_2 > \sigma_3$.

Fig. 13.1

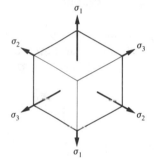

The French engineer Tresca, who proposed this theory, made the assumption that yielding is dependent on the maximum shear stress in the material reaching a critical value. This is taken as the maximum shear stress at yielding in a uniaxial tensile test. The maximum shear stress in the complex stress system will depend on the relative values and signs of the three principal stresses, always being half the difference between the maximum and the minimum. It should be remembered that the minimum can be zero and that a compressive stress is negative in value.

For a general three-dimensional stress system, or in the two-dimensional case with one of the stresses tensile, one compressive and the third zero, the maximum shear stress is

$$\hat{\tau} = (\sigma_1 - \sigma_3)/2$$

Under uniaxial tension there is only one principal stress so that the maximum shear stress is

$$\hat{\tau} = \sigma_1/2$$

and at yield this becomes $\hat{\tau}_Y = \sigma_Y/2$.

Therefore the criterion states that

$$(\sigma_1 - \sigma_3)/2 = \sigma_Y/2$$

or

$$\sigma_1 - \sigma_3 = \sigma_Y \tag{13.1}$$

For the case when two of the principal stresses are of the same type, tension or compression, and the third is zero, then we have

$$\hat{\tau} = (\sigma_1 - 0)/2 = \sigma_1/2$$

and yielding occurs when

$$\sigma_1/2 = \sigma_Y/2 \quad \text{or} \quad \sigma_1 = \sigma_Y \tag{13.2}$$

SHEAR STRAIN ENERGY (VON MISES) CRITERION Huber in 1904 proposed that the total elastic strain energy stored in an element of material could be considered as consisting of energy stored due to change in volume and energy stored due to change in shape, i.e. distortion or shear. It was proposed that the latter contribution of stored strain energy could provide a viable criterion for complex yield conditions. The same criterion was also suggested independently by Maxwell, von Mises and Hencky, but is now generally referred to as the von Mises criterion.

In order to show that the deformation of a material can be separated into change in volume and change in shape, consider the element in Fig. 13.2 subjected to the principal stresses, σ_1, σ_2 and σ_3. These may be

Fig. 13.2

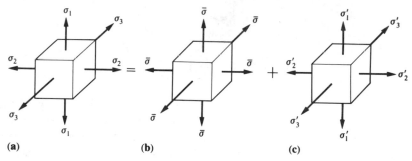

(a)   (b)   (c)

written in terms of the "average" stress in the element as follows:

$$\left.\begin{aligned} \sigma_1 &= \bar{\sigma} + \sigma_1' \\ \sigma_2 &= \bar{\sigma} + \sigma_2' \\ \sigma_3 &= \bar{\sigma} + \sigma_3' \end{aligned}\right\} \tag{13.3}$$

where $\bar{\sigma}$ is the average or mean stress defined as

$$\bar{\sigma} = (\sigma_1 + \sigma_2 + \sigma_3)/3 \tag{13.4}$$

Now, when an element as in Fig. 13.2(b) is subjected to $\bar{\sigma}$ in all directions, this hydrostatic stress will produce a change in volume, but no distortion. Consider the effect of the σ' components of stress. Adding together eqns. 13.3 gives

$$\sigma_1 + \sigma_2 + \sigma_3 = 3\bar{\sigma} + \sigma_1' + \sigma_2' + \sigma_3'$$

but $\bar{\sigma} = \frac{1}{3}(\sigma_1 + \sigma_2 + \sigma_3)$; hence

$$\sigma_1' + \sigma_2' + \sigma_3' = 0 \tag{13.5}$$

But from the stress–strain relationships,

$$\left.\begin{array}{l}\varepsilon_1' = \dfrac{\sigma_1'}{E} - \dfrac{v}{E}(\sigma_2' + \sigma_3') \\[2mm] \varepsilon_2' = \dfrac{\sigma_2'}{E} - \dfrac{v}{E}(\sigma_3' + \sigma_1') \\[2mm] \varepsilon_3' = \dfrac{\sigma_3'}{E} - \dfrac{v}{E}(\sigma_1' + \sigma_2')\end{array}\right\} \tag{13.6}$$

Hence

$$\varepsilon_1' + \varepsilon_2' + \varepsilon_3' = \varepsilon_v' = \frac{(1-2v)}{E}(\sigma_1' + \sigma_2' + \sigma_3') \tag{13.7}$$

and since the sum of the three stresses is zero, eqn. (13.5),

$$\varepsilon_1' + \varepsilon_2' + \varepsilon_3' = \varepsilon_v' = 0 \tag{13.8}$$

Thus the stress components cause no change in volume but only change in shape.

We now turn to the determination of the strain energy quantities in the expression

$$U_T = U_V + U_S$$

where U_T = total strain energy, U_V = volumetric strain energy and U_S = shear or distortion strain energy.

The total strain energy is given by the sum of the energy components due to the three principal stresses and principal strains so that,

$$U_T = \tfrac{1}{2}\sigma_1\varepsilon_1 + \tfrac{1}{2}\sigma_2\varepsilon_2 + \tfrac{1}{2}\sigma_3\varepsilon_3 \tag{13.9}$$

Substituting for the principal strains, from the stress–strain relationships and rearranging gives

$$U_T = \frac{1}{2E}(\sigma_1{}^2 + \sigma_2{}^2 + \sigma_3{}^2) - \frac{v}{2E}(2\sigma_1\sigma_2 + 2\sigma_2\sigma_3 + 2\sigma_3\sigma_1)$$

$$\text{per unit volume} \quad (13.10)$$

The volumetric strain energy can now be determined from the hydrostatic component of stress, $\bar{\sigma}$.

$$U_V = \tfrac{1}{2}\bar{\sigma}\bar{\varepsilon}$$

$$= \tfrac{1}{2}\bar{\sigma}\frac{3\bar{\sigma}}{E}(1-2v)$$

$$= \tfrac{1}{2}\left[\frac{\sigma_1 + \sigma_2 + \sigma_3}{3}\right]\left[\frac{3(1-2v)}{E}\right]\left[\frac{\sigma_1 + \sigma_2 + \sigma_3}{3}\right]$$

$$= \frac{1-2v}{6E}(\sigma_1 + \sigma_2 + \sigma_3)^2 \text{ per unit volume} \tag{13.11}$$

But $U_S = U_T - U_V$; therefore

$$U_S = \frac{1}{2E}[\sigma_1^2 + \sigma_2^2 + \sigma_3^2 - 2\nu(\sigma_1\sigma_2 + \sigma_2\sigma_3 + \sigma_3\sigma_1)]$$

$$- \frac{1-2\nu}{6E}(\sigma_1 + \sigma_2 + \sigma_3)^2$$

which reduces to

$$U_S = \frac{1+\nu}{6E}[(\sigma_1 - \sigma_2)^2 + (\sigma_2 - \sigma_3)^2 + (\sigma_3 - \sigma_1)^2]$$

per unit volume   (13.12)

or alternatively, using the relationship between E, G and ν,

$$U_S = \frac{1}{12G}[(\sigma_1 - \sigma_2)^2 + (\sigma_2 - \sigma_3)^2 + (\sigma_3 - \sigma_1)^2]$$

per unit volume   (13.13)

Now, the shear or distortion strain energy theory proposes that yielding commences when the quantity U_S reaches the equivalent value at yielding in simple tension. In the latter case σ_2 and $\sigma_3 = 0$ and $\sigma_1 = \sigma_Y$; therefore

$$U_S = \frac{\sigma_Y^2}{6G} \text{ per unit volume}$$

(13.14)

and

$$\frac{1}{12G}[(\sigma_1 - \sigma_2)^2 + (\sigma_2 - \sigma_3)^2 + (\sigma_3 - \sigma_1)^2] = \frac{\sigma_Y^2}{6G}$$

or

$$(\sigma_1 - \sigma_2)^2 + (\sigma_2 - \sigma_3)^2 + (\sigma_3 - \sigma_1)^2 = 2\sigma_Y^2$$

(13.15)

In the two-dimensional system, $\sigma_3 = 0$ and

$$\sigma_1^2 + \sigma_2^2 - \sigma_1\sigma_2 = \sigma_Y^2$$

(13.16)

for yielding to occur.

The above analysis has been directly aimed at establishing a yield criterion on an energy basis. However, from eqn. (13.15) one might equally well propose that yielding occurs as a function of the differences between principal stresses. On this hypothesis it is evident that eqn. (13.15) can also be obtained by considering the root mean square of the principal stress differences in the complex stress system in relation to simple tension. Thus

$$[\tfrac{1}{3}\{(\sigma_1 - \sigma_2)^2 + (\sigma_2 - \sigma_3)^2 + (\sigma_3 - \sigma_1)^2\}]^{1/2} = [\tfrac{1}{3}(2\sigma_Y^2)]^{1/2}$$

(13.17)

The right-hand side of the equation is obtained for simple tension by putting $\sigma_1 = \sigma_Y$ and $\sigma_2 = \sigma_3 = 0$. Equation (13.17) reduces to

$$(\sigma_1 - \sigma_2)^2 + (\sigma_2 - \sigma_3)^2 + (\sigma_3 - \sigma_1)^2 = 2\sigma_Y^2$$

which is the same as eqn. (13.15).

Many experiments have been conducted under complex stress conditions to study the behaviour of metals and it has been shown that hydrostatic pressure, and by inference hydrostatic tension, does not cause yielding. Now any complex stress system can be regarded as a combination of hydrostatic stress and a function of the difference of principal stresses, and therefore a yield criterion such as that of Tresca or von Mises which is based on principal stress difference would seem to be the most logical.

YIELD ENVELOPE AND LOCUS

For the case of three principal stresses, all non-zero, the shear strain energy criterion, eqn. (13.15), is represented by a circular cylinder whose longitudinal axis is equally inclined to the three co-ordinate axes σ_1, σ_2, σ_3 (Fig. 13.3). The surface of the cylinder represents the envelope between an elastic stress system within the cylinder and a plastic stress state outside.

Fig. 13.3

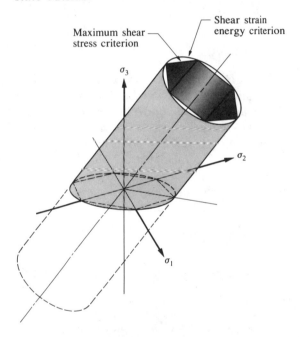

The maximum shear stress criterion for three-dimensional states of stress is represented by a hexagonal cylinder lying within the circular cylinder as shown shaded in Fig. 13.3. Again, the hexagonal envelope divides elastic from plastic or yielded stress states.

If we wish to consider two-dimensional stress states in which, say $\sigma_3 = 0$, then the yield boundary or locus is given by the intersection of the σ_1, σ_2 plane with the two cylinders, as shown by the dashed lines in Fig. 13.3. The yield loci for the above two criteria for ductile materials subjected to two-dimensional principal stress states is illustrated in Fig. 13.4.

A number of experimental studies have been carried out on various ductile metals to establish the appropriate criterion for yielding under various combinations of principal stresses. There are several classical test

Fig. 13.4

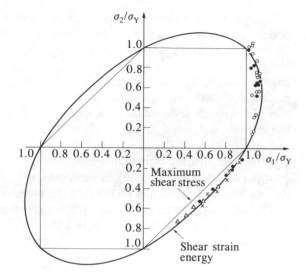

methods for studying complex stress behaviour. These include a thin-walled tube subjected to internal pressure and tensile or compressive axial load, or a tube under combined torsion and axial load or bending. Suitable measurements of strain or deformation are recorded as applied loading is increased to a point where linear elasticity is no longer obtained. The principal stresses may then be calculated and plotted to represent yielding, as shown in Fig. 13.4. It is seen that there is close correlation with the von Mises shear strain energy criterion and that the Tresca maximum shear stress criterion is also satisfactory but somewhat more conservative.

EXAMPLE 13.1

A mild steel shaft of 50 mm diameter is subjected to a bending moment of 1.9 kN-m. If the yield point of the steel in simple tension is 200 MN/m², find the maximum torque that can also be applied according to: (a) the maximum shear stress; (b) the shear strain energy theories of yielding.

SOLUTION

The maximum bending stress occurs at the surface of the shaft and is given by

$$\sigma_x = \frac{32M}{\pi d^3} = \frac{32}{\pi} \times \frac{1900}{125 \times 10^{-6}} = 155 \text{ MN/m}^2$$

The maximum shear stress at the surface is

$$\tau_{xy} = \frac{16T}{\pi d^3} = \frac{16}{\pi} \times \frac{T}{125 \times 10^{-6}} = 40.7 \times 10^3 T$$

(a) Maximum shear stress theory

$$\tau = \frac{\sigma_1 - \sigma_2}{2} = \frac{200 \times 10^6}{2}$$

$$\frac{\sqrt{(\sigma_x^2 + 4\tau_{xy}^2)}}{2} = 200 \times 10^6$$

$$155^2 + [4 \times (0.0407T)^2] = 200^2$$

$$0.001\,66T^2 = 4000$$

$$T = 1.55 \text{ kN-m}$$

(b) Shear strain energy theory

$$\sigma_1^2 + \sigma_2^2 - \sigma_1\sigma_2 = (200 \times 10^6)^2$$

Putting $\sigma_x^2 + 4\tau_{xy}^2 = A$,

$$\tfrac{1}{4}(\sigma_x + \sqrt{A})^2 + \tfrac{1}{4}(\sigma_x - \sqrt{A})^2 - \tfrac{1}{4}(\sigma_x - \sqrt{A})(\sigma_x + \sqrt{A})$$
$$= (200 \times 10^6)^2$$
$$\tfrac{1}{4}[\sigma_x^2 + 2\sigma_x\sqrt{A} + A + \sigma_x^2 - 2\sigma_x\sqrt{A} + A - \sigma_z^2 + A]$$
$$= (200 \times 10^6)^2$$

Simplifying gives

$$\sigma_x^2 + 3\tau_{xy}^2 = (200 \times 10^6)^2$$
$$155^2 + [3 \times (0.001\,66T^2)] = 200^2$$
$$0.001\,66T^2 = 5330$$

Therefore

$$T = 1.79 \text{ kN-m}$$

EXAMPLE 13.2

A thin-walled steel cylinder of 2 m diameter is subjected to an internal pressure of 2.5 MN/m². Using a safety factor of 2 and a yield stress in simple tension of 400 MN/m², calculate wall thickness on the basis of Tresca and von Mises yield criteria. It may be assumed that the radial stress in the wall is negligible.

SOLUTION

The stress system in the wall of the cylinder consists of three principal stresses, circumferential, axial and radial, of which the latter may be neglected and will be taken as zero. Hence, using eqns. (2.9) and (2.10) we have

$$\sigma_1 = \frac{pr}{t} \quad \text{and} \quad \sigma_2 = \frac{pr}{2t}$$

(a) Tresca criterion

Since both axial and circumferential stresses are tension the maximum difference between principal stresses gives

$$\hat{\tau} = \frac{\sigma_1 - 0}{2} = \frac{\sigma_Y}{2} \quad \text{at yielding}$$

Hence

$$\sigma_1 = \sigma_Y \quad \text{therefore} \quad \frac{2.5 \times 1000}{t} = \frac{400}{2}$$

$$t = 12.5 \text{ mm}$$

(b) *Von Mises criterion*

For $\sigma_3 \simeq 0$ we have

$$\sigma_1^2 + \sigma_2^2 - \sigma_1\sigma_2 = \sigma_Y^2$$

$$\frac{p^2r^2}{t^2} + \frac{p^2r^2}{4t^2} - \frac{p^2r^2}{2t^2} = \sigma_Y^2$$

$$\frac{3}{4}\frac{p^2r^2}{t^2} = \sigma_Y^2$$

$$t = \left(\frac{3}{4}\right)^{1/2} \cdot \frac{pr}{\sigma_Y} = \left(\frac{3}{4}\right)^{1/2} \cdot \frac{2.5 \times 1000}{200}$$

$$t = 10.8 \text{ mm}$$

The slightly larger plate thickness given by the Tresca criterion illustrates its more conservative characteristic compared with the von Mises criterion.

FRACTURE CRITERIA: BRITTLE MATERIALS

Brittleness in a material may be defined as an inability to deform plastically. Materials such as glass, some cast irons, concrete and some plastics, when subjected to tensile stress will generally fracture at or just beyond the elastic limit. We are therefore not much concerned with a yield criterion as a fracture criterion for brittle materials under complex principal stress states. The most widely used criterion is that suggested by Rankine known as the *maximum principal stress theory*.

MAXIMUM PRINCIPAL STRESS (RANKINE) CRITERION

This hypothesis, proposed by Rankine, which was also intended for use to predict yielding of a ductile material, states that "failure" (i.e. fracture of a brittle material or yielding of a ductile material) will occur in a complex stress state when the maximum principal stress reaches the stress at "failure" in simple tension. The two-dimensional locus for this theory is illustrated in Fig. 13.5. It will be noticed that in the first and third quadrants the boundary is the same as for the maximum shear stress theory.

MOHR FRACTURE CRITERION

Some materials, such as cast iron, have much greater strength in compression than that in tension. Mohr proposed that, in the first and third quadrants of a "failure" locus, a maximum principal stress theory was appropriate based on the ultimate strength of the material in tension or compression respectively. In the second and fourth quadrants where the two principal stresses are of opposite sign the maximum shear stress theory should apply.

This results in a diagram as shown in Fig. 13.6.

Fig. 13.5

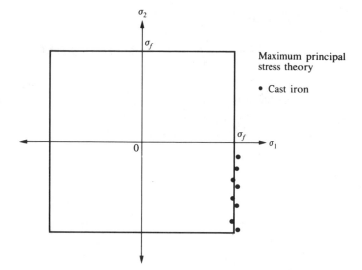

Fig. 13.6

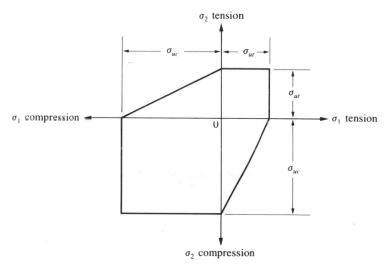

EXAMPLE 13.3

In a cast-iron component the maximum principal stress is to be limited to one-third of the tensile strength. Determine the maximum value of the minimum principal stress using the Mohr theory. What would be the values of the principal stresses associated with a maximum shear stress of 390 MN/m²? The tensile and compressive strengths of the cast iron are 360 MN/m² and 1.41 GN/m² respectively.

SOLUTION

Maximum principal stress = 360/3 = 120 MN/m² (tension). According to Mohr's theory, in the second and fourth quadrant

$$\frac{\sigma_1}{\sigma_{ut}} + \frac{\sigma_2}{\sigma_{uc}} = 1$$

Therefore

$$\frac{120}{360} + \frac{\sigma_2}{-1410} = 1 \quad \text{and} \quad \sigma_2 = -940 \, \text{MN/m}^2$$

The Mohr's stress circle construction for the second part of this problem is shown in Fig. 13.7. If the maximum shear stress is 390 MN/m², a circle is drawn of radius 390 units to touch the two envelope lines. The principal stresses can then be read off as +200 MN/m² and −580 MN/m².

Fig. 13.7

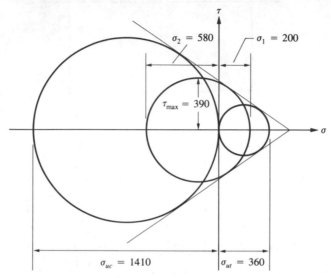

STRENGTH OF LAMINATES

In the previous sections we have been dealing with isotropic materials, so that it has been possible to develop a failure criterion on the basis of one limiting parameter (the yield strength). For fibre-reinforced composites, however, it is not possible to define failure in terms of a single parameter. It was shown in Chapter 12 that unidirectional composites have a longitudinal strength which is many times the transverse and shear strengths. Thus when a multi-axial stress system is applied it is necessary to consider the effect of this in relation to the various strength components of the composite. For laminates the picture is complicated still further by the different orientations of the individual lamina in relation to the applied stress system. However, since a chain is only as strong as its weakest link, the strength of a laminate will be determined by the strength of the individual plies within the laminate.

One of the most popular failure criteria for laminates is the Tsai–Hill[1] criterion. This is based on the von Mises failure criterion which was expanded by Hill to anisotropic bodies and applied to composites by Tsai. The criterion may be expressed as

$$\left(\frac{\sigma_x}{T_x}\right)^2 - \left(\frac{\sigma_x \sigma_y}{T_x^2}\right) + \left(\frac{\sigma_y}{T_y}\right)^2 + \left(\frac{\tau_{xy}}{S_{xy}}\right)^2 = 1 \tag{13.18}$$

where

σ_x = the stress parallel to the fibres;
σ_y = is the stress perpendicular to the fibres;
T_x = is the tensile strength parallel to the fibres;
T_y = is the tensile strength perpendicular to the fibres;
τ_{xy}, S_{xy} = the shear stress and shear strength values.

The Tsai–Hill criterion would be used as follows. Consider a laminate subjected to in-plane stresses which are applied relative to the global X–Y-directions. The strains may then be calculated in the X–Y-directions using the compliance matrix $|a|$ (*see* Ch. 12).

$$\begin{Bmatrix} \varepsilon_X \\ \varepsilon_Y \\ \gamma_{XY} \end{Bmatrix} = |a| \begin{Bmatrix} \sigma_X \\ \sigma_Y \\ \tau_{XY} \end{Bmatrix} \tag{13.19}$$

For example, if there is only a stress in the global X-direction, then

$$\varepsilon_X = a_{11}\sigma_X; \qquad \varepsilon_Y = a_{12}\sigma_X; \qquad \gamma_{XY} = 0$$

For in-plane loading the strains will be the same in all the plies so that these strain values may be used to get the stress in each ply. Using the stiffness matrix $|A|$ for each ply:

$$\begin{Bmatrix} \sigma_X \\ \sigma_Y \\ \tau_{XY} \end{Bmatrix}_{ply} = |A|_{ply} \begin{Bmatrix} \varepsilon_X \\ \varepsilon_Y \\ \gamma_{XY} \end{Bmatrix}_{ply/laminate} \tag{13.20}$$

From the terminology for $|A|$ in Chapter 12, this would give

$$\sigma_X = \bar{Q}_{11}\varepsilon_X + \bar{Q}_{12}\varepsilon_Y + \bar{Q}_{16}\gamma_{XY}$$

$$\sigma_Y = \bar{Q}_{21}\varepsilon_X + \bar{Q}_{22}\varepsilon_Y + \bar{Q}_{26}\gamma_{XY}$$

$$\tau_{XY} = \bar{Q}_{61}\varepsilon_X + \bar{Q}_{62}\varepsilon_Y + \bar{Q}_{66}\gamma_{XY}$$

These are the stresses in each ply in the global X–Y-directions. They would then need to be transferred to the local x–y-directions using the transformation matrix $|T|$.

$$\begin{Bmatrix} \sigma_x \\ \sigma_y \\ \tau_{xy} \end{Bmatrix} = |T| \begin{Bmatrix} \sigma_X \\ \sigma_Y \\ \tau_{XY} \end{Bmatrix} \tag{13.21}$$

At this point the Tsai–Hill criterion could be used to establish whether or not failure would be expected in the ply under consideration. To do this eqn. (13.18) could be applied or the alternative form shown below which is popular because it gives a safety factor S.F.:

$$\text{S.F.} = \frac{T_x}{\sqrt{\left(\sigma_x^2 - \sigma_x\sigma_y + \dfrac{T_x^2}{T_y^2}\sigma_y^2 + \dfrac{T_x^2}{S_{xy}^2}\tau_{xy}^2\right)}} \tag{13.22}$$

CONCEPTS OF STRESS CONCENTRATION

In previous chapters the problems analysed have had stress distributions which were either uniform or varied smoothly and gradually over a significant area. However, in the vicinity of the point of application of a concentrated load there is a rapid variation in stress over a small area, in which the maximum value is considerably higher than the average stress in the full section of the material. This situation is known as a *stress concentration*.

The cause of stress concentration is perhaps most readily understood from consideration of analogous systems such as the flow of a fluid. In a simple strut subjected to an axial compressive force as shown in Fig. 13.8(*a*), the force is transmitted through the strut via the medium of the stresses exerted on every small element of material by its neighbouring elements. The lines of transmission of the stress are similar to the flow lines which would be observed if a fluid entered (at the point of application of the force) a channel of the same cross-section as the strut. The densely-packed flow lines at the entry and exit points are representative of the concentration of stress at those points.

Fig. 13.8

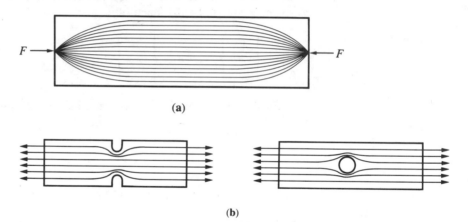

(a)

(b)

Another way in which a stress concentration can be produced is at a geometrical discontinuity in a body, such as a hole, keyway, or other sharp change in sectional dimensions. Figure 13.8(*b*) again uses the flow analogy to illustrate the stress concentration effect which occurs at notches and holes.

Two questions come to mind when considering the above effect. If, as is the case, all points of support and load application disturb the uniformity of stress and cause stress concentration, it is surprising that one can obtain realistic results by, say, the simple bending and torsion theories considered earlier. This problem was studied theoretically by St Venant, who stated the following principle. If the forces acting on a small area of a body are replaced by a statically equivalent system of forces acting on the same area, there will be considerable changes in the local stress distribution, but the effect on the stresses at distances large compared with the area on which the forces act will be negligible. For instance, in a bar gripped at each end and subjected to axial tension, the stress distribution at the ends will vary considerably according to whether gripping is by screw thread, button head, or wedge jaws. However, it has been shown that, at a distance of between one and two diameters from the ends, the stress distribution is quite uniform across the section. Similarly, it is immaterial how the couples are applied at the ends of a beam in pure bending. So long as the length is markedly greater than the cross-sectional dimensions, the assumptions and simple theory of bending will hold good at a distance of approximately one beam depth away from the concentrated force.

The second feature that is of interest is that, although a stress concentration is only effective locally (St Venant), the peak stress at this point is sometimes far in excess of the average stress calculated in the body of the component. Why is it then normal practice under static loading to base design calculations on the main field of stress, and not on the maximum stress concentration value where a load is applied? Consider a point or line application of load, then theoretically the elastic stress would become infinite in the material under the load. This obviously cannot occur in practice since a ductile material will reach a yield point and plastic deformation will occur under the point of application of the load. The effect of the plastic flow is to cause a local redistribution of stress, which relieves the stress concentration slightly so that the peak value of stress does not continue to increase with increasing load at the same rate as in the elastic range. Eventually, with still greater loading, general yielding in the body of the material will tend to catch up and encompass what was the stress concentration area.

Brittle materials have little or no capacity for plastic deformation and therefore the stress concentration is maintained up to fracture. Whether or not there is an accompanying reduction in nominal strength depends largely on the structure of the material. Those such as glass and some cast irons which have inherent internal flaws, which themselves set up stress concentration, show little reduction in strength over the unnotched condition. Others which have a homogeneous stress-free structure will show a considerable decrease in static strength for a severe notch.

From all of the foregoing it would appear that stress concentration does not present too serious a problem for components in service. However, there are two main aspects of material behaviour in which stress concentration plays the major part in causing failure. These are fatigue and brittle fracture (notch brittle reaction of a normally ductile metal), both of which topics are dealt with at length in later chapters.

The theoretical analysis of stress concentration is generally very complex by classical mathematics. Many theoretical solutions are due to Neuber, and a number of individual problems have been solved by other theoreticians and are available in published papers. The development of finite element analysis in recent times has provided many more solutions.

The principal experimental method which has provided many simple and accurate solutions to problems of stress concentration is the technique of photoelasticity.

CONCENTRATED LOADS AND CONTACT STRESSES

CONCENTRATED LOAD ON THE EDGE OF A PLATE

The local distribution of stress at the point of application of a concentrated load normal to the edge of an infinite plate was first studied in 1891 using photoelasticity. This led to the theoretical solutions a year later of Boussinesq and Flamant.

Consider the three systems of forces shown in Fig. 13.9 acting on the edge of an infinitely large plate of thickness b. The resultant force in each

Fig. 13.9

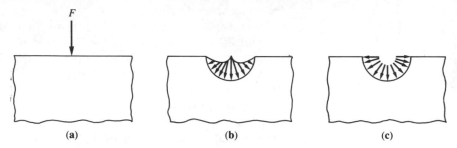

(a)   (b)   (c)

case is the same, and hence the systems are statically equivalent and therefore satisfy the principle of St Venant. Now, case (*a*) is the one we wish to solve, but this will result in practice in a small volume of plastic flow as explained previously. To overcome this difficulty we replace the point load on the straight edge by a radial distribution of forces, as in (*b*) or (*c*), around a small semicircular groove. Experiment has shown that the forces in (*c*) give the better representation of the stress distribution due to a concentrated load on a straight edge. The solution by Flamant on this basis shows that the stress distribution is a simple radial one involving compression only. Using polar co-ordinates, and referring to Fig. 13.10, any element distant r from O at an angle θ to the normal to

Fig. 13.10

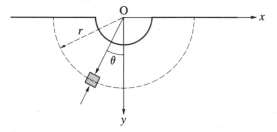

the edge of the plate is subjected to simple radial compression only of a magnitude

$$\sigma_r = -\frac{2F \cos \theta}{\pi b r}$$

CONCENTRATED LOAD BENDING A BEAM

The cross-section of the beam at which the load is acting is subjected to a complex stress condition composed of stress due to simple bending plus the stress due to the concentrated load itself.

Fig. 13.11

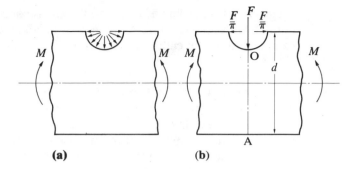

(a)   **(b)**

Considering the radial pressure distribution on the small groove in Fig. 13.11(*a*), then the horizontal components give rise to forces F/π acting parallel to the edge of the beam, so that the system of forces equivalent to the pressure distribution is as shown in Fig. 13.11(*b*). In this problem we are not considering an infinite plate, but a beam of finite depth, and consequently the horizontal forces, F/π, set up longitudinal tension and bending stresses. The former is given simply as load divided by area or $F/\pi \times (1/bd)$. The latter are determined by considering the bending moment about the axis of the beam given by $(F/\pi) \times \frac{1}{2}d$. The bending stresses are therefore

$$\sigma_x = \mp \frac{Fd}{2\pi} \frac{y}{I}$$

The total stress acting across the section OA of the beam is then obtained by the superposition of the various separate quantities:

$$\sigma_{x_{OA}} = \pm \frac{Fl}{4} \frac{y}{I} \mp \frac{Fd}{2\pi} \frac{y}{I} + \frac{F}{\pi bd}$$

$$= \pm \left(\frac{l}{4} - \frac{d}{2\pi}\right) \frac{12Fy}{bd^3} + \frac{F}{\pi bd}$$

This expression is often referred to as the Wilson–Stokes solution.

The distribution of $\sigma_{x_{OA}}$ for long and short spans is compared with the simple bending distribution in Fig. 13.12, and it is seen that the more accurate solution gives rise to maximum longitudinal stresses which are *less* than those from simple bending theory.

Fig. 13.12

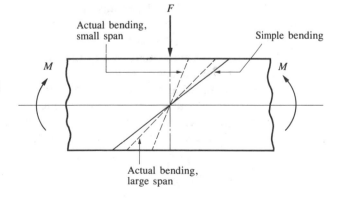

Hence in this problem, although the stress concentration causes high normal compressive stresses, the tensile bending stress which would be expected to be the cause of failure is in fact reduced by the concentrated load.

CONTACT STRESS

Another important problem involving stress concentration is the condition of contact of two bodies under load. Typical examples may be found in the mating of gear teeth, in a shaft in a bearing, and in the balls and rollers in bearings. Solutions are too complex and lengthy to be considered here but a few useful results will be quoted.

Considering first the situation of a ball under loaded contact with a flat surface, Fig. 13.13. The point of contact when unloaded develops into a small spherical surface when under load, which is initially elastic deformation.

Fig. 13.13

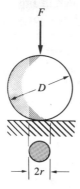

The radius of the circular contact area is given as

$$r = 0.88\left(\frac{FD}{E}\right)^{1/3}$$

where F is the contact force, D is the diameter of the ball and E is the modulus of the two materials (assumed the same in this instance).

The distribution of pressure over the contact area is such that a maximum value occurs at the centre of the circle equal to

$$\sigma_{max} = 0.62\left(\frac{FE^2}{D^2}\right)^{1/3}$$

From the dimensions and pressure on the contact surface the stress distribution can be calculated along the axis normal to contact. If maximum shear stress is taken as the criterion for yielding, it is found that the greatest value occurs not at the surface of contact but at a small depth below the surface in each body. It is generally at this point that failure of the material would originate if the loading were excessive.

For a roller in contact with a plane surface, Fig. 13.14, the contact area is rectangular of length l, and width w, with maximum pressure occurring at the centre of the rectangle.

$$w = 2.15\left(\frac{FD}{El}\right)^{1/2}$$

Fig. 13.14

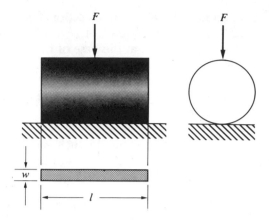

where F is the contact force, D the diameter of the roller and l its length and E the modulus of the two materials, as before. The maximum pressure is given as

$$\sigma_{max} = 0.59\left(\frac{FE}{lD}\right)^{1/2}$$

GEOMETRICAL DISCONTINUITIES

It was previously explained in the introduction that abrupt changes in geometry of a component give rise to stress concentration in a similar manner to those described in previous sections for loading. In most cases, the failure of a component can be attributed to some form of geometrical stress raiser, either from bad design or misfortune.

Typical examples of stress raisers are oil holes, keyways and splines, threads, and fillets at changes of section.

Figure 13.15(a) illustrates a bar under tensile loading into which has been machined two grooves. The uniform stress is σ at a distance from

Fig. 13.15

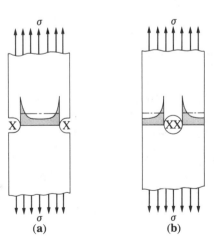

(a) (b)

the discontinuity where the cross-sectional area is A_1. The *average* stress at section XX, where the area is A_2, will be $\sigma(A_1/A_2)$. However, it is seen that at each point X at the base of the grooves there is a peak value of σ which is much higher than the average stress. There is a steep stress gradient to the lower levels of stress in the central region of the bar.

The area under this stress curve must be the same as the area under the average stress line.

STRESS CONCENTRATION FACTOR

The stress concentration set up by such geometrical discontinuities is a function of the shape and dimensions of the discontinuity, and is expressed in terms of the *elastic stress concentration factor* denoted by K_t:

$$K_t = \frac{\text{Maximum boundary stress at the discontinuity}}{\text{Average stress at that cross-section of the body}}$$

This factor is obviously constant within the elastic range of the material. Since this is dependent on the geometrical proportions and shape of the notch and the type of stress system (tension, torsion, etc.), it is readily appreciated that to cover a wide range of parameters would require a great deal more space than part of one chapter. Furthermore, the limitations on theoretical solutions for stress concentration at notches have resulted in many analyses being obtained experimentally (generally photoelastically) and presented as charts of K_t against geometrical proportions. The subject has been treated very thoroughly by Neuber,[2] Peterson,[3] Frocht[4] and others, and therefore attention will be confined here to a few special cases of interest.

CIRCULAR HOLE

The distribution of axial and transverse stress at the hole cross-section is illustrated in Fig. 13.15(b). The peak stress arises at points XX at the edge of the hole and, for a small hole, in a thin infinite plate subjected to tension, the stress concentration factor $K_t = 3$. This is the highest value obtainable and, for a circular hole in a finite-width strip under axial loading K_t, it lies between 2 and 3. The relationship between K_t and geometrical proportions is plotted in Fig. 13.16.

In pure bending of a finite width plate with a transverse hole, K_t is a function of plate thickness as well as of radius of hole r and width of plate w. For a very thick plate and small hole, the K_t against r/w curve is identical with Fig. 13.16 for the tension strip. When the hole is large and the plate thin, K_t varies from 1.85 to 1.1 for decreasing width.

The case of a shaft with a transverse hole subjected to tension, bending or torsion is a common one. It introduces the feature of a three-dimensional stress system as against plane stress in the plate case. The stress concentration factor in a triaxial stress field, although defined in the same way as for a biaxial stress system, will be denoted by K_t'. Frocht has studied photoelastically a circular bar with a transverse hole under tension and has obtained a curve of K_t' against r/d as shown in Fig. 13.17. Also plotted here for comparison is Howland's curve for K_t for a plate under tension, and it is interesting to note that K_t' is noticeably higher.

Fig. 13.16 Stress concentration factor K_t for axial loading of a finite-width plate with a transverse hole (reference 3: by courtesy of John Wiley & Sons, Inc)

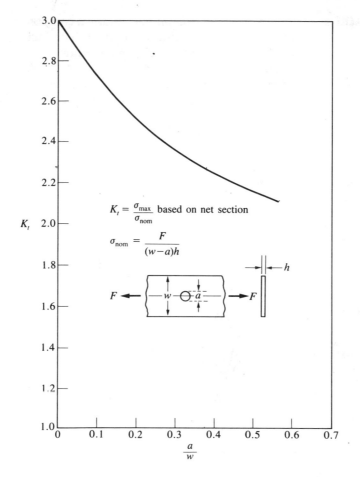

$$K_t = \frac{\sigma_{\max}}{\sigma_{\mathrm{nom}}} \text{ based on net section}$$

$$\sigma_{\mathrm{nom}} = \frac{F}{(w-a)h}$$

K_t (vertical axis)

$\dfrac{a}{w}$ (horizontal axis)

Fig. 13.17 Summary of experimental results for shafts with transverse holes in tension (reference 4: by courtesy of John Wiley & Sons, Inc)

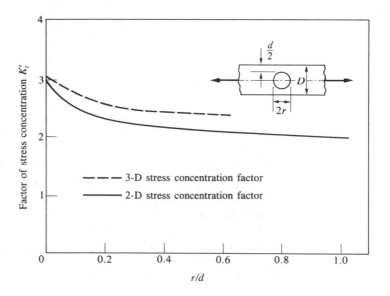

Factor of stress concentration K_t' (vertical axis)

r/d (horizontal axis)

– – – – 3-D stress concentration factor

——— 2-D stress concentration factor

FILLET RADIUS   Almost without exception cylindrical components do not have a uniform diameter from one end to the other. The journal bearings of a crankshaft have to mate into the web, and a motor shaft or railway axle requires shoulders to retain bearings or a wheel. At these changes of section a sharp corner would introduce an intolerable stress concentration which could lead to failure by fatigue. The problem is lessened by the introduction of a fillet radius, to blend one section smoothly into the next. Even a fillet radius will give rise to some stress concentration;

Fig. 13.18 Stress concentration factor K_t for the bending of a shaft with a shoulder fillet (reference 3: by courtesy of John Wiley & Sons, Inc)

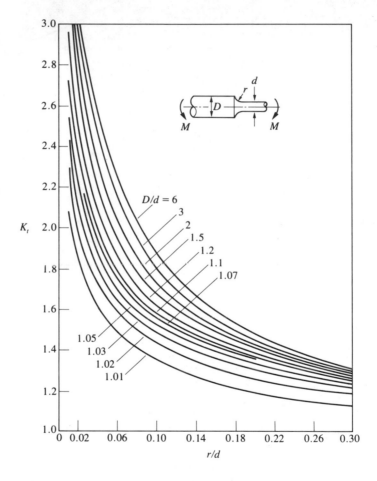

however, this will be considerably less than with the sharp corner. The stress concentration at a fillet radius is not readily amenable to mathematical treatment and solutions have been obtained by photoelasticity and other experimental means.

Values of K_t for various geometrical proportions have been obtained for shafts in tension, bending and torsion. The latter two cases are probably the most common in practice and therefore only the charts for these, in Figs. 13.18 and 13.19, have been included, for reasons of space.

Fig. 13.19 Stress concentration factor K_{ts} for the torsion of a shaft with a shoulder fillet (reference 3: by courtesy of John Wiley & Sons, Inc)

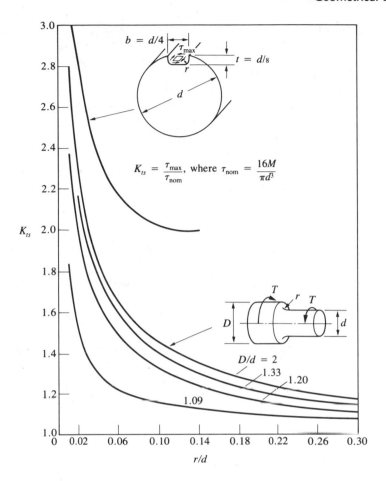

$$K_{ts} = \frac{\tau_{max}}{\tau_{nom}}, \text{ where } \tau_{nom} = \frac{16M}{\pi d^3}$$

KEYWAYS AND SPLINES IN TORSION

Another design feature which requires careful attention is the keyway or spline in a shaft subjected to torsion.

The rectangular keyway of standard form having root fillets has been solved mathematically by Leven, and the curve for K_{ts} against the ratio of fillet radius to shaft diameter is shown in Fig. 13.19. The maximum shear stress occurs at a point 15° from the bottom of the keyway on the fillet radius.

GEAR TEETH

The stress distribution in a loaded gear tooth is a complex problem. Stress concentration at the point of mating of two teeth due to contact load varies in position with rotation of the teeth, and further stress concentration occurs at the root fillets of a tooth, the latter being the more serious. The former can be analysed with the aid of the expressions given in the section under contact stresses. The fillet stress concentration is a function of the load components, (causing bending and direct stress) and the gear tooth geometry.

A very comprehensive investigation was conducted by Jacobson,[5] who studied involute spur-gears (20° pressure angle) photoelastically and produced a series of charts of strength factor against the reciprocal of the number of teeth in the gear, which cover the whole possible range of

Fig. 13.20

spur-gear combinations. Figure 13.20 shows the photoelastic stress pattern and this clearly indicates the areas of stress concentration at the root radius and point of contact.

SCREW THREADS

One of the most common causes of machinery or plant having to be shut down is the fatigue of bolts or studs. This is principally due to the high stress concentration at the root of the thread. The problem has been tackled theoretically by Sopwith, and photoelastically by Hetenyi, and Brown and Hickson.[6] Each of the investigations confirmed that, for a bolt and nut of conventional design, the load distribution along the screw (Fig. 13.21) is far from uniform and reaches a maximum intensity at the plane of the bearing face of the nut. This is due in part to the unmatched and opposing signs of the strains in the screw and nut. This can be overcome to a large extent by altering the design of the nut, principally so that the nut thread is in tension and so matching the strains more evenly with the screw.

Fig. 13.21 Stress distribution in ordinary stud and nut (reference 6: reprinted by permission of the Institution of Mechanical Engineers)

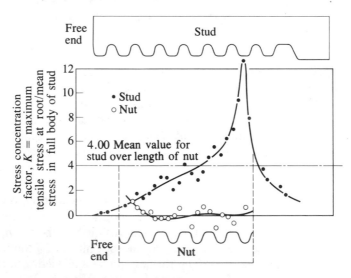

EXAMPLE 13.4

A shaft is stepped from 48 mm diameter to 40 mm diameter through a fillet of 2.4 mm radius. A gear wheel is keyed to the larger diameter of the shaft and the radii at the internal corners of the keyway are each

1.5 mm. What is the maximum torque that can be transmitted if the steel has a shear stress limit of 200 MN/m^2?

SOLUTION

$$D/d = 48/40 = 1.2 \quad \text{and} \quad r/d = 2.4/40 = 0.06$$

From Fig. 13.19, $K_{ts} = 1.5$ at the fillet radius. At the keyway $r/d = 1.5/48 = 0.031$. From Fig. 13.19, $K_{ts} = 2.4$. Now

$$K_{ts} = \frac{\tau_{max}}{\tau_{nom}}, \quad \text{where} \quad \tau_{max} = 200 \text{ MN/m}^2$$

and

$$\tau_{nom} = \frac{16T}{\pi d^3}$$

At the fillet radius

$$\frac{16T}{\pi \times 0.04^3} = \frac{200 \times 10^6}{1.5}$$

$$T = 1676 \text{ Nm}$$

At the keyway

$$\frac{16T}{\pi \times 0.048^3} = \frac{200 \times 10^6}{2.4}$$

$$T = 1810 \text{ Nm}$$

The maximum torque that can be transmitted is 1676 Nm which is governed by the shear stress at the fillet radius.

YIELD AND PLASTIC STRESS CONCENTRATION FACTORS

The usual definition of stress concentration factor, as given in the introduction, is in terms of the maximum stress at the discontinuity. There is of course a biaxial stress condition at the free surface of a notch, and therefore it might be more realistic to express stress concentration in terms of the maximum "equivalent stress," i.e. using one of the yield criteria discussed earlier. If the biaxial stresses can be determined theoretically at the free surface of a notch, then considering the shear strain energy criterion, which seems the most appropriate for both static and fatigue stress conditions in a ductile material, we have for $\sigma_1 > \sigma_2$ and $\sigma_3 = 0$

$$\sigma_{equiv} = \sqrt{(\sigma_1^2 - \sigma_1\sigma_2 + \sigma_2^2)}$$

Then

$$K_e = \frac{\sigma_{equiv}}{\sigma_{nom}}$$

In the elastic range there is a constant ratio between σ_1 and σ_2 for any particular point on the surface; therefore

$$\sigma_1 = k\sigma_2$$

and

$$K_e = \frac{\sigma_2\sqrt{[1 - (1/k) + (1/k^2)]}}{\sigma_{nom}}$$

It is seen from the above that K_e is always less than K_t, except when $k = 1$ and then $K_e = K_t$. This may partly explain why notched strength reduction in fatigue (Ch. 21) is nearly always less than would be expected from considerations of K_t.

If loading is increased to the point where yielding occurs at the notch, there is a redistribution of stress similar to that at a concentrated load. The stress concentration factor is still defined in the same way as in the elastic range but is now denoted by K_p, the *plastic stress concentration factor*, and this is now also a function of the degree of plastic deformation that has occurred. It is found in practice that for ductile materials that, as $\sigma_{max} \rightarrow \sigma_{ult}$, $K_p \rightarrow 1$, i.e. the stress concentration does not reduce the strength of a component loaded statically. However, in fatigue (Ch. 21) stress concentration only achieves a plastic state in a microscopic locality which is contained in a larger elastic stress concentration area. This is the cause in all cases of the initiation of the fatigue crack.

SUMMARY

The latter part of this chapter demonstrates that although designers basically work in the elastic range of materials and indeed employ safety factors to make that more sure, the presence of stress concentration can still result in some local yielding. It is therefore very important to have acceptable yield criteria to apply to complex stress states. For ductile materials it is now well proven that the shear strain energy and maximum shear stress theories both give satisfactory predictions of the onset of yielding. Brittle materials, on the other hand, fracture rather than yield, and this situation can be adequately designed for on the basis of the maximum principal stress theory or the Mohr modified maximum shear stress criterion.

There are plenty of data nowadays on stress concentration factors owing to the primary effects in fatigue failure and fracture mechanics which will be discussed in later chapters. From a design point of view we obviously cannot eliminate stress raisers entirely, but it is possible to minimize their effect by careful attention to the detail of points of load application, appropriate surface hardening and heat treatment and the avoidance of sharp internal corners and sudden changes of section.

REFERENCES

1. Tsai, S. W. and Hahn, H. T. *Introduction to Composite Materials,* Technomic Publishing Co., Westport, Conn., 1980.
2. Neuber, N. *Theory of Notch Stress,* Edwardes, Mich., 1946.
3. Peterson, P. E. *Stress Concentration Factors,* Wiley, New York, 1974.
4. Frocht, M. M. *Photoelasticity,* Vol. I (1941), Vol. II (1948), Wiley, New York.
5. Jacobson, M. A. "Bending stresses in spur gear teeth", *Proc. Inst. Mech. Engrs.,* **169** (1955), 587–604.
6. Brown, A. F. C. and Hickson, V. M. "A photoelastic study of stresses in screw threads", *Proc. Inst. Mech. Engrs.,* **16** (1952–53), 605–612.

PROBLEMS

13.1 A shaft subjected to pure torsion yields at a torque of 1.2 kN-m. A similar shaft is subjected to a torque of 720 N-m and a bending moment M. Determine the maximum allowable value of M according to (*a*) maximum shear stress theory, (*b*) shear strain energy theory.

13.2 Show that the maximum shear stress in a helical spring subjected either to axial force or axial couple is independent of the helix angle of the coils.

   An open-coiled helical spring has ten coils of 50 mm pitch and 76 mm mean diameter made from steel wire of 12.7 mm diameter. If the 0.1% proof stress for the steel is 840 MN/m^2, determine the allowable axial load according to the maximum shear stress criterion. $E = 206$ GN/m^2, $v = 0.3$

13.3 A circular steel cylinder of wall thickness 10 mm and internal diameter 200 mm is subjected to a constant internal pressure of 15 MN/m^2. Determine how much (*a*) axial tensile load, and (*b*) axial compressive load can be applied to the cylinder before yielding commences according to the maximum shear stress theory. The yield stress of the material in simple tension is 240 MN/m^2. Assume that the radial stress in the wall of the cylinder is zero. Sketch the plane yield stress locus for the maximum shear stress theory, and show the two points representing the cases above.

13.4 Part of a supporting bracket for a machine consists of a steel rod of 12 mm diameter fixed at its lower end and containing a right-angle bend which lies in the xy-plane shown in Fig. 13.22. The force P applied at the free end lies in the yz-plane inclined at 30° to the z-axis. The 0.1% proof stress for yielding in simple tension of the steel is 200 MN/m^2. Calculate the value of P which will cause yielding at point A on the outer surface according to the shear–strain energy criterion.

Fig. 13.22

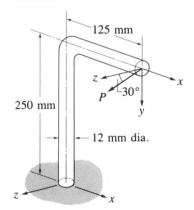

373

Fig. 13.23

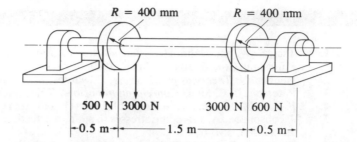

R = 400 mm   R = 400 mm

500 N | 3000 N   3000 N | 600 N

|←0.5 m→|←——— 1.5 m ———→|←0.5 m→|

13.5 In the pulley system shown in Fig. 13.23 the pulleys cause an additional load of 500 N each. Calculate a suitable shaft diameter so as to avoid failure by the maximum shear stress criterion. The tensile yield strength of the shaft material is 248 MN/m². The shaft weight may be neglected and the shaft bearings may be treated as simple supports.

13.6 A cast-iron tube 50 mm and 40 mm outside and inside diameters respectively, is being assembled into a structure. Owing to misalignment it is subjected to a torque about the longitudinal axis of 2.5 kN-m and a tensile force of 50 kN and it fractures. It was discovered that the line of application of the tensile force was parallel to the axis of the tube but offset from it. Calculate the amount of eccentricity which must have occurred to cause failure of the tube according to the maximum principal stress theory. The failure stress of the cast iron in simple tension is 280 MN/m².

13.7 A cast-iron cylinder of 60 mm internal diameter and 5 mm wall thickness is to be used to check the Mohr theory of failure. The tensile and compressive strengths of the material have been measured as 400 MN/m² and 1200 MN/m² respectively. Determine (a) the internal pressure to cause failure, and (b) the axial compressive load to cause failure when combined with an internal pressure of 50 MN/m².

13.8 An aircraft fuselage spacing strut is made from carbon-fibre-reinforced epoxy and has the shape shown in Fig. 13.24. In the manufacture of the strut, six plies are laid in the loading direction and then two plies at 60° and two plies at −60° from the loading axis. In service, the axial load causes a compressive stress of 70 MN/m² in the strut. Use the Tsai–Hill criterion to establish whether or not the composite would be expected to fail at this stress and, if not, determine the stress at which it would fail. The compressive strengths of a unidirectional carbon-fibre composite are 840 MN/m² in the fibre direction and 42 MN/m² in the transverse direction. The shear strength is 56 MN/m². Also for the unidirectional lamina, $E_x = 207$ GN/m², $E_y = 7.7$ GN/m², $G_{xy} = 4.9$ GN/m², $v_{xy} = 0.3$.

Fig. 13.24

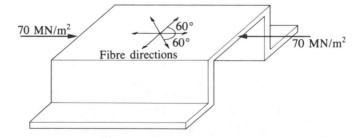

70 MN/m²   60°
60°
Fibre directions   70 MN/m²

13.9 A filament-wound composite-pressure vessel consists of carbon fibre in an epoxy matrix. The vessel is cylindrical with a diameter of 600 mm and a wall thickness of 10 mm. The fibres are arranged with two plies at 45°, two plies at −45° and six plies at 0° to the axis of the cylinder. Use the Tsai–Hill criterion to estimate the internal pressure which would cause the vessel to fail. For a unidirectional composite using the fibre-matrix combination the following data applies: $E_x =$ 207 GN/m^2, $T_x = 1200$ MN/m^2, $E_y = 7.7$ GN/m^2, $T_y = 28$ MN/m^2, $G_{xy} = 4.9$ GN/ m^2, $T_{xy} = 43$ MN/m^2, $v_{xy} = 0.3$.

13.10 A steel beam of width 30 mm and of depth 90 mm is simply-supported at each end of a 2 m span. It is subjected to a concentrated load of 5 kN at mid-span. Determine the approximate depth of compressive yielding beneath the loading point. Taking into account the effect of the concentrated load, compare the maximum resultant tensile stress on the lower surface of the beam with that obtained by simple bending theory. Compressive yield stress = 400 MN/m^2.

13.11 A railway-wagon wheel is 500 mm in diameter and has an approximate contact width on the rail of 40 mm. If the compressive yield stress of the rail steel is 600 MN/m^2 and the modulus is 208 GN/m^2 determine the working load that can be carried per wheel using a load factor of 2.

13.12 A spherical steel pressure vessel is 3 m in diameter and contains a hole of 200 mm diameter to accommodate a safety valve. If the working pressure is 1.4 MN/m^2 determine a suitable value for the shell thickness. The allowable tensile stress for the steel is 350 MN/m^2. (*Hint*: Make a reasonable estimate for K_t using Fig. 13.16.)

13.13 A shaft projects through a roller bearing from where it may be assumed to be cantilevered. It is 50 mm diameter for a length of 100 mm and then is stepped down to 25 mm diameter for a further 100 mm to the free end. At this point a load of 2.68 kN is applied. If the limiting design bending stress is 280 MN/m^2 determine a suitable value for the fillet radius at the change of section. What is the safety factor at the bearing housing? (*See* Fig. 13.18 for relevant data.)

Chapter 14

Variation of stress and strain

In Chapter 12 the conditions of stress and strain at a point in a material were considered. It is now necessary to take the analysis a stage further by examining the variation of stress between adjacent points and deriving suitable expressions for this variation. As in the earlier work, relationships for stresses may be found by considering the equilibrium of a small element of material. The solution of these equations of equilibrium must satisfy the boundary conditions of the problem as defined by the applied forces. However, it is not possible to obtain the individual components of stress directly from the above equations, owing to the statically-indeterminate nature of the problem. It is necessary, therefore, to consider the elastic deformations of the material such that, in a continuous strain field, the displacements are compatible with the stress distribution. These relationships are termed the *equations of compatibility*. From this point it is only required to have a relationship between stress and strain, e.g. Hooke's law, to obtain a complete solution of the stress components in a body.

The equations of equilibrium and compatibility are quite general and may be derived in terms of various co-ordinate systems. The mathematical solution of a problem may often be simplified if an appropriate set of co-ordinates is chosen. With this in mind, and suitable illustrative applications in the following chapter, the various equations will be derived in two-dimensional Cartesian and cylindrical co-ordinates.

EQUILIBRIUM EQUATIONS: PLANE STRESS CARTESIAN CO-ORDINATES

Consider the equilibrium of a small rectangular element of dimensions δx, δy, δz, Fig. 14.1. Owing to the variation of stress through the material, $\sigma_{x_{AB}}$ is a little different from $\sigma_{x_{CD}}$, and likewise for the other

Fig. 14.1

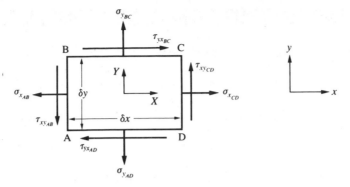

stresses σ_y and τ_{xy}. The variation that must occur over any particular face may be neglected, as it cancels out when the force equilibrium on opposite pairs of faces is considered. On this occasion body forces arising from gravity, inertia, etc., must be taken into account, and these are shown as X and Y per unit volume.

For equilibrium in the x-direction,

$$(\sigma_{x_{CD}} - \sigma_{x_{AB}})\delta y \delta z + (\tau_{yx_{BC}} - \tau_{yx_{AD}})\delta x \delta z + X\,\delta x\,\delta y\,\delta z = 0 \qquad (14.1)$$

Dividing by $\delta x\,\delta y\,\delta z$ gives

$$\frac{\sigma_{x_{CD}} - \sigma_{x_{AB}}}{\delta x} + \frac{\tau_{yx_{BC}} - \tau_{yx_{AD}}}{\delta y} + X = 0$$

In the limit, as $\delta x \to 0$ and $\delta y \to 0$ and the element becomes smaller and smaller, the terms become partial differentials with respect to x and y, and thus

$$\frac{\partial \sigma_x}{\partial x} + \frac{\partial \tau_{yx}}{\partial y} + X = 0 \qquad (14.2)$$

Considering the y-direction,

$$(\sigma_{y_{BC}} - \sigma_{y_{AD}})\,\delta x\,\delta z + (\tau_{xy_{CD}} - \tau_{xy_{AB}})\,\delta y\,\delta z + Y\,\delta x\,\delta y\,\delta z = 0 \qquad (14.3)$$

Dividing by $\delta x\,\delta y\,\delta z$ and for $\delta x \to 0$ and $\delta y \to 0$,

$$\frac{\partial \sigma_y}{\partial y} + \frac{\partial \tau_{xy}}{\partial x} + Y = 0 \qquad (14.4)$$

It is often the case that the only body force is the weight of the component and that it can be neglected in comparison with the applied forces. Then

$$\frac{\partial \sigma_x}{\partial x} + \frac{\partial \tau_{yx}}{\partial y} = 0 \qquad (14.5)$$

$$\frac{\partial \sigma_y}{\partial y} + \frac{\partial \tau_{xy}}{\partial x} = 0 \qquad (14.6)$$

EQUILIBRIUM EQUATIONS: PLANE STRESS CYLINDRICAL CO-ORDINATES

As has been previously stated, there are certain cases such as cylinders, discs, curved bars, etc., in which it is rather more convenient to use r, θ, z co-ordinates. Consider the element ABCD, Fig. 14.2, which is bounded by radial lines OC and OD, subtending an angle $\delta\theta$ at the origin, and circular arcs AB and CD at radii r and $r + \delta r$ respectively. The element is of thickness δz

Fig. 14.2

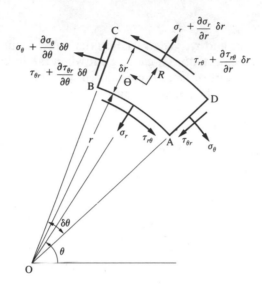

In the preceding section and Fig. 14.1, the stress variation on the element was represented symbolically by the different letter subscripts in order that the first analysis could be written down simply. However, it is quite usual to show stress variation in mathematical symbols, and this has been adopted in Fig. 14.2 for this case. The body forces are shown as R radially and Θ tangentially per unit volume.

Considering equilibrium along the radial centre-line of the element, there will be, in addition to forces from the radial stresses, the resolved components of force from the hoop and shear stresses. Hence

$$\left(\sigma_r + \frac{\partial \sigma_r}{\partial r}\,\delta r\right)(r + \delta r)\,\delta\theta\,\delta z - \sigma_r r\,\delta\theta\,\delta z$$

$$-\left(\sigma_\theta + \frac{\partial \sigma_\theta}{\partial \theta}\,\delta\theta\right)\delta r\,\delta z\,\sin\frac{\delta\theta}{2} - \sigma_\theta\,\delta r\,\delta z\,\sin\frac{\delta\theta}{2}$$

$$+\left(\tau_{\theta r} + \frac{\partial \tau_{\theta r}}{\partial \theta}\,\delta\theta\right)\delta r\,\delta z\,\cos\frac{\delta\theta}{2} - \tau_{\theta r}\,\delta r\,\delta z\,\cos\frac{\delta\theta}{2} + Rr\,\delta\theta\,\delta r\,\delta z = 0$$

$$(14.7)$$

As $\delta\theta \to 0$, $\sin\frac{1}{2}\delta\theta \to \frac{1}{2}\delta\theta$ and $\cos\frac{1}{2}\delta\theta \to 1$. Also, neglecting second- and higher-order terms and dividing by $r\,\delta r\,\delta\theta\,\delta z$, the equation reduces to

$$\frac{\sigma_r}{r} + \frac{\partial \sigma_r}{\partial r} - \frac{\sigma_\theta}{r} + \frac{1}{r}\frac{\partial \tau_{\theta r}}{\partial \theta} + R = 0$$

or

$$\frac{\partial \sigma_r}{\partial r} + \frac{1}{r}\frac{\partial \tau_{\theta r}}{\partial \theta} + \frac{\sigma_r - \sigma_\theta}{r} + R = 0 \tag{14.8}$$

Resolving in the tangential direction,

$$\left(\sigma_\theta + \frac{\partial \sigma_\theta}{\partial \theta} \delta\theta\right) \delta r \, \delta z \cos\frac{\delta\theta}{2} - \sigma_\theta \, \delta r \, \delta z \cos\frac{\delta\theta}{2}$$

$$+ \left(\tau_{\theta r} + \frac{\partial \tau_{\theta r}}{\partial \theta} \delta\theta\right) \delta r \, \delta z \sin\frac{\delta\theta}{2} + \tau_{\theta r} \, \delta r \, \delta z \sin\frac{\delta\theta}{2}$$

$$+ \left(\tau_{r\theta} + \frac{\partial \tau_{r\theta}}{\partial r} \delta r\right)(r + \delta r) \, \delta\theta \, \delta z - \tau_{r\theta} r \, \delta\theta \, \delta z$$

$$+ \Theta r \, \delta r \, \delta\theta \, \delta z = 0 \tag{14.9}$$

In the limit, as $\delta\theta \to 0$, neglecting the appropriate terms and dividing by $r \, \delta r \, \delta\theta \, \delta z$,

$$\frac{1}{r}\frac{\partial \sigma_\theta}{\partial \theta} + \frac{\partial \tau_{r\theta}}{\partial r} + 2\frac{\tau_{\theta r}}{r} + \Theta = 0 \tag{14.10}$$

AXIAL SYMMETRY In certain cases, such as a ring, disc or cylinder, the body is symmetrical about a central axis, z, through O, Fig. 14.2. Then by symmetry the stress components depend on r only, and σ_θ at any particular radius is constant. Also, from the consideration of symmetry, the shear stress components, $\tau_{\theta r}$ must vanish. Eqn. (14.10) no longer exists and the equilibrium equation (14.8) becomes

$$\frac{d\sigma_r}{dr} + \frac{\sigma_r - \sigma_\theta}{r} + R = 0 \tag{14.11}$$

If the body force can be neglected, then

$$\frac{d\sigma_r}{dr} + \frac{\sigma_r - \sigma_\theta}{r} = 0 \tag{14.12}$$

STRAIN IN TERMS OF DISPLACEMENT: CARTESIAN CO-ORDINATES

With reference to a set of fixed axes the movement of an elastic body consists of displacement and rotation of the body combined with strain in the material. Consider a continuous strain field and the displacement of elements OA, of length δx, and OB, of length δy, referred to the axes Ox and Oy, Fig. 14.3. The point O moves to O' having co-ordinates u and v, which are in general functions of x and y. The rate of change of u

Fig. 14.3

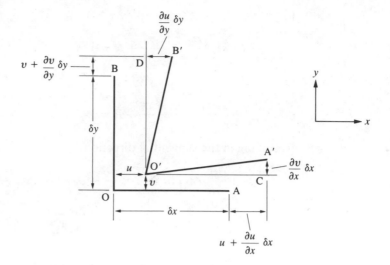

in the x-direction with respect to x will be $\partial u/\partial x$. Therefore, since OA is of length δx, the point A will move to A', where the displacement in the x-direction will be $u + (\partial u/\partial x)\delta x$. The change in length along this axis is thus $(\partial u/\partial x)\delta x$ and the strain is

$$\varepsilon_x = \frac{\dfrac{\partial u}{\partial x}\,\delta x}{\delta x}$$

$$= \frac{\partial u}{\partial x} \tag{14.13}$$

Strain in the y-direction, is obtained from a consideration of the displacement of OB to O'B'. The rate of change of v in the y-direction with respect to y will be $\partial v/\partial y$, and therefore the point B' will have been displaced from B in the y-direction an amount $v + (\partial v/\partial y)\delta y$. The strain occurring will thus be

$$\varepsilon_y = \frac{\dfrac{\partial v}{\partial y}\,\delta y}{\delta y}$$

$$= \frac{\partial v}{\partial y} \tag{14.14}$$

The shearing strain in the element AOB will be given by the change from the original right angle to the new angle A'O'B'. Hence

$$\gamma_{xy} = \angle\text{CO'A'} + \angle\text{DO'B'}$$

For small displacements,

$$\angle\text{CO'A'} \approx \frac{\text{CA'}}{\text{O'C}} \quad \text{and} \quad \text{DO'B'} \approx \frac{\text{DB'}}{\text{O'D}}$$

Now, CA' is the rate of change of v in the x-direction for an amount δx

or $(\partial v/\partial x)\delta x$. Similarly DB′ is the rate of change of u in the y-direction for a length δy, giving $(\partial u/\partial y)\delta y$. Therefore

$$\angle\text{CO′A′}\approx\frac{\dfrac{\partial v}{\partial x}\delta x}{\delta x}\approx\frac{\partial v}{\partial x}$$

and

$$\angle\text{DO′B′}\approx\frac{\dfrac{\partial u}{\partial y}\delta y}{\delta y}\approx\frac{\partial u}{\partial y}$$

Thus, for very small displacements,

$$\gamma_{xy}=\frac{\partial v}{\partial x}+\frac{\partial u}{\partial y} \tag{14.15}$$

In a two-dimensional strain field, the strains in terms of displacements are therefore

$$\varepsilon_x=\frac{\partial u}{\partial x};\qquad \varepsilon_y=\frac{\partial v}{\partial y};\qquad \gamma_{xy}=\frac{\partial v}{\partial x}+\frac{\partial u}{\partial y} \tag{14.16}$$

STRAIN IN TERMS OF DISPLACEMENT: CYLINDRICAL CO-ORDINATES

When dealing with the analysis of elements having a circular geometry (curved bars, discs, etc.) it is often more convenient to consider strain and displacements in terms of cylindrical co-ordinates as was done for the equilibrium equations. Consider the element ABCD, Fig. 14.4,

Fig. 14.4

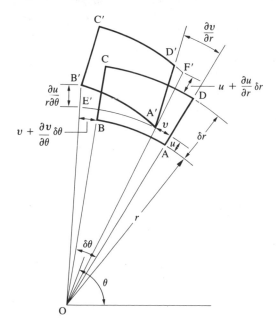

subtending an angle $\delta\theta$, with AB at radius r and CD at $r + \delta r$. This element is displaced to A'B'C'D' so that the radial and tangential movements to A are u and v respectively. The displacement of the point D to D' in the r-direction will be $u + (\partial u/\partial r)\delta r$, where $\partial u/\partial r$ is the rate of change of u with respect to r. The change in length of AD is therefore $(\partial u/\partial r)\delta r$, and hence the strain in the radial direction is

$$\varepsilon_r = \frac{\dfrac{\partial u}{\partial r}\delta r}{\delta r} = \frac{\partial u}{\partial r} \tag{14.17}$$

In the tangential direction there are two effects of displacement on strain. As B moves to B' there is the change of v with respect to θ, giving $(\partial v/\partial\theta)\delta\theta$ as the increase in length and hence a strain of

$$\frac{(\partial v/\partial\theta)\delta\theta}{r\,\delta\theta} \quad \text{or} \quad \frac{1}{r}\cdot\frac{\partial v}{\partial\theta}$$

There is also the tangential strain due to the element moving out to the new radius, $r + u$. This part is

$$\frac{(r+u)\,\delta\theta - r\,\delta\theta}{r\,\delta\theta} = \frac{u}{r}$$

Therefore, the total tangential strain is

$$\varepsilon_\theta = \frac{1}{r}\frac{\partial v}{\partial\theta} + \frac{u}{r} \tag{14.18}$$

The shear strain, $\gamma_{r\theta}$, is given by the difference between $\angle$DAB and $\angle$D'A'B', which is

$$\gamma_{r\theta} = \angle\text{D'A'F'} + \angle\text{B'A'E'}$$

Now, $\angle$D'A'F' is the difference between the tangential displacement of D to D', $(\partial v/\partial r)\delta r$, which as an angle is $\partial v/\partial r$, and the rigid body rotation about O, $\angle$DOF', which is v/r. Therefore

$$\angle\text{D'A'F'} = \frac{\partial v}{\partial r} - \frac{v}{r}$$

$\angle$B'A'E' is in respect of displacement in the radial direction u and is therefore $\partial u/r\,\partial\theta$. Hence

$$\gamma_{r\theta} = \frac{\partial v}{\partial r} - \frac{v}{r} + \frac{1}{r}\frac{\partial u}{\partial\theta} \tag{14.19}$$

The three strains in terms of displacements are

$$\varepsilon_r = \frac{\partial u}{\partial r} \tag{14.20}$$

$$\varepsilon_\theta = \frac{1}{r}\frac{\partial v}{\partial\theta} + \frac{u}{r} \tag{14.21}$$

$$\gamma_{r\theta} = \frac{\partial v}{\partial r} + \frac{1}{r}\frac{\partial u}{\partial\theta} - \frac{v}{r} \tag{14.22}$$

AXIAL SYMMETRY   For problems which are symmetrical about a z-axis through O, there will be no tangential displacement, v, and no shear strain, $\gamma_{r\theta}$, since $\tau_{r\theta}$ is zero. The above equations then reduce to

$$\varepsilon_r = \frac{\partial u}{\partial r} = \frac{du}{dr} \tag{14.23}$$

since σ_θ is constant at any one value of r; and

$$\varepsilon_\theta = \frac{u}{r} \tag{14.24}$$

COMPATIBILITY EQUATIONS: CARTESIAN CO-ORDINATES

Because the above three strain components are in terms of the functions u and v, there must be a relationship between them. This may be obtained by differentiating ε_x twice with respect to y, ε_y twice with respect to x, and γ_{xy} with respect to both x and y. Hence

$$\frac{\partial^2 \varepsilon_x}{\partial y^2} = \frac{\partial^3 u}{\partial x\, \partial y^2}; \qquad \frac{\partial^2 \varepsilon_y}{\partial x^2} = \frac{\partial^3 v}{\partial y\, \partial x^2}; \qquad \frac{\partial^2 \gamma_{xy}}{\partial x\, \partial y} = \frac{\partial^2}{\partial x\, \partial y}\left(\frac{\partial v}{\partial x} + \frac{\partial u}{\partial y}\right)$$

Eliminating u and v between these three equations provides the relationship

$$\frac{\partial^2 \varepsilon_x}{\partial y^2} + \frac{\partial^2 \varepsilon_y}{\partial x^2} = \frac{\partial^2 \gamma_{xy}}{\partial x\, \partial y} \tag{14.25}$$

which is known as a *compatibility equation* in terms of strains. To obtain a compatibility relationship in terms of stresses, it is only necessary to substitute for the strains in eqn. (14.25), using the stress–strain relationships.

For the case of plane stress, $\sigma_z = 0$, the stress–strain relationships are

$$\varepsilon_x = \frac{\sigma_x}{E} - \frac{v\sigma_y}{E}$$

$$\varepsilon_y = \frac{\sigma_y}{E} - \frac{v\sigma_x}{E}$$

$$\gamma_{xy} = \frac{\tau_{xy}}{G} = \frac{2\tau_{xy}(1+v)}{E}$$

Substituting the above in eqn. (14.25),

$$\frac{1}{E}\frac{\partial^2 \sigma_x}{\partial y^2} - \frac{v}{E}\frac{\partial^2 \sigma_y}{\partial y^2} + \frac{1}{E}\frac{\partial^2 \sigma_y}{\partial x^2} - \frac{v}{E}\frac{\partial^2 \sigma_x}{\partial x^2} = \frac{2(1+v)}{E}\frac{\partial^2 \tau_{xy}}{\partial x\, \partial y} \tag{14.26}$$

Considering the equilibrium equations and neglecting the body force, if it is only the weight of the body, then differentiating eqn. (14.5) with respect to x and eqn. (14.6) with respect to y and adding we get

$$\frac{\partial^2 \sigma_x}{\partial x^2} + \frac{\partial^2 \sigma_y}{\partial y^2} = -2\frac{\partial^2 \tau_{xy}}{\partial x\, \partial y} \tag{14.27}$$

Eliminating τ_{xy} between eqns. (14.26) and (14.27),

$$\frac{1}{E}\frac{\partial^2 \sigma_x}{\partial y^2} - \frac{v}{E}\frac{\partial^2 \sigma_y}{\partial y^2} + \frac{1}{E}\frac{\partial^2 \sigma_y}{\partial x^2} - \frac{v}{E}\frac{\partial^2 \sigma_x}{\partial x^2} = -\frac{(1+v)}{E}\left(\frac{\partial^2 \sigma_x}{\partial x^2} + \frac{\partial^2 \sigma_y}{\partial y^2}\right)$$

Simplifying,

$$\frac{\partial^2 \sigma_x}{\partial x^2} + \frac{\partial^2 \sigma_y}{\partial x^2} + \frac{\partial^2 \sigma_x}{\partial y^2} + \frac{\partial^2 \sigma_y}{\partial y^2} = 0$$

or

$$\left(\frac{\partial^2}{\partial x^2} + \frac{\partial^2}{\partial y^2}\right)(\sigma_x + \sigma_y) = 0 \tag{14.28}$$

An analysis similar to that above can be used to show that the compatibility eqn. (14.28) also applies to the case of plane strain.

COMPATIBILITY EQUATIONS: CYLINDRICAL CO-ORDINATES

As in the case of Cartesian co-ordinates there must be a relationship between the strains in cylindrical co-ordinates. This may be obtained by eliminating u and v between the strain-displacement equations.

Differentiating eqn. (14.21) with respect to r and eqn. (14.22) with respect to θ, dividing the latter by r, and subtracting, gives

$$\frac{\partial \varepsilon_\theta}{\partial r} - \frac{1}{r}\frac{\partial \gamma_{r\theta}}{\partial \theta} = \frac{1}{r}\frac{\partial u}{\partial r} - \frac{u}{r^2} - \frac{1}{r^2}\frac{\partial^2 u}{\partial \theta^2}$$

Multiplying by r^2 and substituting for $\partial u/\partial r$, we have

$$r^2\frac{\partial \varepsilon_\theta}{\partial r} - r\frac{\partial \gamma_{r\theta}}{\partial \theta} = r\varepsilon_r - u - \frac{\partial^2 u}{\partial \theta^2} \tag{14.29}$$

Differentiating this equation with respect to r, and eqn. (14.20) with respect to θ twice, and eliminating u gives

$$2r\frac{\partial \varepsilon_\theta}{\partial r} + r^2\frac{\partial^2 \varepsilon_\theta}{\partial r^2} - \frac{\partial \gamma_{r\theta}}{\partial \theta} - r\frac{\partial^2 \gamma_{r\theta}}{\partial \theta\, \partial r} = \varepsilon_r + r\frac{\partial \varepsilon_r}{\partial r} - \varepsilon_r - \frac{\partial^2 \varepsilon_r}{\partial \theta^2}$$

and simplifying,

$$\frac{\partial \gamma_{r\theta}}{\partial \theta} + r\frac{\partial^2 \gamma_{r\theta}}{\partial \theta\, \partial r} = \frac{\partial^2 \varepsilon_r}{\partial \theta^2} + r^2\frac{\partial^2 \varepsilon_\theta}{\partial r^2} + 2r\frac{\partial \varepsilon_\theta}{\partial r} - r\frac{\partial \varepsilon_r}{\partial r} \tag{14.30}$$

This is the compatibility equation in terms of strain. To obtain a similar relationship for stresses it is necessary to use the stress–strain relationships and equilibrium equations in cylindrical co-ordinates.

The stress–strain equations in two dimensions are

$$\varepsilon_r = \frac{\sigma_r}{E} - \frac{v\sigma_\theta}{E}$$

$$\varepsilon_\theta = \frac{\sigma_\theta}{E} - \frac{v\sigma_r}{E}$$

$$\gamma_{r\theta} = \frac{\tau_{r\theta}}{G} = \frac{2(1+v)}{E} \tau_{r\theta}$$

Substituting in eqn. (14.30),

$$\frac{2(1+v)}{E} \frac{\partial \tau_{r\theta}}{\partial \theta} + \frac{2(1+v)}{E} r \frac{\partial^2 \tau_{r\theta}}{\partial \theta \, \partial r} = \frac{\partial^2 \sigma_r}{E \, \partial \theta^2} - \frac{v\partial^2 \sigma_\theta}{E \, \partial \theta^2} + r^2 \frac{\partial^2 \sigma_\theta}{E \, \partial r^2}$$

$$- r^2 \frac{v\partial^2 \sigma_r}{E \, \partial r^2} + \frac{2r}{E} \frac{\partial \sigma_\theta}{\partial r} - \frac{2rv}{E} \frac{\partial \sigma_r}{\partial r} - r \frac{\partial \sigma_r}{E \, \partial r} + \frac{rv}{E} \frac{\partial \sigma_\theta}{\partial r} \quad (14.31)$$

Now the equilibrium equations (14.8) and (14.10) are used to eliminate $\tau_{r\theta}$. Multiply through eqn. (14.8) by r, and differentiate with respect to r; then

$$\frac{\partial \sigma_r}{\partial r} + r \frac{\partial^2 \sigma_r}{\partial r^2} + \frac{\partial^2 \tau_{r\theta}}{\partial r \, \partial \theta} + \frac{\partial \sigma_r}{\partial r} - \frac{\partial \sigma_\theta}{\partial r} = 0$$

Differentiating eqn. (14.10) with respect to θ,

$$\frac{1}{r} \frac{\partial^2 \sigma_\theta}{\partial \theta^2} + \frac{\partial^2 \tau_{r\theta}}{\partial r \, \partial \theta} + \frac{2}{r} \frac{\partial \tau_{r\theta}}{\partial \theta} = 0$$

Multiplying each of the above equations by r, adding and simplifying, we get

$$2 \frac{\partial \tau_{r\theta}}{\partial \theta} + 2r \frac{\partial^2 \tau_{r\theta}}{\partial r \, \partial \theta} = -\frac{\partial^2 \sigma_\theta}{\partial \theta^2} - r^2 \frac{\partial^2 \sigma_r}{\partial r^2} - 2r \frac{\partial \sigma_r}{\partial r} + r \frac{\partial \sigma_\theta}{\partial r} \quad (14.32)$$

Substituting for $\tau_{r\theta}$ in eqn. (14.31) and simplifying,

$$\frac{\partial^2 \sigma_r}{\partial r^2} + \frac{\partial^2 \sigma_\theta}{\partial r^2} + \frac{1}{r} \frac{\partial \sigma_r}{\partial r} + \frac{1}{r} \frac{\partial \sigma_\theta}{\partial r} + \frac{1}{r^2} \frac{\partial^2 \sigma_r}{\partial \theta^2} + \frac{1}{r^2} \frac{\partial^2 \sigma_\theta}{\partial \theta^2} = 0$$

or

$$\left(\frac{\partial^2}{\partial r^2} + \frac{1}{r} \frac{\partial}{\partial r} + \frac{1}{r^2} \frac{\partial^2}{\partial \theta^2} \right)(\sigma_r + \sigma_\theta) = 0 \quad (14.33)$$

This is the equation of compatibility in terms of stresses. The complete analysis of stress distribution in a body may now be made using the equilibrium equations (14.8) and (14.10), the above compatibility equation, and the boundary conditions appropriate to the applied forces or displacements.

For cases of axial symmetry, since stress and displacement are independent of θ, the equation becomes

$$\left(\frac{\partial^2}{\partial r^2} + \frac{1}{r} \frac{\partial}{\partial r} \right)(\sigma_r + \sigma_\theta) = 0 \quad (14.34)$$

Multiplying out, eqn. (14.34) gives

$$\frac{\partial^2 \sigma_r}{\partial r^2} + \frac{1}{r}\frac{\partial \sigma_r}{\partial r} + \frac{\partial^2 \sigma_\theta}{\partial r^2} + \frac{1}{r}\frac{\partial \sigma_\theta}{\partial r} = 0 \tag{14.35}$$

Now, from eqn. (14.12),

$$\sigma_\theta = r\frac{\partial \sigma_r}{\partial r} + \sigma_r \tag{14.36}$$

Therefore

$$\frac{\partial \sigma_\theta}{\partial r} = \frac{\partial \sigma_r}{\partial r} + r\frac{\partial^2 \sigma_r}{\partial r^2} + \frac{\partial \sigma_r}{\partial r}$$

and

$$\frac{\partial^2 \sigma_\theta}{\partial r^2} = 3\frac{\partial^2 \sigma_r}{\partial r^2} + r\frac{\partial^3 \sigma_r}{\partial r^2}$$

Substituting these expressions for σ_θ in eqn. (14.35) and gathering terms together gives

$$r\frac{\partial^3 \sigma_r}{\partial r^3} + 5\frac{\partial^2 \sigma_r}{\partial r^2} + \frac{3}{r}\frac{\partial \sigma_r}{\partial r} = 0 \tag{14.37}$$

which is the general equation for σ_r in an axially-symmetrical stress system with no body force. It can be verified by substitution that one particular solution of this equation is

$$\sigma_r = A - \frac{B}{r^2}$$

and substituting for σ_r and $\partial \sigma_r / \partial r$ in eqn. (14.36) gives

$$\sigma_\theta = A + \frac{B}{r^2}$$

A and B are constants which are determined from the particular boundary conditions of the problem.

These expressions for σ_r and σ_θ will be derived and used for specific examples in the next chapter.

STRESS FUNCTIONS

It is beyond the scope of this text to discuss and make use of stress functions in detail. However, since the stress function method of solution plays an essential part in the mathematical theory of elasticity, it is perhaps of some value to mention it here.

In the last paragraph of the preceding section the principles of obtaining the complete solution for the stresses in a body were stated. A stress function is a mathematical function of the co-ordinate variables x and y, on r and θ. The stress components can be expressed in terms of such a function which, on substitution, will satisfy the equilibrium equations. The compatibility equations can also be written in terms of the stress function.

For a particular problem, the solution of the various partial differential equations is therefore a matter of choosing a suitable mathematical function by which to express the stress components. In Cartesian co-ordinates the stress components may be written as

$$\sigma_x = \frac{\partial^2 \phi}{\partial y^2} \qquad \sigma_y = \frac{\partial^2 \phi}{\partial x^2} \qquad \tau_{xy} = -\frac{\partial^2 \phi}{\partial x\, \partial y} \tag{14.38}$$

where ϕ is the stress function. It may be easily verified that the above three equations completely satisfy the equilibrium equations without body force, namely (14.5) and (14.6). The compatibility equation (14.28) may now be expressed in the form

$$\left(\frac{\partial^2}{\partial x^2} + \frac{\partial^2}{\partial y^2} \right) \phi = 0 \tag{14.39}$$

In cylindrical co-ordinates the stress components in terms of a stress function are as follows:

$$\left. \begin{array}{l} \sigma_r = \dfrac{1}{r} \dfrac{\partial \phi}{\partial r} + \dfrac{1}{r^2} \dfrac{\partial^2 \phi}{\partial \theta^2} \\[4mm] \sigma_\theta = \dfrac{\partial^2 \phi}{\partial r^2} \\[4mm] \tau_{r\theta} = \dfrac{1}{r^2} \dfrac{\partial \phi}{\partial \theta} - \dfrac{1}{r} \dfrac{\partial^2 \phi}{\partial r\, \partial \theta} \end{array} \right\} \tag{14.40}$$

It can also be shown that these expressions satisfy the equilibrium equations (14.8) and (14.10), without body forces R and Θ. The compatibility equation in polar co-ordinates is, from the above,

$$\frac{\partial^2 \sigma_r}{\partial r^2} + \frac{\partial^2 \sigma_\theta}{\partial r^2} + \frac{1}{r} \frac{\partial \sigma_r}{\partial r} + \frac{1}{r} \frac{\partial \sigma_\theta}{\partial r} + \frac{1}{r^2} \frac{\partial^2 \sigma_r}{\partial \theta^2} + \frac{1}{r^2} \frac{\partial^2 \sigma_\theta}{\partial \theta^2} = 0$$

Substituting for σ_r and σ_θ from above and simplifying, we obtain

$$\left(\frac{\partial^2}{\partial r^2} + \frac{1}{r} \frac{\partial}{\partial r} + \frac{1}{r^2} \frac{\partial^2}{\partial \theta^2} \right) \left(\frac{\partial^2 \phi}{\partial r^2} + \frac{1}{r} \frac{\partial \phi}{\partial r} + \frac{1}{r^2} \frac{\partial^2 \phi}{\partial \theta^2} \right) = 0 \tag{14.41}$$

as the compatibility equation expressed in terms of the stress function.

SUMMARY

Many engineering design problems involve complex variations of the stress and strain fields within the component. However, we still must employ the three basic tenets of equilibrium of forces, compatibility of strain and displacements and the stress–strain relationships of elasticity.

Now that we have derived the necessary equations in Cartesian and cylindrical co-ordinates we can proceed in the next chapter to apply these to some design examples.

PROBLEMS

14.1 Show that the equilibrium equation in the radial direction for a varying stress field in spherical co-ordinates r, θ, ψ with body force R is of the following form:

$$R \sin \psi + \frac{\partial \sigma_r}{\partial r} \sin \psi + \frac{1}{r}\left(2\sigma_r \sin \psi - \sigma_\theta \sin \psi - \sigma_\psi \sin \psi\right.$$

$$\left. + \tau_{\psi r} \cos \psi + \frac{\partial \tau_{r\theta}}{\partial \theta} + \frac{\partial \tau_{\psi r}}{\partial \psi} \sin \psi\right) = 0$$

14.2 What are the strain–displacement relationships in spherical co-ordinates for an axi-symmetrical stress field?

Using the equilibrium equation in the previous question, simplified for an axi-symmetrical stress field without body force, show that the displacement at radius r, for a spherical shell, is given by

$$\frac{d^2u}{dr^2} + \frac{2}{r}\frac{du}{dr} - \frac{2u}{r^2} = 0$$

14.3 For a particular problem the strain–displacement equations in cylindrical co-ordinates are

$$\varepsilon_r = \frac{\partial u}{\partial r} \qquad \varepsilon_\theta = \frac{u}{r} \qquad \varepsilon_z = \gamma_{r\theta} = \gamma_{\theta z} = \gamma_{zr} = 0$$

Show that the compatibility equation in terms of the stresses σ_r and σ_θ is

$$rv\frac{\partial \sigma_r}{\partial r} - r(1-v)\frac{\partial \sigma_\theta}{\partial r} + \sigma_r - \sigma_\theta = 0$$

What is the problem?

14.4 Derive compatibility equations from the following strain–displacement relationships:

(a) $\quad \gamma_{xy} = \dfrac{\partial u}{\partial y} + \dfrac{\partial v}{\partial x} \qquad \gamma_{xz} = \dfrac{\partial u}{\partial z} \qquad \gamma_{yz} = \dfrac{\partial v}{\partial z}$

(b) $\quad \varepsilon_z = \dfrac{\partial w}{\partial z} \qquad \gamma_{\theta z} = \dfrac{1}{r}\dfrac{\partial w}{\partial \theta}$

14.5 Commencing with the six strain–displacement relationships in three-dimensional Cartesian co-ordinates, derive the six compatibility equations for three-dimensional states of strain.

Chapter 15

Applications of equilibrium and strain–displacement relationships

INTRODUCTION

It has been explained in principle in the last chapter how the stress components may be determined in a body by use of the equilibrium, the compatibility equations and the particular boundary conditions of the problem. In a majority of cases, the solutions are complex and the stress-function method mentioned in the previous chapter is used.

However there are a few problems in beams and axi-symmetrical bodies in which a simpler analysis is possible using the equilibrium, strain–displacement and stress–strain relationships. As it is important to understand how to apply these principles, the present chapter commences with two simple beam-bending situations. These are followed by important engineering components, namely the thick-walled cylinder used typically in high-pressure chemical engineering and the rotating disc or rotor used in steam and gas turbines.

SHEAR STRESS IN A BEAM

The distribution of transverse shear stress in a beam in terms of the shear force and the geometry of the cross-section was obtained in Chapter 6, using simple bending theory. An alternative approach will now be developed.

Consider the beam in Fig. 15.1, which, for simplicity of solution, is shown simply-supported and carrying a uniformly-distributed load, w per unit length. The origin of Cartesian co-ordinates is taken on the neutral axis, with x positive left to right and y positive downwards.

This is treated as a two-dimensional problem with no variation of stress through the thickness of the beam, and therefore only two equations of

Fig. 15.1

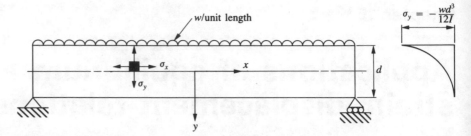

equilibrium are applicable. From page 377

$$\frac{\partial \sigma_x}{\partial x} + \frac{\partial \tau_{xy}}{\partial y} = 0 \tag{15.1}$$

$$\frac{\partial \sigma_y}{\partial y} + \frac{\partial \tau_{xy}}{\partial x} = 0 \tag{15.2}$$

Making use of the exact solution for pure bending

$$\sigma_x = \frac{My}{I} \tag{15.3}$$

then eqn. (15.2) is not required, neither is any strain–displacement relationship. This is because, in the derivation of eqn. (15.3), the geometry of deformation and the stress–strain relationship were included. Substituting for σ_x in eqn. (15.1) gives

$$\frac{\partial \left(\dfrac{My}{I} \right)}{\partial x} + \frac{\partial \tau_{xy}}{\partial y} = 0$$

Therefore

$$\partial \tau_{xy} = -\frac{\partial M}{\partial x} \frac{y}{I} \partial y$$

But $\partial M / \partial x = Q$ the shear force on the section, so that

$$\partial \tau_{xy} = -\frac{Qy}{I} \partial y \tag{15.4}$$

Integrating gives

$$\tau_{xy} = -\frac{Q}{I} \int y \, dy + C = -\frac{Qy^2}{2I} + C \tag{15.5}$$

At the top and bottom free surface of the beam the shear stress must be zero therefore,

$$\tau_{xy} = 0 \quad \text{at } y = \pm \frac{d}{2} \quad \text{from which} \quad C = \frac{Qd^2}{8I}$$

and

$$\tau_{xy} = -\frac{Qy^2}{2I} + \frac{Qd^2}{8I} \tag{15.6}$$

At the neutral axis $y = 0$ and the shear stress has its maximum value

$$\tau_{xy} = \frac{Qd^2}{8I} \tag{15.7}$$

This agrees with the value obtained in Chapter 6.

TRANSVERSE NORMAL STRESS IN A BEAM

A further example on the analysis of beams is that of the distribution of direct stress in the y-direction due to the application of a distributed load w. For the beam in Fig. 15.1 the equilibrium equation which is applicable is

$$\frac{\partial \sigma_y}{\partial y} + \frac{\partial \tau_{xy}}{\partial x} = 0 \tag{15.8}$$

Now, the shear stress, τ_{xy}, was determined in eqn. (15.6), and substituting that value in eqn. (15.8) gives

$$\frac{\partial \sigma_y}{\partial y} + \frac{\partial}{\partial x}\left(-\frac{Qy^2}{2I} + \frac{Qd^2}{8I}\right) = 0$$

But $\partial Q/\partial x = -w$. Therefore

$$\frac{\partial \sigma_y}{\partial y} + \frac{wy^2}{2I} - \frac{wd^2}{8I} = 0 \tag{15.9}$$

or

$$\sigma_y = -\int\left(\frac{wy^2}{2I} - \frac{wd^2}{8I}\right)dy + C$$

$$= -\frac{w}{I}\left(\frac{y^3}{6} - \frac{d^2y}{8}\right) + C \tag{15.10}$$

Using the boundary condition that at the upper surface $y = -\frac{1}{2}d$, the compressive stress is $\sigma_y = -w/b$, where b is the beam thickness; then

$$C = -\frac{w}{b} + \frac{w}{I}\left(-\frac{d^3}{48} + \frac{d^3}{16}\right)$$

Substituting

$$\frac{1}{b} = \frac{d^3}{12I}$$

$$C = -\frac{wd^3}{24I}$$

Therefore

$$\sigma_y = -\frac{w}{I}\left(\frac{y^3}{6} - \frac{d^2y}{8} + \frac{d^3}{24}\right) \tag{15.11}$$

The distribution of stress is illustrated in Fig. 15.1.

A check on this solution may be made by considering the condition at the lower free surface. Here $y = +\frac{1}{2}d$; from which $\sigma_y = 0$, which is correct.

STRESS DISTRIBUTION IN A PRESSURIZED THICK-WALLED CYLINDER

This problem is of considerable practical importance in pressure vessels and gun barrels. It is a further case which can be solved without using a stress-function solution and is an application of the cylindrical co-ordinate system, r, θ, z.

A long hollow cylinder which is subjected to uniformly distributed internal and external pressure is shown in Fig. 15.2(a) and (b). The two

Fig. 15.2

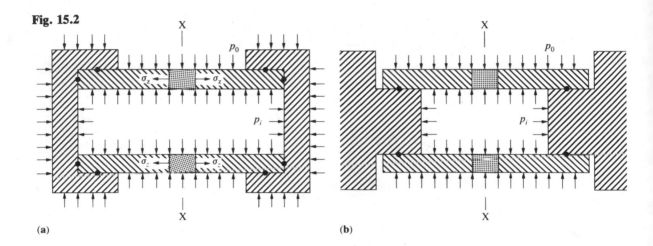

(a)　　　　　　　　　　　　　　　　　　(b)

methods of maintaining the pressure inside the cylinder are either by end caps which are attached to the cylinder as shown in Fig. 15.2(a) or by pistons in each end of the cylinder Fig. 15.2(b). Considering a cross-sectional slice XX as shown in Fig. 15.3, the deformations produced are symmetrical about the longitudinal axis of the cylinder, and the small element of material in the wall supports the stress system shown. This is the same as in Fig. 14.2 for the general stress system except that for axial symmetry $\tau_{r\theta} = 0$ and σ_θ is constant at any particular radius. Hence σ_θ and σ_r are principal stresses and additionally are quite independent of the method of end closure of the cylinder. Considering axial stress σ_z and axial strain ε_z then both of these occur in the case of end cap closures (Fig. 15.2(a)). For closure by pistons (Fig. 15.2(b)) it is evident that $\sigma_z = 0$ and ε_z occurs only due to the Poisson's ratio effect of σ_r and σ_θ. From the symmetry of the system and for a long cylinder, we come to the conclusion that plane cross-sections remain plane when subjected to pressure and therefore axial deformation, w, across the section is independent of r and $\mathrm{d}w/\mathrm{d}r = 0$.

Fig. 15.3

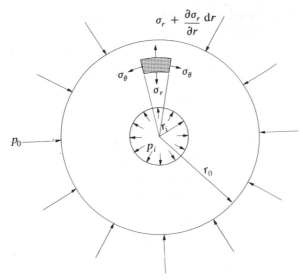

CYLINDER WITH END CAPS

The equations of equilibrium for an element of material are

$$\frac{\mathrm{d}\sigma_r}{\mathrm{d}r} + \frac{\sigma_r - \sigma_\theta}{r} = 0 \quad (\text{eqn. (14.12)}) \tag{15.12}$$

and

$$\frac{\mathrm{d}\sigma_z}{\mathrm{d}z} = 0 \tag{15.13}$$

The strain–displacement equations, from page 383 are

$$\varepsilon_r = \frac{\mathrm{d}u}{\mathrm{d}r} \tag{15.14}$$

$$\varepsilon_\theta = \frac{u}{r} \tag{15.15}$$

$$\varepsilon_z = \frac{\mathrm{d}w}{\mathrm{d}z} \tag{15.16}$$

The stress–strain relationships are

$$\varepsilon_r = \frac{\sigma_r}{E} - \frac{v}{E}(\sigma_\theta + \sigma_z) = \frac{\mathrm{d}u}{\mathrm{d}r} \tag{15.17}$$

$$\varepsilon_\theta = \frac{\sigma_\theta}{E} - \frac{v}{E}(\sigma_z + \sigma_r) = \frac{u}{r} \tag{15.18}$$

$$\varepsilon_z = \frac{\sigma_z}{E} - \frac{v}{E}(\sigma_r + \sigma_\theta) = \frac{\mathrm{d}w}{\mathrm{d}z} \tag{15.19}$$

Differentiating eqn. (15.18) with respect to r gives

$$\frac{E}{r}\left(\frac{\mathrm{d}u}{\mathrm{d}r} - \frac{u}{r}\right) = \frac{\mathrm{d}\sigma_\theta}{\mathrm{d}r} - v\frac{\mathrm{d}\sigma_z}{\mathrm{d}r} - v\frac{\mathrm{d}\sigma_r}{\mathrm{d}r}$$

Substituting for du/dr and u/r from eqns. (15.17) and (15.18) and simplifying,

$$\frac{1+v}{r}(\sigma_r - \sigma_\theta) = \frac{d\sigma_\theta}{dr} - v\frac{d\sigma_z}{dr} - v\frac{d\sigma_r}{dr} \qquad (15.20)$$

Now, since $\varepsilon_z = $ constant, $d\varepsilon_z/dr = 0$ and differentiating eqn. (15.19) gives

$$\frac{d\sigma_z}{dr} = v\left(\frac{d\sigma_r}{dr} + \frac{d\sigma_\theta}{dr}\right) \qquad (15.21)$$

Substituting into eqn. (15.20) for $d\sigma_z/dr$ from eqn. (15.21) and $(\sigma_r - \sigma_\theta)/r$ from eqn. (15.12) and simplifying gives

$$(1 - v^2)\left(\frac{d\sigma_\theta}{dr} + \frac{d\sigma_r}{dr}\right) = 0 \qquad (15.22)$$

From eqns. (15.22) and (15.21) we see that $d\sigma_z/dr = 0$ and therefore σ_z is constant through the wall thickness. Integrating eqn. (15.22) shows that

$$(\sigma_\theta + \sigma_r) = \text{constant} = 2A \qquad (15.23)$$

Eliminating σ_θ between eqns. (15.23) and (15.12) gives

$$\frac{d\sigma_r}{dr} + \frac{2\sigma_r - 2A}{r} = 0 \qquad (15.24)$$

from which, multiplying by r^2,

$$2Ar - 2r\sigma_r - r^2\frac{d\sigma_r}{dr} = 0$$

and

$$2Ar - \frac{d}{dr}(r^2\sigma_r) = 0$$

By integration,

$$Ar^2 - r^2\sigma_r = B$$

Hence

$$\sigma_r = A - \frac{B}{r^2} \qquad (15.25)$$

and from eqn. (15.23);

$$\sigma_\theta = A + \frac{B}{r^2} \qquad (15.26)$$

where A and B are constants which may be found using the boundary conditions.

CYLINDER WITH PISTONS   In this case, Fig. 15.2(b), $\sigma_z = 0$ and there is a condition of plane stress. This solution has been included to show that we can arrive at the same expressions for σ_r and σ_θ by deriving a differential equation for displacement, u.

Putting $\sigma_z = 0$ in eqns. (15.17) and (15.18) and solving for σ_r and σ_θ in terms of u gives

$$\sigma_r = \left(\frac{\mathrm{d}u}{\mathrm{d}r} + \frac{vu}{r}\right)\frac{E}{(1-v^2)} \tag{15.27}$$

$$\sigma_\theta = \left(v\frac{\mathrm{d}u}{\mathrm{d}r} + \frac{u}{r}\right)\frac{E}{(1-v^2)} \tag{15.28}$$

From eqn. (15.27),

$$\frac{\mathrm{d}\sigma_r}{\mathrm{d}r} = \left(\frac{\mathrm{d}^2u}{\mathrm{d}r^2} + v\frac{\mathrm{d}u}{\mathrm{d}r} - \frac{vu}{r}\right)\frac{E}{(1-v^2)} \tag{15.29}$$

Substituting eqns. (15.27), (15.28) and (15.29) into eqn. (15.12) and simplifying gives

$$\frac{\mathrm{d}^2u}{\mathrm{d}r^2} + \frac{1}{r}\frac{\mathrm{d}u}{\mathrm{d}r} - \frac{u}{r^2} = 0 \tag{15.30}$$

This is the differential equation for the radial displacement in the cylinder wall.

The general solution of this equation is

$$u = Cr + \frac{C'}{r} \tag{15.31}$$

Substituting for u and $\mathrm{d}u$ in eqns. (15.27) and (15.28),

$$\sigma_r = \left\{C(1+v) - \frac{C'}{r^2}(1-v)\right\}\frac{E}{(1-v^2)} \tag{15.32}$$

$$\sigma_\theta = \left\{C(1+v) + \frac{C'}{r^2}(1-v)\right\}\frac{E}{(1-v^2)} \tag{15.33}$$

where C and C' are constants.

These equations may be rewritten with different constants as

$$\sigma_r = A - \frac{B}{r^2}$$

and

$$\sigma_\theta = A + \frac{B}{r^2}$$

which are the same as eqns. (15.25) and (15.26).

BOUNDARY CONDITIONS

The next stage is the determination of the constants A and B.

1. Internal and external pressure

The boundary conditions of the problem are: at $r = r_i$, $\sigma_r = -p_i$ (pressure being negative in sign); and at $r = r_0$, $\sigma_r = -p_0$,

$$-p_i = A - \frac{B}{r_i^2} \quad \text{and} \quad -p_0 = A - \frac{B}{r_0^2}$$

from which, eliminating A, we get

$$B = \frac{(p_i - p_0)r_i^2 r_0^2}{r_0^2 - r_i^2} \quad \text{and} \quad A = \frac{p_i r_i^2 - p_0 r_0^2}{r_0^2 - r_i^2}$$

Therefore the radial and hoop stresses become

$$\sigma_r = \frac{p_i r_i^2 - p_0 r_0^2}{r_0^2 - r_i^2} - \frac{(p_i - p_0)r_i^2 r_0^2}{r^2(r_0^2 - r_i^2)} \qquad (15.34)$$

$$\sigma_\theta = \frac{p_i r_i^2 - p_0 r_0^2}{r_0^2 - r_i^2} + \frac{(p_i - p_0)r_i^2 r_0^2}{r^2(r_0^2 - r_i^2)} \qquad (15.35)$$

These equations were first derived by Lamé and Clapeyron in 1833.

Let the radius ratio $r_0/r_i = k$ then eqns. (15.34) and (15.35) may be written as

$$\sigma_r = \frac{p_i - k^2 p_0}{k^2 - 1} - \frac{(p_i - p_0)}{k^2 - 1}\left(\frac{r_0}{r}\right)^2$$

$$\sigma_\theta = \frac{p_i - k^2 p_0}{k^2 - 1} + \frac{(p_i - p_0)}{k^2 - 1}\left(\frac{r_0}{r}\right)^2$$

It is important to note that the stresses depend on the k ratio rather than on the absolute dimensions.

2. Internal pressure only

An important special case of the above is when the external pressure is atmospheric only and can be neglected in relation to the internal pressure. Then with $p_0 = 0$,

$$\sigma_r = \frac{p_i r_i^2}{r_0^2 - r_i^2}\left(1 - \frac{r_0^2}{r^2}\right) = \frac{p_i}{k^2 - 1}\left(1 - \frac{r_0^2}{r^2}\right) \qquad (15.36)$$

$$\sigma_\theta = \frac{p_i r_i^2}{r_0^2 - r_i^2}\left(1 + \frac{r_0^2}{r^2}\right) = \frac{p_i}{k^2 - 1}\left(1 + \frac{r_0^2}{r^2}\right) \qquad (15.37)$$

At the inner surface, σ_r and σ_θ each have their maximum value so that at $r = r_i$,

$$\sigma_r = -p_i \quad \text{(radial compressive stress)}$$

It is appropriate at this point to note that the radial stress shown on the element in Fig. 15.3 in the positive sense i.e. tension, is in fact in the opposite sense, i.e. compression.

The circumferential or hoop stress

$$\sigma_\theta = \frac{r_0^2 + r_i^2}{r_0^2 - r_i^2}p_i$$

$$= \frac{k^2 + 1}{k^2 - 1}p_i$$

At the outer surface, where $r = r_0$,

$$\sigma_r = 0 \quad \text{and} \quad \sigma_\theta = \frac{2p_i}{k^2 - 1}$$

AXIAL STRESS AND STRAIN

Now that expressions have been developed for the radial and circumferential stresses within the cylinder, the next step is to consider what conditions of stress and strain can exist axially along the cylinder. These will depend on the boundary conditions at the ends of the cylinder.

1. Cylinder with end caps but free to change in length

In this case there must be equilibrium between the force exerted on the end cover by the internal pressure and the force of the axial stress integrated across the wall of the vessel. Therefore

$$\sigma_z(\pi r_0^2 - \pi r_i^2) - p_i \pi r_i^2 = 0$$

so that

$$\sigma_z = \frac{p_i r_i^2}{r_0^2 - r_i^2} = \frac{p_i}{k^2 - 1} \tag{15.38}$$

and

$$\varepsilon_z = \frac{\sigma_z}{E} - \frac{v}{E}(\sigma_r + \sigma_\theta) = \frac{(1 - 2v)p_i}{E(k^2 - 1)}$$

2. Pressure retained by piston in each end of cylinder

Since there is no connection between the piston and the cylinder, the axial force due to pressure is reacted entirely by the pistons, and therefore there can be no axial stress in the wall of the cylinder. Thus

$$\sigma_z = 0 \tag{15.39}$$

and

$$\varepsilon_z = -\frac{v}{E}(\sigma_r + \sigma_\theta) = -\frac{2vp_i}{E(k^2 - 1)}$$

3. Cylinder built-in between rigid end supports

For this case $\varepsilon_z = 0$; in other words, plane strain exists.
    Therefore

$$\frac{\sigma_z}{E} - \frac{v}{E}(\sigma_r + \sigma_\theta) = 0$$

$$\sigma_z = v(\sigma_r + \sigma_\theta) \tag{15.40}$$

Substituting for σ_r and σ_θ from eqns. (15.36) and (15.37),

$$\sigma_z = \frac{2vp_i r_i^2}{r_0^2 - r_i^2} = \frac{2vp_i}{k^2 - 1} \tag{15.41}$$

SHEAR STRESS IN THE CYLINDER

Since the radial and circumferential stresses are principal stresses the maximum shear stress in the plane of the cross-section is given by

$$\hat{\tau}_{\theta r} = \frac{\sigma_\theta - \sigma_r}{2}$$

$$= \frac{p_i}{k^2 - 1}\left(\frac{r_0}{r}\right)^2 \tag{15.42}$$

Equations (15.36), (15.37) and (15.38) may be written as

$$\sigma_r = \frac{p_i}{k^2 - 1} - \frac{p_i}{k^2 - 1}\left(\frac{r_0}{r}\right)^2 = \sigma_h - \hat{\tau}$$

$$\sigma_\theta = \frac{p_i}{k^2 - 1} + \frac{p_i}{k^2 - 1}\left(\frac{r_0}{r^2}\right)^2 = \sigma_h + \hat{\tau}$$

$$\sigma_z = \frac{p_i}{k^2 - 1} = \sigma_h$$

where σ_h = hydrostatic stress, and $\hat{\tau}$ = maximum shear stress. Thus the stress distribution in the wall of the cylinder consists of pure shear stress and superimposed hydrostatic stress.

YIELDING IN THE CYLINDER

Yielding will commence at the inner surface and for the Tresca criterion we have

$$\sigma_Y = \sigma_\theta - \sigma_r \quad \text{at} \quad r = r_i$$

Using eqn. (15.42),

$$\sigma_Y = \frac{2k^2 p_i}{k^2 - 1}$$

hence the internal pressure to cause yielding is

$$p_i = \frac{(k^2 - 1)}{2k^2}\sigma_Y \qquad (15.43)$$

The von Mises criterion is employed as follows, using eqn. (13.15)

$$(\sigma_\theta - \sigma_r)^2 + (\sigma_r - \sigma_z)^2 + (\sigma_z - \sigma_\theta)^2 = 2\sigma_Y^2$$

Substituting the expressions for the three principal stresses and simplifying gives

$$\frac{6p_i^2 k^4}{(k^2 - 1)^2} = 2\sigma_Y^2$$

from which

$$p_i = \frac{(k^2 - 1)\sigma_Y}{\sqrt{3}k^2} \qquad (15.44)$$

which is the internal pressure to cause initial yielding at the bore.
The difference between these two criteria is approximately 15%.

STRESS DISTRIBUTIONS FOR σ_θ AND σ_r

To complete the basic analysis of the elastically-deformed thick-walled pressure vessel, the variation of the two principal stresses σ_θ and σ_r is shown plotted through the wall thickness in Fig. 15.4 for internal pressure and a k ratio of 3.

Fig. 15.4

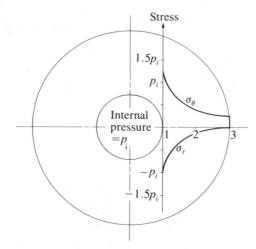

EXAMPLE 15.1

The cylinder of a hydraulic jack has a bore (internal diameter) of 150 mm and is required to operate up to 13.8 MN/m². Determine the required wall thickness for a limiting tensile stress in the material of 41.4 MN/m².

SOLUTION

The given boundary conditions are that at $r = 75 \times 10^{-3}$, $\sigma_r = -13.8 \times 10^6$ and $\sigma_\theta = 41.4 \times 10^6$, since the maximum tensile hoop stress occurs at the inner surface. Therefore

$$-13.8 \times 10^6 = A - \frac{B}{5600 \times 10^{-6}}$$

and

$$41.4 \times 10^6 = A + \frac{B}{5600 \times 10^{-6}}$$

Adding the two equations,

$$2A = 27.6 \times 10^6$$

$$A = 13.8 \times 10^6 \, \text{N/m}^2 \quad \text{and} \quad B = 154.5 \, \text{kN}$$

At the outside surface, $\sigma_r = 0$; therefore

$$0 = A - \frac{B}{r^2} = 13.8 \times 10^6 - \frac{154.5 \times 10^3}{r^2}$$

$$r^2 = 0.011 \, 2 \, \text{m}^2$$

$$r = 0.106 \, \text{m}$$

METHODS OF CONTAINING HIGH PRESSURE

From Fig. 15.4 it will be observed that there is a marked variation in the stress in the wall of a thick cylinder subjected to internal pressure, and

Fig. 15.5 (a) Compound cylinder; (b) wire-wound cylinder; (c) autofret-taged cylinder

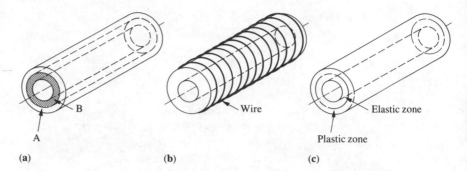

(a) (b) (c)

this situation gets worse when designing for even higher pressures. In order to secure a more uniform stress distribution, one method is to build up the cylinder by "shrinking" one tube on the outside of another (Fig. 15.5(*a*)). The inner tube is subjected to hoop compression by the shrink fit of the external tube, which will therefore be subjected to pressure causing hoop tension. When the compound tube is subjected to working pressure, the resultant stresses are the algebraic sum of that due to the shrinking and that due to the internal pressure. The resultant tensile stress at the inner surface of the inner tube is not so large as if the cylinder were composed of one thick tube. The final tensile stress at the inner surface of the outer tube is larger than if the cylinder consisted of one thick tube. Thus a more even stress distribution is obtained.

In gun-making, it is not an easy matter to turn and bore long tubes to the degree of accuracy required for shrinking. It is usual in this case to wind around the outside of a tube a high-tensile-strength ribbon of a rectangular section with sufficient tension to bring the tube into a state of hoop compression (Fig. 15.5(*b*)). Subsequent internal pressure then has to overcome the hoop compression before tensile stress can be set up in the tube.

A further method for creating hoop compression at the bore of a cylinder is known as *autofrettage*. This consists in applying internal pressure to a single cylinder until yielding and a prescribed amount of plastic deformation occurs at the bore (Fig. 15.5(*c*)). Since the hoop stress falls in magnitude from the inner to the outer surface, a part of the wall from the bore will be in the plastic range of the metal while the remainder will still be elastic. On the release of pressure the elastic material cannot return to its original geometry owing to the permanent deformation at the inside of the vessel. Therefore the material that had been in the plastic range will be subjected to hoop compression, and the elastic outer material to hoop tension. This technique is discussed further in Chapter 16.

STRESSES SET UP BY A SHRINK-FIT ASSEMBLY

A shrink fit between two components is a very important and secure method of assembly. It consists, in the case of two cylindrical objects, of the inner diameter of the outer cylinder being slightly less (by a fraction of a millimetre) than the outer diameter of the inner cylinder. Consequently, when they are at the same temperature the outer cannot be passed over the inner. However, if the outer cylinder is heated and the inner cylinder is cooled then the thermal expansion and contraction can be made sufficient to allow one cylinder to pass over the other. On returning each to room temperature there is "interference" at the mating surface since they cannot regain their original dimensions at the interface. The two components are locked firmly together and a system of radial and circumferential stresses are set up at the interface and through the wall of each cylinder. For elastic conditions the principle of superposition can be used to add together the stresses due to shrink-fit interference to those due to internal pressure or rotation.

For two cylindrical components we require eqns. (15.25) and (15.26), for each component given by

$$\sigma_\theta = A + \frac{B}{r^2}; \quad \sigma_r = A - \frac{B}{r^2} \text{ (inner component)}$$

and

$$\sigma_\theta = C + \frac{D}{r^2}; \quad \sigma_r = C - \frac{D}{r^2} \text{ (outer component)}$$

$$(15.45)$$

where the constants A, B, C and D are determined from the boundary conditions. For shrink-fit stresses only, the boundary conditions are that σ_r is zero at the inside of the inner cylinder and outside of the outer cylinder, and at the mating surface r_m the radial stress in each vessel must be the same, therefore

$$A - \frac{B}{r_m{}^2} = C - \frac{D}{r_m{}^2} \qquad (15.46)$$

Finally, at the mating surface the radial interference δ is the sum of the displacement of the inner cylinder inwards, $-u'$, and the outer cylinder outwards, $+u''$, thus

$$\delta = -u' + u'' = r_m(\varepsilon_\theta'' - \varepsilon_\theta') \qquad (15.47)$$

We next substitute into eqn. (15.47) the expressions for ε_θ' and ε_θ'',

$$\varepsilon_\theta'' = \frac{\sigma_\theta'}{E'} - \frac{v'}{E'}\sigma_r'$$

and

$$\varepsilon_\theta'' = \frac{\sigma_\theta''}{E''} - \frac{v''}{E''}\sigma_r''$$

and thence the relationships (eqns. (15.45)), for σ_θ', σ_θ'', σ_r' and σ_r'' at $r = r_m$.

We now have sufficient equations to solve for the constants A, B, C and D.

The following example illustrates the analytical process.

EXAMPLE 15.2

A bronze bush of 25 mm wall thickness is to be shrunk on to a steel shaft 100 mm in diameter. If an interface pressure of 69 MN/m^2 is required, determine the interference between bush and shaft. Steel: $E = 207$ GN/m^2, $v = 0.28$; bronze: $E = 100$ GN/m^2, $v = 0.29$.

SOLUTION

Using constants A and B for the shaft and C and D for the bush, then the radial stress for the shaft is

$$\sigma_{r_s} = A - \frac{B}{r^2}.$$

At the centre of the shaft $r = 0$ and this might imply that σ_{r_s} was infinite, but this cannot be so and therefore B must be zero; hence $\sigma_{r_s} = A = \sigma_{\theta_s}$ at all points in the shaft. The boundary conditions are

At the interface $\quad \sigma_{r_s} = -69 \times 10^6 = A$

At $r_m = 50$,

$$\sigma_{r_b} = C - \frac{D}{0.002\,5} = -69 \times 10^6$$

At $r_0 = 75$,

$$\sigma_{r_b} = C - \frac{D}{0.005\,6} = 0$$

From which $D = 312 \times 10^3$ and $C = 55.5 \times 10^6$. So at $r_m = 50$ mm, $\sigma_{r_b} = -69$ MN/m^2, $\sigma_{\theta_b} = 180$ MN/m^2.

Now, the interference is

$$\delta = -u_s + u_b = r_m(\varepsilon_{\theta_b} - \varepsilon_{\theta_s})$$

where $r_m = 50$; substituting the values for σ_{θ_b}, σ_{r_b}, σ_{θ_s}, σ_{r_s}

$$\delta = 0.05\{(180 - 0.29(-69))/100 - (-69 - 0.28(-69))/207\}$$

$$= 0.112 \text{ mm}$$

which is the interference required at the nominal interface radius of 50 mm between the shaft and the bush.

STRESS DISTRIBUTION IN A PRESSURIZED COMPOUND CYLINDER

As was explained briefly on page 400 the containment of high internal pressure in, for example, chemical processes can be achieved more effectively by shrinking two or more cylinders one over the other to give a compound or multi-tube vessel. The analysis simply uses the basic thick cylinder equations for σ_r and σ_θ together with the shrink fit and other boundary conditions.

The method and stress distribution is illustrated by the following worked example.

EXAMPLE 15.3

A vessel is to be used for internal pressures up to 207 MN/m². It consists of two hollow steel cylinders which are shrunk one on the other. The inner tube has an internal diameter of 200 mm and a nominal external diameter of 300 mm, while the outer tube is 300 mm nominal and 400 mm for the inner and outer diameters respectively. The interference at the mating surface of the two cylinders is 0.1 mm. Determine the radial and circumferential stresses at the bores and outside surfaces. The axial stress in the cylinders is to be neglected. $E = 207 \, \text{GN/m}^2$. Compare the stress distributions with that for a single steel cylinder, having the same overall dimensions, subjected to the same internal pressure.

SOLUTION

The boundary conditions are: inner tube; $r = 100$ mm, $\sigma_r = -207 \, \text{MN/m}^2$; outer tube; $r = 200$ mm, $\sigma_r = 0$; at the mating surface or interface, $r = 150$ mm nominally, and the radial stresses in the inner and outer tubes are equal, $\sigma_{ri} = \sigma_{ro}$. Also the radial displacement, u_i, of the inner cylinder inwards plus the radial displacement, u_0, of the outer cylinder outwards due to the shrink fit must equal the interference value; therefore

$$-u_i + u_0 = 0.1 \, \text{mm}$$

Using the above conditions and constants A, B and C, D for the inner and outer tubes respectively, we have four equations:

$$-207 \times 10^6 = A - \frac{B}{0.01} \tag{15.48}$$

$$0 = C - \frac{D}{0.04} \tag{15.49}$$

$$A - \frac{B}{0.022\,5} = C - \frac{D}{0.022\,5} \tag{15.50}$$

and

$$-\frac{u_i}{0.15} + \frac{u_0}{0.15} = \frac{0.000\,1}{0.15}$$

or

$$-\varepsilon_{\theta i} + \varepsilon_{\theta 0} = \frac{0.000\,1}{0.15}$$

and substituting for the strains in terms of the stresses,

$$-\sigma_{\theta i} + \nu\sigma_{r_i} + \sigma_{\theta 0} - \nu\sigma_{r_0} = \frac{0.000\,1}{0.15} \times 207 \times 10^9$$

and since $\sigma_{r_i} = \sigma_{r_0}$,

$$\sigma_{\theta 0} - \sigma_{\theta i} = 138 \times 10^6$$

Therefore
$$C + \frac{D}{0.022\,5} - \left(A + \frac{B}{0.022\,5}\right) = 138 \times 10^6 \tag{15.51}$$

The next step is to solve for the constants using eqns. (15.48) to (15.51) and the following values are obtained:

$$A = 28\,\text{MN/m}^2; \qquad B = 2.35\,\text{MN}$$

$$C = 98.2\,\text{MN/m}^2; \qquad D = 3.93\,\text{MN}$$

The required stresses are then computed from the basic equations with the values of the constants above.

Radius (mm)	σ_r (MN/m²)	σ_θ (MN/m²)	
100	−207	+263	} inner cylinder
150	−76.4	+132.4	
150	−76.4	+272.6	} outer cylinder
200	0	+196.4	

If the cylinder had been made of one thick tube of the same overall dimensions as the compound vessel then at the bore $r = 100$ mm, $\sigma_r = -207\,\text{MN/m}^2$

$$-207 \times 10^6 = A - \frac{B}{0.01} \tag{15.52}$$

and at $r = 200$ mm, $\sigma_r = 0$

$$0 = A - \frac{B}{0.04} \tag{15.53}$$

Hence $A = 69\,\text{MN/m}^2$ and $B = 2.76\,\text{MN}$ from which σ_θ and σ_r are as follows:

Radius (mm)	σ_r (MN/m²)	σ_θ (MN/m²)
100	−207	+345
150	−53.5	+191.5
200	0	+138

The values of σ_θ may be compared for the compound cylinder and the monobloc cylinder.

Figure 15.6 shows diagrammatically the distribution of radial and hoop stresses through the wall of the compound and single cylinders, and illustrates the more efficient use of material in the former case.

The stress distribution due to shrinkage only may be obtained directly as the difference between the curves for the single and compound cylinders as plotted above. This approach would only apply if the compound tube was made of the one type of material throughout.

Fig. 15.6
——— Compound cylinder
– – – Single cylinder
—·— Shrinkage stress

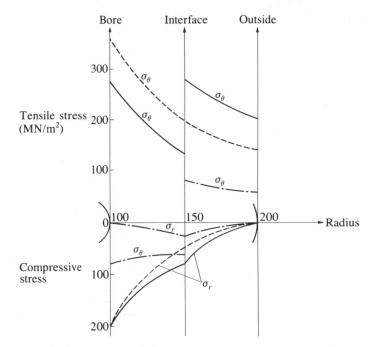

Shrinkage stresses alone could have been obtained by calculation from the four boundary conditions: $\sigma_r = 0$ at $r = 100$ and 200, and $\sigma_{r_i} = \sigma_{r_0}$ and $-u_i + u_0 = 0.1$ at the interface radius $r = 150$, as described earlier.

If the two tubes are of different materials then the appropriate elastic constants have to be used where they occur in the various equations.

STRESS DISTRIBUTION IN A THIN ROTATING DISC

A simplified example of a component such as a gas turbine rotor is a uniformly thin disc which, when rotating at a constant velocity, is subjected to stresses induced by centripetal acceleration. This is a problem which produces deformations symmetrical about the rotating axis. If the disc is thin in section then it is assumed that plane stress exists, so the radial and hoop stresses are constant through the thickness, and there is no stress in the z-direction.

The equation of equilibrium of an element, Fig. 15.7, is that derived in Chapter 14 for the axially-symmetrical stress system, but in this case a body force term must be included which is determined from the centripetal acceleration. Hence, from eqn. (14.11),

$$\frac{\mathrm{d}\sigma_r}{\mathrm{d}r} + \frac{\sigma_r - \sigma_\theta}{r} + R = 0$$

where R is the body force per unit volume.

In order to analyse the rotating disc as a static equilibrium problem, D'Alembert's principle is applied whereby the inward force on the element due to the centripetal acceleration is replaced with an outward

405

Fig. 15.7

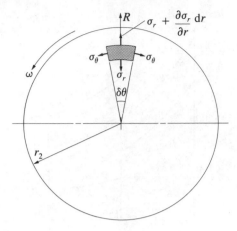

centrifugal force given by

$$F = (\rho r\, \delta\theta\, \delta r\, z)\omega^2 r = Rr\, \delta\theta\, \delta r\, z$$

where ρ is the density, ω is the steady rotational velocity in radians per second and z is the thickness of the disc. Therefore

$$R = \rho\omega^2 r$$

and

$$\frac{\mathrm{d}\sigma_r}{\mathrm{d}r} + \frac{\sigma_r - \sigma_\theta}{r} + \rho\omega^2 r = 0 \tag{15.54}$$

The strain–displacement equations for axial symmetry are

$$\varepsilon_r = \frac{\mathrm{d}u}{\mathrm{d}r} \qquad \varepsilon_\theta = \frac{u}{r}$$

and the stress–strain relationships are

$$\varepsilon_r = \frac{\sigma_r}{E} - \frac{\nu\sigma_\theta}{E} \qquad \varepsilon_\theta = \frac{\sigma_\theta}{E} - \frac{\nu\sigma_r}{E}$$

Using these four equations to obtain σ_r and σ_θ in terms of u,

$$\sigma_r = \left(\frac{\mathrm{d}u}{\mathrm{d}r} + \frac{\nu u}{r}\right)\frac{E}{1 - \nu^2} \tag{15.55}$$

$$\sigma_\theta = \left(\nu\frac{\mathrm{d}u}{\mathrm{d}r} + \frac{u}{r}\right)\frac{E}{1 - \nu^2} \tag{15.56}$$

Substituting in eqn. (15.54) we obtain

$$\frac{\mathrm{d}^2 u}{\mathrm{d}r^2} + \frac{1}{r}\frac{\mathrm{d}u}{\mathrm{d}r} - \frac{u}{r^2} + \left(\frac{1 - \nu^2}{E}\right)\rho\omega^2 r = 0$$

or

$$\frac{\mathrm{d}^2 u}{\mathrm{d}r^2} + \frac{1}{r}\frac{\mathrm{d}u}{\mathrm{d}r} - \frac{u}{r^2} = -\left(\frac{1 - \nu^2}{E}\right)\rho\omega^2 r \tag{15.57}$$

This is a linear differential equation of the second order. The general solution consists of the sum of two separate solutions known as the complementary function and the particular integral. The former is the solution of the left-hand side and the latter is obtained by considering the right-hand side of eqn. (15.57). Thus the complementary function is

$$u = Cr + \frac{C'}{r}$$

and the particular integral is

$$u = -\left(\frac{1 - v^2}{E}\right)\frac{\rho\omega^2 r^3}{8}$$

The complete solution is therefore

$$u = Cr + \frac{C'}{r} - \left(\frac{1 - v^2}{E}\right)\frac{\rho\omega^2 r^3}{8} \tag{15.58}$$

in which C and C' are constants to be determined from the boundary conditions.

Using eqn. (15.58) and substituting for u and du in eqns. (15.55) and (15.56) and simplifying the various constant terms by inserting new ones, A and B, we obtain

$$\sigma_r = A - \frac{B}{r^2} - \left(\frac{3 + v}{8}\right)\rho\omega^2 r^2 \tag{15.59}$$

$$\sigma_\theta = A + \frac{B}{r^2} - \left(\frac{1 + 3v}{8}\right)\rho\omega^2 r^2 \tag{15.60}$$

The constants A and B are found from the appropriate boundary conditions of the problem.

SOLID DISC WITH UNLOADED BOUNDARY

If the disc is continuous from the centre to some outer radius $r = r_2$, then it is apparent that, unless $B = 0$, the stresses would become infinite at $r = 0$. To find A it is only necessary to use the condition that

$$\sigma_r = 0 \quad \text{at} \quad r = r_2$$

from which

$$A = \frac{3 + v}{8}\rho\omega^2 r_2^2$$

and

$$\sigma_r = \frac{3 + v}{8}\rho\omega^2(r_2^2 - r^2) \tag{15.61}$$

$$\sigma_\theta = \frac{3 + v}{8}\rho\omega^2 r_2^2 - \frac{1 + 3v}{8}\rho\omega^2 r^2$$

$$= \frac{\rho\omega^2}{8}[(3 + v)r_2^2 - (1 + 3v)r^2] \tag{15.62}$$

Fig. 15.8

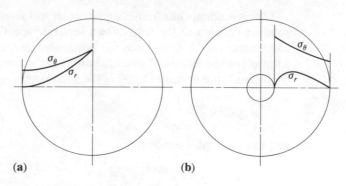

(a) (b)

The distributions of radial and hoop stresses are shown in Fig. 15.8(a). The maximum stress occurs at the centre, where $r = 0$, and then

$$\sigma_r = \sigma_\theta = \frac{3+v}{8}\,\rho\omega^2 r_2^2$$

DISC WITH A CENTRAL HOLE AND UNLOADED BOUNDARIES

In this case the boundary conditions are that the radial stress will be zero at $r = r_1$ and $r = r_2$, the radius of the hole and outside periphery of the disc respectively. Therefore

$$\sigma_r = A - \frac{B}{r_2^2} - \frac{3+v}{8}\,\rho\omega^2 r_2^2 = 0$$

and also

$$\sigma_r = A - \frac{B}{r_1^2} - \frac{3+v}{8}\,\rho\omega^2 r_1^2 = 0$$

Solving these two equations for the constants gives

$$B = \frac{3+v}{8}\,\rho\omega^2 r_1^2 r_2^2 \quad \text{and} \quad A = \frac{3+v}{8}\,\rho\omega^2 (r_1^2 + r_2^2)$$

Therefore

$$\sigma_r = \frac{3+v}{8}\,\rho\omega^2 \left[r_1^2 + r_2^2 - \frac{r_1^2 r_2^2}{r^2} - r^2 \right] \tag{15.63}$$

and

$$\sigma_\theta = \frac{3+v}{8}\,\rho\omega^2 \left[r_1^2 + r_2^2 + \frac{r_1^2 r_2^2}{r^2} - \frac{1+3v}{3+v}\,r^2 \right] \tag{15.64}$$

The maximum value of the hoop stress, σ_θ, is at $r = r_1$, and is given by

$$\sigma_{\theta\,max} = \frac{3+v}{4}\,\rho\omega^2 \left[r_2^2 + \frac{1-v}{3+v}\,r_1^2 \right] \tag{15.65}$$

σ_r is a maximum when $d\sigma_r/dr = 0$ or $r = \sqrt{(r_1 r_2)}$; therefore

$$\sigma_{r\,max} = \frac{3+v}{8}\,\rho\omega^2 (r_2 - r_1)^2 \tag{15.66}$$

The stress distributions of σ_θ and σ_r for the disc with the hole are shown in Fig. 15.8(b).

DISC WITH A LOADED BOUNDARY

In general, rotor discs will have mounted on the outer boundary a large number of blades. These will themselves each have a centrifugal force component which will have to be reacted at the periphery of the disc. Given the mass of each blade, its effective centre of mass and the number of blades we can compute the force due to each blade at a particular value of ω. Multiplying by the number of blades gives the total force which may then be computed as a uniformly distributed load. Dividing this by the thickness of the outer boundary gives the required value of σ_r to use as the boundary condition when evaluating A and B.

DISC SHRUNK ONTO A SHAFT

The concept of a shrink fit between two components was developed on page 400 for the compound cylinder. The same principle is very valuable for locating a rotating disc onto a shaft and avoiding mechanical attachments, e.g. bolting, riveting. The stresses set up by the shrink-fit mechanism are quite independent when the disc is stationary. However, since at the mating surface the radial stress is compressive, when rotation commences there is a superposition of radial tensile stress. It is therefore necessary to ensure that the shrink-fit stress is always in excess of the rotational stress at the mating surfaces so that the disc does not become "free" on the shaft.

ROTATIONAL SPEED FOR INITIAL YIELDING

The design of rotating discs must take account of the limit of rotational speed which would induce initial yielding at some point in the disc.

SOLID DISC

The maximum stress occurs at $r = 0$ and is

$$\sigma_r = \sigma_\theta = \frac{3 + v}{8} \rho \omega^2 r_2^2$$

For the maximum shear stress criterion (Tresca)

$$\sigma_Y = \sigma_\theta \quad \text{(since } \sigma_z = 0\text{)}$$

Therefore

$$\sigma_Y = \frac{(3 + v)}{8} \rho \omega_Y^2 r_2^2$$

$$\omega_Y = \frac{1}{r_2} \sqrt{\frac{8\sigma_Y}{(3 + v)\rho}} \tag{15.67}$$

For the shear strain energy criterion (von Mises)

$$\sigma_Y^2 = \sigma_\theta^2 + \sigma_r^2 - \sigma_\theta \sigma_r$$

$$= \sigma_\theta^2 \quad \text{(since } \sigma_\theta = \sigma_r\text{)}$$

$$\sigma_Y = \sigma_\theta$$

and

$$\omega_Y = \frac{1}{r_2} \sqrt{\frac{8\sigma_Y}{(3+v)\rho}} \qquad (15.68)$$

which is the same value as for the Tresca criterion.

DISC WITH CENTRAL HOLE

The maximum hoop stress occurs at $r = r_1$ and is

$$\sigma_\theta = \frac{3+v}{4} \rho\omega^2 \left\{ r_2^2 + \frac{(1-v)}{(3+v)} r_1^2 \right\}$$

For the Tresca criterion

$$\sigma_Y = \sigma_\theta = \frac{3+v}{4} \rho\omega_Y^2 \left\{ r_2^2 + \frac{(1-v)}{(3+v)} r_1^2 \right\}$$

$$\omega_Y = \left\{ \frac{4\sigma_Y}{\rho[(3+v)r_2^2 + (1-v)r_1^2]} \right\}^{1/2} \qquad (15.69)$$

For the von Mises criterion since $\sigma_r = \sigma_z = 0$ at $r = r_1$ yield occurs when $\sigma_Y = \sigma_\theta$, hence the rotational speed at yield is the same as given by eqn. (15.69).

EXAMPLE 15.4

A steel ring has been shrunk on to the outside of a solid steel disc and shaft. The interface radius is 250 mm and the outer radius of the assembly is 356 mm. If the pressure between the ring and the disc is not to fall below 34.5 MN/m², and the circumferential stress at the inside of the ring must not exceed 207 MN/m², determine the maximum speed at which the assembly can be rotated. What is then the stress at the centre of the disc? $\rho = 7.75$ Mg/m³, $v = 0.28$.

SOLUTION

For the ring at $r = 356$ mm, $\sigma_r = 0$, and at $r = 250$ mm, $\sigma_r = -34.5 \times 10^6$; therefore

$$0 = A - \frac{B}{0.126} - \left(\frac{3+0.28}{8} \times 7.75\omega^2 \times 0.126 \times 10^3 \right)$$

and

$$-34.5 \times 10^6 = A - \frac{B}{0.062\,5} - \left(\frac{3+0.28}{8} \times 7.75\omega^2 \times 0.062\,5 \times 10^3 \right)$$

from which

$$B = (4280 + 0.025\omega^2)10^3 \quad \text{and} \quad A = (34\,000 + 0.6\omega^2)10^3$$

Also when $r = 250$ mm, σ_θ must not exceed 207 MN/m²; therefore

$$207 \times 10^6 = A + \frac{B}{0.062\,5} - \left[\frac{1+(3 \times 0.28)}{8} \times 7.75 \times 10^3 \omega^2 \times 0.062\,5 \right]$$

Substituting for A and B,

$$207 \times 10^3 = 34\,000 + 0.6\omega^2 + 68\,500 + 0.4\omega^2 - 0.111\omega^2$$

From which

$$\omega = 343\,\text{rad/sec} \quad \text{and} \quad N = 3280\,\text{rev/min}$$

For the solid disc, using constants C and D, as shown previously, D must be zero; therefore

$$\sigma_r = C - \left(\frac{3+v}{8}\right)\rho\omega^2 r^2$$

At $r = 250$, $\sigma_r = -34.5 \times 10^6$; therefore

$$-34.5 \times 10^6 = C - \left(\frac{3+0.28}{8} \times 7.75 \times 10^3 \times 117\,500 \times 0.062\,5\right)$$

and

$$C = (-34.5 \times 10^6) + (23.3 \times 10^6) = -11.2 \times 10^6\,\text{N/m}^2$$

But at the centre of the disc, $r = 0$ and $\sigma_r = \sigma_\theta = C$; therefore

$$\sigma_r = \sigma_\theta = -11.2\,\text{MN/m}^2$$

STRESSES IN A ROTOR OF VARYING THICKNESS WITH RIM LOADING

The thin uniform discs analysed in the previous section, although illustrating equilibrium and compatibility concepts, do not represent a very realistic design configuration. Because the hoop stress σ_θ is highest at the bore and reduces towards the periphery it is more economical to vary the thickness in similar proportions. A varying cross-section for a turbine rotor might be as shown in Fig. 15.9. A simple method of solution was devised by Donath for steam turbines which consisted of dividing up the cross-section into a number of constant thickness rings as shown.

Fig. 15.9

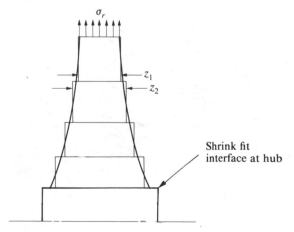

The equations (15.59) and (15.60) for σ_r and σ_θ can be applied to each ring observing the required conditions of equilibrium and compatibility at each. Let

$$S \text{ (sum)} = \sigma_\theta + \sigma_r = 2A - 4\frac{(1+v)}{8}\rho\omega^2 r^2 \tag{15.70}$$

and

$$D \text{ (difference)} = \sigma_\theta - \sigma_r = 2\frac{B}{r^2} + 2\frac{(1-v)}{8}\rho\omega^2 r^2 \tag{15.71}$$

Replacing ωr by the tangential velocity V at radius r gives

$$S = \frac{(1+v)}{2}\rho\{-V^2 + C_1\} \tag{15.72}$$

$$D = \frac{(1-v)}{4}\rho\left\{V^2 + \frac{C_2}{V^2}\right\} \tag{15.73}$$

Donath then constructed a chart of a series of S and D curves for various values of C_1 and C_2 and for a particular density and Poisson's ratio. A typical Donath chart is illustrated in Fig. 15.10.

Fig. 15.10

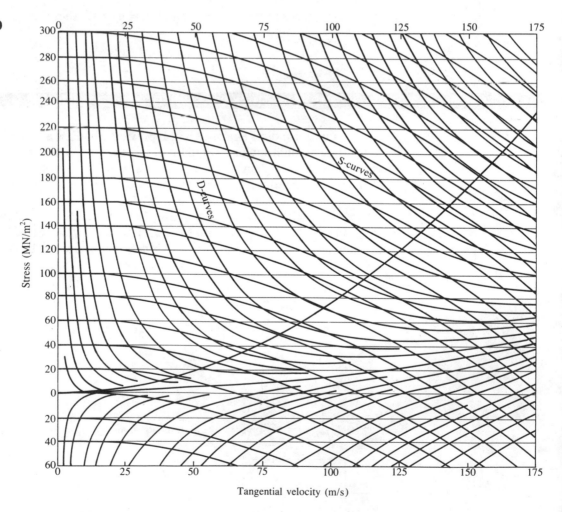

Tangential velocity (m/s)

If S and D values are known at any radius then σ_θ and σ_r can be found since

$$\sigma_\theta = \frac{S+D}{2} \quad \text{and} \quad \sigma_r = \frac{S-D}{2} \tag{15.74}$$

The procedure is as follows having divided the disc into several rings of constant thickness:

1. Calculate the rim loading due to blades and thence σ_{r_0} at the rim.
2. Assume a value of σ_{θ_0} at the rim and then calculate S and D at the rim.
3. Using the chart and the tangential velocity at the rim will give starting positions on the appropriate S and D curves.
4. Proceed along the S and D lines from the rim to the first interface between rings at the correct tangential velocity for that interface radius. Use these new S and D values to calculate σ_{r_1}.
5. At this interface we must satisfy equilibrium for each side of the interface hence

$$\sigma_{r_1} \cdot z_1 = \sigma_{r_2} \cdot z_2 \tag{15.75}$$

where z_1 and z_2 are the thicknesses of the adjacent rings.
   Now $\Delta\sigma_r = \sigma_{r_1} - \sigma_{r_2}$

$$\Delta\sigma_r = \sigma_{r_1}\left(1 - \frac{\sigma_{r_2}}{\sigma_{r_1}}\right) = \sigma_{r_1}\left(1 - \frac{z_1}{z_2}\right) \tag{15.76}$$

and we must also satisfy compatibility so that circumferential strains must be equal on each side of the step hence, $\varepsilon_{\theta_1} = \varepsilon_{\theta_2}$ and

$$\sigma_{\theta_1} - \nu\sigma_{r_1} = \sigma_{\theta_2} - \nu\sigma_{r_2} \tag{15.77}$$

$$\Delta\sigma_\theta = \nu\Delta\sigma_r$$

$$\Delta\sigma_\theta = \nu\sigma_{r_1}\left(1 - \frac{z_1}{z_2}\right) \tag{15.78}$$

Adding and subtracting eqns. (15.76) and (15.78),

$$\Delta\sigma_\theta + \Delta\sigma_r = \Delta S = (1 + \nu)\Delta\sigma_r \tag{15.79}$$

$$\Delta\sigma_\theta - \Delta\sigma_r = \Delta D = (\nu - 1)\Delta\sigma_r \tag{15.80}$$

6. The new S and D values for the other side of the interface are calculated using eqns. (15.76), (15.79) and (15.80).
7. Follow these curves on the chart to the next interface velocity V and calculate σ_r using the S and D values. Repeat calculations (5) and (6).
8. Continue for the remaining steps until reaching the centre or the inner radius. The final values should be $\sigma_r = \sigma_\theta$ for a solid disc and $\sigma_r = 0$ or a shrink-fit pressure for a disc with a central hole.
9. The correct values will probably not be obtained on the first run and an adjustment will then be made to the σ_θ value at the rim and the process repeated until the conditions at the bore or centre are correct.

10. The distribution of σ_r and σ_θ may now be plotted by calculation from the values off the S and D curves at the radii for which each ring thickness equals the disc thickness.

The above analysis may seem rather outdated with the current availability of computers, but the object here is to demonstrate the principle of analysis and the Donath chart is merely a graphical aid. The following simplified worked example will help to clarify the solution.

EXAMPLE 15.5

A steel rotor disc of 800 mm diameter and varying thickness, as shown in Fig. 15.9, rotates at 2860 rev/min. The outer periphery is subjected to radial and circumferential stresses of 17 and 33 MN/m² respectively. Evaluate the magnitudes and distributions of radial and circumferential stresses in the disc. The interface radii are 50, 100, 175 and 275 mm and the ring segments are of width 100, 65, 43 and 30 mm respectively. $v = 0.3$.

SOLUTION

The first step is to calculate the velocities at the periphery and at each ring interface and these are given in the table below.
At the periphery

$$S = 33 + 17 = 50\,\text{MN/m}^2$$

and

$$D = 33 - 17 = 16\,\text{MN/m}^2$$

which gives the starting-points on the chart, Fig. 15.10, at a velocity of 120 m/s. The remainder of the solution is shown in tabulated form. The final values of σ_r and σ_θ are determined from the S and D values half-way between each interface where the ring thickness equals the disc thickness. In this example further iteration is not required and the negative value of σ_r is due to a shrink fit.

V (m/s)	z_1	z_2	z_1/z_2	$(1 - z_1/z_2)$	S	D	σ_r	$\Delta\sigma_r$	ΔS	ΔD	σ_θ	σ_r
120	—	30	—	—	50	16	17	—	—	—	33	17
											40	30
82.4	30	43	0.7	0.3	90	0	45	13.5	17.5	−9.5		
											47	39
52.4	43	65	0.66	0.34	95	5	45	15.3	19.9	−10.7		
											49	27
30	65	100	0.65	0.35	80	38	21	7.3	9.5	−5.1		
											72	0
15	100	—	—	—	74	140	−33	—	—	—	107	−33

SUMMARY

The application of the equilibrium and strain–displacement relationships has been demonstrated at length in relation to two important engineering units, namely the thick-walled pressure vessel and the rotating turbine rotor. Various design problems have been examined such as containing very high pressure by using a compound vessel. This leads to the concept of shrink fitting and the associated initial stresses prior to pressurizing. Numerical constants in the equations for radial and hoop stresses have to be determined from the boundary conditions which must be accurately assessed. The axial condition in the cylinder wall depends on the method of end closure and sealing and this is an important design consideration.

Stresses set up by rotation are a major design feature of turbine rotors and compressors. The analysis commenced with the thin uniform disc to enable a full understanding of the plane stress solution for σ_θ and σ_r obtained from the equilibrium, compatibility and stress–strain equations, together with a variety of boundary conditions. It became a fairly straightforward process then to develop an analysis for the varying thickness rotor.

Finally, we must remember that elastic design relies on the application of yield criteria limits such as those of Tresca and von Mises to enable us to determine a limiting pressure for a cylinder and limiting speed for a rotor. However, it will be seen in Chapter 16 that there can be good reasons for developing limited amounts of plastic deformation in a cylinder or rotor to induce favourable residual stresses which allow enhanced performance. In *all* cases the basic principles of equilibrium of forces, geometry of deformation and the elastic or plastic stress–strain relationship, together with the appropriate boundary conditions, must be followed.

PROBLEMS

15.1 A beam of depth d and length l is simply-supported at each end and carries a uniformly distributed load over the whole span. Show that the maximum vertical direct stress σ_y is $\frac{4}{3}(d/l)^2$ times the maximum bending stress σ_x at mid-span, and therefore in most cases may be considered insignificant in relation to the latter.

15.2 Solve Problem 13.3 using the thick-walled cylinder relationships and compare the difference in axial loads.

15.3 Determine the k ratio for a thick-walled cylinder subjected to an internal pressure of $80 \, \text{MN/m}^2$ if the circumferential stress is not to exceed $140 \, \text{MN/m}^2$. What are the maximum shear stresses at the inside and outside surfaces?

15.4 In a pressure test on a hydraulic cylinder of 120 mm external diameter and 60 mm internal diameter the hoop and longitudinal strains are measured by means of strain gauges on the outer surface and found to be 266×10^{-6} and 69.6×10^{-6} respectively for an internal pressure of $100 \, \text{MN/m}^2$. Determine the actual hoop stress at the outer surface and compare this result with the calculated value. Determine also the safety factor for the cylinder according to the maximum shear stress theory. The properties of the cylinder material are as follows: $\sigma_y = 280 \, \text{MN/m}^2$; $E = 208 \, \text{GN/m}^2$; $v = 0.29$.

15.5 A cylinder of internal radius a and external radius b is sealed at atmospheric pressure and put into the sea. If the pressure due to the water is 10 atmospheres, calculate the maximum hoop stress in the cylinder in units of bars. Assume that the volume of the cylinder does not change.

15.6 Derive expressions for the radial, circumferential and axial strains at the inner and outer surfaces of a thick-walled cylinder of radius ratio k with closed ends subjected to internal pressure p.

15.7 One method of determining Poisson's ratio for a material is to subject a cylinder to internal pressure and to measure the axial, ε_z, and circumferential, ε_θ, strains on the outer surface. Show that

$$\nu = \frac{\varepsilon_\theta - 2\varepsilon_z}{2\varepsilon_\theta - \varepsilon_z}$$

The axial and circumferential strains on the outer surface of a closed-ended cylinder of diameter ratio 3 subjected to internal pressure were found to be 1.02×10^{-4} and 4.1×10^{-4} respectively. Calculate the internal pressure and the hoop strain at the bore. It may be assumed that

$$\sigma_z = \frac{\sigma_r + \sigma_\theta}{2} \qquad E = 207 \text{ GN/m}^2$$

15.8 Show that the pressure generated at the interface between two cylinders when they are shrunk together is given by

$$p = \left(\frac{E\delta}{r} \right) \left[\frac{k_2^2 + 1}{k_2^2 - 1} + \frac{k_1^2 + 1}{k_1^2 - 1} \right]^{-1}$$

where δ is the interference between the outer and inner cylinders and r is the nominal interface radius. The suffices 1 and 2 refer to the inner and outer cylinders respectively.

15.9 A gun barrel is formed by shrinking a tube of 224 mm external diameter and 168 mm internal diameter upon another tube of 126 mm internal diameter. After shrinking, the radial pressure at the common surface is 13.8 MN/m². Determine the hoop stresses at the inner and outer surfaces of each tube. Plot diagrams to show the variation of the hoop and radial stresses with radius for both tubes.

15.10 A steel tube has an internal diameter of 25 mm and an external diameter of 50 mm. Another tube, of the same steel, is to be shrunk over the outside of the first so that the shrinkage stresses just produce a condition of yield at the inner surface of each tube. Determine the necessary difference in diameters of the mating surfaces before shrinking and the required external diameter of the outer tube. Assume that yielding occurs according to the maximum shear stress criterion and that no axial stresses are set up due to shrinking. Yield stress in simple tension or compression = 414 MN/m², $E = 207 \text{ GN/m}^2$.

15.11 A steel bush is to be shrunk on to a steel shaft so that the internal diameter is extended 0.152 mm above its original size. The inside and outside diameters of the sleeve are 203 mm and 305 mm respectively. Find (i) the normal pressure intensity between the bush and the shaft, (ii) the hoop stresses at the inner and outer surfaces of the bush. $E = 207 \text{ GN/m}^2$, $\nu = 0.28$.

15.12 A solid steel shaft of 0.2 m diameter has a bronze bush of 0.3 m outer diameter shrunk on to it. In order to remove the bush the whole assembly is raised in temperature uniformly. After a rise of 100 °C the bush can just be moved along the shaft. Neglecting any effect of temperature in the axial direction, calculate the original interface pressure between the bush and the shaft. $E_{steel} = 208 \text{ GN/m}^2$; $\nu_{steel} = 0.29$; $\alpha_{steel} = 12 \times 10^{-6}$ per deg C; $E_{bronze} = 112 \text{ GN/m}^2$; $\nu_{bronze} = 0.33$; $\alpha_{bronze} = 18 \times 10^{-6}$ per deg C.

15.13 Commencing from the relationship derived in Problem 14.2 show that the radial and circumferential stresses in a thick-walled spherical shell may be expressed as

$$\sigma_r = A - 2\frac{B}{r^3} \quad \text{and} \quad \sigma_\theta = A + \frac{B}{r^3}$$

Determine the maximum shear stress at the inner and outer surfaces of a spherical shell, having a k ratio of 1.5, for an internal pressure of 7 MN/m^2.

15.14 A thin disc of inner and outer radii 150 and 300 mm respectively rotates at 150 rad/sec. Determine the maximum radial and hoop stresses. $v = 0.304$, $\rho = 7.7 \text{ Mg/m}^3$.

15.15 A thin uniform disc with a central hole is pressed on a shaft in such a manner that when the whole is rotated at n revolutions per minute the pressure at the common surface is p. Derive an expression for the hoop stress in the disc at the periphery, if the inside radius of the disc is r_1 and the outside radius r_2.

15.16 A solid steel disc 457 mm in diameter and of small constant thickness has a steel ring of outer diameter 610 mm and the same thickness shrunk on to it. If the interference pressure is reduced to zero at a rotational speed of 3000 rev/min, calculate the difference in diameters of the mating surfaces of the disc and ring before assembly and the interface pressure. $v = 0.29$, $\rho = 7.7 \text{ Mg/m}^3$, $E = 207 \text{ GN/m}^2$.

15.17 A steel rotor disc of uniform thickness 50 mm has an outer rim of diameter 750 mm and a central hole of diameter 150 mm. There are 200 blades each of weight 0.22 kg at an effective radius of 430 mm pitched evenly around the periphery. Determine the rotational speed at which yielding first occurs according to the maximum shear stress criterion. Yield stress in simple tension for the steel is 700 MN/m^2, $v = 0.29$, $\rho = 7.3 \text{ Mg/m}^3$, $E = 207 \text{ GN/m}^2$.

15.18 A disc is to be designed having uniform strength, that is, the radial and hoop stresses are the same at any point in the disc. Show that the required profile of thickness variation is given by

$$z = z_0 e^{-\rho \omega^2 r^2 / 2\sigma}$$

where z_0 is the thickness at $r = 0$, σ is the uniform stress, and e is the base of Napieran logarithms.

Chapter 16

Elementary plasticity

INTRODUCTION

Engineering design is primarily concerned with maintaining machines and structures working within their elastic range. The analyses of Chapter 13 were specifically related to the assessment of the yield boundaries for components and the influence of stress concentration in possibly causing material to exceed the elastic range locally. However, it would be imprudent if designers knew nothing of what would happen to components that were, say, grossly overloaded to the point where marked yielding and plastic deformation occurred. Another important aspect is that in some circumstances enhanced performance can be achieved by prior plastic deformation resulting in favourable residual stresses as, for example, in a thick-walled pressure vessel or rotor disc. A further application of plasticity relates to shaping metals, and although this is generally a subject outside the scope of this text we shall look at the simple elements of beams and shafts plastically deformed.

It is evident that the same principles must apply as for elastic deformation, namely equilibrium of forces, compatibility of deformations and a stress–strain relationship. It is the nature of this latter material behaviour which particularly dictates the final solution. Elastic-plastic stress-strain relationships are illustrated in Fig. 16.1(a), (b) and (c). The first is a typical curve for a real strain-hardening material and its linear to non-linear development causes some complication in analysis. Because of this, semi-idealized behaviour is often assumed as in Fig. 16.1(b) in which strain-hardening occurs linearly from initial yield, while in Fig. 16.1(c) strain-hardening is ignored and we have a linear-elastic-non-hardening plastic relationship.

PLASTIC BENDING OF BEAMS: PLASTIC MOMENT

In considering the behaviour of beams subjected to pure bending which results in fibres being stressed beyond the limit of proportionality, the

Fig. 16.1

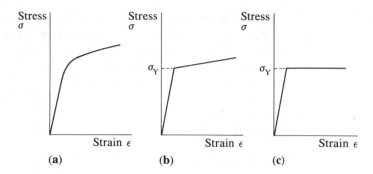

(a)　　　　　(b)　　　　　(c)

following assumptions will be made:

1. That the fibres are in a condition of simple tension or compression.
2. That any cross-section of the beam will remain plane during bending as in elastic bending.

The axis, of cross-section, passing through the centroid remains a neutral axis during inelastic bending; thus a cross-section will bend about the neutral axis and the stresses will be greatest in the extreme fibres of the beam.

In elastic bending of a beam there is a linear stress distributed over a cross-section and the extreme fibres reach the yield stress, when the bending moment for this condition is given by

$$M_Y = \sigma_Y \frac{I}{y} \tag{16.1}$$

where I is the second moment of area of the cross-section about the neutral axis, and y is the distance from the neutral axis to an extreme fibre. The value of the yield bending moment M_Y will be found for beams of various cross-section.

RECTANGULAR SECTION

From eqn. (16.1),

$$M_Y = \sigma_Y \frac{bd^2}{6} \tag{16.2}$$

and the stress distribution corresponding to this condition is shown in Fig. 16.2(a), all the fibres of the beam being in the elastic condition.

Fig. 16.2

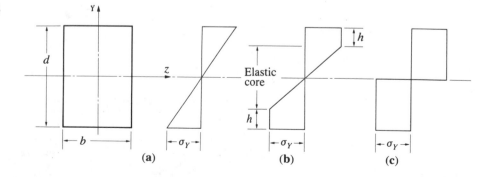

(a)　　　　　(b)　　　　　(c)

When the bending moment is increased above the value given in eqn. (16.2), some of the fibres near the top and bottom surfaces of the beam begin to yield and the appropriate stress diagram is given in Fig. 16.2(b). With further increase in bending moment, plastic deformation penetrates deeper into the beam. The total bending moment is obtained by consideration of both the plastic stress near the top and bottom of the beam and the elastic stress in the core of the beam. This moment is called the *elastic–plastic bending moment*. Therefore

$$M = \sigma_Y bh(d - h) + \sigma_Y \frac{b(d - 2h)^2}{6}$$

$$= \frac{\sigma_Y bd^2}{6} \left[\frac{6hd - 6h^2 + d^2 - 4dh + 4h^2}{d^2} \right]$$

$$= \frac{\sigma_Y bd^2}{6} \left[1 + 2\frac{h}{d} \left(1 - \frac{h}{d} \right) \right] \tag{16.3}$$

At a distance $(\tfrac{1}{2}d - h)$ from the neutral axis, the stress in the fibres has just reached the value σ_Y; then, if R is the radius of curvature, we have

$$\sigma_Y = \frac{E(\tfrac{1}{2}d - h)}{R}$$

or

$$\frac{1}{R} = \frac{\sigma_Y}{E(\tfrac{1}{2}d - h)} \tag{16.4}$$

The values of M and $1/R$ calculated from eqns. (16.3) and (16.4) when plotted give the graph shown in Fig. 16.3. The connection between these quantities is linear up to a value of $M = M_Y$. Beyond this point the relationship is non-linear and the slope decreases with increase in depth, h, of the plastic state. When h becomes equal to $\tfrac{1}{2}d$, the stress distribution becomes that shown in Fig. 16.2(c) and the highest value of bending moment is reached.

This *fully plastic moment* is given by eqn. (16.3), putting $h = \tfrac{1}{2}d$, as

$$M_p = \tfrac{3}{2}\sigma_Y \frac{bd^2}{6} = \sigma_Y \frac{bd^2}{4} \tag{16.5}$$

$$= \tfrac{3}{2}M_Y \tag{16.6}$$

Fig. 16.3

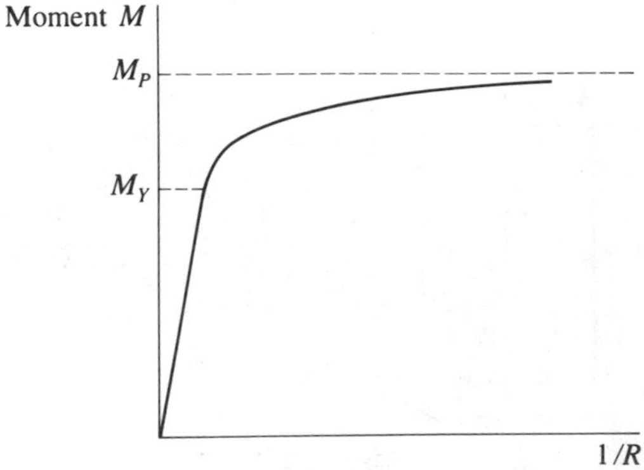

420

The value of M_p is shown in Fig. 16.3 and is the horizontal asymptote of the curve. Plastic collapse of the beam is shown, by eqn. (16.6), to occur at a bending moment of one and a half times that at initial yielding of the extreme fibres of the beam.

I-SECTION

When yielding is about to occur at the extreme fibres, the beam is still in the elastic condition and, for the dimensions in Fig. 16.4(*a*),

$$M_Y = \sigma_Y\left(\frac{bd^3}{12} - \frac{b_1d_1^3}{12}\right)\frac{2}{d} \tag{16.7}$$

Fig. 16.4

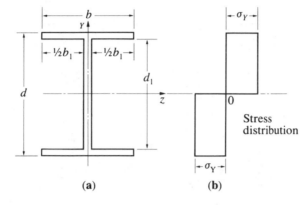

(a)   (b)

In the fully plastic condition, the stress diagram is shown in Fig. 16.4(*b*), and from eqn. (16.5) the fully plastic moment is given by

$$M_p = \sigma_Y\left(\frac{bd^2}{4} - \frac{b_1d_1^2}{4}\right) \tag{16.8}$$

and the ratio M_p/M_Y which is termed the *shape factor* is

$$\frac{M_p}{M_Y} = \frac{\left(\dfrac{bd^2}{4} - \dfrac{b_1d_1^2}{4}\right)d}{\left(\dfrac{bd^3}{12} - \dfrac{b_1d_1^3}{12}\right)2}$$

or

$$\frac{M_p}{M_Y} = \frac{3}{2}\frac{\left(1 - \dfrac{b_1d_1^2}{bd^2}\right)}{\left(1 - \dfrac{b_1d_1^3}{bd^3}\right)} \tag{16.9}$$

In an I-beam, 100 mm × 300 mm, with flanges and web 14 mm and 9 mm thick respectively,

$$\frac{M_p}{M_Y} = \frac{3}{2}\frac{\left(1 - \dfrac{91 \times 272^2}{100 \times 300^2}\right)}{\left(1 - \dfrac{91 \times 272^3}{100 \times 300^3}\right)} = 1.16$$

421

This shape factor is fairly representative of standard rolled I-section beams, and the fully plastic moment is only 16% greater than that at which initial yielding occurs.

ASYMMETRICAL SECTION

In the two previous cases the neutral axis in bending of the section coincided with an axis of symmetry. If the cross-section is asymmetrical about the axis of bending, then the position of the neutral axis must be determined. Fig. 16.5(*a*) shows a T-bar section in which YY is the only axis of symmetry and ZZ passes through the centroid of the section.

Fig. 16.5

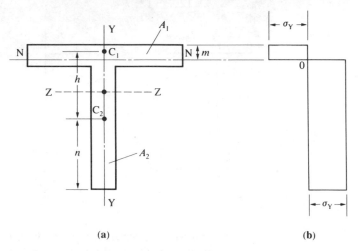

(a)   (b)

In the fully plastic condition the beam is bent about the neutral axis NN. If A_1 and A_2 are the areas of the cross-section above and below NN respectively, then since there can be no longitudinal resultant force in the beam during bending without end load,

$$A_1\sigma_Y = A_2\sigma_Y$$

or $A_1 = A_2 = \frac{1}{2}A$, where A is the total area of the cross-section. Thus for the fully plastic state the neutral axis divides the cross-section into two equal areas and the stress diagram is shown in Fig. 16.5(*b*).

If C_1 is the centroid of the area A_1, C_2 the centroid of the area A_2, and h the distance between C_1 and C_2, then the fully plastic moment is given by

$$M_p = \frac{1}{2}A\sigma_Y h = \frac{1}{2}\sigma_Y A h \tag{16.10}$$

This equation applies for any shape of cross-section.

EXAMPLE 16.1

The flange and web of the T-bar section in Fig. 16.5 are each 12 mm thick, the flange width is 100 mm, and the overall depth of the section is 100 mm. The centroid of the section is at a distance of 70.6 mm from the bottom of the web, and the second moment of area I_z, of the section about a line through the centroid and parallel to the flange is $2.03 \times 10^6 \text{ mm}^4$. Determine the value of the shape factor.

Let m be the distance of the neutral axis NN from the top of the flange (Fig. 16.5); then

$$A_1 = 100m \quad \text{and} \quad A_2 = 100(12 - m) + (88 \times 12)$$
$$100m = 100(12 - m) + (88 \times 12)$$
$$m = 11.3 \text{ mm}$$

If n is the distance of the centroid of area A_2 from the bottom of the web, then

$$(100 \times 0.7 \times 88.35) + (12 \times 88 \times 44) = [(100 \times 0.7) + (12 \times 88)]n$$
$$n = 46.8 \text{ mm}$$

Therefore h, the distance between C_1 (the centroid of A_1) and C_2 (the centroid of A_2), is

$$h = 88.7 - 46.8 + \frac{11.3}{2} = 47.55 \text{ mm}$$

so that

$$M_p = \frac{12 \times 188}{2} \sigma_Y \times 47.55 = 53\,636\sigma_Y$$
$$M_Y = \frac{2.03 \times 10^6}{70.6} \sigma_Y = 28\,754\sigma_Y$$

and the shape factor

$$\frac{M_p}{M_Y} = 1.87$$

PLASTIC COLLAPSE OF BEAMS

The fully plastic bending moment developed in the preceding section was due to the application of pure bending. A beam would therefore become fully plastic at *all* cross-sections along the whole length once M_p was reached. However, in practice pure bending rarely occurs and bending-moment distribution varies depending on the loading conditions. The point of maximum bending moment along the beam will be the first cross-section which becomes fully plastic as the load magnitude increases. Cross-sections adjacent to the fully plastic section will have commenced yielding to various depths. For a beam simply supported at each end and carrying a central concentrated load the shape of the plastic zone associated with the central fully plastic cross-section is illustrated in Fig. 16.6(a). The boundary between elastic and plastic material is parabolic in shape.

Fig. 16.6

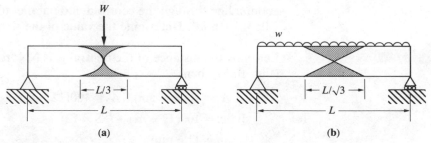

(a) (b)

The plastic zones for distributed loading are triangular in shape, as shown in Fig. 16.6(*b*). When a cross-section such as those shown reaches the fully plastic state it cannot carry any higher loading and the beam forms a hinge at that cross-section. This is termed a *plastic hinge* about which rotation of the two halves of the beam occurs, as shown in Fig. 16.7. When one or more plastic hinges occur such that the beam or structure becomes a *mechanism* then this situation is described as *plastic collapse*.

Fig. 16.7 (a) Elastic; (b) plastic

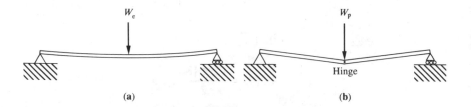

(a) (b)

In the example of Fig. 16.7 the maximum bending moment is at the centre and is $WL/4$. Therefore plastic collapse occurs for the *single* hinge formation and

$$M_p = \frac{W_p L}{4} \quad \text{or} \quad W_p = \frac{4M_p}{L} \tag{16.11}$$

The next example is a cantilever propped at the free end and carrying a concentrated load at mid-span, as shown in Fig. 16.8(*a*). The resultant bending-moment diagram may be obtained from the superposition of the two load cases namely due to the central load W and the prop load F. The resultant diagram shows two peak values of M at A and B. Now the beam does not become a mechanism and collapse until M_p is reached at *both* A and B. From the geometry of the resultant bending-moment diagram at B,

$$M_p + \tfrac{1}{2}M_p = \frac{W_p L}{4}$$

or

$$W_p = \frac{6M_p}{L} \tag{16.12}$$

Finally, we shall consider the case of a beam fixed at each end carrying a

Fig. 16.8

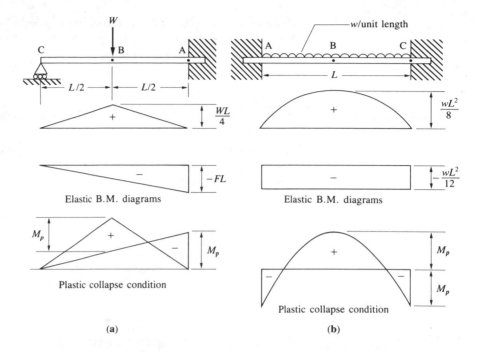

uniformly distributed load (Fig. 16.8(b)). It is clear that plastic collapse cannot occur until *three* hinges have formed at A, B and C. The resultant bending moment is again obtained by superposition and from the diagram the required condition for collapse is

$$M_p + M_p = w_p L^2/8$$

from which

$$w_p = \frac{16M_p}{L^2} \tag{16.13}$$

The simple examples above illustrate the principles of the concept which may be summarized as follows:

(a) Elastic behaviour occurs in a structural member until a plastic hinge is formed at a section.

(b) If rotation at this hinge results in diffusion of the load to other parts of the structure or supports then additional load may be carried until another plastic hinge is formed.

(c) As each hinge forms the moment remains constant at the fully plastic value irrespective of additional load or deformation.

(d) When there is no remaining stable portion able to carry additional load then collapse will occur.

(e) The structure as a whole, or in part, will form a simple mechanism at collapse.

(f) The collapse load may be calculated by statical equilibrium if the locations of the hinges can be identified.

The above principles may of course, be applied also to the investigation of the collapse of plane frameworks, which leads on to a fuller treatment of what is termed *limit analysis* of structures.

EXAMPLE 16.2

The beam illustrated in Fig. 16.9 is made of I-section mild steel having a shape factor of 1.15 and a yield stress of 240 MN/m². Using a load factor against collapse of 2 find the required section modulus.

Fig. 16.9

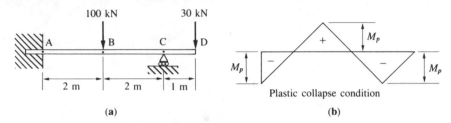

(a)   (b)

SOLUTION

The elastic bending-moment diagram is such that peaks (of different magnitude) will occur at A, B and C. Plastic hinges will form when these peaks reach the value of M_p. Collapse will occur when the three hinges are formed as shown in Fig. 16.9(b). From the geometry of the figure we have that between A and C

$$2M_p = \frac{WL}{4} = \frac{100 \times 4}{4}$$

$$M_p = 50 \text{ kN-m}$$

Also

$$\frac{M_p}{M_Y} = 1.15$$

Hence

$$M_Y = \frac{2 \times 50}{1.15} = 86.96 \text{ kN-m}$$

having incorporated the load factor of 2.

Now the elastic section modulus

$$Z = \frac{M_Y}{\sigma_Y} = \frac{86.96}{240} = 0.362 \text{ m}^3$$

For the hinge to form at C due to the load at D we shall have

$$M_p = WL = 30 \times 1 = 30 \text{ kN-m}$$

This value would require a section modulus of

$$Z = \frac{2 \times 30}{1.15 \times 240} = 0.217 \text{ m}^3$$

Therefore the larger section modulus is required.

PLASTIC TORSION OF SHAFTS: PLASTIC TORQUE

In the following discussion it will be assumed that we have an ideal stress–strain relationship for the material as shown in Fig. 16.10, that a plane cross-section of the shaft remains plane when in the plastic state, and that a radius remains straight. The shearing strain γ at a distance r from the axis of the shaft will be given by $\gamma = r\theta/L$.

Fig. 16.10

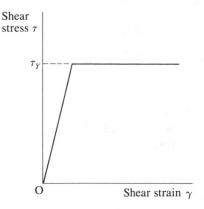

When the shaft has a torque applied in the elastic range, the shear stress increases from zero at the shaft axis to a maximum value at the surface of the shaft, and

$$T = \frac{\tau\pi}{2}r^3$$

for a solid circular shaft. When the shear stress at the surface of the shaft has reached the value τ_Y the torque required to give this stress is

$$T_Y = \frac{\tau_Y\pi}{2}r^3 \tag{16.14}$$

If the torque is increased beyond this value, then plasticity occurs in fibres at the surface of the shaft and the stress diagram is as shown in Fig. 16.11. The torque carried by the elastic core is

$$T_1 = \frac{\tau_Y\pi}{2}r_Y^3 \tag{16.15}$$

where r_Y is the interface radius between elastic and plastic material, and that carried by the plastic zone is

$$T_2 = \int_{r_Y}^{r_0} 2\pi r^2 \tau_Y \, dr = \frac{2\pi}{3}\tau_Y(r_0^3 - r_Y^3) \tag{16.16}$$

and the total torque, T, is

$$T_1 + T_2 = \tfrac{1}{2}\tau_Y\pi r_Y^3 + \tfrac{2}{3}\pi\tau_Y(r_0^3 - r_Y^3)$$

Therefore

$$T = \tfrac{2}{3}\pi r_0^3 \tau_Y\left[1 - \frac{r_Y^3}{4r_0^3}\right] \tag{16.17}$$

427

Fig. 16.11

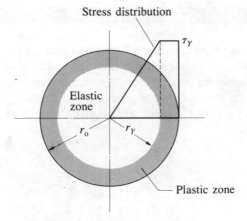

Stress distribution

Elastic zone

r_o

r_Y

τ_Y

Plastic zone

When the fully plastic condition is reached, the shear stress at all points in the cross-section is τ_Y, and it follows from eqn. (16.16) that the fully plastic torque is given by

$$T_p = \frac{2\pi}{3}\,\tau_Y r_0^{\,3} \tag{16.18}$$

and the ratio T_p/T_Y is

$$\frac{T_p}{T_Y} = \frac{4}{3}$$

When the fibres at the outer surface of the shaft are about to become plastic, the angle of twist is given by

$$\theta_Y = \frac{\tau_Y L}{Gr_0} \tag{16.19}$$

and when the shaft is in the elastic–plastic condition, the angle of twist of the elastic core is given by

$$\theta = \frac{\tau_Y L}{Gr_Y} \tag{16.20}$$

Since we have assumed that radii remain straight, then the outer plastic region has the same angle of twist. From eqns. (16.19) and (16.20) it follows that

$$\frac{\theta_Y}{\theta} = \frac{r_Y}{r_0} \tag{16.21}$$

It is evident that as the shaft approaches the fully plastic state, the angle of twist tends to infinity.

Equation (16.17) may be expressed in the form

$$T = \tfrac{2}{3}\pi r_0^{\,3}\tau_Y\left[1 - \frac{1}{4}\left(\frac{\theta_Y}{\theta}\right)^3\right] \tag{16.22}$$

EXAMPLE 16.3

A mild steel shear coupling in a metal-working process is 40 mm in diameter and 250 mm in length. It is subjected to an overload torque of 1800 N-m which is known to have caused shear yielding in the shaft.

Determine the radial depth to which plasticity has penetrated and the angle of twist. $\tau_Y = 120 \, \text{MN/m}^2$, $G = 80 \, \text{GN/m}^2$.

SOLUTION

Using eqn. (16.17) for the elastic–plastic torque

$$1800 = \tfrac{2}{3}\pi \times 0.02^3 \times 120 \times 10^6 \left[1 - \frac{r_Y^3}{4 \times 0.02^3} \right]$$

from which $r_Y = 15$ mm.

Hence depth of plastic deformation is 5 mm.

The shear strain at $r_Y = 15$ mm is

$$\gamma = \frac{120 \times 10^6}{80 \times 10^9} = 0.0015$$

but

$$\gamma L = r_Y \theta$$

Therefore

$$0.0015 \times 0.25 = 0.015\theta$$

and

$$\theta = 0.025 \, \text{rad} = 1.43°$$

PLASTICITY IN A PRESSURIZED THICK-WALLED CYLINDER

The problems analysed in the previous sections only involved a uniaxial stress condition, and hence a simple tensile or shear yield stress was sufficient to define the onset of plastic deformation. In two- or three-dimensional stress systems it is necessary to use a yield criterion of the type discussed in Chapter 13 in order to determine the initiation of plastic flow. A thick-walled cylindrical pressure vessel is a good example of this type of situation in that there will be radial and hoop stresses and generally also axial stress. The process of inducing plastic deformation partly through the wall thickness is known as *autofrettage* and its purpose was described in Chapter 15.

The following analysis will only consider an ideal elastic–plastic material, since the problem for a strain-hardening material is beyond the scope of this text. Furthermore, in order to simplify the mathematics the maximum shear stress (Tresca) theory of yielding will be adopted.

For a thick cylinder under internal pressure, the maximum shear stress occurs at the inner surface (*see* Fig. 15.4) and therefore as the pressure is increased, plastic deformation will commence first at the bore and penetrate deeper and deeper into the wall, until the whole vessel reaches the yield condition.

At a stage when plasticity has penetrated partly through the wall, the vessel might be regarded as a compound cylinder with the inner tube plastic and the outer elastic. If the elastic–plastic interface is at a radius a and the radial pressure there is p_a, then from eqns. (15.36) and (15.37),

$$\sigma_r = \frac{p_a a^2}{r_0^2 - a^2}\left(1 - \frac{r_0^2}{a^2}\right) \tag{16.23}$$

$$\sigma_\theta = \frac{p_a a^2}{r_0^2 - a^2}\left(1 + \frac{r_0^2}{a^2}\right) \tag{16.24}$$

and

$$\tau_{max} = \frac{\sigma_\theta - \sigma_r}{2} = \frac{p_a r_0^2}{r_0^2 - a^2} \tag{16.25}$$

But at the interface, yielding has just been reached; therefore

$$\tau_{max} = \frac{\sigma_Y}{2}$$

and

$$\frac{p_a r_0^2}{r_0^2 - a^2} = \frac{\sigma_Y}{2}$$

Therefore

$$p_a = \frac{\sigma_Y}{2r_0^2}(r_0^2 - a^2) \tag{16.26}$$

From this value of p_a the stress conditions in the elastic zone can be determined using eqns. (15.36) and (15.37). It is now necessary to consider the equilibrium of the plastic zone in order to find the internal pressure required to cause plastic deformation to a depth of $r = a$. The equilibrium equation is (15.12).

$$r\frac{d\sigma_r}{dr} + \sigma_r - \sigma_\theta = 0 \tag{16.27}$$

Hence

$$r\frac{d\sigma_r}{dr} - 2\tau_{max} = 0$$

or

$$\frac{d\sigma_r}{dr} = \frac{\sigma_Y}{r}$$

Integrating this equation gives

$$\sigma_r = \sigma_Y \log_e r + C \tag{16.28}$$

Now, at $r = a$, $\sigma_r = -p_a$; therefore

$$-p_a = \sigma_Y \log_e a + C$$

or

$$C = -p_a - \sigma_Y \log_e a$$

Substituting for C in eqn. (16.28),

$$\sigma_r = \sigma_Y \log_e r - p_a - \sigma_Y \log_e a \qquad (16.29)$$

$$= -\sigma_Y \log_e \frac{a}{r} - p_a \qquad (16.30)$$

Therefore, using eqn. (16.26),

$$\sigma_r = -\sigma_Y \log_e \frac{a}{r} - \frac{\sigma_Y}{2r_0^2}(r_0^2 - a^2) \qquad (16.31)$$

which gives the distribution of radial stress in the plastic zone; and at

$$r = r_i, \qquad \sigma_r = -p_i$$

therefore

$$p_i = +\sigma_Y \log_e \frac{a}{r_i} + \frac{\sigma_Y}{2r_0^2}(r_0^2 - a^2) \qquad (16.32)$$

where p_i is the internal pressure to cause yielding to a depth of $r = a$. The hoop stress is

$$\sigma_\theta = \sigma_Y + \sigma_r$$

Therefore

$$\sigma_\theta = \sigma_Y\left(1 - \log_e \frac{a}{r}\right) - \frac{\sigma_Y}{2r_0^2}(r_0^2 - a^2) \qquad (16.33)$$

The internal pressure, p_{max}, required to cause yielding right through the wall is found by putting $a = r_0$ in eqn. (16.32):

$$p_{max} = +\sigma_Y \log_e \frac{r_0}{r_i} \qquad (16.34)$$

and from eqns. (16.31) and (16.33)

$$\sigma_r = -\sigma_Y \log_e \frac{r_0}{r} \qquad (16.35)$$

$$\sigma_\theta = \sigma_Y\left(1 - \log_e \frac{r_0}{r}\right) \qquad (16.36)$$

The stress distribution of σ_r and σ_θ for the cases considered above are illustrated in Fig. 16.12 in terms of σ_Y for a vessel where $r_0 = 2r_i$.

Fig. 16.12

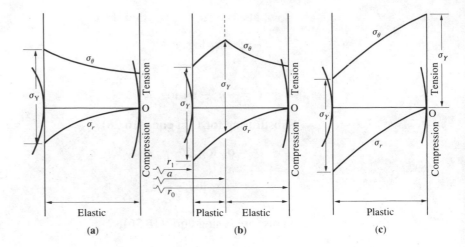

Elastic   Plastic   Elastic   Plastic

(a)   (b)   (c)

EXAMPLE 16.4

A thick-walled cylindrical pressure vessel has to be autofrettaged prior to its use for a high-pressure chemical process. The radius ratio is 2.5 and 20% of the wall thickness has to be brought into the plastic range of the alloy steel which has a yield stress of 400 MN/m². Calculate the internal pressure required to achieve the specified plastic deformation. What are the values of hoop stress at the bore, elastic–plastic interface and the outer surface caused by that internal pressure? Determine the reserve factor against yielding right through the wall thickness.

SOLUTION

To determine a, the radius of the elastic–plastic interface, in terms of r_i we know that 20% of the wall thickness is in the plastic range so that

$$a = r_i + 0.2(r_0 - r_i) = 1.3r_i$$

From eqn. (16.32),

$$p_i = 400 \log_e 1.3 + \frac{400}{2}\left[1 - \left(\frac{1.3r_i}{2.5r_i}\right)^2\right]$$

$$= 104.8 + 146 = 250.8 \text{ MN/m}^2$$

At the bore

$$\sigma_r = -250.8 \text{ MN/m}^2$$

and

$$\sigma_\theta = 400 - 250.8 = 149.2 \text{ MN/m}^2$$

At the interface from eqn. (16.31) putting $r = a$,

$$\sigma_r = -\frac{400}{2}\left[1 - \left(\frac{1.3r_i}{2.5r_i}\right)^2\right] = -146 \text{ MN/m}^2$$

and

$$\sigma_\theta = 400 - 146 = 254 \text{ MN/m}^2$$

The value of σ_r at the interface is also the value of p_a in eqn. (16.26). Therefore at $r = r_0$

$$\sigma_r = 0$$

and from eqn. (16.24)

$$\sigma_\theta = \frac{146 \times 2}{2.69} = 108.5 \text{ MN/m}^2$$

The shape of the stress distribution is similar to that in eqn. 16.15(b). For yielding right through the wall we use eqn. (16.34)

$$p_{max} = 400 \log_e 2.5 = 366.4 \text{ MN/m}^2$$

The reserve factor on pressure is

$$\frac{366.4}{250.8} = 1.46$$

PLASTICITY IN A ROTATING THIN DISC

In a rotating disc with a central hole the maximum hoop stress occurs on the surface of the hole, and both hoop and radial stresses are positive throughout the disc, the former being the greater of the two. Therefore maximum shear stress is given by

$$\tau_{max} = \frac{\sigma_\theta - 0}{2} \tag{16.37}$$

since $\sigma_z = 0$.

Yielding occurs first at the hole and gradually spreads outwards with increasing rotational speed. Using the maximum shear stress criterion, the speed ω_Y at which yielding first commences is given by

$$\tau_{max(r=r_1)} = \frac{\sigma_\theta}{2} = \frac{\sigma_Y}{2} \tag{16.38}$$

From eqn. (15.64) putting $r = r_1$,

$$\sigma_\theta = \frac{\rho \omega_Y^2}{4} [r_2^2(3 + v) + r_1^2(1 - v)] = \sigma_Y \tag{16.39}$$

Therefore initial yielding commences at a speed of

$$\omega_Y = \left\{ \frac{4\sigma_Y}{\rho[r_2^2(3 + v) + r_1^2(1 - v)]} \right\}^{1/2} \tag{16.40}$$

Next we determine the rotational speed ω at which there is a plastic zone from $r = r_1$ to $r = c$. The equilibrium equation is

$$r\frac{d\sigma_r}{dr} + \sigma_r - \sigma_\theta + \rho\omega^2 r^2 = 0$$

and, since $\sigma_\theta = \sigma_Y$ in the plastic zone,

$$r\frac{d\sigma_r}{dr} + \sigma_r = \sigma_Y - \rho\omega^2 r^2 \qquad (16.41)$$

Integrating this equation,

$$r\sigma_r = r\sigma_Y - \frac{\rho}{3}\omega^2 r^3 + K$$

Therefore

$$\sigma_r = \sigma_Y - \frac{\rho}{3}\omega^2 r^2 + \frac{K}{r} \qquad (16.42)$$

At $\quad r = r_1,\ \sigma_r = 0$

therefore

$$0 = \sigma_Y - \frac{\rho}{3}\omega^2 r_1^2 + \frac{K}{r_1}$$

so that

$$K = -r_1\sigma_Y + \frac{\rho}{3}\omega^2 r_1^3$$

Substituting for K in eqn. (16.42)

$$\sigma_r = \sigma_Y - \frac{\rho}{3}\omega^2 r^2 - \frac{r_1}{r}\sigma_Y + \frac{\rho}{3}\omega^2\frac{r_1^3}{r} \qquad (16.43)$$

At $r = c$, eqn. (16.43) becomes

$$\sigma_r = \frac{\sigma_Y}{c}(c - r_1) - \frac{\rho\,\omega^2}{3\,c}(c^3 - r_1^3) \qquad (16.44)$$

But this value of σ_r must be the same as σ_r at $r = c$ for the elastic zone. It is firstly necessary to determine the constants A and B in eqns. (15.59) and (15.60). The boundary conditions are: at $r = r_2$, $\sigma_r = 0$; and at $r = c$, $\sigma_\theta = \sigma_Y$. Therefore

$$0 = A - \frac{B}{r_2^2} - \frac{3 + v}{8}\rho\omega^2 r_2^2$$

and

$$\sigma_Y = A + \frac{B}{c^2} - \frac{1 + 3v}{8}\rho\omega^2 c^2$$

from which

$$A = \frac{c^2\sigma_Y + \dfrac{\rho}{8}\omega^2\{(1+3v)c^4 + (3+v)r_2^4\}}{c^2 + r_2^2}$$

$$B = \frac{c^2r_2^2\sigma_Y + \dfrac{\rho}{8}\omega^2 c^2 r_2^2\{(1+3v)c^2 - (3+v)r_2^2\}}{c^2 + r_2^2}$$

Therefore, for the elastic zone at $r = c$,

$$\sigma_r = \frac{c^2\sigma_Y + \dfrac{\rho}{8}\omega^2\{(1+3v)c^4 + (3+v)r_2^4\}}{c^2 + r_2^2}$$

$$- \frac{r_2^2\sigma_Y + \dfrac{\rho}{8}\omega^2 r_2^2\{(1+3v)c^2 - (3+v)r_2^2\}}{c^2 + r_2^2} - \left(\frac{3+v}{8}\right)\rho\omega^2 c^2 \tag{16.45}$$

Equating eqns. (16.44) and (16.45) and simplifying,

$$\omega = \left[\frac{12\sigma_Y\{2cr_2^2 - r_1(c^2 + r_2^2)\}}{\rho\{3(3+v)cr_2^4 - (1+3v)(2r_2^2 - c^2)c^3 - 4r_1^3(r_2^2 + c^2)\}}\right]^{1/2} \tag{16.46}$$

where ω is the angular speed to cause plasticity to a radial depth of $r = c$.

If ω_p is the speed at which the disc becomes fully plastic, then substituting $c = r_2$ in eqn. (16.46) and simplifying

$$\omega_p = \left\{\frac{3\sigma_Y}{\rho(r_2^2 + r_1 r_2 + r_1^2)}\right\}^{1/2} \tag{16.47}$$

The stress distributions of σ_r and σ_θ for various degrees of plastic deformation are shown in Fig. 16.13 for $r_2 = 10r_1$. A more complete analysis of this problem has been made by Heyman (*Proc. Instn. Mech. Engrs*, **172**, 1958).

Fig. 16.13

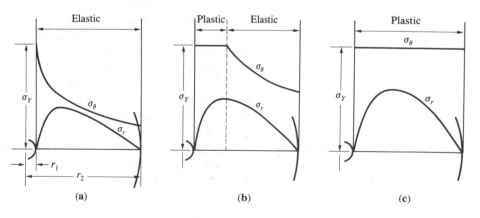

(a)   (b)   (c)

RESIDUAL STRESS DISTRIBUTIONS AFTER PLASTIC DEFORMATION

BENDING OR TORSION

When a beam is bent or a shaft twisted beyond the elastic limit permanent deformation occurs which does not disappear when the load is removed. Those regions which have suffered permanent deformation prevent those which are elastically strained from recovering their initial dimension when the load is removed. Consequently the interaction produces what are termed *residual stresses*.

In a beam of rectangular cross-section, we will assume for simplicity that the beam has been bent to the fully plastic condition such that the stress distribution diagrams are the two rectangles Ocdb and Oeka, shown in Fig. 16.14(a). Assuming that when the material is stretched beyond the yield point and then unloaded it will be linear elastic during unloading, then the bending stresses which are superposed while the beam is being unloaded follow the linear law represented by the line a_1b_1. The shaded areas represent the stresses which remain after the beam is unloaded and are thus the residual stresses produced in the beam by plastic deformation.

Fig. 16.14

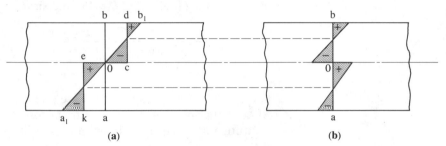

(a)   (b)

The rectangular and triangular stress distributions represent bending moments of the same magnitude; hence the moment of the rectangle Ocdb about the axis eOc is equal to the moment of the triangle Obb_1 about the same axis. Since bd represents σ_Y it follows for equal moments about the axis eOc that the stress represented by bb_1 is equal to $1\frac{1}{2}\sigma_Y$, and thus the maximum residual tension and compression stresses after loading and unloading are db_1, and $a_1k = \frac{1}{2}\sigma_Y$. Near the neutral axis the residual stresses are equal to σ_Y.

In Fig. 16.14(b) the residual stresses are replotted on a conventional base of a cross-section, in order to show these residual tensile and compressive stresses more clearly. Diagrams showing the residual stresses in a partially yielded beam are shown in Fig. 16.15. In this case the material has yielded to such an extent that the stress distribution diagrams are represented by the areas Ocdb and Oeka in Fig. 16.15(a). Again assuming the material to follow Hooke's law during unloading, the bending stress during this operation will follow the linear law represented by a_1b_1, such that the moments of Ocdb and Obb_1 about the neutral axis are equal, and the shaded areas represent the residual stresses. These stresses are shown replotted on a conventional cross-section in Fig. 16.15(b).

Fig. 16.15

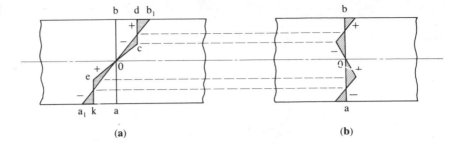

(a)   (b)

The solution for residual stresses after plastic torsion follows the same arguments as above (*see* Ex. 16.6, p. 439).

AXIALLY SYMMETRICAL COMPONENTS

If plastic deformation is caused in one part of a body and not in the remainder, then, on removal of external load, there still exists a stress system in the body, owing to the strain gradient and hence interaction between parts of the body not being able to return to the unstrained state. Residual stresses have important implications in engineering practice.

In the pressurized thick-walled cylinder (pp. 429–32) it was assumed that plastic deformation had partly penetrated the wall. On release of pressure the elastic outer zone tries to return to its original dimensions, but is partly prevented by the permanent deformation of the inner plastic material. Hence the latter is put into hoop compression and the former is in hoop tension, such that equilibrium exists. The residual stress distributions are then as shown diagrammatically in Fig. 16.16. They may be calculated by using eqns. (15.36) and (15.37) to determine the elastic unloading stresses from the internal pressure p_i (eqn. (16.32)) and superposing these on to the loaded stress distribution. The residual compressive hoop stress at the bore has to be nullified first on repressurizing, thus allowing a greater elastic range of hoop stress and therefore a greater internal pressure.

When a rotating disc is stopped after partial plasticity has occurred, a similar condition of residual stresses is obtained as for the thick cylinder above. Compressive hoop stress is obtained at the central hole, which may be calculated using the elastic stress equations and the appropriate rotational speed, followed by superposition on to the loaded stress

Fig. 16.16

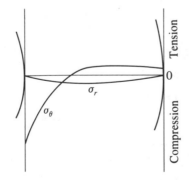

Fig. 16.17

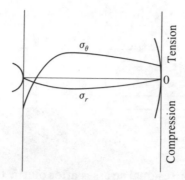

pattern. The above action, known as *overspeeding,* helps to increase the elastic stress range, and hence the speed limit available under working conditions. The residual stress distributions for this case are shown in Fig. 16.17 for the same conditions as in Fig. 16.13(b).

EXAMPLE 16.5

After the autofrettage process is completed in Example 16.4 what are the residual stresses at the bore, interface and outer surface of the vessel assuming elastic unloading?

SOLUTION

The elastic unloading stress *range* may be found from the elastic equations for σ_θ and σ_r using $p_i = 250.8$ MN/m^2.

At $r = r_i$,

$$\sigma_\theta = \frac{250.8}{k^2 - 1}(1 + k^2)$$

$$= 250.8 \times \frac{7.25}{5.25} = 346.3 \text{ MN/m}^2$$

Therefore the residual stress is

$$\sigma_\theta{}^R = +149.2 - 346.3$$
$$= -197.1 \text{ MN/m}^2$$

At $r = a$,

$$\sigma_\theta = \frac{250.8}{5.25}(1 + 3.69) = 224 \text{ MN/m}^2$$

The residual stress

$$\sigma_\theta{}^R = +254 - 224$$
$$= +30 \text{ MN/m}^2$$

At $r = r_0$, $\sigma_\theta = 95.5$ MN/m^2, and the residual stress

$$\sigma_\theta{}^R = +108.5 - 95.5$$
$$= +13 \text{ MN/m}^2$$

At $r = r_i$ and $r = r_0$ the residual radial stresses are zero and at $r = a$ the radial stress due to unloading elastically from $p_i = 250.8 \, \text{MN/m}^2$ is $\sigma_r = -128.5 \, \text{MN/m}^2$. Therefore the residual stress is

$$\sigma_r^R = -146 - (-128.5) = -17.5 \, \text{MN/m}^2$$

The shape of the residual stress distribution is similar to that of Fig. 16.16.

EXAMPLE 16.6

Determine the residual shear stress distribution after elastic unloading for the plastically-deformed shear coupling in Example 16.3. What is the permanent twist?

SOLUTION

The elastic unloading torque of 1800 N-m gives rise to a shear stress range at the outer surface of

$$\tau = \frac{16 \times 1800}{\pi \times 0.04^3} = 143 \, \text{MN/m}^2$$

The residual shear stress at the outer surface of the coupling is the value of 143 MN/m² above less the yield stress value of 120 MN/m² which gives 23 MN/m².

The residual shear stress is zero at $r = 0$ and where the unloading stress range equals 120 MN/m², therefore from Fig. 16.18(a)

$$\frac{120}{143} = \frac{r}{0.02}$$

$$r = 16.8 \, \text{mm}$$

Fig. 16.18

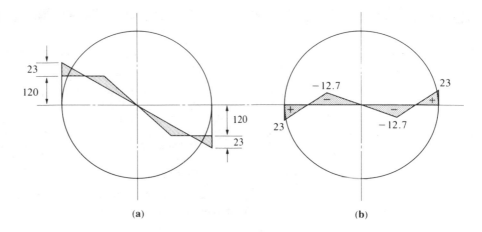

(a)   (b)

The other peak occurs at $r = 15$ mm so that

$$\tau_{15} = 143 \times \frac{0.015}{0.02}$$

$$= 107.3 \text{ MN/m}^2$$

and the *residual* shear stress is $120 - 107.3 = 12.7 \text{ MN/m}^2$.

The complete residual shear stress diagram is shown in Fig. 16.18(*b*).

The elastic unloading twist is given by

$$\theta = \frac{1800 \times 32}{\pi \times 0.4^4} \times \frac{0.25}{80 \times 10^9} = 0.0224 \text{ rad}$$

Permanent angle of twist $= 0.025 - 0.0224$
$$= 0.0026 \text{ rad}.$$

SUMMARY

The occurrence of plastic deformation in engineering components by design or accident is not common, but nevertheless an understanding of the behaviour is important. Since equilibrium and strain-displacement equations are independent of the type or state of the material, the principal difference from elastic analysis is the form of the stress–strain relationship. For convenience this is often assumed to be elastic–ideally plastic, i.e. non-strain-hardening, although a linear plastic strain-hardening relationship is a straightforward extension of the former.

Plastic bending and torsion represent elementary illustrations of the principles of analysis. Plasticity in pressurized cylinders and rotating discs can be seen to have important practical implications for performance through the influence of residual stresses and strains. The introduction of this latter concept is an opportunity to emphasize the role, both advantageous and detrimental, played by residual stresses in engineering components.

PROBLEMS

16.1 Determine the ratio of the fully plastic to the maximum elastic moment for a beam of elastic–perfectly plastic material subjected to pure bending for the following shapes of cross-section: (*a*) solid circular; (*b*) solid square about a diagonal axis; (c) thin-walled circular tube; (*d*) thin-walled square tube about a centroidal axis parallel to one of the sides.

16.2 A steel beam of I-section, as shown in Fig. 16.19, and of length 5 m is simply-supported at each end and carries a uniformly-distributed load of 114 kN/m over the full span. Steel reinforcing plates 12 mm thick are welded to each flange and are made of elastic–ideally plastic material. Calculate the plate width such that yielding has just spread through each reinforcing plate at mid-span under the given load. Determine the positions along the reinforcing plates at which the outer surfaces have just reached the yield point. Yield stress $= 300 \text{ MN/m}^2$, second moment of area $= 80 \times 10^{-6} \text{ m}^4$.

Fig. 16.19

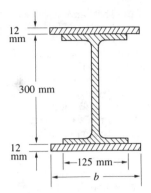

16.3 The flange and web of a T-section are each 12 mm thick, the flange width is 100 mm and the overall depth is 100 mm. The beam is simply-supported over a length of 2 m and it is subjected to a point load W at mid-span. Calculate the maximum value of W if the beam is to be designed such that yielding is permitted to penetrate the web to a depth of 20 mm. The yield strength of the beam material is 300 MN/m².

16.4 A short column of 0.05 m square cross-section is subjected to a compressive load of 0.5 MN parallel to but eccentric from the central axis. The column is made from elastic-perfectly plastic material which has a yield stress in tension or compression of 300 MN/m². Determine the value of the eccentricity which will result in the cross-section becoming just fully plastic.

16.5 Prove that for a beam simply-supported at each end and carrying a concentrated load at mid-span, which has developed full plasticity, the shape of the boundary between elastic and plastic material is parabolic.

16.6 A horizontal cantilever of length L is simply-supported at the same level at the free end and is subjected to a uniformly-distributed load w over the full span. Determine the location and magnitude of the collapse load.

16.7 Part of a small bridge deck is represented as shown in Fig. 16.20. What will be the mode of collapse and the value of the collapse load?

Fig. 16.20

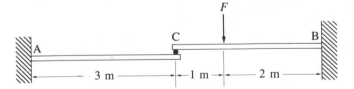

16.8 A steel shaft 100 mm in diameter and 1 m long is in an elastic–plastic state under a torque of 30 kN-m. Determine the diameter of the elastic core of the shaft and the angle of twist. What is the value of the fully plastic torque for the shaft?
$\tau_Y = 120$ MN/m², $G = 80$ GN/m².

16.9 A solid circular shaft is subjected to pure torsion and the material is elastic–perfectly plastic with a yield stress in shear of 152 MN/m². When the shear stress at one-third of the radius from the centre of the shaft reaches the yield stress, determine the shear strain on the outer surface. Also find the ratio of the torque carried in the above conditions to the maximum elastic torque for the shaft.
$G = 83$ GN/m².

16.10 A solid cylindrical composite shaft 1 m long consists of a copper core of 50 mm diameter surrounded by a well-fitting steel sleeve having an external diameter of 62 mm. If the steel has an elastic–perfectly plastic stress–strain relationship, determine the torque that can be applied to the shaft to cause yielding to develop just through to the inner surface of the sleeve. Neglect any stresses set up by the fit between the two parts of the shaft and assume there is no slipping at the copper–steel interface. G (steel) = 83 GN/m², G (copper) = 45 GN/m². τ_Y (steel) = 124 MN/m², τ_Y (copper) = 76 MN/m².

16.11 A thick-walled cylinder of radius ratio 2:1 and made of an elastic–perfectly plastic material is subjected to internal pressure. Plot a diagram of (p_a/p_{max}) against a, where p_a is the internal pressure to cause yielding to a depth a through the wall, and p_{max} is the pressure which results in yielding right through the wall.

16.12 A metal disc of uniform thickness is 300 mm diameter and has a central hole of 50 mm diameter. Determine the increase in rotational speed over that for initial yielding at the hole necessary to cause plastic deformation throughout the disc. $v = 0.3$.

16.13 Calculate the residual stress at the outer surfaces of the column in Problem 16.4 after elastic unloading from the fully plastic condition.

16.14 A rectangular beam 30 mm wide and 50 mm deep is simply-supported over a length of 2 m. If it is subjected to a uniformly distributed load of 8 kN/m calculate the depth of penetration of plastically deformed material in the beam. If the load is removed calculate the force necessary at mid-span to straighten the beam. The yield stress of the beam material is 240 MN/m².

16.15 If the composite shaft in Problem 16.10. has the torque removed, calculate the residual shear stress at the outer surface of the steel sleeve and plot the residual stress distribution.

16.16 A thin circular disc with a central hole has inner and outer diameters of 51 and 304 mm respectively. It is required to have a residual compressive hoop stress of 77 MN/m² at the hole when the disc is stationary. Assuming ideal elastic–plastic conditions, yield according to the maximum shear stress theory and elastic unloading, determine the rotational speed necessary to effect the required residual stress. By how much is this speed greater than the speed for initial yielding? Also find the depth of the plastic zone. $\sigma_Y = 340$ MN/m², $\rho = 7.83$ Mg/m³, $v = 0.3$.

Chapter 17

Thin plates and shells

INTRODUCTION

The purpose of plates in engineering is to cover, generally, a rectangular or circular area and to support concentrated or distributed loading normal to the plane of the plate. A typical example is a pressure diaphragm, as a safety or control device, supported around its circular periphery and subjected to uniform pressure on one face and perhaps a central point load on the opposite face.

Some simple examples of thin shells were studied in Chapter 2 to illustrate statically-determinate problems.

The engineering applications of thin shells include storage tanks for liquids or solids and pressure vessels for a variety of chemical processes, rocket motor casings, boiler drums, etc.

"Thin" is a relative term which indicates that the thickness of the material is small compared with the overall geometry, a ratio of 10:1 or greater being the usual criterion.

Solutions for "thick" plates and shells are more complex and may be found in specialized texts.

A further factor which affects the nature of the analysis is the range of deformation, again in qualitative terms referred to as "small", or "large". This chapter will only consider *small* elastic deformations of plates and shells.

ASSUMPTIONS FOR SMALL DEFLECTION OF THIN PLATES

(a) No deformation in the middle plane of the plate, i.e. a neutral surface.
(b) Points in the plate lying initially on a normal to the middle plane of the plate remain on the normal during bending.
(c) Normal stresses in the direction transverse to the plate can be disregarded.

Assumption (*a*) does not of course hold if there are external forces acting in the middle plane of the plate. Assumption (*b*) disregards the effect of shear force on deflection.

The deflection, *w*, is a function of the two co-ordinates in the plane of the plate, the elastic constants of the material, and the loading.

RELATIONSHIPS BETWEEN MOMENTS AND CURVATURES FOR PURE BENDING

GENERAL CASE IN RECTANGULAR COORDINATES

The first important step in the analysis of plates is similar to that for beams and is to relate the bending moments to curvature from which slope and deflection are determined.

Fig. 17.1

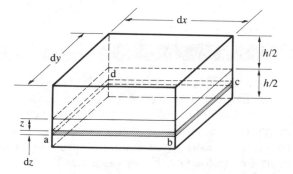

Consider an element of material as shown in Fig. 17.1 cut from a plate subjected to pure bending as in Fig. 17.2. The bending moments M_x and M_y *per unit length* are positive as drawn acting on the middle of the plate. This plane is undeformed and constitutes the *neutral surface*. The material above it is in a state of biaxial compression, and below it, in biaxial tension. The curvatures of the mid-plane in sections parallel to the *xz*- and *yz*-planes are denoted by $1/R_x$ and $1/R_y$ respectively. At a depth *z* below the neutral surface the strains in the *x*- and *y*-directions of a lamina such as *abcd* are

$$\varepsilon_x = \frac{z}{R_x} \quad \text{and} \quad \varepsilon_y = \frac{z}{R_y} \tag{17.1}$$

using the same approach as for beams in Chapter 6.

Fig. 17.2

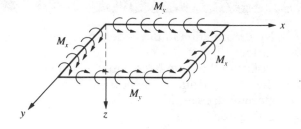

The stress–strain relationships are

$$\varepsilon_x = \frac{\sigma_x}{E} - \frac{v\sigma_y}{E} \quad \text{and} \quad \varepsilon_y = \frac{\sigma_y}{E} - \frac{v\sigma_x}{E}$$

Combining with eqn. (17.1) and rearranging gives

$$\sigma_x = \frac{Ez}{1-v^2}\left(\frac{1}{R_x} + \frac{v}{R_y}\right)$$

$$\sigma_y = \frac{Ez}{1-v^2}\left(\frac{1}{R_y} + \frac{v}{R_x}\right)$$

(17.2)

Equations (17.2) show that bending stresses are a function of plate curvatures and are proportional to distance from the neutral surface.

Next, we equate the required equilibrium between the internal moments, due to the bending stresses acting on the sides of the element, and the applied moments M_x and M_y per unit length.

$$\left. \begin{array}{l} \displaystyle\int_{-h/2}^{h/2} \sigma_x z \; \mathrm{d}y \; \mathrm{d}z = M_x \; \mathrm{d}y \\[3mm] \displaystyle\int_{-h/2}^{h/2} \sigma_y z \; \mathrm{d}x \; \mathrm{d}z = M_y \; \mathrm{d}x \end{array} \right\}$$

(17.3)

Substituting from eqns. (17.2) for σ_x and σ_y in eqns. (17.3) and integrating gives

$$\left. \begin{array}{l} \displaystyle M_x = \frac{Eh^3}{12(1-v^2)}\left(\frac{1}{R_x} + \frac{v}{R_y}\right) = D\left(\frac{1}{R_x} + \frac{v}{R_y}\right) \\[3mm] \displaystyle M_y = \frac{Eh^3}{12(1-v^2)}\left(\frac{1}{R_y} + \frac{v}{R_x}\right) = D\left(\frac{1}{R_y} + \frac{v}{R_x}\right) \end{array} \right\}$$

(17.4)

where

$$D = \frac{Eh^3}{12(1-v^2)} \text{ is termed the } \textit{flexural rigidity}.$$

Since the principal curvatures are given by

$$\frac{1}{R_x} = -\frac{\partial^2 w}{\partial x^2} \quad \text{and} \quad \frac{1}{R_y} = -\frac{\partial^2 w}{\partial y^2}$$

where w is the deflection in the z-direction, the relationships between applied moments and curvatures are

$$\left. \begin{array}{l} \displaystyle M_x = -D\left(\frac{\partial^2 w}{\partial x^2} + v\frac{\partial^2 w}{\partial y^2}\right) \\[3mm] \displaystyle M_y = -D\left(\frac{\partial^2 w}{\partial y^2} + v\frac{\partial^2 w}{\partial x^2}\right) \end{array} \right\}$$

(17.5)

**SYMMETRICAL
BENDING OF
CIRCULAR PLATES
IN CYLINDRICAL
CO-ORDINATES**

Fig. 17.3

When the loading on the surface of a circular plate is symmetrical about a perpendicular central axis, the deflection surface is also symmetrical about that axis. Any diametral section may be used to indicate the deflection curve and the associated slope ψ and deflection w at any radius r, as shown in Fig. 17.3.

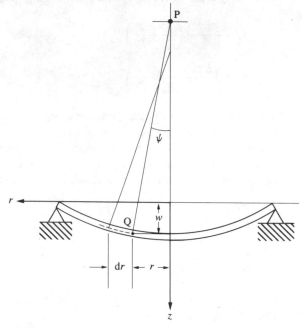

The curvature of the plate in the diametral plane rz is

$$\frac{1}{R_r} = -\frac{d^2w}{dr^2} \tag{17.6}$$

and for small values of w (positive downwards) the slope at any point is

$$\psi = -\frac{dw}{dr}$$

The second principal radius of curvature R_θ is in the plane perpendicular to rz and is represented by lines such as PQ which form a conical surface so that

$$\frac{1}{R_\theta} = \frac{\psi}{r} \tag{17.7}$$

Now we shall consider an element of the plate subjected to bending moments along the edges M_r per unit length and M_θ per unit length respectively, as shown in Fig. 17.4. This element can be analysed in the

Fig. 17.4

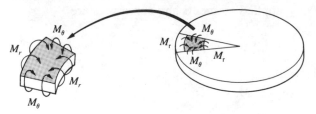

same manner as on page 445 for rectangular coordinates. Thus eqns. (17.4) can be expressed for the circular plate as follows in terms of slope ψ,

$$
\left.
\begin{aligned}
M_r &= \frac{Eh^3}{12(1-v^2)}\left(\frac{\mathrm{d}\psi}{\mathrm{d}r} + v\frac{\psi}{r}\right)\\[2mm]
M_\theta &= \frac{Eh^3}{12(1-v^2)}\left(\frac{\psi}{r} + v\frac{\mathrm{d}\psi}{\mathrm{d}r}\right)
\end{aligned}
\right\}
\tag{17.8}
$$

These equations can then be interpreted in terms of curvatures or deflection by further substitution to give

$$
\left.
\begin{aligned}
M_r &= -D\left(\frac{\mathrm{d}^2 w}{\mathrm{d}r^2} + \frac{v}{r}\frac{\mathrm{d}w}{\mathrm{d}r}\right)\\[2mm]
M_\theta &= -D\left(\frac{1}{r}\frac{\mathrm{d}w}{\mathrm{d}r} + v\frac{\mathrm{d}^2 w}{\mathrm{d}r^2}\right)
\end{aligned}
\right\}
\tag{17.9}
$$

RELATIONSHIPS BETWEEN BENDING MOMENT AND BENDING STRESS

We can determine bending stress as a function of bending moment by eliminating the curvatures between eqns. (17.2) and (17.4) so that

$$
\sigma_x = \frac{12M_x z}{h^3} \quad \text{and} \quad \sigma_y = \frac{12M_y z}{h^3}
\tag{17.10}
$$

Similar expressions apply for the bending stresses in circular plates as a function of M_r and M_θ.

RELATIONSHIPS BETWEEN LOAD, SHEAR FORCE AND BENDING MOMENT

If an element, Fig. 17.5, is taken from the plate then it must be in equilibrium under the action of uniformly distributed loading p per unit area and the resulting shear forces Q per unit length and bending moments M per unit length. Owing to symmetry there are no shear forces

Fig. 17.5

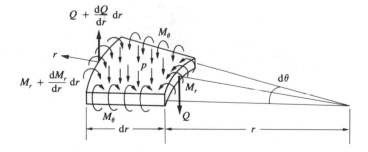

on the radial sides of the element. For vertical equilibrium,

$$Qr \, d\theta + pr \, dr \, d\theta - \left(Q + \frac{dQ}{dr} dr\right)(r + dr) \, d\theta = 0$$

from which

$$\frac{dQ}{dr} + \frac{Q}{r} = p \tag{17.11}$$

For moment equilibrium,

$$\left(M_r + \frac{dM_r}{dr} dr\right)(r + dr) \, d\theta - M_r r \, d\theta - 2M_\theta \, dr \sin\frac{d\theta}{2} + Qr \, d\theta \, dr = 0$$

which may be simplified to

$$r\frac{dM_r}{dr} + M_r - M_\theta + Qr = 0 \tag{17.12}$$

RELATIONSHIPS BETWEEN DEFLECTION, SLOPE AND LOADING

Next we substitute eqns. (17.8) into eqn. (17.12) and after simplifying obtain

$$\frac{d^2\psi}{dr^2} + \frac{1}{r}\frac{d\psi}{dr} - \frac{\psi}{r^2} = -\frac{Q}{D} \tag{17.13}$$

which relates the slope at any radius to the shear force.

If eqns. (17.9) are substituted into eqn. (17.12) then

$$\frac{d^3w}{dr^3} + \frac{1}{r}\frac{d^2w}{dr^2} - \frac{1}{r^2}\frac{dw}{dr} = \frac{Q}{D} \tag{17.14}$$

expresses the variation of deflection with radius.

Equations (17.13) and (17.14) can be expressed in a form which makes a solution by integration rather more obvious, as follows:

$$\frac{d}{dr}\left[\frac{1}{r}\frac{d}{dr}(r\psi)\right] = -\frac{Q}{D} \tag{17.15}$$

and

$$\frac{d}{dr}\left[\frac{1}{r}\frac{d}{dr}\left(r\frac{dw}{dr}\right)\right] = \frac{Q}{D} \tag{17.16}$$

The shear force Q is a function of the applied loading p and may be related by simple statical equilibrium or by integrating eqn. (17.11). Multiplying through by $r \, dr$ gives

$$r \, dQ + Q \, dr = pr \, dr \quad \text{or} \quad d(Qr) = pr \, dr$$

$$Qr = \int_0^r pr \, dr$$

Substituting in eqn. (17.16)

$$r\frac{d}{dr}\left[\frac{1}{r}\frac{d}{dr}\left(r\frac{dw}{dr}\right)\right]=\frac{1}{D}\int_0^r pr\,dr \qquad (17.17)$$

If we know that p is equal to $f(r)$ or is constant then eqn. (17.17) can be integrated to find the deflection at any radius.

The constants of integration are evaluated from the boundary conditions for the particular problem being solved.

Some typical loading and boundary situations will now be considered.

PLATE SUBJECTED TO UNIFORM PRESSURE

In this problem the right-hand side of eqn. (17.17) reduces to $pr^2/2$, since p is constant; therefore

$$\frac{d}{dr}\left[\frac{1}{r}\frac{d}{dr}\left(r\frac{dw}{dr}\right)\right]=\frac{pr}{2D} \qquad (17.18)$$

Integrating,

$$\frac{1}{r}\frac{d}{dr}\left(r\frac{dw}{dr}\right)=\frac{pr^2}{4D}+C_1$$

Multiplying both sides by r and integrating again,

$$r\frac{dw}{dr}=\frac{pr^4}{16D}+\frac{C_1r^2}{2}+C_2$$

$$\frac{dw}{dr}=\frac{pr^3}{16D}+\frac{C_1r}{2}+\frac{C_3}{r} \qquad (17.19)$$

Finally

$$w=\frac{pr^4}{64D}+\frac{C_1r^2}{4}+C_2\log_e r+C_3 \qquad (17.20)$$

CLAMPED PERIPHERY

For a plate of radius a the boundary conditions are $dw/dr=0$ at $r=0$ and $r=a$, and $w=0$ at $r=a$ (Fig. 17.6(a)).

Fig. 17.6

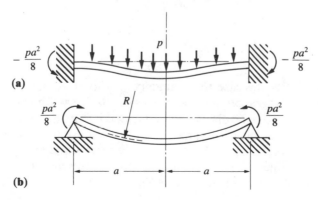

(a)

(b)

Hence $C_2 = 0$ and $\dfrac{pa^3}{16D} + C_1\dfrac{a}{2} = 0$ so that $C_1 = -\dfrac{pa^2}{8D}$

From eqn. (17.20),

$$0 = \frac{pa^4}{64D} - \frac{pa^4}{32D} + C_3 \quad \text{or} \quad C_3 = \frac{pa^4}{64D}$$

The deflection curve is thus

$$w = \frac{p}{64D}(a^2 - r^2)^2 \tag{17.21}$$

The maximum deflection occurs at the centre, so that

$$w_{max} = \frac{pa^4}{64D} \tag{17.22}$$

Bending stress may now be determined from the moment–slope relationships, eqns. (17.8) and (17.19):

$$M_r = \frac{p}{16}[a^2(1 + v) - r^2(3 + v)]$$

$$M_\theta = \frac{p}{16}[a^2(1 + v) - r^2(1 + 3v)]$$

At the periphery $r = a$ and

$$M_r = -\frac{pa^2}{8} \quad \text{and} \quad M_\theta = -v\frac{pa^2}{8}$$

At the centre $r = 0$ and

$$M_r = M_\theta = \frac{pa^2}{16}(1 + v) \tag{17.23}$$

From eqns. (17.10) we have

$$\sigma_r = \frac{12M_r z}{h^3} \quad \text{and} \quad \sigma_\theta = \frac{12M_\theta z}{h^3}$$

The maximum stresses occur at $r = a$ and $z = \pm h/2$, hence

$$\sigma_{r_{max}} = \pm\frac{6M_{r_{max}}}{h^2} = \pm\frac{3}{4}\frac{pa^2}{h^2} \tag{17.24}$$

$$\sigma_{\theta_{max}} = \pm\frac{3}{4}vpa^2/h^2$$

SIMPLY-SUPPORTED PERIPHERY

For this case the boundary conditions to determine three constants of integration are: $M_r = 0$, $r = a$; $dw/dr = 0$, $r = 0$; and $w = 0$, $r = a$. The problem will instead be tackled by an alternative method using the principle of superposition. We can use the solution from the fixed edge case combined with that for a plate simply-supported and subjected to edge moments equal but opposite in sense to the fixing moment, as illustrated in Fig. 17.6(b).

From eqn. (17.4) since, for $M_x = M_y$, $R_x = R_y$, then

$$\frac{1}{R} = \frac{M}{D(1 + v)}$$

The deflection at the centre of a spherical surface of radius a and curvature $1/R$ is

$$w = \frac{a^2}{2R}$$

Therefore

$$w = \frac{Ma^2}{2D(1 + v)} \tag{17.25}$$

But from page 450 the fixing moment is $M_r = -pa^2/8$; therefore

$$w = \frac{pa^4}{16D(1 + v)} \tag{17.26}$$

The resultant deflection at the centre for the simply-supported plate is, by superposition of eqns. (17.22) and (17.26),

$$w = \frac{pa^4}{64D} + \frac{pa^4}{16D(1 + v)} = \frac{5 + v}{64(1 + v)D} pa^4 \tag{17.27}$$

The maximum bending moment occurs at the centre, and by superposition of eqn (17.23) and $M = pa^2/8$ for the pure bending contribution,

$$M_r = M_\theta = \frac{pa^2}{16}(3 + v) \tag{17.28}$$

The maximum stress occurs at the centre and is again obtained using eqn. (17.10) thus

$$\sigma_{r_{max}} = \frac{3}{8}\frac{pa^2}{h^2}(3 + v)$$

PLATE WITH CENTRAL CIRCULAR HOLE

EDGE MOMENTS A ring of moments is applied to the inner and outer boundaries as shown in Fig. 17.7. Since there is no shear force at the inner boundary, eqn.

Fig. 17.7

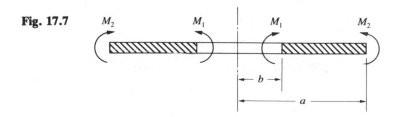

(17.15) reduces to

$$\frac{d}{dr}\left[\frac{1}{r}\frac{d}{dr}\left(r\frac{dw}{dr}\right)\right]=0 \tag{17.29}$$

Integrating twice and simplifying gives

$$\frac{dw}{dr}=\frac{C_1 r}{2}+\frac{C_2}{r} \tag{17.30}$$

and

$$w=\frac{C_1 r^2}{4}+C_2\log_e r+C_3 \tag{17.31}$$

To find C_1 and C_2 we use the boundary conditions: at $r=b$, $M_r=M_1$; and at $r=a$, $M_r=M_2$; and by substituting eqn. (17.30) and its differential into eqn. (17.9) and solving, we obtain

$$C_1=-\frac{2(a^2 M_2-b^2 M_1)}{(1+v)(a^2-b^2)D}$$

$$C_2=-\frac{a^2 b^2(M_2-M_1)}{(1-v)(a^2-b^2)D}$$

Since $w=0$ at $r=a$ then, from eqn. (17.31),

$$C_3=\frac{a^2(a^2 M_2-b^2 M_1)}{2(1+v)(a^2-b^2)D}+\frac{a^2 b^2(M_2-M_1)\log_e a}{(1-v)(a^2-b^2)D}$$

The deflection at any radius may now be determined by substituting the values of C_1, C_2 and C_3 into eqn. (17.31).

EDGE FORCES   In this case uniformly distributed transverse forces Q_0 are applied at the inner and outer edges as shown in Fig. 17.8. The shear force per unit length at radius r is

$$Q=\frac{2\pi b Q_0}{2\pi r}=\frac{bQ_0}{r}$$

Substituting in eqn. (17.15) and integrating gives

$$\frac{dw}{dr}=\frac{bQ_0}{4D}(2\log_e r-1)+\frac{C_1 r}{2}+\frac{C_2}{r} \tag{17.32}$$

$$w=\frac{bQ_0 r^2}{4D}(\log_e r-1)+\frac{C_1 r^2}{4}+C_2\log_e r+C_3 \tag{17.33}$$

Fig. 17.8

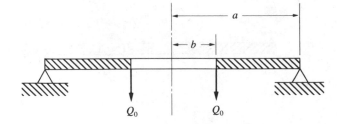

$$Q_0 \qquad Q_0$$

For a simply-supported periphery the boundary conditions are: at $r = a$, $w = 0$ and $M_r = 0$; at $r = b$, $M_r = 0$.

By substituting eqn. (17.32) and its differential into eqn. (17.9) and solving, we obtain

$$C_1 = -\frac{bQ_0}{2D}\left(\frac{1-v}{1+v} + \frac{2(a^2\log_e a - b^2\log_e b)}{a^2 - b^2}\right)$$

$$C_2 = \frac{bQ_0}{2D}\left(\frac{(1+v)}{(1-v)}\frac{a^2b^2}{a^2-b^2}\log_e\frac{b}{a}\right)$$

Then, from eqn. (17.33),

$$C_3 = \frac{a^2bQ_0}{4D}\left(1 + \frac{1}{2}\frac{(1-v)}{(1+v)} - \frac{b^2}{a^2-b^2}\log_e\frac{b}{a} - \frac{2(1+v)}{(1-v)}\frac{b^2}{a^2-b^2}\log_e\frac{b}{a}\log_e a\right)$$

These constants may now be substituted into eqns. (17.32) and (17.33) to give the slope and deflection at any radius.

SOLID PLATE WITH CENTRAL CONCENTRATED FORCE

SIMPLY-SUPPORTED EDGE

Considering the problem solved in the previous section and equating the total load around the inner periphery to the concentrated force F so that

$$2\pi bQ_0 = F$$

and taking the limiting case when b is infinitely small, $b^2\log_e(b/a)$ tends to zero and the above constants of integration become

$$C_1 = -\frac{F}{4\pi D}\left\{\frac{1-v}{1+v} + 2\log_e a\right\}; \qquad C_2 = 0;$$

$$C_3 = \frac{Fa^2}{8\pi D}\left\{1 + \frac{1}{2}\left(\frac{1-v}{1+v}\right)\right\}$$

Substituting these values into eqn. (17.33),

$$w = \frac{F}{8\pi D}\left[\frac{3+v}{2(1+v)}(a^2 - r^2) + r^2\log_e\frac{r}{a}\right] \qquad (17.34)$$

which is the deflection at any radius for a solid plate simply supported and subjected to a concentrated force at the centre.

FIXED EDGE

For the fixed edge case we find the slope at the edge for the simply-supported plate by differentiating eqn. (17.34); then this slope has to be made zero by the superposition of the appropriate ring of edge moments. The deflection due to these moments may then be superposed onto the deflection given by eqn. (17.34) to obtain a resultant.

EXAMPLE 17.1

A cylinder head valve of diameter 38 mm is subjected to a gas pressure of 1.4 MN/m². It may be regarded as a uniform thin circular plate simply supported around the periphery by the seat, as shown in Fig. 17.9. Assuming that the valve stem applies a concentrated force at the centre of the plate, calculate the movement of the stem necessary to lift the valve from its seat. The flexural rigidity of the valve is 260 N-m, and Poisson's ratio for the material is 0.3.

Fig. 17.9

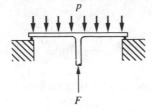

SOLUTION

We have already derived solutions for a simply supported plate subjected to uniform loading, p, and a concentrated force, F, at the centre; hence the deflection at the centre is equal to the sum of the deflections due to the two separate load components. Therefore

$$w_{max} = \frac{Fa^2}{16\pi D}\frac{(3+v)}{(1+v)} + \frac{pa^4}{64D}\frac{(5+v)}{(1+v)}$$

but when the valve lifts from its seat $F = -\pi a^2 p$; therefore

$$w_{max} = -\frac{(7+3v)}{(1+v)}\frac{pa^4}{64D}$$

$$= -\frac{7.9 \times 1.4 \times 10^6 \times 0.019^4}{1.3 \times 64 \times 260}$$

$$= -0.0665 \text{ mm}$$

OTHER FORMS OF LOADING AND BOUNDARY CONDITION

The solutions that have been obtained on pages 449–53 can be used to advantage with the principle of superposition to analyse a number of other plate problems.

CONCENTRIC LOADING

A plate which is uniformly loaded transversely along a circle of radius b, as illustrated in Fig. 17.10(a), can be split into the two components shown at (b), and the separate solutions which have been obtained previously can be superposed using the appropriate boundary conditions, namely that there must be continuity of slope at radius b.

Fig. 17.10

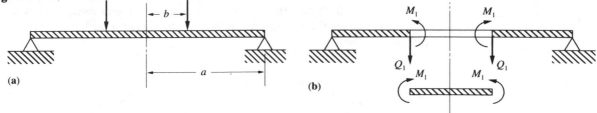

(a)

(b)

DISTRIBUTED LOADING ON PLATE WITH CENTRAL HOLE	The situation illustrated in Fig. 17.11(*a*) can be simulated by taking the deflection of the solid plate subjected to uniform loading and superposing that due to the appropriate moment and shear force at radius *b*, as in Fig. 17.11(*b*).

Fig. 17.11

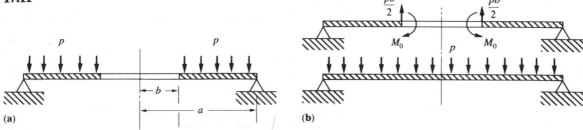

(a)

(b)

FIXED INNER BOUNDARY	This edge condition can be associated with several types of loading and merely entails applying the appropriate moment to give zero slope and shear force as a function of the applied loading. A variety of configurations and loadings are illustrated in Fig. 17.12, all

Fig. 17.12

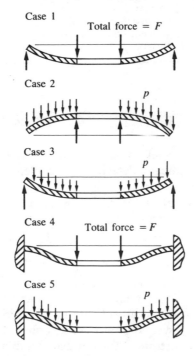

Case 1 — Total force = F

Case 2 — p

Case 3 — p

Case 4 — Total force = F

Case 5 — p

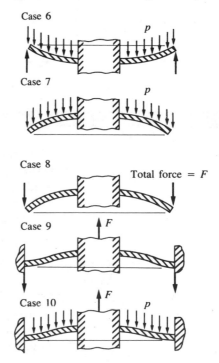

Case 6 — p

Case 7 — p

Case 8 — Total force = F

Case 9 — F

Case 10 — F, p

of which can be dealt with easily by superposition of the required components which give the appropriate boundary conditions. However, in all these cases the maximum deflection can be represented by the following relations:

$$w_{max} = c' \frac{pa^4}{Eh^3} \quad \text{or} \quad w_{max} = c' \frac{Fa^2}{Eh^3}$$

where c' is a factor involving the ratio a/b and Poisson's ratio.

The maximum stresses can also be expressed by formulae as follows:

$$\sigma_{max} = c'' \frac{pa^2}{h^2} \quad \text{or} \quad \sigma_{max} = \frac{c''F}{h^2}$$

where c'' is also a factor as defined above. Values of c' and c'' for $v = 0.3$ and a/b in the range $1\frac{1}{4}$ to 5 for the cases in Fig. 17.12 are given in Table 17.1.

Table 17.1
Coefficients c' and c'' for the plate cases shown in Fig. 17.12

$a/b =$	1.25		1.5		2		3		4		5	
Case	c''	c'	c''	c'	c''	c'	c''	c'	c''	c'	c''	c'
1	1.10	0.341	1.26	0.519	1.48	0.672	1.88	0.734	2.17	0.724	2.34	0.704
2	0.66	0.202	1.19	0.491	2.04	0.902	3.34	1.220	4.30	1.300	5.10	1.310
3	0.592	0.184	0.976	0.414	1.440	0.664	1.880	0.824	2.08	0.830	2.19	0.813
4	0.194	0.00504	0.320	0.0242	0.454	0.0810	0.673	0.172	1.021	0.217	1.305	0.238
5	0.105	0.00199	0.259	0.0139	0.480	0.0575	0.657	0.130	0.710	0.162	0.730	0.175
6	0.122	0.00343	0.336	0.0313	0.74	0.1250	1.21	0.291	1.45	0.417	1.59	0.492
7	0.135	0.00231	0.410	0.0183	1.04	0.0938	2.15	0.293	2.99	0.448	3.69	0.564
8	0.227	0.00510	0.428	0.0249	0.753	0.0877	1.205	0.209	1.514	0.293	1.745	0.350
9	0.115	0.00129	0.220	0.0064	0.405	0.0237	0.703	0.062	0.933	0.092	1.13	0.114
10	0.090	0.00077	0.273	0.0062	0.71	0.0329	1.54	0.110	2.23	0.179	2.80	0.234

AXI-SYMMETRICAL THIN SHELLS

The analysis of thin shells of revolution subjected to uniform pressure was treated as a statically-determinate problem in Chapter 2. The fundamental relationship between the principal membrane stresses in the wall and the principal curvatures of the shell to the applied pressure and wall thickness was shown to be

$$\frac{\sigma_1}{r_1} + \frac{\sigma_2}{r_2} = \frac{p}{t} \tag{17.35}$$

The simple applications of eqn. (17.35) to the cylinder and sphere under internal pressure were also dealt with in Chapter 2, together with an example on the self-weight of a concrete dome. The following example on liquid storage illustrates a further use of a thin shell.

EXAMPLE 17.2

The water storage tank illustrated in Fig. 17.13, of 20 mm uniform wall thickness, consists of a cylindrical section which is supported at the top edge and joined at the lower end to a spherical portion. An angle-section reinforcing ring of 5000 mm² cross-sectional area is welded into the lower joint as shown.

Calculate the maximum stresses in the cylindrical and spherical portions of the tank and the hoop stress in the reinforcing ring when the water is at the level shown. Density of water is 9.81 kN/m³.

Fig. 17.13

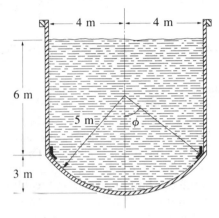

SOLUTION

The total force due to the weight of water in the tank is

$$W = 9.81 \times 10^3 \left[(\pi \times 6 \times 4^2) + \left(\frac{2\pi}{3} \times 5^3 \right) + \left(\frac{\pi \times 3^3}{3} \right) - (\pi \times 5^2 \times 3) \right]$$

$$= 3.5 \, \text{MN}$$

The axial stress in the cylindrical part of the tank is

$$\sigma_a = \frac{3.5}{2\pi \times 4 \times 0.02} = 6.95 \, \text{MN/m}^2$$

and the hoop stress is

$$\sigma_{h_{max}} = \frac{6 \times 9.81 \times 10^3 \times 4}{0.02} = 11.8 \, \text{MN/m}^2$$

The maximum stress in the spherical portion of the tank occurs at the bottom, where the pressure is $9.81 \times 10^3 \times 8 \, \text{N/m}^2$; therefore

$$\sigma_{max} = \frac{9.81 \times 10^3 \times 8 \times 5}{2 \times 0.02} = 9.81 \, \text{MN/m}^2$$

For the reinforcing ring, the tangential force per unit length at the edge of the spherical part is $W/(2\pi \times 4 \sin \phi)$. The inward radial component of that force is

$$\frac{W}{2\pi \times 4 \sin \phi} \cos \phi$$

and therefore the compressive force in the ring is

$$\frac{W \cot \phi}{2\pi \times 4} \times 4$$

and the compressive stress is

$$
\begin{aligned}
\sigma_c &= \frac{W \cot \phi}{2\pi} \times \frac{1}{5000 \times 10^{-6}} \\
&= \frac{3.5 \times 10^6 \times \frac{3}{4}}{2\pi \times 5 \times 10^{-3}} = 83 \text{ MN/m}^2
\end{aligned}
$$

LOCAL BENDING STRESSES IN THIN SHELLS

Whenever there is a change in geometry of the shell, particularly for discontinuities in the meridian such as in the above example, the membrane stresses cause displacements which give rise to local bending in the wall. The resulting bending stresses may be significant in comparison with the membrane stresses. This was the reason for introducing the reinforcing ring in the above problem.

To illustrate the method of analysing local bending we will consider the elementary situation of a cylindrical vessel with hemispherical ends of the same thickness subjected to internal pressure as illustrated in Fig. 17.14.

Fig. 17.14

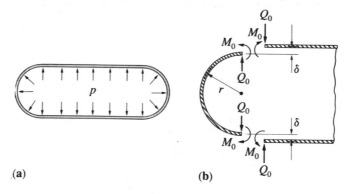

(a) (b)

The membrane stresses for the cylinder are

$$\sigma_1 = \frac{pr}{t} \quad \text{and} \quad \sigma_2 = \frac{pr}{2t}$$

and for the hemisphere,

$$\sigma = \frac{pr}{t}$$

as calculated in Chapter 2.

The corresponding radial displacements for the cylinder and hemisphere respectively, are

$$u_c = \frac{r}{E}(\sigma_1 - v\sigma_2) = \frac{pr^2}{2tE}(2 - v)$$

$$u_h = \frac{r}{E}(\sigma_1 - v\sigma_2) = \frac{pr^2}{2tE}(1 - v)$$

The difference in deformation radially is

$$\delta = \frac{pr^2}{2tE}[(2 - v) - (1 - v)] = \frac{pr^2}{2tE} \tag{17.36}$$

In order to overcome this difference, shear and moment reactions are set up at the joint as shown in Fig. 17.14.

Since the cylindrical section is symmetrical with respect to its axis we may consider the reactions Q_0 and M_0 per unit length acting on a strip of unit width as illustrated in Fig. 17.15. The inward bending of the strip

Fig. 17.15

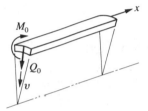

due to Q_0 sets up compressive circumferential strain. If the radial displacement is v then

$$\varepsilon_\theta = \frac{v}{r} \quad \text{and} \quad \sigma_\theta = \frac{Ev}{r}$$

The circumferential force due to this stress on the edge of the strip per unit length is Evt/r. The outward radial component of this force is

$$R = 2\frac{Evt}{r}\sin\frac{d\theta}{2} = \frac{Evt}{r}d\theta$$

$$= \frac{Evt}{r^2} \tag{17.37}$$

This force opposes the deflection of the strip and is distributed along the strip, being proportional to the deflection v at any point. This is a special case of the bending of a beam on an elastic foundation for which the deflection curve is

$$EI\frac{d^4v}{dx^4} = -w$$

where w is the distributed loading function. For the present problem EI is replaced by D, since the strip is restrained from distortion by adjacent material as for plates; therefore

$$D\frac{d^4v}{dx^4} = -\frac{Et}{r^2}v = -4D\beta^4v \tag{17.38}$$

The solution of this equation is

$$v = e^{\beta x}(A \cos \beta x + B \sin \beta x) + e^{-\beta x}(C \cos \beta x + D' \sin \beta x) \qquad (17.39)$$

where A, B, C and D' are constants determined by the boundary conditions, and

$$\beta = \left[\frac{Et}{4Dr^2}\right]^{1/4} = \left[\frac{3(1 - v^2)}{r^2 t^2}\right]^{1/4}$$

As $x \to \infty$, $v \to 0$ and $M \to 0$, which gives $A = B = 0$. At $x = 0$, $M = M_0$ and $Q = Q_0$; therefore

$$D \frac{d^2 v}{dx^2} = -M_0$$

and

$$D \frac{d^3 v}{dx^3} = -Q_0$$

from which

$$C = \frac{1}{2\beta^3 D}(Q_0 - \beta M_0) \quad \text{and} \quad D' = \frac{M_0}{2\beta^2 D}$$

Substituting into eqn. (17.39), the deflection curve for the strip becomes

$$v = \frac{e^{-\beta x}}{2\beta^3 D}[Q_0 \cos \beta x - \beta M_0(\cos \beta x - \sin \beta x)] \qquad (17.40)$$

This is a rapidly damped oscillatory curve, and thus bending of the cylinder and head is local to the joint.

When the cylinder and head are of the same material and wall thickness then the deflections and slopes at the joint produced by Q_0 are equal and $M_0 = 0$. Therefore the boundary condition is at $x = 0$, $v = \delta/2$, and from eqn. (17.40),

$$\frac{\delta}{2} = \frac{Q_0}{2\beta^3 D}$$

or

$$Q_0 = \delta \beta^3 D = \frac{pr^2}{2tE} \frac{Et}{4\beta r^2} = \frac{p}{8\beta} \qquad (17.41)$$

The deflection curve becomes

$$v = \frac{e^{-\beta x} p \cos \beta x}{16\beta^4 D} = \frac{pr^2}{4tE} e^{-\beta x} \cos \beta x \qquad (17.42)$$

By differentiating this equation twice the bending moment and hence the bending stress can be calculated for any cross-section. These have to be added to the membrane stresses to get the resultant stress.

If the wall thickness of the head and cylinder are different then $M_0 \neq 0$ and the boundary conditions would be that (a) the sum of the edge deflections must be zero, and (b) the rotation of the edges must be the same.

The above solution is equally applicable for other shapes of head.

EXAMPLE 17.3

SOLUTION

Calculate the local bending stresses in the vessel shown in Fig. 17.14 if $p = 1\,\text{MN/m}^2$, $r = 500\,\text{mm}$, $t = 10\,\text{mm}$, $v = 0.3$.

$$\beta = \left[\frac{3(1 - 0.3^2)}{0.5^2 \times 0.01^2}\right]^{1/4} = 18.2\,\text{m}^{-1}$$

$$Q_0 = \frac{p}{8\beta} = \frac{1 \times 10^6}{8 \times 18.2} = 6.87\,\text{kN/m}$$

Since the bending moment in the strip is $M = -D(\text{d}^2v/\text{d}x^2)$, and from eqn. (17.40) with $M_0 = 0$,

$$v = \frac{Q_0}{2\beta^3 D}\,\text{e}^{-\beta x}\cos\beta x$$

Therefore

$$M = -\frac{Q_0}{\beta}\,\text{e}^{-\beta x}\sin\beta x$$

This expression takes the largest value for $\beta x = \pi/4$, which gives

$$M_{max} = 0.121\,\text{kN-m/m}$$

This gives rise to a maximum bending stress of

$$\text{(axial)}\quad \sigma_b = \frac{6M_{max}}{t^2} = \frac{6 \times 0.121 \times 10^3}{0.01^2} = 7.26\,\text{MN/m}^2$$

The membrane stress is

$$\text{(axial)}\quad \sigma_m = \frac{pr}{2t} = \frac{1 \times 0.5}{2 \times 0.01} = 25\,\text{MN/m}^2$$

The total axial stress is $\sigma_a = 7.26 + 25 = 32.26\,\text{MN/m}^2$.

The bending of the strip also produces circumferential stresses: (a) due to prevention of strip from distorting, as in plates $= \pm 6vM/t^2$; (b) stresses due to shortening of the circumference $= -Ev/r$.

Using the above values for v and M and summing gives

$$\text{(hoop)}\quad \sigma_b = \frac{Q_0\text{e}^{-\beta x}}{\beta t^2}\left(6v\sin\beta x - \frac{12(1 - v^2)}{2\beta^2 tr}\cos\beta x\right)$$

$$= 22.6\,\text{e}^{-\beta x}(0.3\sin\beta x - 0.55\cos\beta x)\,\text{MN/m}^2$$

The maximum value of this expression is 1.58 MN/m^2, which is small compared with the hoop membrane stress.

$$\text{(hoop)} \quad \sigma_m = \frac{pr}{t} = \frac{1 \times 0.5}{0.01} = 50 \text{ MN/m}^2$$

Hence local bending does not have a serious influence in this particular case.

BENDING IN A CYLINDRICAL STORAGE TANK

An upright cylindrical storage tank, of radius r, uniform wall thickness t, and height h, is filled to the top with liquid of density ρ. The base of the tank is built in to its foundation, Fig. 17.16, and we need to design for the maximum bending moment due to discontinuity in the shell at the base. Since $t \ll h$ or r the shell may be regarded as infinitely long.

Fig. 17.16

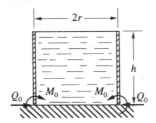

The governing equation is basically the same as eqn. (17.38) with an additional term which defines the variation of pressure loading due to the liquid:

$$D\frac{d^4v}{dx^4} + 4D\beta^4 v = -\rho(h - x) \tag{17.43}$$

The particular integral part of the solution is $-\rho(h-x)/4D\beta^4$ and the complete solution is

$$v = e^{\beta x}(C_1 \cos \beta x + C_2 \sin \beta x) + e^{-\beta x}(C_3 \cos \beta x + C_4 \sin \beta x)$$
$$-\frac{\rho(h-x)}{4D\beta^4} \tag{17.44}$$

The boundary conditions are as follows:

1. The height of the cylinder can be regarded as "infinite", so that $M \to 0$ and $v \to 0$, giving $C_1 = C_2 = 0$.
2. At $x = 0$, $v = 0$ and $dv/dx = 0$; hence

$$C_3 = \frac{\rho h}{4D\beta^4} \quad \text{and} \quad C_4 = \frac{\rho}{4D\beta^4}\left(h - \frac{1}{\beta}\right)$$

Putting these values in eqn. (17.44) gives the deflection curve

$$v = -\frac{\rho h}{4D\beta^4}\left[1 - \frac{x}{h} - e^{-\beta x}\cos \beta x - e^{-\beta x}\left(1 - \frac{1}{\beta h}\right)\sin \beta x\right] \tag{17.45}$$

But $M = -D(\mathrm{d}^2v/\mathrm{d}x^2)$; so that differentiating eqn. (17.45) twice and substituting gives

$$M = \frac{\rho h}{2\beta^2} \left[-\mathrm{e}^{-\beta x} \sin \beta x + \left(1 - \frac{1}{\beta h}\right) \mathrm{e}^{-\beta x} \cos \beta x \right] \qquad (17.46)$$

Now, the maximum value of M occurs at the discontinuity, where $x = 0$; therefore

$$M_{max} = \frac{\rho h}{2\beta^2} \left(1 - \frac{1}{\beta h}\right) = \frac{\rho}{2\beta^3} (\beta h - 1)$$

The resultant bending stresses can now be calculated as in Example 17.3.

SUMMARY

The basic principles for the analysis of plates, have been developed involving equilibrium, geometry of deformation and the stress–strain relationships. These are essentially the two-dimensional development from the simple bending theory for beams. Thereafter, each plate solution is dependent on the particular boundary conditions of loading and support.

The design of thin shells depends principally on the magnitude of the general system of membrane stresses. However, attention must also be given to the effect of local bending stresses at regions of discontinuity in the shell, the severity of which will depend on local geometry and stiffness.

BIBLIOGRAPHY

Roark, R. J. and Young, W. C. *Formulas for Stress and Strain,* McGraw-Hill, New York, 1975.

Timoshenko, S. and Woinowsky-Kreiger, S. *Theory of Plates and Shells,* McGraw-Hill, New York, 1970.

PROBLEMS

17.1 Show that for a flat circular steel plate subjected to a uniform pressure on one surface, the maximum stress when the periphery is freely supported is 1.65 times that when the periphery is clamped. $v = 0.3$.

17.2 Calculate the ratio of (i) the radial stresses, and (ii) the tangential stresses at the edge and centre of a flat circular steel plate clamped at its periphery and subjected to a uniform pressure p. $v = 0.29$.

17.3 A circular thin steel diaphragm having an effective diameter of 200 mm is clamped around its periphery and is subjected to a uniform gas pressure of 180 kN/m². Calculate a minimum thickness for the diaphragm if the deflection at the centre is not to exceed 0.5 mm. $E = 208$ GN/m², $v = 0.287$.

17.4   A circular aluminium plate 6 mm thick has an outer diameter of 250 mm and a concentric hole of 50 mm diameter. The edge of the hole is subjected to a bending moment of magnitude 900 N-m/m. Determine the deflection of the inner edge relative to the outer. $E = 70\ \text{GN/m}^2$, $\nu = 0.3$.

17.5   The end plate of a tube is made from 5 mm thick steel plate as in Fig. 17.17. If a 30 mm diameter rod welded to the end plate is subjected to a force of 10 kN what would be the movement of the rod? Calculate also the maximum stresses in the end plate. $E = 207\ \text{GN/m}^2$, $\nu = 0.29$.

Fig. 17.17

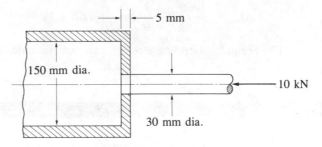

5 mm

150 mm dia.

10 kN

30 mm dia.

17.6   A circular plate 500 mm diameter and 2.5 mm thick is clamped around its edge and is subjected to a concentrated load of 900 N at its centre. Calculate the radial and tangential bending stresses at the fixed edge. $\nu = 0.29$.

17.7   A pressure transducer uses a probe to measure the deflection of a thin circular diaphragm, clamped at its periphery and having a rigid seat at its centre to eliminate curvature in line with the probe as shown in Fig. 17.18. Calculate the thickness of the diaphragm if the probe can measure 0–0.25 mm and the range of the pressure transducer is to be 0–0.5 MN/m². Young's modulus and Poisson's ratio for the diaphragm material are $207\ \text{GN/m}^2$ and 0.3 respectively.

Fig. 17.18

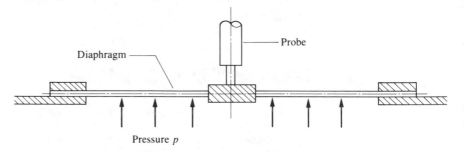

Probe

Diaphragm

Pressure p

17.8   A 200 mm diameter circular plate is clamped around its periphery and subjected to a uniform pressure of 15 MN/m². If a rod supports the plate at its centre so that the central deflection of the plate is zero calculate the force in the rod.

17.9   A circular plate is to be simply-supported at its periphery and subjected to a load of 15 kN distributed around a circle which has the same centre as the plate. If the ratio of the plate diameter to the loading circle diameter is 2 calculate the thickness of the plate so that its maximum stress does not exceed 200 MN/m². $\nu = 0.3$.

17.10   A circular steel plate of 304 mm diameter and 12 mm thick is clamped around the edge. A concentric ring of loading of 20 kN is applied uniformly on a circle of 152 mm diameter. Calculate the deflection at the centre of the plate. $E = 208\ \text{GN/m}^2$, $\nu = 0.29$.

17.11   Determine the membrane stresses in a concrete hemispherical dome of radius 4 m and thickness 200 mm. The concrete has a density of 2.31 Mg/m³.

17.12   A conical water-storage tank has an included angle of 60° as shown in Fig. 17.19 and a vertical depth of water of 3 m. If the wall thickness is 5 mm determine the

Fig. 17.19

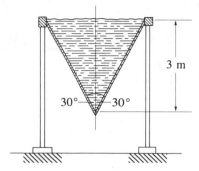

location and magnitude of the maximum hoop and meridional stresses. The water loading is 9.81 kN/m³.

17.13 A toroidal pressure vessel has an interior diameter of 1 m and an exterior diameter of 2 m as shown in Fig. 17.20, with a wall thickness of 4 mm. Derive expressions for the principal membrane stresses and determine values at the horizontal section of symmetry for an internal pressure of 150 kN/m².

Fig. 17.20

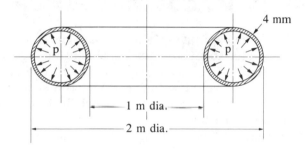

17.14 A steel pipe of outer diameter 200 mm and wall thickness 6 mm is to have a steel ring shrunk on to the outside. The inside diameter of the ring is 199 mm and it is 80 mm² in cross-sectional area. Determine the maximum local bending stress in the pipe due to the shrink pressure of the ring. $E = 208$ GN/m² and $v = 0.3$.

17.15 A stainless steel drum of outer diameter 400 mm and 6 mm wall thickness is closed at each end by steel discs which house central bearings. The drum has to rotate in service at 1000 rev/min. Determine the local bending stress in the drum where it is fixed to the discs. $E = 208$ GN/m², $v = 0.3$, $\rho = 7700$ kN/m³.

17.16 An upright cylindrical steel tank is built in to a concrete base. It is 3 m in diameter and has a wall thickness of 12 mm. If the height of water during a test is 2 m determine the maximum bending stress. $\rho = 1000$ kg/m³ $v = 0.3$.

Chapter 18

Finite element method

The finite element method for analysing structural parts has now been around for about 30 years, but although it is generally accepted as an extremely valuable tool, many engineers do not know how to go about using it and very few engineers understand it. One of the main reasons for this is that the subject has generally been surrounded by a high level of research activity. Coupled with this is the fact that because of the amount of calculations which the method involves, it tended to go directly from its embryonic stage to an advanced computing stage. There never seemed to be an intermediate stage at which it could have been conveniently slotted into a curriculum on solid mechanics. And yet when one looks at the basic principles of the method, it is not an advanced topic and is probably out of place at this late stage in a book on solid mechanics. Basically the finite element method involves the application of the three basic conditions introduced in Chapter 4, i.e. *equilibrium of forces, compatibility of displacements* and *stress–strain relationships*. Finite element methods could, therefore, have been dealt with at Chapter 5. However, it should also be pointed out that due to the versatility of the finite element method it can be applied to very complex problems and the matrix methods involved in the solution of these can provide material for a whole textbook in their own right. In this chapter it is proposed to provide a simple introduction to the finite element method and remove some of the mystery which surrounds the subject by illustrating its application to a number of typical problems.

PRINCIPLE OF FINITE ELEMENT METHOD

If a truss of the type shown in Fig. 18.1(*a*) is being analysed then it is a straightforward exercise because it is formed from discrete members. The

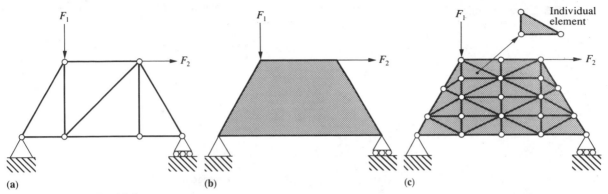

Fig. 18.1

paths of force transmission through the truss to the ground are readily apparent and the forces may be determined using equilibrium equations as illustrated in Chapter 1. If the truss had been statically indeterminate then both equilibrium equations and deformation compatibility would be necessary, but the method of solution is well established. If, however, a plate of the same shape as the truss (Fig. 18.1(*b*)) is to be analysed then this is not so straightforward. The reason is that the plate is an elastic continuum and since the force transmission paths are not readily apparent the problem does not lend itself to simple mathematical analysis. Although, for the purpose of analysis, it might be tempting to consider the truss as being equivalent to the plate, this would not give accurate results since it ignores the restraining effect which all points in a continuum will experience and exert on neighbouring points. However, if the continuum was considered to be subdivided into a large number of triangular panels (Fig. 18.1(*c*)) it should be possible to develop a picture of the stress distribution in the whole plate by analysing each of the small panels in turn. To do this it would of course be necessary to (*a*) analyse the *equilibrium* of each of the triangular panels in relation to its neighbours and (*b*) have available equations for the *geometry of deformation* and the *stress–strain relationships* for a triangular panel. This subdivision of a continuum into a large number of discrete elements is the basis of the finite element method of stress analysis. The triangular panels referred to in this example are the "elements", but this is only one type of element. Others include a spring element (one-dimensional), a plane rectangular element (two-dimensional) and solid elements (three-dimensional) as shown in Fig. 18.2.

The accuracy of the solution depends on the number of sub-divisions (elements); the more there are the greater the accuracy. However, although the analysis of each individual element is straightforward the analysis of a large number of elements becomes extremely tedious. For this reason finite element solutions to problems are generally carried out on computers and there are many commercial software packages available. To some extent this has led to the current situation where many engineers are put off by the apparent complexity of the subject and they leave it to the experts who tend to attach a certain mystique to the subject through the use of computer jargon!

Fig. 18.2 Types of finite elements

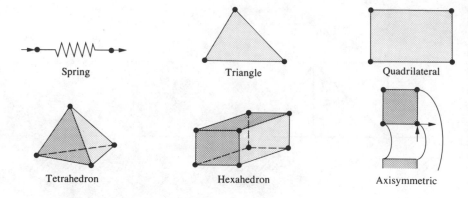

Spring   Triangle   Quadrilateral

Tetrahedron   Hexahedron   Axisymmetric

As an introduction to finite element methods it is convenient to consider the matrix analysis of skeletal structures using the stiffness method. An introduction to matrix algebra is given in Appendix B.

ANALYSIS OF SPRING ELEMENTS

A SINGLE SPRING ELEMENT

When a pin-jointed tie or strut is part of a structure, its ends will be able to move due to displacement of the structure and deformation of the member. This may be modelled by a spring element as shown in Fig. 18.3.

Fig. 18.3

The points of attachment of the element to other parts of the structure are called *nodes* and are indicated by the points 1 and 2 in Fig. 18.3. Here, F and u are the force and displacement values and their suffix indicates the node to which they apply. In situations where the spring element is used to model a tie or strut of length and area A, the stiffness of the spring k_1 will be given by

$$k_1 = \frac{AE}{L} \tag{18.1}$$

where E is Young's modulus for the material.

For the simple system illustrated in Fig. 18.3, using the sign convention that forces and displacements are positive in the x-direction, then the forces may be related to the displacements by the following equations:

$$F_1 = k_1(u_1 - u_2) = k_1 u_1 - k_1 u_2 \tag{18.2}$$
$$F_2 = k_1(u_2 - u_1) = -k_1 u_1 + k_1 u_2 \tag{18.3}$$

Equations (18.2) and (18.3) may be written in matrix form as

$$\begin{Bmatrix} F_1 \\ F_2 \end{Bmatrix} = \begin{bmatrix} k_1 & -k_1 \\ -k_1 & k_1 \end{bmatrix} \begin{Bmatrix} u_1 \\ u_2 \end{Bmatrix} \tag{18.4}$$

In shorthand form this may be written as

$$\{F\} = |K^e| \{u\} \tag{18.5}$$

where $|K^e|$ is referred to as the *stiffness matrix* for the spring element. An important property of the stiffness matrix for an element (and, as will be seen later, for a complete structure) is that it is symmetrical.

AN ASSEMBLY OF SPRING ELEMENTS

Consider now a system consisting of two spring elements as shown in Fig. 18.4.

Using eqn. (18.4) the force–displacement equation for each element may be written as

$$\begin{Bmatrix} F_1 \\ F_2 \end{Bmatrix} = \begin{bmatrix} k_1 & -k_1 \\ -k_1 & k_1 \end{bmatrix} \begin{Bmatrix} u_1 \\ u_2 \end{Bmatrix}$$

$$\begin{Bmatrix} F_2 \\ F_3 \end{Bmatrix} = \begin{bmatrix} k_2 & -k_2 \\ -k_2 & k_2 \end{bmatrix} \begin{Bmatrix} u_2 \\ u_3 \end{Bmatrix}$$

Fig. 18.4

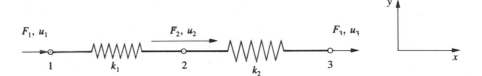

If each of these two equations is expanded so that they are in an equivalent form

$$\begin{Bmatrix} F_1 \\ F_2 \\ F_3 \end{Bmatrix} = \begin{bmatrix} k_1 & -k_1 & 0 \\ -k_1 & k_1 & 0 \\ 0 & 0 & 0 \end{bmatrix} \begin{Bmatrix} u_1 \\ u_2 \\ u_3 \end{Bmatrix} \quad \text{and}$$

$$\begin{Bmatrix} F_1 \\ F_2 \\ F_3 \end{Bmatrix} = \begin{bmatrix} 0 & 0 & 0 \\ 0 & k_2 & -k_2 \\ 0 & -k_2 & k_2 \end{bmatrix} \begin{Bmatrix} u_1 \\ u_2 \\ u_3 \end{Bmatrix}$$

As the spring elements are in series, these matrices may be added to give

$$\begin{Bmatrix} F_1 \\ F_2 \\ F_3 \end{Bmatrix} = \begin{bmatrix} k_1 & -k_1 & 0 \\ -k_1 & k_1 + k_2 & -k_2 \\ 0 & -k_2 & k_2 \end{bmatrix} \begin{Bmatrix} u_1 \\ u_2 \\ u_3 \end{Bmatrix} \tag{18.6}$$

or

$$\{F\} = |K|\{u\}$$

where $|K|$ is the stiffness matrix for the *structure*, i.e. the assembly of two spring elements. If the diagonal terms in the *element* stiffness matrix $|K^e|$ are referred to as direct stiffness and the off-diagonal terms in $|K^e|$ are indirect stiffness then Rockey *et al.*[1] have summarized the rules for the formation of $|K|$ as follows:

(a) The term in location *ii* consists of the sum of the direct stiffnesses of all the elements meeting at node *i*.
(b) The term in location *ij* consists of the sum of the indirect stiffness relating to nodes *i* and *j* of all the elements joining node *i* to node *j*.

This approach to a solution is the basis of the finite element method and it is illustrated in the following example of a statically-indeterminate force system.

EXAMPLE 18.1

Three dissimilar materials are friction welded together and placed between rigid end supports as shown in Fig. 18.5. If forces of 50 kN and

Fig. 18.5

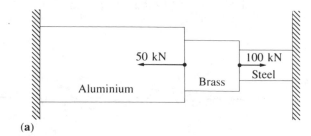

(b)

100 kN are applied as indicated calculate the movement of the interfaces between the materials and the forces exerted on the end supports.

Table 18.1

For aluminium	For brass	For steel
Area = 400 mm²	Area = 200 mm²	Area = 70 mm²
Length = 280 mm	Length = 100 mm	Length = 100 mm
$E = 70$ GN/m²	$E = 100$ GN/m²	$E = 200$ GN/m²

SOLUTION

This system may be represented by a model consisting of three spring elements as shown.

$$k_1 = \frac{A_1 E_1}{L_1} = \frac{400 \times 70 \times 10^3}{280} = 100 \text{ kN/mm};$$

$$k_2 = 200 \text{ kN/mm}$$

$$k_3 = 140 \text{ kN/mm}$$

The force–deformation relationships for each element may then be written as

$$\begin{Bmatrix} F_1 \\ F_2 \end{Bmatrix} = \begin{bmatrix} 100 & -100 \\ -100 & 100 \end{bmatrix} \begin{Bmatrix} u_1 \\ u_2 \end{Bmatrix} \qquad \begin{Bmatrix} F_2 \\ F_3 \end{Bmatrix} = \begin{bmatrix} 200 & -200 \\ -200 & 200 \end{bmatrix} \begin{Bmatrix} u_2 \\ u_3 \end{Bmatrix}$$

$$\begin{Bmatrix} F_3 \\ F_4 \end{Bmatrix} = \begin{bmatrix} 140 & -140 \\ -140 & 140 \end{bmatrix} \begin{Bmatrix} u_3 \\ u_4 \end{Bmatrix}$$

Using the two rules for the formation of the overall stiffness matrix,

$$\begin{Bmatrix} F_1 \\ F_2 \\ F_3 \\ F_4 \end{Bmatrix} = \begin{bmatrix} 100 & -100 & 0 & 0 \\ -100 & (100+200) & -200 & 0 \\ 0 & -200 & (200+140) & -140 \\ 0 & 0 & -140 & 140 \end{bmatrix} \begin{Bmatrix} u_1 \\ u_2 \\ u_3 \\ u_4 \end{Bmatrix}$$

This yields the equations

$$F_1 = 100u_1 - 100u_2 \tag{18.7}$$

$$F_2 = -100u_1 + 300u_2 - 200u_3 \tag{18.8}$$

$$F_3 = -200u_2 + 340u_3 - 140u_4 \tag{18.9}$$

$$F_4 = -140u_3 + 140u_4 \tag{18.10}$$

Using the boundary conditions that $u_1 = u_4 = 0$ and letting $F_2 = -50$ kN, $F_3 = 100$ kN then eqns (18.8) and (18.9) may be solved simultaneously to give $u_2 = 0.048$ mm and $u_3 = 0.323$ mm. Using eqns. (18.7) and (18.10) then gives $F_1 = 4.8$ kN and $F_4 = -45.2$ kN. These forces will both act to the right on the supports.

ANALYSIS OF FRAMEWORKS

Although the spring elements considered so far have been collinear this need not be the case. The spring element could, for example, be used to analyse the forces and deformation of a framework as shown in Fig. 18.6.

However, as the spring elements in the model are at different angles to one another it is necessary to express the forces and deformations for each element (calculated for its own *local* co-ordinate system x, y) to the

Fig. 18.6 (a) Framework; (b) model using spring elements

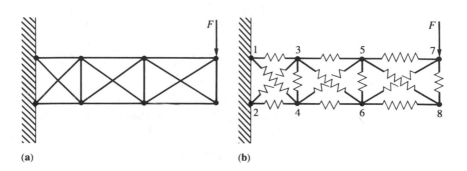

(a)

(b)

Fig. 18.7

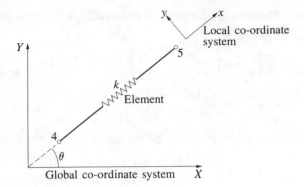

deformation of the whole framework relative to the global co-ordinate system X, Y. For example, consider member 4–5 in Fig. 18.7.

In the general case it is necessary to consider forces and displacements in both local co-ordinate directions at each node. For a pin-jointed member there will only be forces in the x-direction so the F_y components will be zero. Thus equation (18.4) may be expanded to the form

$$\begin{Bmatrix} F_{x4} \\ F_{y4} \\ F_{x5} \\ F_{y5} \end{Bmatrix} = \begin{bmatrix} k & 0 & -k & 0 \\ 0 & 0 & 0 & 0 \\ -k & 0 & k & 0 \\ 0 & 0 & 0 & 0 \end{bmatrix} \begin{Bmatrix} \delta_{x4} \\ \delta_{y4} \\ \delta_{x5} \\ \delta_{y5} \end{Bmatrix} \qquad (18.11)$$

Referring to Fig. 18.7 the forces in the local co-ordinate directions may be related to the forces in the global co-ordinate directions as follows:

$$F_{x4} = F_{X4} \cos \theta + F_{Y4} \sin \theta$$
$$F_{y4} = -F_{X4} \sin \theta + F_{Y4} \cos \theta$$
$$F_{x5} = F_{X5} \cos \theta + F_{Y5} \sin \theta$$
$$F_{y5} = -F_{X5} \sin \theta + F_{Y5} \cos \theta$$

Using $s = \sin \theta$ and $c = \cos \theta$ as a convenient shorthand, these equations may be expressed in matrix form

$$\begin{Bmatrix} F_{x4} \\ F_{y4} \\ F_{x5} \\ F_{y5} \end{Bmatrix} = \begin{bmatrix} c & s & 0 & 0 \\ -s & c & 0 & 0 \\ 0 & 0 & c & s \\ 0 & 0 & -s & c \end{bmatrix} \begin{Bmatrix} F_{X4} \\ F_{Y4} \\ F_{X5} \\ F_{Y5} \end{Bmatrix}$$

or

$$\{F\}_L = |T| \{F\}_G \qquad (18.12)$$

where L refers to the local co-ordinate system and G refers to the global co-ordinate system. Here, $|T|$ is called the transformation matrix. A similar analysis will show that the local displacements are related to the global displacements by the following equation:

$$\{\delta\}_L = |T| \{\delta\}_G \qquad (18.13)$$

Now, by combining eqns. (18.5), (18.12) and (18.13) we may write

$$|T|\{F\}_G = |K^e||T|\{\delta\}_G$$

Premultiplying each side by $|T|^{-1}$ this becomes

$$\{F\}_G = |T|^{-1}|K^e||T|\{\delta\}_G \tag{18.14}$$

or

$$\{F\}_G = |K^e|_G\{\delta\}_G \tag{18.15}$$

where $|K^e|_G$ is the stiffness matrix for the element in global coordinates. Performing the matrix multiplication $|T|^{-1}|K^e||T|$ gives the value of $|K^e|_G$ as

$$|K^e|_G = k\begin{bmatrix} c^2 & cs & -c^2 & -cs \\ cs & s^2 & -cs & -s^2 \\ -c^2 & -cs & c^2 & cs \\ -cs & -s^2 & cs & s^2 \end{bmatrix} \tag{18.16}$$

where $k = AE/L$.

The use of this method of analysing frameworks using spring elements is illustrated in the following example.

EXAMPLE 18.2

Determine the vertical and horizontal displacements at the loading point in the framework shown in Fig. 18.8. The value of AE for each of the members is 200 MN.

Fig. 18.8

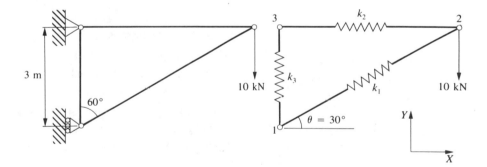

SOLUTION

Table 18.2

Member	Length (m)	θ	$c\ (=\cos\theta)$	$s\ (=\sin\theta)$	k (MN/m)
1–2	6	30	0.866	0.5	33.33
2–3	5.196	0	1	0	38.5
3–1	3	90	0	1	66.7

Then, from eqn. (18.16) the stiffness matrix for each of the elements in global co-ordinates will be

$$|K_1^e|_G = \begin{bmatrix} 25 & 14.4 & -25 & -14.4 \\ 14.4 & 8.33 & -14.4 & -8.33 \\ -25 & -14.4 & 25 & 14.4 \\ -14.4 & -8.33 & 14.4 & 8.33 \end{bmatrix}$$

$$|K_2^e|_G = \begin{bmatrix} 38.5 & 0 & -38.5 & 0 \\ 0 & 0 & 0 & 0 \\ -38.5 & 0 & 38.5 & 0 \\ 0 & 0 & 0 & 0 \end{bmatrix}$$

$$|K_3^e|_G = \begin{bmatrix} 0 & 0 & 0 & 0 \\ 0 & 66.7 & 0 & -66.7 \\ 0 & 0 & 0 & 0 \\ 0 & -66.7 & 0 & 66.7 \end{bmatrix}$$

If these matrices are expanded with rows and columns of zeros so that they are in an equivalent form, they may then be added to give the stiffness matrix for the whole structure. This will give equation (18.17) as follows

$$\begin{Bmatrix} F_{X1} \\ F_{Y1} \\ F_{X2} \\ F_{Y2} \\ F_{X3} \\ F_{Y3} \end{Bmatrix} = \begin{bmatrix} 25 & 14.4 & -25 & -14.4 & 0 & 0 \\ 14.4 & 75 & -14.4 & -8.33 & 0 & -66.7 \\ -25 & -14.4 & 63.5 & 14.4 & -38.5 & 0 \\ -14.4 & -8.33 & 14.4 & 8.33 & 0 & 0 \\ 0 & 0 & -38.5 & 0 & 38.5 & 0 \\ 0 & -66.7 & 0 & 0 & 0 & 66.7 \end{bmatrix} \begin{Bmatrix} \delta_{X1} \\ \delta_{Y1} \\ \delta_{X2} \\ \delta_{Y2} \\ \delta_{X3} \\ \delta_{Y3} \end{Bmatrix}$$

(18.17)

Recognizing that $\delta_{X1} = \delta_{X3} = \delta_{Y3} = 0$ and also $F_{Y1} = 0$ then this set of simultaneous equations may be solved for δ_{Y1}, δ_{X2} and δ_{Y2}. Hence

$\delta_{Y1} = -0.15$ mm (i.e. downwards)

$\delta_{X2} = 0.448$ mm (i.e. to the right)

$\delta_{Y2} = -2.12$ mm (i.e. downwards)

The values of the forces at the supports may also be determined from eqn. (18.17). This gives $F_{X1} = 17.3$ kN, $F_{Y1} = 10$ kN and $F_{X3} = -17.3$ kN which agrees with the values obtained from a simple equilibrium analysis of the structure.

The forces in each of the elements may be determined by reverting to the local co-ordinate system for each element. This gives[2]

$$F_{ij} = k_{ij}[c\ s]_{ij} \begin{Bmatrix} (\delta_{Xj} - \delta_{Xi}) \\ (\delta_{Yj} - \delta_{Yi}) \end{Bmatrix}_G$$

(18.18)

For example, for member 1–2, Fig. 18.8, the force is given by

$$F_{12} = 33.33[0.866 \quad 0.5] \begin{Bmatrix} 0.448 \\ -1.97 \end{Bmatrix} 10^{-3}$$

$$= 20 \text{ kN}$$

Similarly $F_{23} = 17.3$ kN, and these values agree with the values obtained by taking a free-body diagram at the loading point.

The framework in this example was deliberately kept simple in order that the steps in the solution could be illustrated and the calculation performed manually. For a large plane or three-dimensional structure the individual steps in the solution are identical to those illustrated but, although the calculations are straightforward, they are so numerous that they are best left to a computer. Even on a computer the time taken to obtain a solution can be relatively long if a large complex structure is being analysed. This is where it can be beneficial to have some understanding of the nature of the calculations being performed so that data input can be rationalized to streamline the solution procedure. For example one good way to achieve this in a structural analysis program is to ensure that the difference between the node numbers on each element is kept to the minimum. This has the effect of condensing the data into a band along the main diagonal of the overall stiffness matrix. This then reduces the subsequent computation and provides a valuable saving in computing time.

ANALYSIS OF CONTINUA

Although the use of the simple linear spring element is a convenient way to introduce finite element methods, it is quite limited in its application. The major advantage of the finite element method is its ability to model complex two-dimensional and three-dimensional solids. In these cases the elements used may be of the types shown earlier in Fig. 18.2. However, the approach to a solution is still similar to the method illustrated for the linear spring element. In essence the solid continuum is modelled by an array of plane or three-dimensional elements which are joined to each other at their node points. The system of external loads acting on the actual solid must then be replaced by an equivalent system of forces acting at the node points. The type and number of elements used can be decided by the analyst. In general the accuracy of the solution will be greater if the number of elements is large. However, computer time (and cost) also increases with the number of elements chosen so it is generally wise only to use a dense concentration of elements in the critical areas of the solid which are likely to be of particular interest. Typical examples are shown in Fig. 18.9.

A computer program is then used to obtain the distribution of forces and displacements in the solid based on the stiffness matrix for the particular type of element chosen. Before extending the analysis of the spring element to other more practical element shapes it is convenient to express the steps outlined on pages 468–75 in a more generalized form. In this analysis there is no necessity to employ global *and* local coordinate systems.

Fig. 18.9

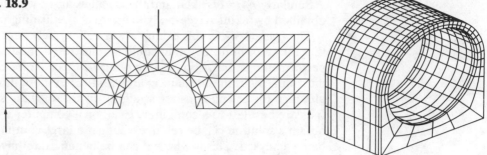

GEOMETRY OF DEFORMATION

In general, two-dimensional displacement patterns may be expressed in terms of two linear polynomials:

$$\delta_x = \alpha_1 + \alpha_2 x + \alpha_3 y \tag{18.19}$$

$$\delta_y = \beta_1 + \beta_2 x + \beta_3 y \tag{18.20}$$

where α_i and β_j are constants which may be determined from the co-ordinates of the nodal points of the element. The strains in the element may be determined from Chapter 14.

$$\varepsilon_x = \frac{\partial(\delta_x)}{\partial x} = \alpha_2 \tag{18.21}$$

$$\varepsilon_y = \frac{\partial(\delta_y)}{\partial y} = \beta_3 \tag{18.22}$$

$$\gamma_{xy} = \frac{\partial(\delta_y)}{\partial x} + \frac{\partial(\delta_x)}{\partial y} = \beta_2 + \alpha_3 \tag{18.23}$$

In matrix form, the strains in the element may therefore be related to the deformation of the nodes

$$\{\varepsilon\} = |B|\{\delta\} \tag{18.24}$$

where

$$\{\delta\} = \begin{Bmatrix} \delta_x \\ \delta_y \end{Bmatrix}$$

$$\{\varepsilon\} = \begin{Bmatrix} \varepsilon_x \\ \varepsilon_y \\ \gamma_{xy} \end{Bmatrix}$$

and $|B|$ is a matrix dependent on the geometric size and shape of the element. For the spring element of length L, this would be given by

$$|B| = \begin{bmatrix} \dfrac{1}{L} & -\dfrac{1}{L} \\ -\dfrac{1}{L} & \dfrac{1}{L} \end{bmatrix}$$

STRESS–STRAIN RELATIONSHIPS

For a two-dimensional plane stress element, the stresses and strains are related by the following equations:

$$\sigma_x = \left(\frac{E}{1-v^2}\right)(\varepsilon_x + v\varepsilon_y)$$

$$\sigma_y = \left(\frac{E}{1-v^2}\right)(\varepsilon_y + v\varepsilon_x)$$

$$\tau_{xy} = \frac{E}{2(1+v)}\gamma_{xy}$$

In matrix form these equations may be expressed as

$$\begin{Bmatrix} \sigma_x \\ \sigma_y \\ \tau_{xy} \end{Bmatrix} = \frac{E}{1-v^2} \begin{bmatrix} 1 & v & 0 \\ v & 1 & 0 \\ 0 & 0 & \frac{1}{2}(1-v) \end{bmatrix} \begin{Bmatrix} \varepsilon_x \\ \varepsilon_y \\ \gamma_{xy} \end{Bmatrix} \qquad (18.25)$$

or in shorthand form

$$\{\sigma\} = |D|\{\varepsilon\} \qquad (18.26)$$

where $|D|$ expresses the material stiffness properties of the element. For the spring element of modulus E this would be given by

$$D = \begin{bmatrix} E & 0 \\ 0 & E \end{bmatrix}$$

EQUILIBRIUM OF FORCES

The final step in the solution procedure is to relate the forces on each nodal point to the stresses on the element. In general this will be a simple force balance depending on the shape of the element. Therefore the forces in the x- and y-directions at each node may be written as

$$\{F\} = |A|\{\sigma\} \qquad (18.27)$$

where $|A|$, like $|B|$ is related to the shape and size of the element. For the spring element of area $\bar{A}$ this would be given by

$$|A| = \begin{bmatrix} \bar{A} & 0 \\ 0 & \bar{A} \end{bmatrix}$$

Now, combining eqns (18.27), (18.26) and (18.24) the forces at each node may be written as

$$\{F\} = |A||D||B|\{\delta\} \qquad (18.28)$$

The three matrices $|A|$, $|D|$ and $|B|$ may then be multiplied out to give

$$\{F\} = |K^e|\{\delta\}$$

which is equivalent to eqn (18.5), where $|K^e|$ was referred to as the stiffness matrix for the element. Note that for the spring element, the product $|A||D||B|$ gives

$$\begin{bmatrix} k & -k \\ -k & k \end{bmatrix}$$

as in eqn (18.4). Then, as shown in the analysis for the spring elements, the stiffness matrices for each element may be assembled into a stiffness matrix for the whole structure. Once this master matrix has been assembled it may then be transformed and solved to give the force and displacement distribution in the structure.

Having now considered the general approach to defining the characteristics of an element this approach will·be applied to a triangular element.

STIFFNESS MATRIX FOR A TRIANGULAR ELEMENT

Figure 18.10 shows a two-dimensional triangular element typical of that used in plane stress problems.

Fig. 18.10

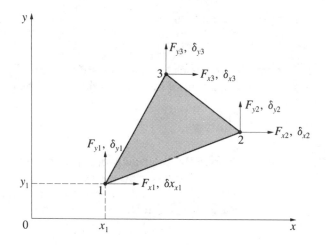

Geometry of deformation

From eqns. (18.19) and (18.20), at node 1

$$\delta_{x1} = \alpha_1 + \alpha_2 x_1 + \alpha_3 y_1$$

$$\delta_{y1} = \beta_1 + \beta_2 x_1 + \beta_3 y_1$$

Similar expressions may be written for δ_{x2}, δ_{y2}, δ_{x3} and δ_{y3}, and if each set of equations is written in matrix form so that they may be added then we get

$$\begin{Bmatrix} \delta_{x1} \\ \delta_{y1} \\ \delta_{x2} \\ \delta_{y2} \\ \delta_{x3} \\ \delta_{y3} \end{Bmatrix} = \begin{bmatrix} 1 & x_1 & y_1 & 0 & 0 & 0 \\ 0 & 0 & 0 & 1 & x_1 & y_1 \\ 1 & x_2 & y_2 & 0 & 0 & 0 \\ 0 & 0 & 0 & 1 & x_2 & y_2 \\ 1 & x_3 & y_3 & 0 & 0 & 0 \\ 0 & 0 & 0 & 1 & x_3 & y_3 \end{bmatrix} \begin{Bmatrix} \alpha_1 \\ \alpha_2 \\ \alpha_3 \\ \beta_1 \\ \beta_2 \\ \beta_3 \end{Bmatrix} \tag{18.29}$$

or in shorthand

$$\{\delta\} = |C| \{\alpha\beta\}$$

This may be rearranged to give the coefficients $\{\alpha\beta\}$ in terms of the displacements

$$|C|^{-1}\{\delta\} = \{\alpha\beta\} \tag{18.30}$$

By matrix manipulation of $|C|$, $|C|^{-1}$ may be written as

$$|C|^{-1} = \frac{1}{2a}\begin{bmatrix} a_1 & 0 & a_2 & 0 & a_3 & 0 \\ b_1 & 0 & b_2 & 0 & b_3 & 0 \\ c_1 & 0 & c_2 & 0 & c_3 & 0 \\ 0 & a_1 & 0 & a_2 & 0 & a_3 \\ 0 & b_1 & 0 & b_2 & 0 & b_3 \\ 0 & c_1 & 0 & c_2 & 0 & c_3 \end{bmatrix}$$

where

$$a_1 = (x_2y_3 - x_3y_2) \qquad b_1 = (y_2 - y_3) \qquad c_1 = (x_3 - x_2)$$
$$a_2 = (x_3y_1 - x_1y_3) \qquad b_2 = (y_3 - y_1) \qquad c_2 = (x_1 - x_3)$$
$$a_3 = (x_1y_2 - x_2y_1) \qquad b_3 = (y_1 - y_2) \qquad c_3 = (x_2 - x_1)$$
$$2a = (a_1 + a_2 + a_3) = 2(\text{area of element})$$

The next step is to get an expression for the strains in the element as a function of its geometry. Expressing eqns (18.21)–(18.23) in matrix form

$$\begin{Bmatrix} \varepsilon_x \\ \varepsilon_y \\ \gamma_{xy} \end{Bmatrix} = \begin{bmatrix} 0 & 1 & 0 & 0 & 0 & 0 \\ 0 & 0 & 0 & 0 & 0 & 1 \\ 0 & 0 & 1 & 0 & 1 & 0 \end{bmatrix} \begin{Bmatrix} \alpha_1 \\ \alpha_2 \\ \alpha_3 \\ \beta_1 \\ \beta_2 \\ \beta_3 \end{Bmatrix} \tag{18.31}$$

or in shorthand form

$$\{\varepsilon\} = |H|\{\alpha\beta\} \tag{18.32}$$

Then combining eqns. (18.30) and (18.32)

$$\{\varepsilon\} = |H|\,|C|^{-1}\{\delta\} \tag{18.33}$$

Comparing this with eqn. (18.24), hence,

$$|B| = |H|\,|C|^{-1}$$

So, multiplying the respective matrices (*see* Appendix B) gives,

$$|B| = \frac{1}{2a}\begin{bmatrix} b_1 & 0 & b_2 & 0 & b_3 & 0 \\ 0 & c_1 & 0 & c_2 & 0 & c_3 \\ c_1 & b_1 & c_2 & b_2 & c_3 & b_3 \end{bmatrix} \tag{18.34}$$

It may be seen that the terms in this matrix are known since they are a function of the co-ordinates of the nodes.

Stress–strain relationships

The stress–strain relationships depend on the material of the element, not on its shape so the equations are as developed on page 477. Thus the material stiffness matrix is

$$|D| = \begin{bmatrix} d_{11} & d_{12} & 0 \\ d_{21} & d_{22} & 0 \\ 0 & 0 & d_{33} \end{bmatrix} \tag{18.35}$$

where, for plane stress,

$$d_{11} = d_{22} = E/(1 - v^2)$$
$$d_{12} = d_{21} = Ev/(1 - v^2)$$
$$d_{33} = E/2(1 + v)$$

Equilibrium of forces

The effect of the external stress system on the triangular element is as shown in Fig. 18.11.

Fig. 18.11

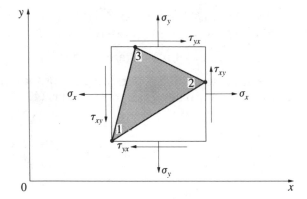

Assuming the thickness of the element to be h, the force on the left-hand side of the element is $-\sigma_x h(y_3 - y_1)$. This force is assumed to be shared equally by nodes 1 and 3, so making the earlier substitution of $b_2 = (y_3 - y_1)$ then

$$F'_{x1} = F'_{x3} = -\tfrac{1}{2}\sigma_x h b_2 \tag{18.36}$$

Note, however, that this is just one component of the forces at nodes 1 and 3. There will be other components due to the direct stress on the right-hand side of the element and the shear stresses on the top and bottom of the element.

Considering the right-hand side of the element, over nodes 1–2 the force to be shared is $\sigma_x h(y_2 - y_1)$. Hence

$$F''_{x1} = -\tfrac{1}{2}\sigma_x h b_3 \tag{18.37}$$

The force at node 1 due to the shear stresses will comprise of contributions from face 1–3 and face 1–2.

$$F'''_{x1} = -\tfrac{1}{2}\tau_{xy} h(c_2 + c_3) = \tfrac{1}{2}\tau_{xy} h c_1 \tag{18.38}$$

Combining eqns. (18.36), (18.37) and (18.38) gives

$$F_{x1} = \tfrac{1}{2}h(b_1\sigma_x + c_1\tau_{xy}) \tag{18.39}$$

Similar expressions may be obtained for the y-direction at node 1 and the x- and y-directions at nodes 2 and 3. The overall interrelationships between nodal forces and applied stresses may then be written in matrix form.

$$\begin{Bmatrix} F_{x1} \\ F_{y1} \\ F_{x2} \\ F_{y2} \\ F_{x3} \\ F_{y3} \end{Bmatrix} = \frac{h}{2} \begin{bmatrix} b_1 & 0 & c_1 \\ 0 & c_1 & b_1 \\ b_2 & 0 & c_2 \\ 0 & c_2 & b_2 \\ b_3 & 0 & c_3 \\ 0 & c_3 & b_3 \end{bmatrix} \begin{Bmatrix} \sigma_x \\ \sigma_y \\ \tau_{xy} \end{Bmatrix} \tag{18.40}$$

or in shorthand form

$$\{F\} = |A|\{\sigma\}$$

We are now in a position to determine the stiffness matrix for the triangular element since the matrices $|A|$, $|B|$ and $|D|$ are available from eqns (18.40), (18.34) and (18.35).

$$|K^e| = |A|\,|D|\,|B|$$

So multiplication of these matrices gives

$$|K^e| = \frac{h}{4a} \begin{bmatrix} k_{11} & k_{12} & k_{13} & k_{14} & k_{15} & k_{16} \\ k_{21} & k_{22} & k_{23} & k_{24} & k_{25} & k_{26} \\ k_{31} & k_{32} & k_{33} & k_{34} & k_{35} & k_{36} \\ k_{41} & k_{42} & k_{43} & k_{44} & k_{45} & k_{46} \\ k_{51} & k_{52} & k_{53} & k_{54} & k_{55} & k_{56} \\ k_{61} & k_{62} & k_{63} & k_{64} & k_{65} & k_{66} \end{bmatrix} \tag{18.41}$$

where $a =$ area of element and

$$k_{11} = d_{11}b_1{}^2 + d_{33}c_1{}^2 \qquad\qquad k_{23} = k_{32} = d_{12}b_2c_1 + d_{33}b_1c_2$$
$$k_{22} = d_{22}c_1{}^2 + d_{33}b_1{}^2 \qquad\qquad k_{24} = k_{42} = d_{22}c_1c_2 + d_{33}b_1b_2$$
$$k_{33} = d_{11}b_2{}^2 + d_{33}c_2{}^2 \qquad\qquad k_{25} = k_{52} = d_{12}b_3c_1 + d_{33}b_1c_3$$
$$k_{44} = d_{22}c_2{}^2 + d_{33}b_2{}^2 \qquad\qquad k_{26} = k_{62} = d_{22}c_1c_3 + d_{33}b_1b_3$$
$$k_{55} = d_{11}b_3{}^2 + d_{33}c_3{}^2 \qquad\qquad k_{34} = k_{43} = (d_{12} + d_{33})b_2c_2$$
$$k_{66} = d_{22}c_3{}^3 + d_{33}b_3{}^2 \qquad\qquad k_{35} = k_{53} = d_{11}b_2b_3 + d_{33}c_2c_3$$
$$k_{12} = k_{21} = (d_{12} + d_{33})b_1c_1 \qquad k_{36} = k_{63} = d_{12}b_2c_3 + d_{33}b_3c_2$$
$$k_{13} = k_{31} = d_{11}b_1b_2 + d_{33}c_1c_2 \qquad k_{45} = k_{54} = d_{21}b_3c_2 + d_{33}b_2c_3$$
$$k_{14} = k_{41} = d_{12}b_1c_2 + d_{33}c_1b_2 \qquad k_{46} = k_{64} = d_{22}c_2c_3 + d_{33}b_3b_2$$
$$k_{15} = k_{51} = d_{11}b_1b_3 + d_{33}c_1c_3 \qquad k_{56} = k_{65} = (d_{21} + d_{33})b_3c_3$$
$$k_{16} = k_{61} = d_{12}b_1c_3 + d_{33}c_1b_3$$

The stiffness matrix for a triangular element is thus a 6×6 matrix. It can be appreciated that when a structure is modelled by a large number of triangular elements, the structure stiffness matrix becomes extremely large and can only be handled by a computer. However, anyone intending to make use of finite elements should, at least once, compile a master stiffness matrix for a simple three- or four-element model. This will serve to illustrate the simple steps in the solution and will reinforce the value of a computer to solve such problems. The following example is a simple three-element model, but the extent of the computation involved gives an idea of what is required when solving a typical practical problem with, say, 1000 elements.

EXAMPLE 18.3

The plate in Fig. 18.12(a) is 4 mm thick and is subjected to a load of 120 kN at a point 0.1 m from the left-hand support. It is to be analysed using a three-element model as shown in Fig. 18.12(b). Construct the stiffness matrix for the whole plate. $E = 200 \, \text{GN/m}^2$, $v = 0.3$

Fig. 18.12

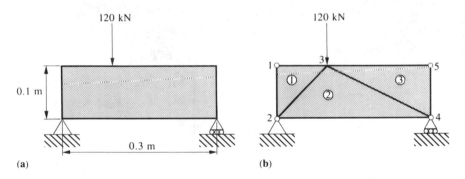

(a)　　　　　　　　　　　　　　　(b)

SOLUTION

For each of the three elements,

$$d_{11} = d_{22} = E/(1 - v^2) = 220 \times 10^6 \, \text{kN/m}^2$$

$$d_{12} = d_{21} = Ev/(1 - v^2) = 66 \times 10^6 \, \text{kN/m}^2$$

$$d_{33} = E/2(1 + v) = 77 \times 10^6 \, \text{kN/m}^2$$

Element 1　　　$a_1 = \frac{1}{2}(0.1 \times 0.1) = 0.005 \, \text{m}^2$　　　$c_1 = 0.1 - 0 = 0.1 \, \text{m}$

$b_1 = 0 - 0.1 = -0.1 \, \text{m}$　　　　　　　$c_2 = 0 - 0.1 = -0.1 \, \text{m}$

$b_2 = 0.1 - 0.1 = 0$　　　　　　　　　　$c_3 = 0 - 0 = 0$

$b_3 = 0.1 - 0 = 0.1 \, \text{m}$

Using eqn. (18.41), we obtain eqn. (18.42) as follows:

$$
\begin{Bmatrix} F_{x1} \\ F_{y1} \\ F_{x2} \\ F_{y2} \\ F_{x3} \\ F_{y3} \end{Bmatrix} = \frac{h}{4a_1}
\begin{bmatrix}
2.97 & -1.43 & -0.77 & 0.66 & -2.2 & 0.77 \\
-1.43 & 2.97 & 0.77 & -2.2 & 0.66 & -0.77 \\
-0.77 & 0.77 & 0.77 & 0 & 0 & -0.77 \\
0.66 & -2.2 & 0 & 2.2 & -0.66 & 0 \\
-2.22 & 0.66 & 0 & -0.66 & 2.2 & 0 \\
0.77 & -0.77 & -0.77 & 0 & 0 & 0.77
\end{bmatrix}
\begin{Bmatrix} \delta_{x1} \\ \delta_{y1} \\ \delta_{x2} \\ \delta_{y2} \\ \delta_{x3} \\ \delta_{y3} \end{Bmatrix}
$$

$$(18.42)$$

Element 2   Similarly for element 2, we obtain eqn. (18.43) as follows:

$$
\begin{Bmatrix} F_{x2} \\ F_{y2} \\ F_{x3} \\ F_{y3} \\ F_{x4} \\ F_{y4} \end{Bmatrix} = \frac{h}{4a_2}
\begin{bmatrix}
5.28 & 2.86 & -4.62 & -1.98 & -0.66 & -0.88 \\
2.86 & 9.57 & -2.31 & -13.2 & -0.55 & 3.63 \\
-4.62 & -2.31 & 6.93 & 0 & -2.31 & 2.31 \\
-1.98 & -13.2 & 0 & 19.9 & 1.98 & -6.6 \\
-0.66 & -0.55 & -2.31 & 1.98 & 2.97 & -1.43 \\
-0.88 & 3.63 & 2.31 & -6.6 & -1.43 & 2.97
\end{bmatrix}
\begin{Bmatrix} \delta_{x2} \\ \delta_{y2} \\ \delta_{x3} \\ \delta_{y3} \\ \delta_{x4} \\ \delta_{y4} \end{Bmatrix}
$$

$$(18.43)$$

Note that for element 2 the local numbering is $1 \rightarrow$ node 2, $2 \rightarrow$ node 4, $3 \rightarrow$ node 3. Thus the terms in the above matrix have had to be rearranged to be compatible with the other two matrices.

Element 3   For element 3 we obtain eqn. (18.44) as follows:

$$
\begin{Bmatrix} F_{x3} \\ F_{y3} \\ F_{x4} \\ F_{y4} \\ F_{x5} \\ F_{y5} \end{Bmatrix} = \frac{h}{4a_3}
\begin{bmatrix}
2.2 & 0 & 0 & 1.32 & -2.2 & -1.32 \\
0 & 0.77 & 1.54 & 0 & -1.54 & -0.77 \\
0 & 1.54 & 3.08 & 0 & -3.08 & -1.54 \\
1.32 & 0 & 0 & 8.8 & -1.32 & -8.8 \\
-2.2 & -1.54 & -3.08 & -1.32 & 5.28 & 2.86 \\
-1.32 & -0.77 & -1.54 & -8.8 & 2.86 & 9.57
\end{bmatrix}
\begin{Bmatrix} \delta_{x3} \\ \delta_{y3} \\ \delta_{x4} \\ \delta_{y4} \\ \delta_{x5} \\ \delta_{y5} \end{Bmatrix}
$$

$$(18.44)$$

If all the terms in the matrices of eqns. (18.42) and (18.44) are factored so that they are expressed as a function of $h/4a_2$ then using the rules of matrix addition eqns. (18.42), (18.43) and (18.44), may be added to give the following:

$$
\begin{Bmatrix} F_{x1} \\ F_{y1} \\ F_{x2} \\ F_{y2} \\ F_{x3} \\ F_{y3} \\ F_{x4} \\ F_{y4} \\ F_{x5} \\ F_{y6} \end{Bmatrix}
= \frac{h}{4a_2}
\begin{bmatrix}
k^{(1)}_{11} & k^{(1)}_{12} & k^{(1)}_{13} & k^{(1)}_{14} & k^{(1)}_{15} & k^{(1)}_{16} & 0 & 0 & 0 & 0 \\
 & k^{(1)}_{22} & k^{(1)}_{23} & k^{(1)}_{24} & k^{(1)}_{25} & k^{(1)}_{26} & 0 & 0 & 0 & 0 \\
 & & \begin{smallmatrix}k^{(1)}_{33}\\+k^{(2)}_{11}\end{smallmatrix} & \begin{smallmatrix}k^{(1)}_{34}\\+k^{(2)}_{12}\end{smallmatrix} & \begin{smallmatrix}k^{(1)}_{35}\\+k^{(2)}_{15}\end{smallmatrix} & \begin{smallmatrix}k^{(1)}_{36}\\+k^{(2)}_{16}\end{smallmatrix} & k^{(2)}_{13} & k^{(2)}_{14} & 0 & 0 \\
 & & & \begin{smallmatrix}k^{(1)}_{44}\\+k^{(2)}_{22}\end{smallmatrix} & \begin{smallmatrix}k^{(1)}_{45}\\+k^{(2)}_{25}\end{smallmatrix} & \begin{smallmatrix}k^{(1)}_{46}\\+k^{(2)}_{26}\end{smallmatrix} & k^{(2)}_{23} & k^{(2)}_{24} & 0 & 0 \\
 & & & & \begin{smallmatrix}k^{(1)}_{55}\\+k^{(2)}_{55}\\+k^{(3)}_{11}\end{smallmatrix} & \begin{smallmatrix}k^{(1)}_{56}\\+k^{(2)}_{56}\\+k^{(3)}_{12}\end{smallmatrix} & \begin{smallmatrix}k^{(2)}_{53}\\+k^{(3)}_{13}\end{smallmatrix} & \begin{smallmatrix}k^{(2)}_{54}\\+k^{(3)}_{14}\end{smallmatrix} & k^{(3)}_{15} & k^{(3)}_{16} \\
 & & & & & \begin{smallmatrix}k^{(1)}_{66}\\+k^{(2)}_{66}\\+k^{(3)}_{22}\end{smallmatrix} & \begin{smallmatrix}k^{(2)}_{63}\\+k^{(3)}_{23}\end{smallmatrix} & \begin{smallmatrix}k^{(2)}_{64}\\+k^{(3)}_{24}\end{smallmatrix} & k^{(3)}_{25} & k^{(3)}_{26} \\
 & & & & & & \begin{smallmatrix}k^{(2)}_{33}\\+k^{(3)}_{33}\end{smallmatrix} & \begin{smallmatrix}k^{(2)}_{34}\\+k^{(3)}_{34}\end{smallmatrix} & k^{(3)}_{35} & k^{(2)}_{36} \\
 & & & \text{Symmetrical} & & & & \begin{smallmatrix}k^{(2)}_{44}\\+k^{(3)}_{44}\end{smallmatrix} & k^{(3)}_{45} & k^{(3)}_{46} \\
 & & & & & & & & k^{(3)}_{55} & k^{(3)}_{56} \\
 & & & & & & & & & k^{(3)}_{66}
\end{bmatrix}
\begin{Bmatrix} \delta_{x1} \\ \delta_{y1} \\ \delta_{x2} \\ \delta_{y2} \\ \delta_{x3} \\ \delta_{y3} \\ \delta_{x4} \\ \delta_{y4} \\ \delta_{x5} \\ \delta_{y5} \end{Bmatrix}
$$

Inserting the numerical values gives the following equation:

$$
\begin{Bmatrix} F_{x1} \\ F_{y1} \\ F_{x2} \\ F_{y2} \\ F_{x3} \\ F_{y3} \\ F_{x4} \\ F_{y4} \\ F_{x5} \\ F_{y5} \end{Bmatrix}
= \frac{h}{4a_2}
\begin{bmatrix}
8.91 & -4.3 & -2.31 & 1.98 & -6.6 & 2.31 & 0 & 0 & 0 & 0 \\
 & 8.91 & 2.31 & -6.6 & -1.98 & -2.31 & 0 & 0 & 0 & 0 \\
 & & 7.59 & 2.86 & -4.62 & -4.29 & -0.66 & -0.88 & 0 & 0 \\
 & & & 16.2 & -4.29 & -13.2 & -0.55 & 3.63 & 0 & 0 \\
 & & & & 16.83 & 0 & -2.31 & 4.29 & -3.3 & -1.98 \\
 & & & & & 23.3 & 4.29 & -6.6 & -2.31 & -1.16 \\
 & & & & & & 7.6 & -1.43 & -4.62 & -2.31 \\
 & & & & & & & 16.2 & -1.98 & -13.2 \\
 & \text{Symmetrical} & & & & & & & 17.92 & 4.29 \\
 & & & & & & & & & 14.4
\end{bmatrix}
\begin{Bmatrix} \delta_{x1} \\ \delta_{y1} \\ \delta_{x2} \\ \delta_{y2} \\ \delta_{x3} \\ \delta_{y3} \\ \delta_{x4} \\ \delta_{y4} \\ \delta_{x5} \\ \delta_{y5} \end{Bmatrix}
$$

When solving this set of equations it is worth noting that they are not a conventional set of simultaneous equations. This is because only some of the 'F' terms are known and some of the 'δ' terms are known from the boundary conditions. In this case, for example, it is known that $\delta_{x2} = \delta_{y2} = \delta_{y4} = 0$. The usual way to satisfy such boundary conditions[3] is to multiply by a very large number the matrix stiffness values which will be multiplied by δ_{x2}, δ_{y2} and δ_{y4}. As a consequence of this the values of

F_{x2}, F_{x3} and F_{y4} need not be known (let them equal 1) and the solution of the equations will provide all the 'δ' values with $\delta_{x2} = \delta_{y2} = \delta_{y4} \approx 0$. The values of '$\delta$' can then be substituted back to get the unknown forces.

Solution of the above equation gives

$$\delta_{x1} = 5.35 \times 10^{-2}\ \text{mm} \qquad \delta_{x4} = 9.6 \times 10^{-2}\ \text{mm}$$

$$\delta_{y1} = -0.9 \times 10^{-2}\ \text{mm} \qquad \delta_{x5} = 4.6 \times 10^{-2}\ \text{mm}$$

$$\delta_{x3} = 4.4 \times 10^{-2}\ \text{mm} \qquad \delta_{y5} = -0.0135 \times 10^{-2}\ \text{mm}$$

$$\delta_{y3} = -9.7 \times 10^{-2}\ \text{mm}$$

Also $F_{y2} = 80$ kN and $F_{y4} = 40$ kN which can be confirmed from equilibrium and moment equations on the whole plate.

It should be realized, however, that these values of deflections are not likely to be accurate due to the simplicity of the three-element model. As mentioned earlier, the accuracy of the solution will increase with the number of elements used. If this problem is solved on a computer using a mesh with 600 elements as shown in Fig. 18.13, then it is found that the deflections predicted by the three-element model are incorrect by a factor of about 20. With the 600-element model it is found that $\delta_{x3} = 0.85$ mm and $\delta_{y3} = -1.56$ mm.

Fig. 18.13 Plate divided into 600 finite elements

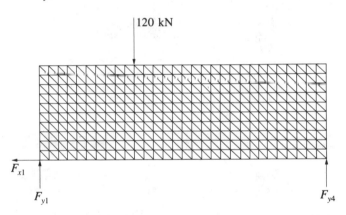

120 kN

F_{x1}

F_{y1}

F_{y4}

SUMMARY

This chapter has been written as a very basic introduction to the finite element method. The approach has been to show that in the first place it is possible to develop quite sophisticated methods of analysis for structures by treating the ties and struts as simple spring elements. Then, having established the basis of the solution method for the one-dimensional spring element it is shown how this can be extended for use with two- and three-dimensional elements. In each case it is apparent that the underlying basis of the finite element method is the application of the conditions of *equilibrium of forces*, *compatibility of displacements* and *stress–strain relationships*. To illustrate this a plane stress triangular element has been analysed in full and a similar analysis can be applied to the other types of finite elements referred to.

Finally it should be stressed that the reader who is unfamiliar with the use of matrices should not be put off by their use in the finite element method. Matrix algebra is used simply as a convenient method of dealing with the large number of simultaneous equations which arise. The reader is encouraged to attempt a few numerical problems of the type shown in the chapter in order to appreciate the simplicity of the mathematics and to gain a better insight to the subtleties of the solution method.

REFERENCES

1. Rockey, K. C., Evans H. R, Griffiths, D. W. and Nethercott, D. A., *The Finite Element Method,* Granada, London, 1975.
2. Budynas R. C. *Advanced Strength of Materials and Applied Stress Analysis,* McGraw-Hill, New York, 1977.
3. Paulsen W. C. "Finite element stress analysis", *Machine Design,* September, 1971, p. 46.

BIBLIOGRAPHY

Holand, I. and Bell, K. *Finite Element Method in Stress Analysis,* Tapir Forlag, Trondheim, Norway, 1970.
Livesley, R. K. *Finite Elements. An Introduction for Engineers,* Cambridge University Press, 1980.
Ross, C. T. F. *Computational Methods in Structural and Continuum Mechanics,* Ellis Horwood, Chichester, 1982.
Zienkiewicz, O. C. *The Finite Element Method in Engineering Science,* McGraw-Hill, New York, 1971.

PROBLEMS

18.1 Three dissimilar cylindrical rods are jointed together as indicated in Fig. 18.14. The brass cylinder is fixed to the rigid support at the top and the aluminium

Fig. 18.14

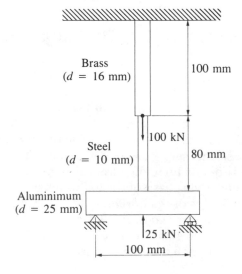

cylinder may be regarded as simply-supported with a point load at mid-span. Using the stiffness matrix approach, calculate the movement at each interface and the forces at the rigid supports when 25 kN and 100 kN forces are applied as indicated. $E_b = 100 \, \text{GN/m}^2$, $E_s = 200 \, \text{GN/m}^2$, $E_a = 70 \, \text{GN/m}^2$.

18.2 For the simple pin-jointed framework shown in Fig. 18.15 use the stiffness matrix approach to calculate the vertical deflection at the 20 kN load and the forces in each of the members. For each member the product $AE = 240 \, \text{MN}$.

Fig. 18.15

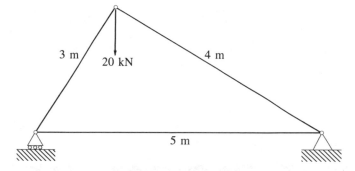

18.3 Use the method of stiffness matrices to determine the vertical deflection at the loading point in the plane pin-jointed framework shown in Fig. 18.16. The product of cross-sectional area and modulus for each member is 400 MN.

Fig. 18.16

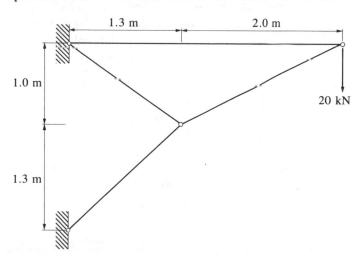

18.4 The cantilever beam shown in Fig. 18.17 is to be represented by three triangular finite elements. Construct the master stiffness matrix for the beam.

Fig. 18.17

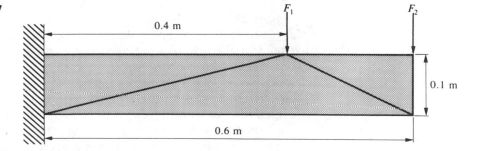

Tension, compression, torsion and hardness

INTRODUCTION

For the design engineer it is just as essential to have an understanding of material behaviour, to aid in the appropriate selection of type and condition, as it is to be able to calculate the stresses and strains which will have to be withstood by the material of a component or structure. This textbook is primarily devoted to the latter aspect and there are many excellent texts which deal fully with the engineering properties of materials. These remaining four chapters are intended therefore to provide a sufficient introduction to enable the reader to appreciate the significance of material response to stress–strain and environmental conditions.

This first chapter concentrates on the principal laboratory tests which are used to characterize materials from which British Standards and manufacturers' specifications of material properties are derived.

STRESS–STRAIN RESPONSE IN A TENSION TEST

The principal concepts of elastic and plastic uniaxial tensile stress–strain behaviour were introduced in Chapter 3 and it is therefore only necessary briefly to reiterate certain key aspects.

Figure 19.1(a) shows the typical shape of a flat or round bar specimen for tension testing (B.S. 18: Part 2: 1971). The enlarged ends are for gripping in the jaws of a testing machine and the reduced parallel portion contains the *gauge length*, across the ends of which is mounted an *extensometer*. This is a sensitive instrument which measures the very small longitudinal deformations that occur in the *elastic range*.

If a second instrument is used to measure the lateral contraction of the gauge length as the load is increased then the ratio of lateral to longitudinal strains can be determined, which is Poisson's ratio v.

Fig. 19.1 (a) Tensile specimens; (b) elastic stress–strain curve

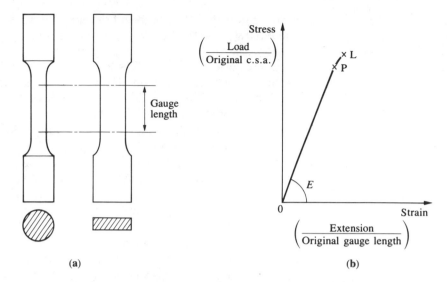

(a)

(b)

Figure 19.1(b) is a stress–strain graph for the elastic range which is derived from the loads applied by the testing machine and the extensions of the gauge length measured by the extensometer. For metals it is a linear relationship, obeying Hooke's law, the slope of which is the Young's modulus of elasticity E. The end point of linearity is termed the *limit of proportionality P,* and a closely adjacent point is the *elastic limit L.*

When straining of a ductile metal is continued beyond the elastic limit, yielding commences and the material is in the plastic range. There are two characteristically different types of transition from elastic to plastic behaviour. These are illustrated in Fig. 19.2(*a*) and (*b*): the first is principally found in low- and medium-carbon steels and the second in alloy steels and non-ferrous metals. In the former, the point Y_u, at which there is a sudden drop in load with further strain, is termed the *upper yield point.* This is followed by Y_L, the *lower yield point,* from which there is a marked extension at almost constant load.

The upper yield point of low- and medium-carbon steels is a complex phenomenon, which is a function of strain rate, temperature, type of testing machine and geometry of the specimen. In some cases the upper

Fig. 19.2 Nominal and true tensile stress–strain curves

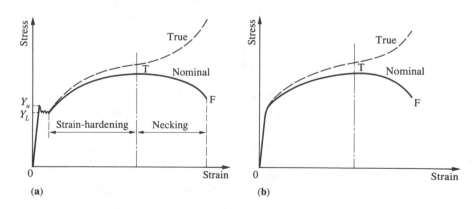

(a)

(b)

yield point does not appear and the curve departs from the elastic range immediately into the horizontal lower yield range. The latter, however, may be regarded as a material property and is used as the yield point stress for design purposes. It is this rather unique shape of stress–strain curve to which one approximates when using an ideal elastic–plastic relationship in plastic bending or torsion theory as in Chapter 16, since mild steel is widely used for structural work.

Many materials, particularly the light alloys, do not exhibit a clearly-defined elastic limit, limit of proportionality or yield point, and several methods of indicating these stresses are in use. The most widely used is that involving a measure of permanent set and is called the *offset method*.

Figure 19.3 shows the stress–strain relationship for a material stressed beyond the limit of proportionality and then unloaded. The slope during the unloading stage is practically similar to that during the elastic range of loading. The stress for any given amount of inelastic deformation is easily obtained from the stress–strain diagram. The amount x is set off on the strain or extension axis and a line AB is drawn parallel to the straight line portion of the loading curve, and the intersection at B gives the stress σ_p for a permanent set of $x\%$ strain.

Fig. 19.3 Determination of proof stress

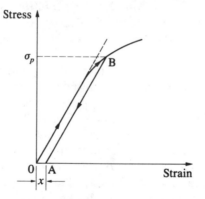

The elastic limit is taken to be that stress at which there is a permanent set of 0.02%; this is generally called the 0.02% elastic limit. The proof stress is that stress at which there is a permanent set of 0.1% and is called the 0.1% proof stress. The yield stress is that stress at which there is a permanent set of 0.2% of the gauge length.

On a 50 mm gauge length, the extensions corresponding to the above stresses are 0.01, 0.05 and 0.1 mm respectively, and thus extensometer measurements are required for the accurate production of the load–extension diagram.

Once the initial yield region has been passed in Figs. 19.2(a) and (b), the stress–strain curves have a common form. An increasing stress is required to cause continued straining, and this behaviour is known as *work-hardening* or *strain-hardening* and the metal does in fact become harder.

The property of work-hardening is quite important and will be discussed further later. The specimen has a limit to which it can be work-hardened uniformly in a simple tension test, and this is reached when the slope of the nominal stress–strain curve becomes zero as at T, in Figs. 19.2(a) and (b). This point is known as the *tensile strength* (in the past, the ultimate tensile strength U.T.S.) and is given by the maximum load carried by the specimen divided by the original cross-sectional area. This is a very important quantity as it is sometimes used in design in conjunction with a safety factor, and is always quoted when comparing metals and describing their mechanical properties. It is also an approximate guide to hardness and fatigue strength.

Up to the point T, the parallel gauge length of the specimen has reduced in cross-section quite uniformly; however, at or close to the tensile strength T one section is slightly weaker due to inhomogeneity than the rest and begins to thin more rapidly, forming a waist or *neck* in the gauge length. Further, extension is now concentrated in the neck and a reducing load is required for continued straining, and thus the nominal stress–strain curve also falls off, until fracture occurs at F.

The lower stress–strain curves in Fig. 19.2 were based on nominal values of stress, that is, the load at all points was divided by the original cross-sectional area. This is a convenient arrangement for most practical purposes and is the generally adopted procedure. However, as the specimen is strained, so the cross-sectional area reduces, and hence the true stress is higher than the nominal stress. The true stress is obtained by dividing the load by the current area of the specimen corresponding to that load. The current area of the bar may be obtained by direct measurement at various stages during the test or by calculation.

The shape of the true stress curve is shown in Fig. 19.2. It is only strictly valid up to the point where necking commences, since the change in geometry at the neck sets up a complex stress system which cannot be determined simply from the load divided by the area of the neck.

The foregoing has dealt principally with the property of "strength" or stress; another property which is of almost equal importance is that of *ductility*, or the ability of a material to withstand plastic deformation. In a tensile test this is expressed in two ways, by the percentage elongation of the gauge length after fracture and by the percentage reduction in cross-sectional area referred to the neck or minimum section at fracture.

The second quantity is expressed algebraically as

$$\text{Reduction in area} = \frac{A_0 - A_F}{A_0} \times 100\%$$

where A_0 and A_F are the original and fracture areas respectively.

Elongation is the increase in the gauge length divided by the original gauge length, or algebraically

$$\text{Elongation} = \frac{L_F - L_0}{L_0} \times 100\%$$

where L_0 and L_F are the original and final gauge lengths respectively. Elongation is only partly a material property since it is also dependent on the geometrical form of the test piece.

Fig. 19.4 Strain distribution along a tensile test piece

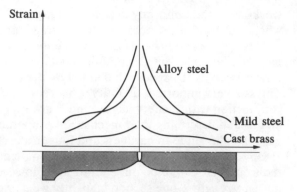

The distribution of strain over the gauge length of a specimen can vary considerably from one metal to another, as illustrated in Fig. 19.4 depending on the grain size and microstructure.

The elongation at fracture includes the necked region, and this is relatively independent of original gauge length and is more a function of the material and shape of cross-section.

Two basic types of fracture can be obtained in tension, depending on the material, temperature, strain rate, etc. and these are termed *brittle* and *ductile*. The main features of the former are that there is little or no plastic deformation, the plane of fracture is normal to the tensile stress and separation of the crystal structure occurs. In the second type, ductile failure is preceded by a considerable amount of plastic deformation and fracture is by shear or sliding of the crystal structure on microscopic or general planes at about 45° to the tensile stress. Probably the most well-known type of failure is called the *cup and cone* in the cylindrical bar.

A cylindrical bar, even though relatively ductile, may not produce a sufficient neck to set up marked triaxiality of stress, and failure is found to occur on a single shear plane right across the specimen. When the two parts of a fractured flat bar are placed together it is seen that there is a gap in the middle region, showing the initiation of fracture there, while the outer regions continued to extend.

STRESS–STRAIN RESPONSE IN A COMPRESSION TEST

The mechanical properties of a ductile metal are generally obtained from a tension test. However, compression behaviour is of interest in the metal-forming industry, since most processes, rolling, forging, etc. involve compressive deformations of the metal, and also often of the forming equipment (rolling mill).

In compression an elastic range is exhibited as in tension and the elastic modulus, proportional limit and yield point or proof stresses have closely corresponding values for the two types of deformation. The real problem arises in a compression test when the metal enters the plastic range. The test piece has to be relatively short ($L/D \not> 4$) to avoid the possibility of

Fig. 19.5 Deformation during a compression test

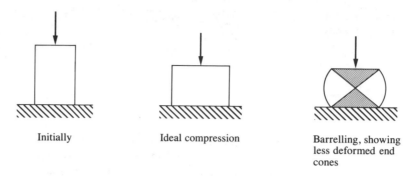

Initially   Ideal compression   Barrelling, showing less deformed end cones

instability and buckling. The axial compression is accompanied by lateral expansion, but this is restrained at the ends of the specimen owing to the friction between the machine platens and the end faces, and consequently on a shortish specimen marked barrelling occurs as in Fig. 19.5. This causes a non-uniformity of stress distribution, and conical sections of material at each end are strained and hardened to a lesser degree than the central region. The effect on the load–compression curve, after the smaller values of plastic strain have been achieved, is a fairly rapid rise in the load required to overcome friction and cause further compression.

Owing to the barrelling effect, only an average stress can be computed from the load–compression curve, based on an average area determined from considerations of constant volume.

Various methods have been attempted to overcome the effects of barrelling, none of which is completely successful. The most satisfactory appears to be the technique of using several cylinders of the same metal having different diameter-to-length ratios. Incremental compression tests are conducted on the set of cylinders at a series of loads of increasing magnitude, measuring the strain for each of the cylinders at each load. Extrapolation of curves of D/L against strain with load as parameter to a value of $D/L = 0$, representing an infinitely long specimen where barrelling would be negligible, enables the true compressive stress–strain curve to be determined, Fig. 19.6. Failure of a ductile metal in compression only occurs owing to excessive barrelling causing axial splitting around the periphery.

Fig. 19.6 Compressive stress–strain curves for various diameter length ratios

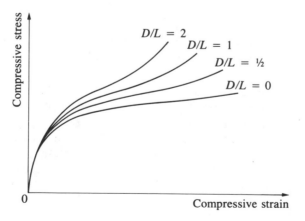

Fig. 19.7 Modes of compression failure in various materials

Concrete

Flake iron

Timber

For brittle materials, such as flake cast-iron, concrete, etc. which would not normally be used in tension, the compression test is used to give quantitative mechanical properties. Although end friction still occurs, which affects the stress values somewhat, owing to the absence of ductility in these materials the barrelling condition is barely achieved. Fracture takes place on planes of maximum shear stress as illustrated in Fig. 19.7.

STRESS–STRAIN RESPONSE IN A TORSION TEST

The usual method of obtaining a relationship between shear stress and shear strain for a material is by means of a torsion test. This may be conducted on a circular-section solid or tubular bar. By applying a torque to each end of the test piece by a testing machine and measuring the angular twist over a specified gauge length, a torque–twist diagram can be plotted. This is the equivalent in torsion to the load–extension diagram in tension. It was shown theoretically in Chapter 15 that, under elastic conditions in torsion, the applied torque is proportional to the angle of twist on the assumption that shear stress is proportional to shear strain. This is found to be true experimentally, and the linear torque–twist relationship obtained enables the shear or rigidity modulus to be determined, since

$$G = \frac{T}{\theta} \frac{L}{J}$$

where the symbols are as specified in Chapter 5.

The torsion test is not like the tension or compression tests in which the stress is uniform across the section of the specimen. In torsion there is a stress gradient across the cross-section, and hence at the end of the elastic range yielding commences in the outer fibres first while the core is still elastic, whereas in the direct stress tests yielding occurs relatively evenly throughout the material. With continued twisting into the plastic range, more and more of the cross-section yields until there is penetration to the axis of the bar (*see* Section 16.4). The torque–twist diagram, Fig. 19.8 appears of much the same form as a load–extension diagram, and work-hardening will occur at a gradually decreasing rate as straining proceeds, but of course there is no fall off in the curve, as in tension, since necking cannot take place. In fact, ductile metals can

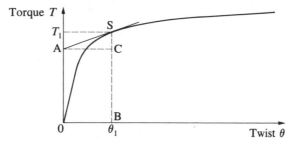

Fig. 19.8 Torque-twist curve and Nadai construction

absorb extremely high values of shear strain (200%) before failure occurs.

Although the shear-stress/shear-strain relationship can be determined easily in the elastic range, difficulties are introduced for a solid bar in the plastic range due to the stress variation mentioned above. One solution to this problem is to conduct the torsion test on a thin-walled tubular specimen in which the shear stress in the plastic range may be assumed to be constant through the wall thickness, and is given by

$$\tau = \frac{T}{2\pi r^2 t}$$

where t is the wall thickness and r is the mean radius. Shear strain is obtained in the plastic range from the same assumptions that apply in the elastic range; hence

$$\gamma = \frac{r\theta}{L}$$

(The tangent of the angle γ must be used for large strains.)

For a solid bar in torsion Nadai developed a construction for the shear–stress/shear-strain curve from the torque–twist curve. The shear strain γ is obtained from the relationship above and the shear stress on the outer surface is given by

$$\tau = \frac{1}{2\pi r^3} \left\{ \theta \frac{\mathrm{d}T}{\mathrm{d}\theta} + 3T \right\}$$

Referring to the torque–twist diagram in Fig. 19.8, at the point S the torque is T_1 and the angle of twist per unit length is θ_1. Drawing a tangent to the curve at S, the intercept on the ordinate is A, and from A drawing a horizontal line to cut SB at C, then

$$\theta \frac{\mathrm{d}T}{\mathrm{d}\theta} + 3T = CS + 3BS$$

Hence the shear stress τ can be computed from the $T-\theta$ curve for various values of θ or γ_0/r_0.

The relationship between the yield stress in simple tension, and that in pure shear is of interest in that it can be found from the von Mises expression for yield criterion, eqn. (13.15).

Putting $\sigma_3 = 0$ and $\sigma_1 = -\sigma_2 = \tau$ for pure shear we have

$$6\tau^2 = 2\sigma_Y^2$$

or

$$\tau = \sigma_Y/\sqrt{3}$$

Hence the yield stress in torsion is 0.577 times the yield stress in simple tension. It can also be shown that increments of plastic shear strain are equal to $\sqrt{3}$ times the increments of plastic tensile strain for a material so that it is possible to construct the *plastic* shear stress–strain curve from the *plastic* tensile stress–strain curve.

Fracture in torsion for ductile metals generally occurs in the plane of maximum shear stress perpendicular to the axis of the bar, whereas for brittle materials failure occurs along a 45° helix to the axis of the bar due to tensile stress across that plane.

PLASTIC OVERSTRAIN AND HYSTERESIS

The behaviour of materials in their plastic range is of particular interest in relation to the analytical treatment in Chapter 16 of plasticity of engineering components.

If a metal is taken into its plastic range in tension, compression or torsion, and at some point the load is removed, then the unloading line, although having slight curvature, approximates to the slope of the original elastic range. On reapplication of load the line diverges only slightly from that for unloading, and yielding occurs only when the load has reached the previous point of commencement of unloading. This effect is shown diagrammatically in Fig. 19.9 and is known as *overstraining*. It should be emphasized that this word does not mean damage to the static properties of the metal although it may convey that impression.

Overstrain is in fact a very useful property in enabling a higher yield range to be obtained, although at the expense of elasticity and proportionality. The latter can be regained by a rest period or mild heat treatment (boiling water). Instances of the usefulness of overstraining on components are mentioned on pages 418, 437, and 438.

Fig. 19.9 Hysteresis loops during overstrain

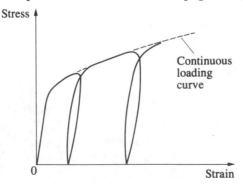

In Fig. 19.9 the "loops" formed by the unloading and reloading lines during overstrain are caused by mechanical hysteresis, that is, there is a lag between stress relaxation and strain recovery, and similarly on reloading. In unidirectional loading (tension only) hysteresis loops are generally quite narrow and in some metals, virtually non-existent. Hysteresis is also the term used to describe the loop obtained when reversed loading is conducted on a metal, i.e. through yield in tension followed by compression or vice versa. Figure 19.10 shows the form of a tension–compression loop.

Fig. 19.10 Hysteresis loop for reversed loading

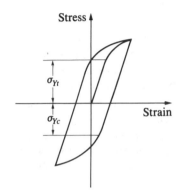

HARDNESS MEASUREMENT

Hardness of materials is a concept which is not directly employed in engineering design as for example yield stress may be. However, it is a "property" which is intimately related to strength and ductility, as discussed in the next section, and in that context it gives valuable quality-control information.

The term *hardness* of a material may be defined in several different ways. These are principally in relation to the resistance to deformation such as indentation, abrasion, scratching and machining. Although the last three are of importance in certain circumstances, they have a limited application in practice, and therefore discussion will be restricted to hardness as measured by resistance to indentation. One of the first recorded tests of this type was made by de Réaumur (1722) in which a piece of material was indented by a tool made of the same material and the volume of the resulting indentation measured. Since then there have been several variations of the principle of this type of test used and widely adopted in engineering practice. The most common are the Brinell, Vickers and Rockwell methods and these are discussed in the following sections.

BRINELL METHOD   Brinell published the details of the indentation hardness test he devised in 1901. The principle involved is that a hardened steel ball is pressed under a specified load into the surface of the metal being tested (B.S. 240: 1961). The hardness, which is quoted as a number, is then defined as

$$\text{Hardness number} = \frac{\text{Load applied to indenter in kg}}{\text{Contact area of indentation in mm}^2}$$

or Brinell Hardness Number (B.H.N.) $= P/A$. It is noted that the number has in fact units of pressure. The contact area A is given by

$$A = \pi Dh$$
$$= \frac{\pi D}{2}[D - \sqrt{(D^2 - d^2)}]$$

where h = depth of impression in mm, D = ball diameter in mm d = surface projected diameter of impression in mm.

If the hardness number of a metal is determined at several different values of load it is found that the number is not constant. The results give all or part of a curve of the form shown in Fig. 19.11 depending on the condition of the material. The curve is attributed to two separate effects. The rising portion with increase in load is caused by the non-proportional effect of work-hardening on the size of the impression. Thus a soft metal will show a marked apparent rise in hardness while a heavily cold-worked material will show none.

Fig. 19.11 Variation of hardness with load in the Brinell test

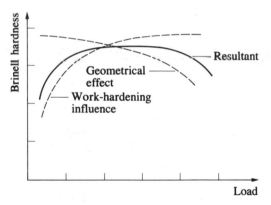

The falling part of the curve in Fig. 19.11 with increasing indenter load is caused by geometrical dissimilarity between the spherical areas of successive impressions. This feature is very important and is worthy of further analysis. Consideration of Fig. 19.12 shows that similarity can only be obtained for different loads if different ball sizes are used, since the total angle subtended by the centre of a ball and the indentation must be equal in each case. Hence the condition for similarity is that

$$\frac{d_1}{D_1} = \frac{d_2}{D_2} = \text{constant}$$

For a given angle of indentation the mean pressure is $P/\frac{1}{4}\pi d^2$, but since $d = \text{constant} \times D$ it follows that for similarity $P/D^2 = \text{constant}$.

The highest value on the curve of Fig. 19.11 is known as the *optimum hardness number* and is the figure quoted for a material when hardness is

Fig. 19.12 Indentation geometry in the Brinell test

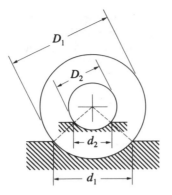

required. Brinell obtained the optimum hardness for steels using a 10 mm diameter ball at a load of 3000 kg, and these conditions have become a British Standard for hardness testing. Using the above values it is seen that $P/D^2 = 30$, and it is therefore possible to obtain comparable hardness numbers on a metal for different sizes of ball if the load is chosen to satisfy the above relationship. For softer metals and thin sheet it is found that other values are required for the constant equal to P/D^2 in order to give optimum hardness. Values of $P/D^2 = 10$ (non-ferrous alloys), 5 (copper, aluminium) and 1 (lead, tin) have been adopted as standard. The peak of the hardness curve generally occurs for a value of d/D between 0.25 and 0.5 and this fact can be used to assist in choosing the correct value of P/D^2. It is important when quoting a hardness

Table 19.1

Material	Condition	Hardness Brinell	Hardness Vickers	Tensile strength, (MN/m^2)
Pure aluminium	Annealed	25	(25)	54
	Cold rolled (hard)	55	(55)	185
Duralumin alloy (3.5–4.5 Cu, 0.4–0.8 Mg, 0.4–0.7 Mn, 0.7 Si)	Solution- and precipitation-treated	120	(120)	433
6% Al–Zn, 4% Mg alloy	Solution- and precipitation-treated	180	(181)	587
Pure copper	Annealed	42	(42)	221
	Cold rolled (hard)	119	(119)	371
Brass (60–40)	Cold drawn	178	(179)	659
Mild steel (0.19 C)	Annealed	127	(127)	451
	Cold rolled	(192)	192	595
Si–Mn spring steel	Quenched and tempered	(415)	435	1180
4% Ni, 1.5% Cr, 0.3% C steel	Quenched and tempered	(434)	460	1640
Ball-bearing steel	—	—	700	—
Tungsten carbide	Sintered	—	1200	—

Note: Equivalent values are in parentheses.

number to state the conditions used, ball size, load, etc. It is also essential to space successive impressions adequately and keep them clear of the edge of the material, owing to the plastic deformation caused in the area around the indentation ("ridging" for hard metals, "sinking" for soft metals).

Another very useful feature of the Brinell method is that an empirical relationship has been found to exist between hardness number and tensile strength for steels. Thus, $K \times$ B.H.N. (kg/mm^2) = tensile strength (MN/m^2), where K lies between 3.4 and 3.9 for the majority of steels. Hence, for checking of correct heat treatment, or a fractured component, it is only necessary to carry out a Brinell test to obtain an approximate value for the tensile strength.

The size of the Brinell indentation is such that the test is generally employed for checking raw stock or unmachined components rather than finished products. Typical hardness values for various materials are given in Table 19.1.

VICKERS METHOD

This test was devised about 1920 and employs a square-based diamond pyramid as the indenting tool (B.S. 427: 1961). The angle between opposite faces of the pyramid is 136° and this was chosen so that close correlation can be obtained between Vickers and optimum Brinell Hardness Numbers. The angle of 136° corresponds to the geometry of an impression given by a d/D ratio of 0.375.

In the Vickers test, hardness number is defined in the same way as for Brinell, that is, indenter load, kg, divided by the contact area of the impression, mm^2. If l is the average length of the diagonal of the impression, the contact area is given by

$$\frac{l^2}{2 \sin \frac{1}{2}(136)} = \frac{l^2}{1.854}$$

Therefore the Vickers Pyramid Number

$$\text{V.P.N.} = 1.854 \frac{P}{l^2}$$

There are two features of this test which are essentially different and advantageous over the Brinell method. Firstly, there is geometrical similarity between impressions under different indenter loads, and hardness number is virtually independent of load as shown in Fig. 19.13, except at very low loads where there is often a higher hardness owing to a "skin" effect on the test piece. The standard loads recommended in B.S. 427 are 1, 2.5, 5, 10, 20, 30, 50 and 100 kg.

The second advantage of the Vickers test is that the upper limit of hardness number is controlled by the diamond, therefore allowing values up to 1500 to be determined which is far in excess of that possible with the steel ball in the Brinell test.

The extremely small size of the impression necessitates a very good surface finish on the test sample, but means that it is advantageous in checking the hardness of finished components without leaving a noticeable mark.

Fig. 19.13 Variation of hardness with load in the Vickers test

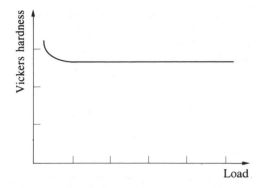

ROCKWELL METHOD

This test was introduced in the U.S.A. at about the same time as the Vickers test in England. It is quite popular as it has a wide range of versatility, is rapid and useful for finished parts.

Two types of penetrator are employed for different purposes, a diamond cone with rounded point for hard metals and a $\frac{1}{16}$ in (1.6 mm) diameter hardened steel ball for metals of medium and lower hardness values.

In this test hardness is defined in terms of the depth of the impression rather than the area, and the hardness number is read directly from an indicator on the machine having three scales, A, B and C.

The procedure for applying load to the specimen is rather different from the Vickers and Brinell and is illustrated in Fig. 19.14. Initially a

Fig. 19.14 Load application in the Rockwell test

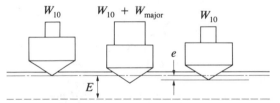

Reference line representing zero hardness

minor load of 10 kg is applied, which is followed by a major load, being an additional 50, 90 or 140 kg depending on the indenter and type of metal. The major load is now removed, but the minor load is retained while the hardness number is read, where

$$H_R = E - e$$

and H_R = Rockwell number; e = depth of penetration due to the major load only but while the minor load is operating; E = arbitrary constant which is dependent on the type of penetrator. The various relevant details of the test may be found in B.S. 891: 1962.

COMPARISON OF HARDNESS VALUES

Owing to the wide use of the Brinell, Vickers and Rockwell methods and the varying preferences for any one of these tests, there are occasions when the same material or component is hardness tested by different methods in different laboratories. This has led to a demand for some correlation between hardness values determined by the three tests. It has been shown that there is no general relationship between the hardness scales, and empirical formulae only hold good for materials of closely similar composition and condition.

However, based on experimental results, the British Standards Institution has issued a table (B.S. 860: 1967) of *approximately* comparative values for the three tests, but it is emphasized that it is not intended that the table shall be used as a conversion system for standard values from one hardness scale to another.

HARDNESS OF VISCOELASTIC MATERIALS

Hardness measurements are also necessary for assessing non-metallic materials such as plastics, rubbers and composites. Indentation methods can be used, but owing to the much lower levels of hardness compared with metals very low indenter loads are used. Material thickness must be adequate in relation to the depth of indentation and a solid mounting plate must be used. The strain-time dependence of viscoelastic materials also has to be taken into account. In particular the depth of penetration of the indenter will increase with time under load and the geometry of the indent may change due to recovery after load removal. In general it has been found that conventional Brinell and Vickers type tests are not successful with plastics and rubbers because these methods require a clear impression on the indent after the indenter is removed. Using a ball indenter the image of the indent tends to be imprecise; with a pyramid indenter the indent is sharper, but in both cases the image is difficult to see due to the poor reflection of light from the surface of the plastic or rubber. For plastics the most successful results are obtained by using a conical or ball indenter (B.S. 2782: Methods 365, B, D) or a pyramid indenter and measuring the *depth of penetration* of the indenter during load application. For rubber, a rigid ball indenter is used (B.S. 903: Part A26). Hardness testing of rubber is in fact very common in industry because there is a known relationship between the hardness number (I.R.H.D. – International Rubber Hardness Degrees) and Young's modulus for vulcanized rubber.

FACTORS INFLUENCING STRENGTH, DUCTILITY AND HARDNESS

The chemical composition of a material and the heat treatment to which it has been subjected have a great effect on the strength and ductility of the material. The mechanical properties of steel are very largely influenced by the amount of carbon in the steel. Its strength increases with increase in carbon content, but the ductility decreases as illustrated in Fig. 19.15.

A steel may be hardened by heating it to a high temperature and then rapidly cooling it in a cold liquid. The strength is greatly increased by this

Fig. 19.15 Effect of carbon content on mechanical properties

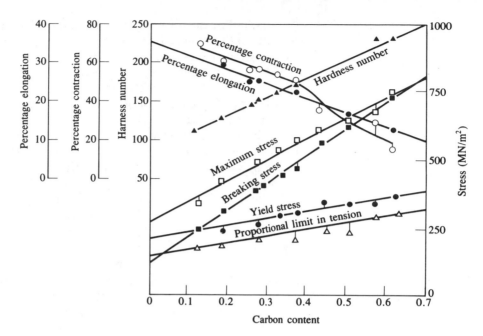

Fig. 19.16 Effect of tempering temperature on mechanical properties

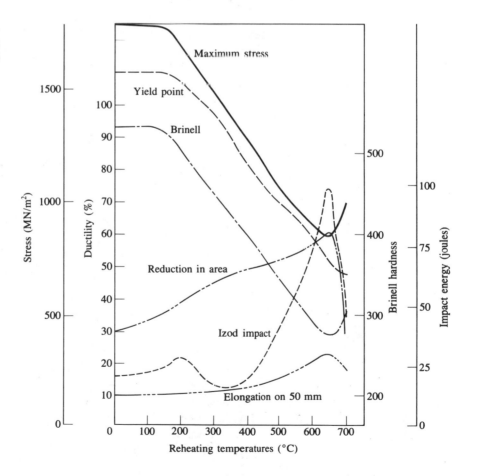

process, but its ductility is reduced and it is in a brittle state. It may be brought to the required degree of strength and ductility by further heat treatment, or tempering. The effect of tempering on a nickel-chrome steel for use in highly stressed components in automobiles and aircraft is shown in Fig. 19.16. The steel was hardened by heating to 820 °C soaked for $\frac{3}{4}$ hour and quenched in oil. It was then tempered for 1 hour at a range of temperatures and cooled in oil.

Similar influences on strength and ductility are obtained for non-ferrous alloys by the appropriate types of heat treatment.

The operating temperature of materials can range quite widely under atmospheric extremes quite apart from special engineering situations such as internal combustion, or liquid petroleum gas (LPG) storage. It is therefore necessary for engineers to have data in this context. A few brief comments follow herewith and a more detailed treatment of high-temperature creep behaviour is given in Chapter 22.

At low temperatures, for example the strength of a mild steel is increased; the material, however, may become very brittle under certain circumstances, its ductility being almost negligible (see Ch. 20.)

With increase in temperature, the elastic limit decreases, and is accompanied by a corresponding decrease in the modulus of elasticity. The tensile strength of the material decreases until a temperature of approximately 150 °C is reached. From this temperature the strength increases, reaching a maximum value in the neighbourhood of 300 °C, after which it decreases as the temperature is further increased. Table 19.2 shows the effect of high temperature on the tensile strengths of various steels.

Table 19.2

Steel	Tensile tests, cold				Tensile strength, hot (°C), (MN/m^2)						
	Elastic limit, (MN/m^2)	Tensile strength, (MN/m^2)	Elongation on 50 mm, (%)	Reduction of area, (%)	600	650	700	750	800	850	900
3% Nickel-chrome	745	874	23.0	62.5	525	365	209	173	142	107	70
Stainless steel	726	746	24.0	58.1	374	264	158	278	94	87	121
Silicon-chrome	790	975	21.0	40.0	650	534	392	275	165	111	60
Chrome steel	650	835	24.5	55.0	560	464	294	201	108	111	113
Cobalt-chrome	650	896	13.0	22.0	695	472	383	210	155	90	125
High-speed steel	711	943	15.0	24.0	634	411	330	260	151	119	130
High-nickel chrome	650	1050	27.0	45.0	660	592	523	442	371	300	232

The ductility decreases with increase in temperature until a temperature of about 150 °C is reached. After this temperature the ductility increases with further increase of temperature. The increase is not regular, however, but takes place in an erratic manner.

Table 19.3 shows the effect of temperature on a cast- and wrought-aluminium alloy respectively.

Table 19.3

Temperature (°C)	Tensile strength, (MN/m²)		
	Wrought	Sand-cast	Die-cast
20	433	286	371
200	332	248	340
250	301	216	302
300	201	201	240
350	124	136	139

One must conclude by emphasizing the importance of the interdependence of strength, ductility and hardness.

In broad terms a high hardness value is associated with high tensile strength and medium to low ductility, while the converse is that a low hardness number relates with lower tensile strength and higher ductility. Hence hardness measurement is widely used as a quality-control test for materials during manufacture.

SUMMARY

The properties discussed in this chapter represent the basic information that the designer needs to have available in relation to the selection of materials for "steady-load" components and structures.

The tensile, compression or shear moduli together with the yield (or proof) stress and Poisson's ratio are needed for linear elastic design. The plastic properties of tensile or compressive strength and ductility are required to give a measure of reserve safety in the event of exceeding the yield level. The hardness of the material gives a check on its condition.

However, we must not get a false sense of security, since it will be seen in the next three chapters that although we may have selected what appears to be a very suitable material in the required condition for "static" working conditions other factors can have very serious consequences during the service life of a component or structure.

BIBLIOGRAPHY

Nadai, A. *Theory of Flow and Fracture of Solids,* McGraw-Hill, New York 1950.
McClintock, F. A. and Argon, A. S. *Mechanical Behaviour of Materials,* Addison-Wesley Publishers, Reading 1966.
Dieter, G. E. *Mechanical Metallurgy,* McGraw-Hill, New York 1976.
Pascoe, K. J. *Introduction to the Properties of Engineering Materials,* Van Nostrand, New York 1978.
Ashby, M. F. and Jones, D. R. *Engineering Materials,* Pergamon Press, Oxford 1980.

19.1 A tensile test has been carried out on a mild steel specimen 10 mm thick and 50 mm wide rectangular cross-section. An extensometer was attached over a 100 mm gauge length and load extension readings were obtained as follows:

Load (kN)	16	32	64	96	128	136	144	152	158
Extension (mm)	0.016	0.032	0.064	0.096	0.128	0.137	0.147	0.173	0.605

Load (kN)	154	168	208	222	226	216	192	185.4
Extension (mm)	1.181	2.42	7.25	12.0	16.8	22.0	24.0	Fracture

Plot load-extension diagrams for the elastic range and the plastic range and determine: (i) Young's modulus; (ii) proportional limit stress; (iii) yield point stress; (iv) tensile strength; (v) percentage elongation.

19.2 An aluminium alloy specimen of 1.2 mm thickness and 25 mm width cross-section and a parallel gauge length of 50 mm is tested in tension giving the following data:

Load (kN)	1.8	3.6	5.4	6.4	7.2	7.6	8.0	8.4	8.8
Extension (mm)	0.044 3	0.088 6	0.133	0.155	0.181	0.198	0.219	0.246	0.281

Load (kN)	9.2	10.8	12.0	12.4
Extension (mm)	0.332	0.645	1.05	1.94

Determine values for: (i) Young's modulus; (ii) 0.1% proof stress; (iii) 0.5% proof stress; (iv) tensile strength.

19.3 The following load-compression data has been obtained on four copper cylinders of 12 mm diameter and lengths 24, 12, 6, 4 mm. Construct the true compressive stress–strain curve for the material.

d_0/h_0	Load kN	13.5	22.5	29.0	34.0	40.0
3	% Compression	3.3	11.5	19.5	26.7	37.1
2	% Compression	3.8	12.6	21.4	29.5	41.5
1	% Compression	4.0	13.8	23.3	32.2	45.8
$\frac{1}{2}$	% Compression	4.3	14.4	24.2	33.6	47.9

19.4 The torque–twist data given below was obtained on a solid cylindrical steel bar of 25 mm diameter. Use the Nadai construction to develop the shear-stress–shear-strain curve for the material.

Torque (kN-m)	7.3	9.3	10.8	12.3	14.2	15.5
Twist (rad/m)	0.2	1.0	2.0	4.0	8.0	12.0

19.5 A Wallace-type of micro-hardness testing machine measures the depth of penetration of a standard Vickers indenter. During a test on polypropylene, using an applied weight of 300 g the instrument records 2.66×10^{-2} mm. From this information calculate the hardness of the plastic.

19.6 A nickel-chrome alloy steel component fails in service and an investigation is required. A Brinell hardness test is carried out on the broken part giving a value of 320 B.H.N. An Izod impact sample cut from the part gives a fracture energy of 50 J. The design maximum stress for the component was 1075 MN/m^2. What is the main contributing factor for the failure? (*Hint*: Examine Fig. 19.16.)

Chapter 20

Fracture mechanics

The material properties discussed in the previous chapter principally relate to the quality control of materials and to initial material selection by a designer. We now need to recognize that in spite of carefully employing the design stress and strain analysis procedures of the earlier chapters, there are other factors to be taken into account in design. If we do not do so then it is possible for the engineering component or structure to fail and by this more is meant than, say, exceeding the yield stress and causing a little plastic deformation. We are now referring to the case of material fracture which may lead to catastrophic disaster of an engineering structure, for example an aeroplane, and loss of life.

The "factors" mentioned above include straining rate, fluctuating stresses, stress concentration, metallurgical flaws, high and low temperatures, corrosion and other special effects. This chapter and the remaining two will attempt to cover these topics in a manner which will provide an initial insight which can be built on as required through the specific texts in these areas.

FRACTURE CONCEPTS

Fracture is concerned with the *initiation* and *propagation* of a crack or cracks in the material until the extent of cracking is such that the applied loading can no longer be sustained by the component or structure. It is generally accepted that initiation of a crack is relatively difficult to design against and, in fact, most components or structures will contain some crack-like flaw/defect by the time manufacturing is completed in spite of rigorous inspection procedures. We, therefore, must try to design for non-propagation or, at next best, controlled propagation. In the latter case normal in-service inspections should enable the presence of growing cracks to be detected. This tends to be the case in respect of the phenomenon known as *fatigue* which is fully discussed in the next chapter. Cracks which propagate in an uncontrolled manner at very high velocity through materials is a situation of great danger in service and is the theme of this chapter.

From a metallurgical viewpoint there are only two paths for a crack passing through a metal, either trans-crystalline or inter-crystalline. The latter only occurs in a few particular circumstances, e.g. creep, stress, corrosion. The former is the more general mechanism of which there are two types related to the crystallographic planes known as *shear* and *cleavage*. Shear is the result of certain crystal planes sliding over one another, termed *slip,* and is associated with a great deal of macroscopic plastic deformation. Cleavage occurs on different crystallographic planes caused by a normal (tensile) stress and involves negligible plastic deformation. Shear fractures have a dull appearance, sometimes described as fibrous, but cleavage fractures reveal smooth reflecting planes described as bright and crystalline or granular. Relating these two modes broadly to stress–strain characteristics it is found that the ductile shear mode tends to be above-yield stress fracture with high energy absorption and hence *toughness.* On the other hand, a fully brittle cleavage mode would be associated with a low toughness, low energy absorption and below-yield stress fracture. Of course there are the possibilities of mixed-mode fractures, depending on a combination of a number of factors, between the two extremes above.

One of the more important influences on the mode of fracture is the state of stress. In engineering components, as compared to simple laboratory uniaxial stress tests, a complex stress system generally exists. In a triaxial stress state where $\sigma_1 > \sigma_2 > \sigma_3$, the maximum shear stress is $\frac{1}{2}(\sigma_1 - \sigma_3)$, but as $\sigma_3 \to \sigma_1$, $\tau \to 0$. In the extreme case of hydrostatic tension and compression, $\sigma_1 = \sigma_2 = \sigma_3$ and $\tau = 0$. It is evident that shear cannot occur and hence cleavage fracture will result. The introduction of a discontinuity or notch into a piece of material causes a stress concentration and triaxiality of stress to a degree which depends on notch geometry and loading condition.

Most metals exhibit some temperature dependence of fracture over a range from, say, $-100\,°C$ to $+100\,°C$. Toughness is reduced by the lowering of fracture temperature, and so this aspect of the working environment of engineering structures and components must be taken into account.

Unstable fracture manifested itself first as a serious engineering problem of nominally ductile low-carbon steels from the mid-1930s to the mid-1950s. Large welded structures such as ships, bridges and storage tanks failed in a catastrophic and apparently brittle manner. From this grew an extensive research programme into what was then called *brittle fracture.* Although factors such as stress concentration at "notches", weld defects and low temperature contributed to the initiation of a crack, the principal controlling factors on fast propagation of the crack are the ability of the material to absorb energy, i.e. the toughness, and the existence of crack arrest barriers. In the latter context riveted or bolted plate structures were better than the "continuous" all-welded structure if the welds were not of high quality.

The past 40 years have seen the development of ultra-high-strength alloys for rocket motors and space vehicles. Some of these low-ductility materials were found to be susceptible to unstable fracture from small defects owing to low toughness. This resulted in the development of the theory of linear elastic fracture mechanics (L.E.F.M.) which was accompanied by the establishment of special tests to measure the fracture resistance of materials. The analytical and experimental techniques of fracture mechanics now have a major influence on design for crack growth which is (i) unstable, (ii) intermittent/cyclic (fatigue) and (iii) time dependent in metallic and non-metallic materials which are homogeneous or fibre–matrix composites.

LINEAR ELASTIC FRACTURE MECHANICS

Linear elastic fracture mechanics (L.E.F.M.) developed from the early work of Griffith[1] who sought to explain why the observed strength of a material is considerably less than the theoretical strength based on the forces between atoms. He concluded that real materials must contain small defects and cracks which reduce their strength. These cracks cause stress concentrations but they cannot be allowed for by calculation of a linear elastic stress concentration factor K_t. This is because an elliptical defect, Fig. 20.1, has its stress concentration factor defined by the equation

$$K_t = 1 + 2\left(\frac{a}{b}\right) \tag{20.1}$$

As $b \rightarrow 0$ the defect becomes a crack, but $K_t \rightarrow \infty$ which would suggest that a material with a crack would not be able to withstand any applied forces. This is contrary to what is observed so Griffith developed a concept to explain how a stable crack could exist in a material. He postulated that a crack only becomes unstable if an increment of crack growth results in more stored energy being released than can be absorbed by the creation of the new crack surface.

Fig. 20.1 Elliptical defect in a stressed plate

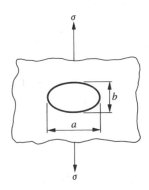

Based on this premise and with subsequent refinements, principally by Irwin,[2] L.E.F.M. has developed as an analytical approach to fracture. It relates the stress distribution in the vicinity of a crack tip to other parameters such as the nominal stress applied to the structure and the size, shape and orientation of the crack. Thus it permits representation of the material fracture properties, often in terms of a single parameter.

There have been two main approaches: (i) energy; (ii) stress intensity factor.

STRAIN ENERGY RELEASE RATE

For a through crack of length, $2a$, in an infinite body of unit width, the surface energy U_s stored in the material due to the formation of the crack is given by

$$U_s = (2a)2\gamma \tag{20.2}$$

where γ = surface energy per unit area. In the context of the fracture of brittle materials this term is replaced by $\gamma = \frac{1}{2}G$, where G is energy absorbed per unit area of crack (note that G refers to the area of crack which will be half the new surface area).

Thus eqn. (20.2) may be written as

$$U_s = 2aG \tag{20.3}$$

From the concept of elastic strain energy introduced in Chapter 3 the elastic energy U_e released by the formation of the crack is given by

$$U_e = \frac{1}{2} \int_a \sigma(x) \cdot \Delta(x, a) \, \mathrm{d}x \tag{20.4}$$

where $\sigma(x)$ is the stress distribution in the vicinity of the crack, and $\Delta(x, a)$ is the vertical opening of the crack.

It can be shown[3] that for the through crack of length $2a$ in an infinite plate, Fig. 20.2,

$$U_e = \frac{\pi\sigma^2 a^2}{4E}(1 + v)(k + 1) \tag{20.5}$$

Fig. 20.2 Sharp crack in a stressed infinite plate

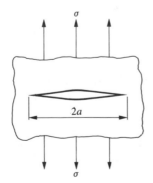

Fig. 20.3 Variation of energy with crack length

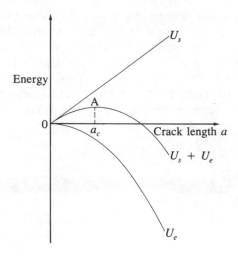

where v is Poisson's ratio, and $k = (3 - 4v)$ for plane strain and $(3 - v)/(1 + v)$ for plane stress.

Thus the surface energy which is stored in the material is increasing linearly with crack length, whereas the energy released by the formation of the crack increases with (crack length).[2] This is illustrated in Fig. 20.3.

The net energy in the presence of the crack is thus the mathematical summation of the surface energy U_s and the energy released U_e. Griffith proposed that the threshold between a stable crack and an unstable crack occurs when an increment of crack growth causes more energy to be released than can be absorbed in the material. In mathematical terms this means that the incremental growth of total energy per incremental growth of crack length has become negative. Thus the critical condition is $dU/da = 0$ which occurs at point A in Fig. 20.3 and hence the critical crack length a_c is defined. From eqns. (20.3) and (20.5)

$$\frac{d}{da}\left(2aG_C - \frac{\pi\sigma^2 a^2}{4E}(1 + v)(k + 1)\right) = 0$$

$$2G_C = \frac{2\pi\sigma^2 a}{4E}(1 + v)(k + 1)$$

This then reduces to

$$\left.\begin{array}{l} (EG_C)^{1/2} = \sigma(\pi a)^{1/2} \quad \text{(plane stress)} \\[2mm] \left(\dfrac{EG_C}{1 - v^2}\right)^{1/2} = \sigma(\pi a)^{1/2} \quad \text{(plane strain)} \end{array}\right\} \tag{20.6}$$

These equations are an expression of the conditions for fast fracture in a brittle material. It should be noted that G_C is a material property which is referred to as the *critical strain energy release rate, toughness* or *crack extension force*, and has the units J/m^2.

A high value means that it is hard to propagate cracks in the material as for example in copper for which $G_C \simeq 10^3$ kJ/m^2. This may be compared with a value for glass of approximately 0.01 kJ/m^2.

STRESS INTENSITY FACTOR

Although Griffith put forward the original concept of L.E.F.M. he restricted his work to brittle materials (e.g. glass) and it was Irwin who developed the technique for metals. He examined the equations that had been developed for the stresses in the vicinity of an elliptical crack in a large plate as illustrated in Fig. 20.4. The equations for the elastic stress distribution at the crack tip are as follows:

$$\left. \begin{aligned} \sigma_x &= \frac{K}{(2\pi r)^{1/2}} \cos\left(\frac{\theta}{2}\right)\left\{1 - \sin\left(\frac{\theta}{2}\right)\sin\left(\frac{3\theta}{2}\right)\right\} \\ \sigma_y &= \frac{K}{(2\pi r)^{1/2}} \cos\left(\frac{\theta}{2}\right)\left\{1 + \sin\left(\frac{\theta}{2}\right)\sin\left(\frac{3\theta}{2}\right)\right\} \\ \tau_{xy} &= \frac{K}{(2\pi r)^{1/2}} \sin\left(\frac{\theta}{2}\right)\cos\left(\frac{\theta}{2}\right)\cos\left(\frac{3\theta}{2}\right) \end{aligned} \right\} \tag{20.7}$$

and

$$\sigma_z = \frac{2vK}{(2\pi r)^{1/2}} \cos\left(\frac{\theta}{2}\right) \quad \text{(plane strain)}$$

or

$$\sigma_z = 0 \quad\quad\quad\quad \text{(plane stress)}$$

Fig. 20.4 Stress distribution at crack tip in an infinite plate

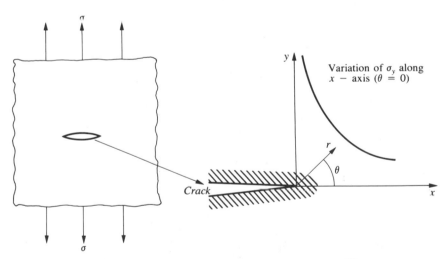

He observed that the stresses are proportional to $(\pi a)^{1/2}$, where 'a' is the half length of the crack. On this basis, Irwin defined a *stress intensity factor K* as

$$K = \sigma(\pi a)^{1/2} \tag{20.8}$$

The stress intensity factor is a means of characterizing the elastic stress distribution near the crack tip but in itself has no physical reality. It has units of MN-m$^{-3/2}$ and should not be confused with the elastic stress concentration factor K_t.

In order to extend the applicability of L.E.F.M. beyond the case of a central crack in an infinite plate, K is usually expressed in the more general form

$$K = Y\sigma(\pi a)^{1/2} \tag{20.9}$$

where Y is a geometry factor and a is the half-length of a central crack or the full length of an edge crack.

The methods for evaluating stress intensity factors for particular loading situations are formidable and outside the scope of this book. The main theoretical methods include: boundary collocation, conformal mapping, numerical methods and analysis of stress functions. The reader should refer to other texts[4,5] for details of these methods.

Figure 20.5 shows some crack configurations of practical interest and the expressions for K are as follows:

(a) Central crack of length $2a$ in a sheet of finite width

$$K = \sigma(\pi a)^{1/2}\left\{\frac{W}{\pi a} \cdot \tan\left(\frac{\pi a}{W}\right)\right\}^{1/2} \tag{20.10}$$

(b) Edge cracks in a plate of finite width

$$K = \sigma(\pi a)^{1/2}\left\{\frac{W}{\pi a}\tan\left(\frac{\pi a}{W}\right) + \frac{0.2W}{\pi a}\sin\left(\frac{\pi a}{W}\right)\right\}^{1/2} \tag{20.11}$$

Fig. 20.5 Typical crack configurations: (a) finite width plate; (b) double edge crack; (c) single edge crack; (d) internal penny crack; (e) elliptical surface crack; (f) three-point bending

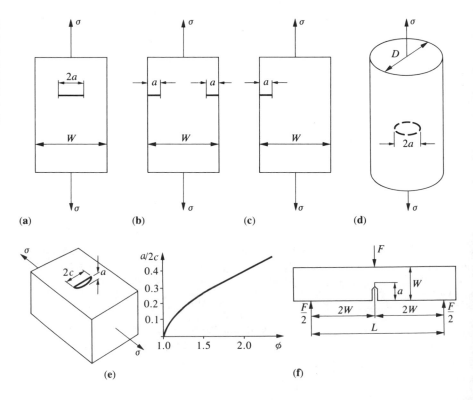

514

(c) Single edge cracks in a plate of finite width

$$K = \sigma(\pi a)^{1/2}\left\{1.12 - 0.23\left(\frac{a}{W}\right) + 10.6\left(\frac{a}{W}\right)^2 - 21.7\left(\frac{a}{W}\right)^3 + 30.4\left(\frac{a}{W}\right)^4\right\}$$

(20.12)

Note: In most cases (a/W) is very small so $Y = 1.12$.

(d) Penny-shaped internal crack

$$K = \sigma(\pi a)^{1/2}\left(\frac{2}{\pi}\right)$$

(20.13)

assuming $a \ll D$

(e) Semi-elliptical surface flaw

$$K = \sigma(\pi a)^{1/2}\left(\frac{1.12}{\phi^{1/2}}\right)$$

(20.14)

(f) Three-point bending

$$K = \frac{3FL}{2BW^{3/2}}\left\{1.93\left(\frac{a}{W}\right)^{1/2} - 3.07\left(\frac{a}{W}\right)^{3/2} + 14.53\left(\frac{a}{W}\right)^{5/2}\right.$$
$$\left. - 25.11\left(\frac{a}{W}\right)^{7/2} + 25.8\left(\frac{a}{W}\right)^{9/2}\right\}$$

(20.15)

or

$$K = \frac{F}{BW^{1/2}} \cdot f_1\left(\frac{a}{W}\right)$$

(20.16)

Thus the basis of the L.E.F.M. design approach is that:

(a) All materials contain cracks or flaws.
(b) The stress intensity value K may be calculated for the particular loading and crack configuration.
(c) Failure is predicted if K exceeds the critical value for the material.

The *critical stress intensity factor* is sometimes referred to as the *fracture toughness* and will be designated K_C. By comparing eqns. (20.8) and (20.6) it may be seen that K_C is related to G_C by the following equations:

$$(EG_C)^{1/2} = K_C \quad \text{(plane stress)}$$

(20.17)

$$\left(\frac{EG_{IC}}{1 - v^2}\right)^{1/2} = K_{IC} \quad \text{(plane strain)}$$

(20.18)

MODES OF CRACK TIP DEFORMATION

So far it has been assumed that the loading plane is symmetrical with respect to the crack plane. This is probably the most common situation and is referred to as the opening mode (designated as mode I), Fig.

Fig. 20.6
    Cracking modes:
    (*a*) opening mode (I)
    (*b*) shearing mode (II)
    (*c*) tearing mode (III)

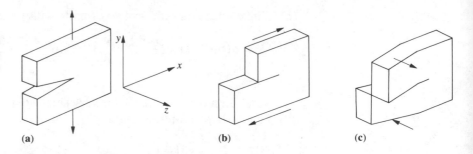

(a)　　　　　　　　　　(b)　　　　　　　　　　(c)

20.6(*a*). Therefore to be strictly correct the stress intensity factor should have the suffix I. There are in fact three deformation possibilities and the other two are shown in Fig. 20.6(*b*) and (*c*).

(a) Opening mode (mode I) having symmetry about the (x, y)- and (x, z)-planes.
(b) Sliding or shear mode (mode II) having anti-symmetry about the (x, z)-plane and symmetry about the (x, y)-plane.
(c) Tearing mode (mode III) having anti-symmetry about the (x, y)- and (x, z)-planes.

Thus the stress intensity factor should be referred to as K_I, K_{II} or K_{III} depending on the deformation mode.

EXPERIMENTAL DETERMINATION OF CRITICAL STRESS INTENSITY FACTOR

From the previous sections it is apparent that the stress intensity factor K for a particular loading situation may be calculated from a knowledge of the nominal stress and the size, shape and orientation of the crack. The basis of L.E.F.M. is that fast fracture will occur when K reaches the critical value for the material. This critical value K_C may be determined experimentally from standardized tests[6,7] on samples of the material.

EFFECT OF SIZE

It has been found that the value of the critical stress intensity factor K_C at which unstable crack growth occurs under static loading depends on the specimen thickness as illustrated in Fig. 20.7.

The limiting value of K_C is observed in plane strain (which is the maximum constraint condition). It is designated K_{IC} and is usually referred to as the fracture toughness. It is important not to confuse K_I with K_{IC}. As shown earlier, K_I depends on the configuration of the system, but K_{IC} is a material property and is independent of the configuration of the system. Since K_{IC} is the least value of stress intensity factor for a material, it is this value which is used in design calculations. Therefore the standardized test conditions are carefully chosen to ensure that it is K_{IC} which is determined and, after testing, checks are made on the validity of the test conditions.

Fig. 20.7 Variation of fracture toughness with plate thickness

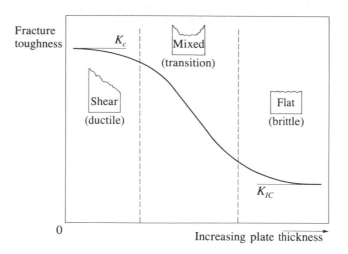

TEST METHODS

The two most commonly used test methods involve bending or tensile loading of the test specimens shown in Fig. 20.8. The specimens contain a carefully-machined notch which may have a plane front or a chevron profile. A chevron notch has been found to keep the crack in-plane and so the machining operation is not quite so critical as with a plane notch.

Fig. 20.8 Details of tensile and bending fracture toughness specimen

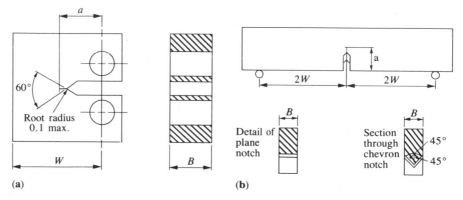

In order to get a sharp crack the machined notch is extended by applying a cyclically varying force to the test piece. This cyclic stressing, in which the peak value of stress is chosen to ensure that K_{max} is less than 70% of the critical value K_{IC}, causes the crack to grow by a fatigue mechanism (*see* Ch. 21). The sharp crack thus produced should be at least 1.25 mm long. At this point the test specimen is removed from the fatigue machine and subjected to a static test. Typical forms of test are illustrated in Fig. 20.9.

The force is measured directly from the load cell on the testing machine and this is recorded automatically as the vertical axis on an $X-Y$ plotter. The horizontal axis is obtained from a displacement (clip) gauge attached to the test piece as indicated in Fig. 20.9.

From the graph so obtained a value of force, termed F_Q, is obtained so that an interim value of stress intensity factor K_Q may be calculated. If the material is perfectly elastic up to fracture then the peak force is taken

Fig. 20.9 Fracture toughness testing

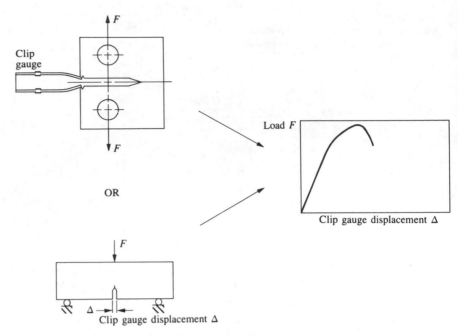

as F_Q. Assuming that the material exhibits some non-linearity as in Fig.
20.10, then the procedure for obtaining F_Q is to draw a line OQ with a
slope of 95% of the initial slope of the force–displacement graph. The
value of force at the point where this line intersects the F–Δ
characteristic is taken as F_Q.

The fracture surface is then examined so that the crack length "a" may
be determined as the average of a_1, a_2 and a_3 where a_1 and a_3 are
mid-way between the centre-line of the specimen and its edge as shown in
Fig. 20.11.

Using the values of F_Q and "a" thus obtained the following equations

Fig. 20.10 Determination
of F_Q

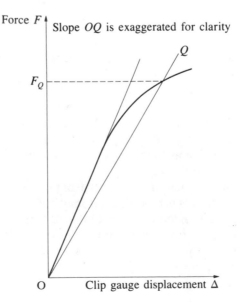

Fig. 20.11 Crack length measurements

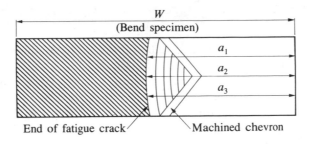

are used to calculate K_Q:

$$\text{Bending} \quad K_Q = \frac{F_Q}{BW^{1/2}} f_1\left(\frac{a}{W}\right) \tag{20.19}$$

$$\text{Tension} \quad K_Q = \frac{F_Q}{BW^{1/2}} f_2\left(\frac{a}{W}\right) \tag{20.20}$$

The values of $f_1(a/W)$ and $f_2(a/W)$ depend on the particular specimen geometry. (*See* Table 20.2 on page 523.)

Having thus obtained an interim value of stress intensity factor K_Q, checks are made that the following conditions have been satisfied.

$$B, a \geqslant 2.5\left(\frac{K_Q}{\sigma_Y}\right)^2; \quad 0.45 < \left(\frac{a}{W}\right) < 0.55; \quad F_Q \leqslant 1.1 F_{max}$$

If these conditions are upheld then the test is a valid one and K_Q is taken to be K_{IC}. Otherwise the result must be discarded. If desired G_{IC} may then be calculated from eqn. (20.18). Table 20.1 gives typical values of K_{IC} and G_{IC} for a range of materials.

Table 20.1
Typical values of K_{IC} and G_{IC}

Material	K_{IC} (MN m$^{-3/2}$)	G_{IC} (kJ/m^2)
Mild steel	100–200	50–95
High-strength steel	30–150	5–110
Cast iron	6–20	0.2–3.0
Titanium alloys	30–120	7–120
Aluminium alloys	22–45	7–30
CFRP (uniaxial fibres)*	20–45	2–30
GFRP (uniaxial fibres)*	10–100	3–60
GFRP (laminate)	10–60	5–100
Wood*	8–13	6–20
Glass	0.3–0.7	0.002–0.01
Acrylic (PMMA)	1.0–2.0	1.3–1.6
Polycarbonate	1.0–3.5	2.0–5.0
Concrete	0.2–0.4	0.03–1
Epoxy	0.5–0.7	0.08–0.34

* Perpendicular to fibre direction.

EXAMPLE 20.1

A pressure vessel is to be fabricated from plate steel which may be either: (i) a maraging (18% nickel) steel with $\sigma_Y = 1900 \text{ MN/m}^2$, $K_{IC} = 82 \text{ MN m}^{-3/2}$; or (ii) a medium-strength steel with $\sigma_Y = 1000 \text{ MN/m}^2$, $K_{IC} = 50 \text{ MN m}^{-3/2}$. Which of these two steels has the better tolerance to defects and compare their fracture toughness if they are to have the same defect tolerance? A factor of safety of 2 should be used for the design stress.

SOLUTION

Assuming that eqn. (20.9) is appropriate for the steel plates then

(i) For the maraging steel the critical defect size is given by

$$a_c = \frac{K_{IC}^2}{\pi \sigma_d^2} \quad \text{where} \quad \sigma_d = \frac{\sigma_Y}{2}$$

$$= \frac{82^2}{\pi (950)^2} = 2.4 \text{ mm}$$

So the critical crack length is 4.8 mm.

(ii) For the medium-strength steel

$$a_c = \frac{50^2}{\pi (500)^2} = 3.18 \text{ mm}$$

So the critical crack length is 6.36 mm.

Hence the medium-strength steel is more tolerant of defects in the material. In order that the maraging steel would have the same tolerance its K_{IC} value would need to be

$$K_{IC} = \sigma_d (\pi a_c)^{1/2} = 950 (\pi \times 3.18 \times 10^{-3})^{1/2}$$
$$= 95 \text{ MN m}^{-3/2}$$

FRACTURE MECHANICS FOR DUCTILE MATERIALS

PLASTIC ZONE CORRECTION

The stress field equations given on page 513 show that the elastic stresses would become very large in the vicinity of the crack tip where $r \ll a$. In practice these large stresses do not occur because in a ductile material this region becomes plastically deformed. It would appear therefore that in such materials the formation of a plastic zone at the crack tip would invalidate the use of linear elastic fracture mechanics.

However, Irwin has shown that when yielding occurs at the crack tip, L.E.F.M. techniques may still be applied if an equivalent crack length is used, i.e. a physical crack length plus an allowance for the extent of the plastic zone. This zone is generally represented by a circular boundary of radius r_p at the crack tip and the equivalent crack length then becomes

$$a' = (a + r_p) \tag{20.21}$$

Fig. 20.12 Stress distribution at crack tip due to local yielding

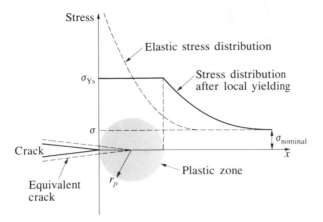

This effect is illustrated in Fig. 20.12 where the modification to the stress system due to local yielding is indicated.

For plane stress situations r_p has been shown to be given by

$$r_p = \frac{1}{2\pi}\left(\frac{K}{\sigma_Y}\right)^2 \tag{20.22}$$

For plane strain cases the extent of the plastic zone is less as indicated in Fig. 20.13. In this case the radius of the plastic zone is given approximately by the following expression:

$$r_p = \frac{1}{6\pi}\left(\frac{K}{\sigma_Y}\right)^2 \tag{20.23}$$

Fig. 20.13 Variation of crack tip plastic zone across thickness of material

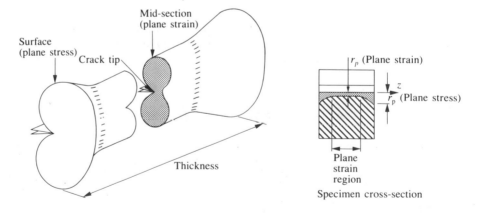

EXAMPLE 20.2

A wide sheet of aluminium alloy has a central crack 25 mm long. If the fracture stress for the sheet is 200 MN/m² and the yield stress of the material is 400 MN/m², calculate the fracture toughness of the material (i) using L.E.F.M., and (ii) using the plastic zone correction.

SOLUTION

(i) Using L.E.F.M.

$$K_{IC} = \sigma_{max}(\pi a)^{1/2}$$

$$= 200\left(\pi\left(\frac{0.025}{2}\right)\right)^{1/2}$$

$$K_{IC} = 39.6 \text{ MN m}^{-3/2}$$

(ii) Using correction for the plastic zone

$$K_{IC} = \sigma_{max}(\pi(a + r_p))^{1/2}$$

$$= \sigma_{max}\left(\pi a\left(1 + \frac{1}{2}\left(\frac{\sigma_{max}}{\sigma_Y}\right)^2\right)\right)^{1/2}$$

$$= 200\left(\pi\left(\frac{0.025}{2}\right)\left(1 + \frac{1}{2}\left(\frac{200}{400}\right)^2\right)\right)^{1/2}$$

$$K_{IC} = 42 \text{ MN m}^{-3/2}$$

The difference between the values of K_{IC} calculated by the elastic and plastic zone correction methods will increase as the ratio (σ_{max}/σ_Y) increases, i.e. when the size of the plastic zone increases. In order to ensure that the plastic zone dimensions are small compared with other dimensions of the test piece, it is found that the more ductile the material the larger must be the test specimen in order to meet the criteria laid down for the validity of the test. In some cases this can lead to specimen handling problems as well as difficulty in getting a testing machine with enough capacity to break the specimen. For these reasons it would be advantageous to have an alternative fracture test method utilizing smaller specimens of tough materials.

This need is also borne out by the fact that for materials which exhibit extensive plasticity before fracture, even the use of the corrected crack length approach becomes inappropriate. As a result of this a number of alternative test and analysis methods have been developed and the more important of these will now be described.

CRACK OPENING DISPLACEMENT (C.O.D.)

This method was proposed by Dugdale[8] and developed by Wells.[9] It is based on the fact that due to plasticity at the crack tip, the crack opens in the direction of the applied stress as illustrated in Fig. 20.14.

Fig. 20.14 Increased crack opening due to yielding at crack tip

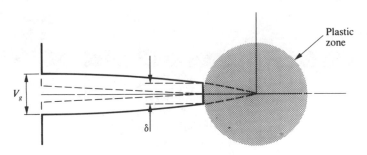

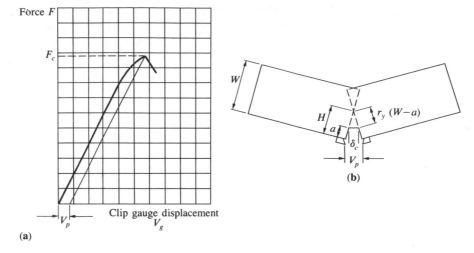

Fig. 20.15 (a) Load–displacement graph; (b) plastic hinge in C.O.D. specimen

(a)

(b)

The C.O.D. test is performed using the same type of specimen and procedure as described on page 517 for determining K_{IC}.[10] The clip gauge measures the sample opening V_g which is a measure of the resistance of the material to fracture initiation. The objective is to determine the critical crack opening at the onset of crack extension. A typical test record is shown in Fig. 20.15(a) and this indicates how the plastic component V_p of the clip gauge displacement is obtained.

For this type of characteristic the critical displacement is taken as the value corresponding to the maximum applied force F_c. Occasionally other shapes of characteristics are obtained and the British Standard[11] should be referred to for details of how to obtain V_p in these cases. The clip gauge displacement V_p must be converted to the crack opening displacement δ using

$$\delta = \frac{K^2(1 - v^2)}{2\sigma_Y E} + \frac{0.4(W - a)V_p}{0.4W + 0.6a + z} \tag{20.24}$$

where

$$K = \frac{F}{BW^{1/2}} f_1\left(\frac{a}{W}\right) \quad \text{and} \quad z = \text{thickness of knife edge}$$

We obtain $f_1(a/W)$ from Table 20.2.

Table 20.2

$\dfrac{a}{W}$	0.45	0.46	0.47	0.48	0.49	0.5	0.51	0.52	0.53	0.54	0.55
$f_1\left(\dfrac{a}{W}\right)$*	9.1	9.37	9.66	9.96	10.28	10.61	10.96	11.33	11.71	12.12	12.55
$f_2\left(\dfrac{a}{W}\right)$†	8.34	8.57	8.81	9.06	9.32	9.6	9.9	10.21	10.54	10.89	11.26

* Three-point bending (for support span $= 4W$)
† Tension.

A basic requirement of toughness tests for critical C.O.D. is that they should be carried out at full thickness. These tests do not have to satisfy criteria related to plane strain conditions because they seek to determine a toughness value relevant to the particular thickness of interest. For application to welded structures it is necessary for tests to be carried out on material representing different regions of the welded joints.

From elastic–plastic analysis of the crack tip region using several simplifying assumptions, the critical crack tip opening displacement has been related to the critical values of fracture toughness.[12] The most commonly used relations are

$$\delta_c = \frac{K_C^2}{\lambda E \sigma_Y} \qquad \text{(plane stress)} \qquad (20.25)$$

$$\delta_c = \frac{K_{IC}^2(1 - v^2)}{\lambda E \sigma_Y} \qquad \text{(plane strain)} \qquad (20.26)$$

where λ is a constant constraint factor which theoretical analyses have shown to be in the range 1–2 and which experimental measurements have shown to be approximately unity for both plane stress and plane strain situations.

EXAMPLE 20.3

A pressure vessel has a diameter of 2 m and a wall thickness of 10 mm. During routine inspection it is discovered that there is a crack 6.5 mm deep on the outside surface of the cylinder. The steel used in the vessel is known to be tough and a C.O.D. test on a sample gives the following results:

Sample width, $B = 25$ mm; $V_p = 0.35$ mm

Sample depth, $w = 50$ mm; $F_c = 65$ kN

Crack length, $a = 25$ mm; $z = 2.5$ mm

Calculate the maximum internal pressure to which the vessel could be subjected. $E = 207$ GN/m^2, $\sigma_Y = 500$ MN/m^2, $v = 0.3$.

SOLUTION

From eqn. (20.24)

$$\delta = \frac{K^2(1 - v^2)}{2\sigma_Y E} + \frac{0.4(w - a)V_p}{0.4W + 0.6a + z}$$

where

$$K = \frac{F}{Bw^{1/2}} f_1\left(\frac{a}{w}\right)$$

From Table 20.2, at $(a/w) = 0.5$, $f_1(a/w) = 10.61$

$$K = \frac{0.065 \times 10.61}{0.025(0.05)^{1/2}} = 123.4 \text{ MN m}^{-3/2}$$

From eqn. (20.24)

$$\delta_c = \frac{(123.4)^2(0.91)}{2 \times 500 \times 207 \times 10^3} + \frac{0.4 \times 25 \times 0.35 \times 10^{-3}}{0.4(50) + 0.6(25) + 2.5}$$

$$= 0.16 \times 10^{-3}\,\text{m}$$

Using eqn. (20.25), assuming plane stress conditions and $\lambda = 1$,

$$K_C = (E\sigma_Y\delta_c)^{1/2} = (207 \times 10^3 \times 500 \times 0.16 \times 10^{-3})^{1/2}$$

$$= 128.7\,\text{MN m}^{-3/2}$$

Using eqn. (20.9) with $Y = 1.16$ for the crack geometry,

$$\text{Hoop stress, } \sigma_\theta = \frac{K_C}{1.16(\pi a)^{1/2}} = \frac{128.7}{1.16(\pi \times 6.5 \times 10^{-3})^{1/2}} = 776\,\text{MN/m}^2$$

$$\sigma_\theta = \frac{pd}{2t}$$

$$p = \frac{2t\sigma_\theta}{d} = \frac{2 \times 10 \times 776}{2000} = 7.8\,\text{MN/m}^2$$

J-CONTOUR INTEGRAL

During the period when the C.O.D. method was being developed in the U.K. an alternative approach was being developed in the U.S.A.[13] This has become known as the J-contour integral approach. It is based on the finding that for a two-dimensional crack situation, the sum of the strain energy density and the work terms along a path completely enclosing the crack tip are independent of the path taken. This is shown in Fig. 20.16.

The energy line integral J is defined for either the elastic or elastic–plastic behaviour as follows:

$$J = \int_p w\,\mathrm{d}y - T\left(\frac{\partial u}{\partial x}\right)\mathrm{d}s \tag{20.27}$$

Fig. 20.16 Crack-tip co-ordinate system and arbitrary line integral contour

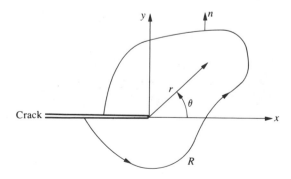

where p = any contour path surrounding the crack tip (*note*: the integral is evaluated anti-clockwise starting at lower surface of crack and proceeding along any path to the top surface);

w = strain energy density $- \int_0^\varepsilon \sigma \, d\varepsilon$;

T = traction vector defined according to the outward normal n along path p;

u = displacement vector;

s = arc length.

From a more physical viewpoint J may be interpreted as the potential energy difference between two identically loaded bodies having crack sizes (a) and $(a + da)$. In this context

$$J = -\frac{1}{B}\frac{\partial U}{\partial a}$$

where B = material width and U = strain energy or work done (area under the load-displacement curve).

The J-contour integral provides a means for describing the severity of conditions at a crack tip in a non-linear elastic material. Although it is a work absorption rate, J is equivalent to Griffith's strain energy release rate concept (note $J = G$ for the linear elastic situation) and it is related to the stress intensity factor used in L.E.F.M. As with these other approaches, the critical value of J has been found to be dependent on thickness and test specimen geometry. However, recommended test procedures have been developed and using the suggested specimen thickness $B \geqslant 25 \, J_{IC}/\sigma_Y$ (or $50 \, J_{IC}/\sigma_Y$), the specimen thickness of, for example, a tough structural steel would be only about 10–20 mm compared with a required thickness of approximately 100 mm to conduct a valid L.E.F.M. test on such a material.

The most commonly used test methods for obtaining J_{IC} are based on compact tension or notch bend specimens as illustrated in Fig. 20.17(a) and (b). A special test-piece geometry is used to permit load-line displacements to be measured as shown in Fig. 20.17(c).

From these test specimens, J is calculated from

$$J = \frac{2U}{B(w - a)} \tag{20.28}$$

where U = the total energy under the load-displacement diagram (using

Fig. 20.17 Test pieces used to determine J

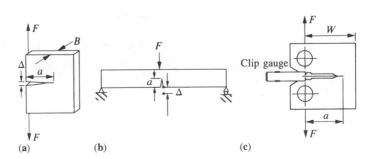

(a)　　　　(b)　　　　(c)

Fig. 20.18 Procedure for J_{IC} measurement: (a) load test pieces to various displacements; (b) measure crack extension; (c) calculate J for each test piece and plot J v. da; (d) construct two curves for J_{IC} measurement

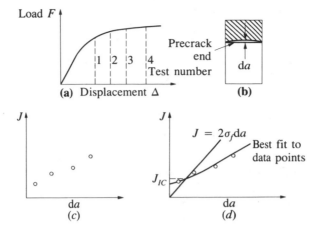

the load line displacement Δ), w = the specimen width and a = the crack depth (so $(w - a)$ is the remaining ligament of the test piece).

When using these test methods it is important not to overlook the fact that the definition of J as a measure of work in fracturing the specimen is just a convenient simplification of the original J-contour integral. It so happens that for the notched bend and compact tension specimens, the total energy represented by the load-displacement diagram is directly proportional to the rate of release of energy as the crack extends a small amount. It is this energy release rate which really justifies the linking of the work done to J. The steps in a typical test procedure are illustrated in Fig. 20.18 and summarized below.

(a) Load the test piece to different displacement values with a testing machine in displacement control.
(b) Unload and mark the extent of crack growth.
(c) Fracture each test piece at a low temperature (e.g. $-150\,°C$) and measure the crack extension.
(d) Calculate J values from the load versus load-line displacement record using eqn. (20.28) with U being the area under the graph at the displacement of interest.
(e) Plot a curve of J versus crack extension da.
(f) Draw a straight line $J = 2\sigma_f da$ to intersect the best fit line through the data points. Here, σ_f is taken as $\frac{1}{2}(\sigma_Y + \sigma_u)$, where σ_u is the tensile strength of the material.
(g) The critical value of J is at the intersection of the two lines.

The most common method of application of the J-contour integral approach is to convert experimentally determined values of J_{IC} to equivalent G_{IC} or K_{IC} values. As would be expected due to the similarity of the tests, it has also been shown that J is related to the crack opening displacement (C.O.D.). The following equations are frequently used to relate the various fracture toughness parameters.

$$\frac{K_{IC}^2}{E'} = G_{IC} = J_{IC} = \lambda \sigma_Y \delta_c \qquad (20.29)$$

where

$$E' = E \quad \text{for plane stress}$$
$$E' = E/(1 - v^2) \quad \text{for plane strain}$$

EXAMPLE 20.4

A bend test piece of the type shown in Fig. 20.8 is made from mild steel. The width B is 20 mm and the depth w is 25 mm. At the point of crack growth the crack length "a" is 15.3 mm and the area under the load-deflection graph is 14.7 J. If the modulus of the mild steel is 207 GN/m^2, calculate its fracture toughness.

SOLUTION

From eqn. (20.28),

$$J = \frac{2U}{B(w - a)}$$
$$= \frac{2 \times 14.7}{0.02(25 - 15.3)10^{-3}} = 151.5 \text{ kJ/m}^2$$

from eqn. (20.29),

$$K_{IC} = \sqrt{EJ}$$
$$= \sqrt{207 \times 10^9 \times 151.5 \times 10^3} \text{ Nm}^{-3/2}$$
$$= 177 \text{ MN m}^{-3/2}$$

R-CURVES   Another method which has been developed to extend the techniques of L.E.F.M. into the regime of elastic–plastic fracture involves the use of crack extension resistance curves or R-curves. An R-curve is a plot of the crack growth resistance (R) in a material as a function of the actual or effective crack extension (Δa). Here, R represents the energy absorbed (dU_s) per increment of crack growth (da). It is therefore given by the value of dU_s/da prior to unstable crack growth at the critical point; R has the same units as the stress intensity factor K.

On page 511 it was assumed that dU_s/da, i.e. R was independent of the crack length since U_s varied linearly with crack length a in Fig. 20.3. This is approximately true for cracks under plane strain conditions, but in situations involving larger proportions of plane stress failure, R is no longer independent of crack length.

Figure 20.19 shows a typical variation of R (calculated using the effective crack length) with crack extension. ASTM E561 provides specific instructions for the measurement of R. If curves of G (corrected to allow for the size of the plastic zone) are superimposed on the

Fig. 20.19 Typical $G{-}R$ curves

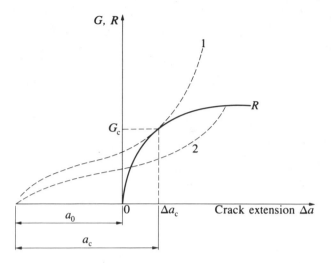

R-curves then the point of instability occurs at the point of tangency between the two types of curves. The value of G should be calculated from $G = Y^2\sigma_Y^2(a + r_p)/E$.

The solid curve shows the resistance to crack extension (R-curve). The dotted lines show the driving force for crack extension. Both dotted lines are defined by $EG = Y^2\sigma_Y^2 a$, but for curve 1 the stress is greater than for curve 2. The higher stress causes unstable crack growth when the crack length reaches the value a_c, shown in Fig. 20.19.

TOUGHNESS MEASUREMENT BY IMPACT TESTING

In addition to the more recent methods of fracture toughness testing for design by fracture mechanics analysis, the traditional quality-control method of measuring the energy absorption of a material during fracture has been by impact bend tests.

The two forms of the test most widely used are the Charpy and Izod. The principle of the former is shown in Fig. 20.20. The test piece is a square bar of material, 10 mm × 10 mm × 55 mm, containing a notch cut in the middle of one face. In this country the notch is a 45° vee, 2 mm deep, with a root radius of 0.25 mm. The original notch and that still used frequently in Europe is of keyhole form. The test piece is simply-supported at each end on anvils 40 mm apart. A heavy pendulum is supported at one end in a bearing on the frame of the machine, and a striker is situated at the other end. The pendulum in its initially raised position has an available energy of 300 J and on release swings down to strike the specimen immediately behind the notch, bending and fracturing it between the supports. A scale and pointer indicate the energy absorbed during fracture.

Fig. 20.20 (a) Charpy-type impact testing machine; (b) location of test piece

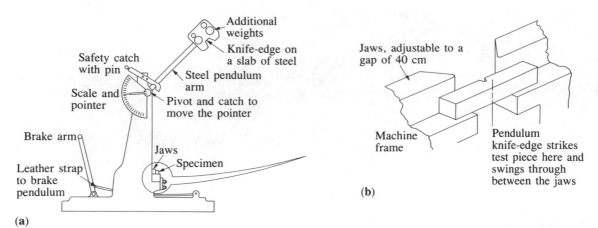

(a)

(b)

In the Izod test, Fig. 20.21 the specimen is of circular section, 11.43 mm in diameter and 71 mm long, or square section, 10 mm × 10 mm × 75 mm, and the Izod notch is a vee as described above for the Charpy test, 3.33 mm and 2 mm deep for the round and square specimens respectively. The specimen is supported as a vertical cantilever "built-in" to jaws up to the notched cross-section. A pendulum and striker having an initial energy of 166 J is arranged to swing and strike the free end of the test piece with a velocity of 3–4 m/s on the same side as the notch and 22 mm above it. The energy absorbed by the test piece is again recorded on a scale.

Fig. 20.21 Arrangement of test piece in the Izod test

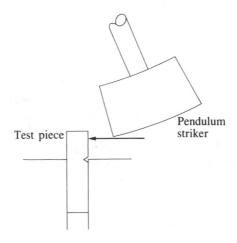

It is seen that in both tests the notch is on the tension side of bending, thus initiating fracture. The Charpy and Izod tests have been adopted as British Standards and details of alternative types of specimen and other conditions are given in B.S. 131: 1982. There is no direct correlation between energy values given in each of these tests; however, experimental results on a wide range of steels show that there is a linear relationship over the range from 20 to 95 J.

Some typical values of strength, ductility and toughness for steels are given in Table 20.3. The most notable feature is the effect of the type of heat treatment given, whether normalizing, quenching or quenching and tempering. The reason for this is the difference in metallurgical structure in each case influencing the ductility of the material, and the ease with which a crack can propagate through the notched bar. The quenched structure is hard and brittle and can absorb little energy. In the quenched and tempered structure, a higher tempering temperature gives greater ductility and therefore higher impact energy. The importance of temperature effects on fracture mechanics can be easily demonstrated in the Charpy or Izod impact tests by prior heating or cooling of the specimen.

Table 20.3

Material	Condition	Tensile strength (MN/m^2)	Reduction in area, (%)	Izod impact energy at room temperature (J)
0.1 C, 0.3 Mn steel	Annealed	377	65	54.2
0.21 C, 0.82 Mn steel	Annealed	505	58	40.6
0.5 C steel	Normalized	787	63	29.8
Ni–Cr–Mo steel	Quenched 840 °C, tempered 650 °C	895	64	114.0
	Quenched 840 °C, tempered 500 °C	1472	47	29.8
3.0 Ni, 1.0 Cr steel	Quenched 840 °C	1668	35	19.0
	Quenched and tempered 550 °C	865	60	93.5
Stainless steel (18–8)	Cold rolled	987	—	46.1

If the test temperature is varied from "high" to "low", then a metal such as mild steel, which exhibits the property of brittle fracture, will have a range of test temperature in which the mechanism of fracture changes from shear to cleavage. This is known as the *transition temperature* range, and within this range will be determined, according to some criterion, a transition temperature.

Typical transition curves are shown in Fig. 20.22(*a*) and (*b*) in which plastic deformation at, or energy to, fracture is plotted against test temperature. The first shows what is known as bimodal behaviour in which there are two distinct branches, one at high energy and temperature and the other at low energy and temperature joined by a

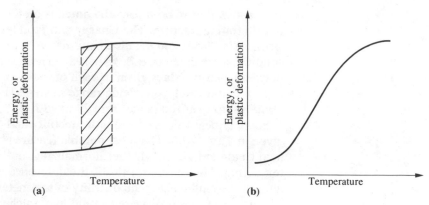

Fig. 20.22 Brittle-ductile energy v. temperature transition curves

narrow region (10–20 °C) of scattered points at upper and lower energy values. The second diagram shows continuous behaviour where there is a gradual fall from high to low energy with decrease in temperature. The transition range in this case can be as much as 100 °C from complete shear to entire cleavage.

RELATIONSHIP BETWEEN IMPACT TESTING AND FRACTURE MECHANICS

The major advantage of the Charpy and Izod types of impact tests is that they are quick and convenient tests to perform when compared with the much more demanding test schedule essential to obtain F–M parameters. Hence there have been a number of attempts in recent years to correlate K_{IC} with the Charpy impact strength C_v. No general rule has yet emerged but the research has met with some success. For high-strength steels a relationship has been found to exist between K_{IC} and C_v for the upper shelf region, and a variation of this can be applied to the transition temperature region.[14] Typical results are shown in Fig. 20.23 and it is felt that the type of relationships observed may have more general applicability to the lower strength steels, particularly for the upper shelf C_v values.

Fig. 20.23 Impact strength v. critical stress intensity for maraging steel

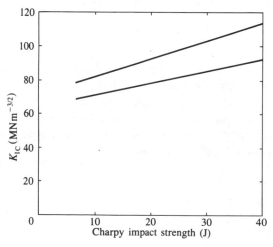

Thus, although Charpy testing says nothing about stress levels which has in the past made it impossible to use Charpy toughness values in design calculations, the correlation between K and C_v has encouraged attempts to provide a criterion for material selection based on C_v.

Finally, there has been extensive work done on the use of Charpy tests to obtain fracture mechanics parameters for polymers. In particular a very useful relationship has been developed[15] between the energy absorbed in breaking the pre-cracked specimen C_v (in J) and the critical strain energy release rate G_{IC} (in J/m^2). This has the form

$$C_v = G_{IC}BD\phi \qquad\qquad (20.30)$$

where B = specimen breadth, D = specimen depth and ϕ = a geometry factor. This permits G_{IC}, which is a material property, to be determined and hence K_{IC} (from eqn. (20.18)) although it should be noted that for polymers the modulus E is not a constant.

SUMMARY

The development of fracture mechanics on the basis that materials contain crack-life defects from which fast fracture may occur, has led to the emergence of new design concepts. In essence these assume that the critical value of the stress intensity factor K_{IC} is a material property which may be used to calculate the maximum defect size for a given stress or the maximum permissible stress for a given intrinsic defect size. Linear elastic fracture mechanics may be applied with confidence in situations where fracture occurs under essentially elastic conditions. The theory can also be extended to include materials which exhibit a relatively small amount of yielding prior to fracture, and the quantity $(K_{IC}/\sigma_Y)^2$ provides a measure of the size of the plastic zone at the crack tip. However, for very tough materials which exhibit gross yielding prior to fracture the use of L.E.F.M. is no longer valid. In such cases recourse may be made to methods such as crack-opening-displacement (C.O.D.), J-contour integral or resistance curves which have emerged in the wake of the success achieved by L.E.F.M.

The traditional toughness tests such as the Charpy and Izod tests still have an important role as quality-control tests, but have the disadvantage that they do not provide information which can be used in design calculations.

REFERENCES

1. Griffith, A. A. "The phenomena of rupture and flow in solids", *Phil. Trans. Roy. Soc., Lond.* **A221** (1921), 163–97.
2. Irwin, G. R. "Fracture mechanics", *J. Appl. Phys.*, **24** (1957) 361.
3. Parker, A. P. *The Mechanics of Fracture and Fatigue*, E. F. Spon, London, 1981.
4. Ibid.

5. Paris, P. C. and Sih, G. C. "Fracture toughness and its applications", *ASTM STP* 381, 1965.
6. B.S. 5447: 1977. *Plane Strain Fracture Toughness of Metallic Materials.*
7. A.S.T.M. Test Method E399-74. *Standard Method of Test for Plane Strain Fracture Toughness of Metallic Materials.*
8. Dugdale, D. S. "Yielding of steel sheets containing slits", *J. Mech. Phys. Solids,* **8** (1960) 100–8.
9. Wells, A. A. "Unstable crack propagation in metals", *Royal Aero. Soc. Symposium on Crack Propagation,* Cranfield, 1961, p. 210–30.
10. B.S. 5762: 1979. *Crack Opening Displacement (C.O.D.) Testing.*
11. Ibid.
12. See ref. 3.
13. Rice, J. R. *Fracture,* Vol. II, ed. by H. Liebowitz, Academic Press, New York, 1968.
14. Rolfe, S. T. and Barsom, J. M. *Fracture and Fatigue Control in Structures,* Prentice-Hall, New Jersey, 1977.
15. Plati, E. and Williams, J. G., *Polymer,* **16** (1975), 915.

BIBLIOGRAPHY

A General Introduction to Fracture Mechanics I. Mech. E., London, 1978.
Knott, J. F. *Fundamentals of Fracture Mechanics,* Butterworths, London, 1973.
Broek, D., *Elementary Engineering Fracture Mechanics,* Nijhoff, The Hague, 1982.
Chell, G. G. *Developments in Fracture Mechanics,* Vol. I, Applied Science Publications, London, 1979.
Ewalds, H. L. and Wanhill, R. J. H. *Fracture Mechanics* Arnold, London, 1984.

PROBLEMS

20.1 Describe how the critical strain energy release rate, G_{IC}, might be obtained by experiment.

20.2 Three fracture toughness samples of an aluminium alloy have identical external dimensions and a thickness of 25 mm. During three tests on the samples the following information was obtained:

Test	Sample crack length (mm)	Applied load (kN)	Sample elongation (mm)
1	20	185	Sample fractured
2	19.5	120	0.26
3	20.5	120	0.263

If the Young's modulus and Poisson's ratio values for the aluminium are 70 GN/m² and 0.3 respectively, calculate the fracture toughness of the material.

20.3 Calculate the critical defect size for each of the following steels assuming they are each to be subjected to a stress of $\frac{1}{2}\sigma_Y$. Comment on the results obtained.

Steel	Yield strength, σ_Y (MN/m²)	Fracture toughness (MN m$^{-3/2}$)
Mild steel	207	200
Low-alloy steel	500	160
Medium-carbon steel	1000	280
High-carbon steel	1450	70
18% Ni (maraging) steel	1900	75
Tool steel	1750	30

20.4 In a fracture toughness test involving three-point bending, the following information was recorded: support span = 180 mm; specimen thickness = 22.8 mm; specimen width = 44.72 mm; crack length = 21.92 mm; fracture force = 19.8 kN. Establish whether or not these data are suitable for measuring K_{IC} for the material. The yield stress of the material is 350 MN/m².

20.5 The load–deflection graph from a tensile fracture toughness test on a 50 mm thick steel specimen is shown in Fig. 20.24. Also shown (to scale) is the fracture surface of the broken specimen. If the yield stress for the steel is 1650 MN/m², establish whether the test is valid for the determination of K_{IC}.

Fig. 20.24

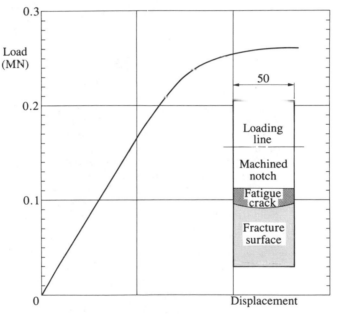

20.6 A sheet of glass 0.5 m wide and 18 mm thick is found to contain a number of surface cracks 3 mm deep and 10 mm long. If the glass is placed horizontally on two supports, calculate the maximum spacing of the supports to avoid fracture of the glass due to its own weight. For glass $K_{IC} = 0.3$ MN m$^{-3/2}$ and density = 2600 kg/m³.

20.7 The accident report on a steel pressure vessel which fractured in a brittle manner when an internal pressure of 19 MN/m² had been applied to it shows that the vessel had a longitudinal surface crack 8 mm long and 3.2 mm deep. A subsequent fracture mechanics test on a sample of the steel showed that it had a K_{IC} value of 75 MN m$^{-3/2}$. If the vessel diameter was 1 m and its wall thickness was 10 mm, determine whether the data reported are consistent with the observed failure.

20.8 A simple lifting crane is illustrated in Fig. 20.25. If the tie bar is found to have 3 mm long cracks at the pin-joint hole, calculate the maximum vertical force at the pulley to avoid fracture of the tie bar. The fracture toughness of the tie-bar steel is $75\ \text{MN m}^{-3/2}$. Comment on the suitability of the equation used to calculate K_{IC}.

Fig. 20.25

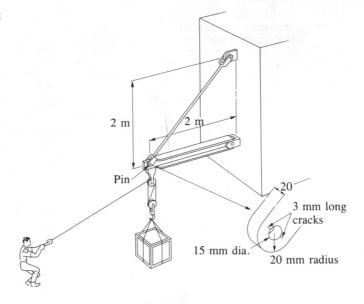

20.9 An aluminium alloy plate with a yield stress of $450\ \text{MN/m}^2$ fails in service at a stress of $110\ \text{MN/m}^2$. The conditions are plane stress and there is some evidence of ductility at the fracture. If a surface crack 20 mm long is observed at the fracture plane calculate the size of the plastic zone at the crack tip. Calculate also the percentage error likely if L.E.F.M. was used to obtain the fracture toughness of this material.

20.10 A large medium-carbon steel crane hook is thought to contain penny-shaped internal cracks. If the non-destructive test equipment used on the hook is not capable of detecting cracks smaller than 20 mm diameter, determine the fracture toughness required from this steel if the safety factor on stress is to be 2. The yield stress of the steel is $1050\ \text{MN/m}^2$.

20.11 As a result of a C.O.D. test on a medium-carbon steel, the following data were recorded: sample width = 25.03 mm; $V_p = 0.33$ mm; sample depth = 49.98 mm; $F_c = 80$ kN; crack length = 24.00 mm; $z = 2.5$ mm. Calculate (a) the fracture toughness of the material and (b) the percentage error in using L.E.F.M. to calculate this value. For the steel, $E = 207\ \text{GN/m}^2$, $\sigma_Y = 950\ \text{MN/m}^2$, $v = 0.3$.

20.12 In a three-point bend test on a mild steel sample the load at the point of crack growth and the final crack length were noted as 70 kN and 26 mm respectively. At the point of crack growth the area under the load–deflection graph was 32.6 J. If the specimen width (W) and thickness (B) were 49.98 mm and 25.05 mm respectively, calculate the fracture toughness of the mild steel. If a C.O.D. gauge (with $z = 2.5$ mm) had been attached to this specimen calculate the plastic component of the gauge reading which you would have expected to record. For the mild steel, $\sigma_Y = 240\ \text{MN/m}^2$, $E = 207\ \text{GN/m}^2$ and $v = 0.3$.

20.13 A bench-top impact testing machine for plastics is shown in Fig. 20.26. The pendulum weighs 4 kg and its centre of gravity is at A. When there is no specimen in position it is found that the pendulum swings through to an angle $\theta = 129°$. When there is a specimen in position the pendulum swings through to $\theta = 89°$. Calculate (a) the windage and friction losses in the machine and (b) the energy absorbed in breaking the specimen.

Fig. 20.26

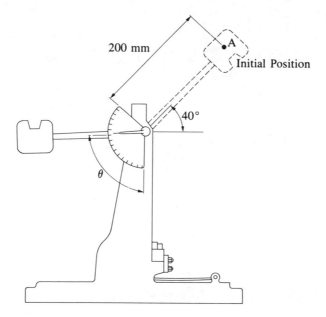

Chapter 21

Fatigue

In the early part of the nineteenth century the failure of some mechanical components subjected to nominal stresses well below the tensile strength of the material aroused some interest among a few engineers of that time. The fact that puzzled these early engineers was that a component such as a bolt or a shaft made from a ductile material such as mild steel could fracture suddenly in what appeared to be a brittle manner. There was no obvious defect in workmanship or material, and the only feature common to these failures was the fact that the stresses imposed were not steady in magnitude, but varied in a cyclical manner. This phenomenon of failure of a material when subjected to a number of varying stress cycles became known as *fatigue*, since it was thought that fracture occurred due to the metal weakening or becoming "tired". The first real attack on this problem was made by the German engineer Wöhler in 1858. Since then a great deal of research has been conducted on fatigue of metals, and in more recent times other materials also, and although this work has resulted in an ever-increasing understanding of the problem there is as yet no complete solution. What has been established is that the early theories that the metal becomes "crystalline" or brittle under the action of the cyclic load are erroneous. It is now well known that a fatigue failure starts on a microscopic scale as a minute crack or defect in the material and this gradually grows under the action of the stress fluctuations until complete fracture occurs.

It has been estimated from time to time that at least 75% of all machine and structural failures have been caused by some form of fatigue. It is therefore evident that every engineer should be aware of this phenomenon, and have some idea of its mechanics and what can be done to minimize or avoid the risk of this type of failure. It is astonishing how many design engineers ignore warnings about the possibility of fatigue failures and include in their design geometrical shapes which cause stress concentrations and initiate fatigue failures. Indeed even when the danger of fatigue failure is recognized it is not always possible to avoid owing to the many stages of manufacture which a component may go through and which are outside the direct control of the designer. Not infrequently the designer may specify non-destructive testing of components to search for flaws that would initiate fatigue failures, and when none are found the test engineer may use a metal stamp to mark the component as having passed inspection. Then in service fatigue cracks initiate from the stamp mark! The number of failures which have occurred in this way is quite considerable.

Figure 21.1(*a*) shows a typical situation where fatigue failures can arise due to geometrical configurations, and Fig. 21.1(*b*) illustrates the appearance of the fracture surface in such cases.

Fig. 21.1 Fatigue cracks in an engine crankshaft

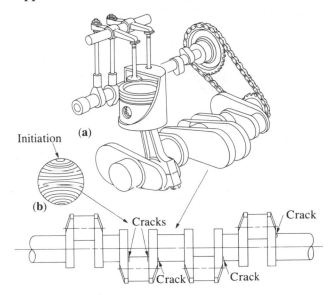

FORMS OF STRESS CYCLE

Throughout the working life of a component subjected to cyclical stress the magnitude of the upper and lower limits of cycles may vary considerably as shown in Fig. 21.2. However, it has been general practice when considering fatigue behaviour to assume or employ a sinusoidal cycle having constant upper and lower stress limits throughout the life.

Fig. 21.2 Variable stress spectrum

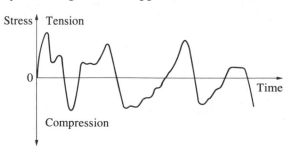

Fig. 21.3 Stress parameters

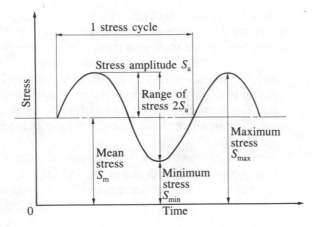

Figure 21.3 shows a general type of stress cycle, which is termed *fluctuating*, in which an alternating stress is imposed on a mean stress. This cycle can consist of any combination of upper and lower limits, within the static strength, which are both positive, or both negative.

When one stress limit in a cycle is positive and the other negative as shown in Fig. 21.4(*a*) it is known as a reversed cycle. There are two particular cases of the fluctuating and reversed cycles which arise frequently in engineering practice. The first is a fluctuating cycle in which the mean stress is half the maximum, the minimum being zero, as shown in Fig. 21.4(*b*). The second is a symmetrical reversed cycle as shown in Fig. 21.4(*c*) in which the mean stress is zero and the upper and lower limits are equal positive and negative.

Fig. 21.4 Typical stress–time cycles

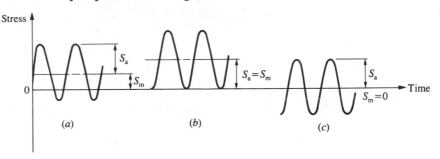

The relationship between the various stress values is of some importance. The mean stress, S_m,* is half the algebraic sum of the maximum stress, S_{max}, and the minimum stress, S_{min}. The range of stress, $2S_a$, is the algebraic difference between S_{max}, and S_{min}. The ratio of the minimum to the maximum stress is termed the *stress ratio*, R. Hence

$$S_m = \frac{S_{max} + S_{min}}{2} \tag{21.1}$$

$$2S_a = S_{max} - S_{min} \tag{21.2}$$

$$R = \frac{S_{min}}{S_{max}} \tag{21.3}$$

* The symbol S is being used for stress to conform with B.S. 3518: Part I: 1984; however, σ is also a recommended symbol (I.S.O.).

It should be noted that the foregoing has only considered stresses as being positive or negative in a general sense. In practice, fatigue can be generated in direct stress due to axial loading or bending or shear stress due to cyclic torsion or any combinations of these.

TEST METHODS

Fatigue failures occur most often in moving machinery parts – for example, shafts, axles, connecting-rods, valves, springs, etc. However, the wings and fuselage of an aeroplane or the hull of a submarine are also susceptible to fatigue failures because in service they are subjected to variations of stress. As it is not always possible to predict where fatigue failures will occur in service and because it is essential to avoid premature fractures in articles such as aircraft components, it is common to do full-scale testing on aircraft wings, fuselage, engine pods, etc. This involves supporting the particular aeroplane section or submarine hull or car chassis in jigs and applying cyclically varying stresses using hydraulic cylinders with specially controlled valves.

Laboratory tests are also carried out on particular materials to establish their fatigue characteristics and to study factors such as their susceptibility to stress concentrations. The earliest and still widely used method of fatigue testing of laboratory specimens is by means of rotation bending. A cylindrical bar is arranged either as a cantilever (Fig. 21.5) or a beam in pure bending. It is then rotated while subjected to a bending moment and hence each fibre of the bar suffers cycles of reversed bending stress. A bending fatigue test can also be arranged, without rotation, by alternating bending in one plane. An advantage of this test over the former is that a mean bending stress can be introduced.

Axial load fatigue tests, although being more difficult to perform, have the significant advantage that they subject a volume of material to a uniform stress condition as opposed to the stress gradient in bending.

Fig. 21.5 Rotating bending fatigue test

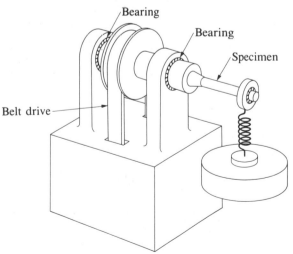

There are three main principles of operation for axial load fatigue machines. Firstly, there is the system whereby an electromagnet in series with the specimen is rapidly energized and de-energized, thus applying cyclical force. To reduce the amount of power input to the electromagnet to achieve the required load amplitude, a resonant vibration system is employed using springs. Another method of achieving a resonant condition of springs in series with a specimen is by mechanical means using a small out-of-balance rotating mass. The third arrangement for producing fluctuating forces on a specimen is by means of hydraulic pulsation.

Frequency is an important factor in fatigue testing, owing to the large number of cycles it is necessary to achieve at lower stress ranges, and the consequent time factor in obtaining fatigue data. Bending fatigue machines generally run in the range from 30 to 80 Hz. The electrical excitation push–pull machines operate between 50 and 300 Hz, while the mechanical excitation is usually restricted to about 50 Hz. The hydraulic machines have a frequency, often variable, in the range from about 1 to 50 Hz. Frequency of an actual test is generally controlled by the stiffness of the specimen, i.e. high frequency can only be achieved for high stiffness and low amplitude of deformation. The testing capacity of axial load fatigue machines varies from 20 to 2000 kN.

Cyclic torsion or combined bending and torsion fatigue machines generally operate on the principle of direct mechanical displacement of the specimen by a variable eccentric, crank and connecting-rod system. Depending on the capacity of the machine, frequencies vary from 16 to 50 Hz.

FATIGUE DATA

S–N CURVES

The most readily obtainable information on fatigue behaviour is the relationship between the applied cyclic stress S and the number of cycles to failure N. When plotted in graphical form the result is known as an S–N curve. Figure 21.6 shows the three ways of plotting the variables: (a) S v. N; (b) S v. $\log_{10} N$, (c) $\log_{10} S$ v. $\log_{10} N$. The number of cycles to failure at any stress level is termed the *endurance*, and this may vary

Fig. 21.6 Typical methods of presenting fatigue curves

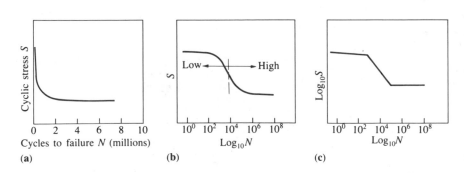

between a few cycles at high stress and as much as 100 million cycles at low cyclic stresses, for a complete $S-N$ curve. It is immediately apparent that a linear scale for endurance N over the whole range is impractical. It is for this reason that the semi-logarithmic or double logarithmic plots (b) and (c) are employed. The former, (b), is the most widely used method of presentation.

HIGH-ENDURANCE FATIGUE

This relates to endurances from about 10^4 cycles to "infinity" (or 50×10^6 in terms of a laboratory test). The $S-N$ curve in Fig. 21.6(b) falls rapidly from 10^4 cycles followed by a "knee" after which, depending on the material, the curve either becomes parallel to the N-axis or continues with a steadily decreasing slope. Most steels and ferrous alloys exhibit the former types of curve, and the stress range at which the curve becomes horizontal is termed the *fatigue limit*. Below this value it appears that the metal cannot be fractured by fatigue. In general, non-ferrous metals do not show a fatigue limit and fractures can still be obtained even after several hundred million cycles of stress. It is usual, therefore, to quote what is termed an *endurance limit* for these metals, that is the stress range to give a specific large number of cycles, usually 50×10^6. Typical $S-N$ curves for an aluminium alloy and a steel tested in air for a plain condition, i.e. no stress concentration, under axial loading, are given in Fig. 21.7.

Fig. 21.7 ($\times$) Aluminium alloy 24S-T3 reversed axial stress fatigue curve; ($\bullet$) mild steel reversed axial stress fatigue curve

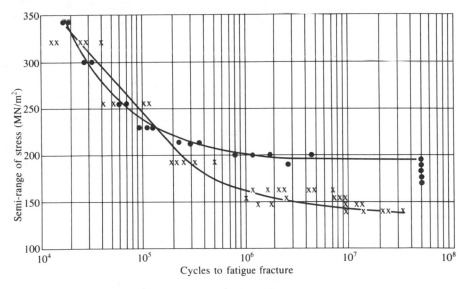

LOW-ENDURANCE FATIGUE

For many years the low-endurance region (<3000 cycles) of fatigue was ignored because the bulk of engineering cyclic stress situations had to have a working life ranging from several millions to infinity before failure. Hence the important information was contained in the lower stress–higher cycles to failure part of the curve. However, low cycles to failure does not necessarily mean a short lifetime because it is a function of cyclic frequency. Hence an aircraft fuselage is only pressurized once every flight and so it may take years to accumulate 1000 cycles and a

Fig. 21.8 Plastic strain hysteresis loop

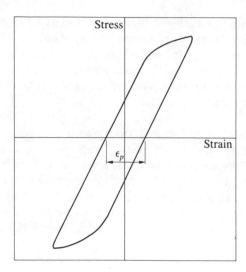

pressure vessel may work for 25 years before achieving 1000 cycles of cleaning and inspection.

Failure at low endurances results from stresses and strains which are high and will result in marked plastic deformation (hysteresis) in every cycle, Fig. 21.8. Due to strain-hardening (or softening) effects it becomes of some significance whether the cycle is of controlled load (stress) amplitude or controlled strain amplitude, a problem which does not arise at long endurance and hence low stress. A good example of strain cycling, which has come into prominence with the nuclear power and rocket era, is the effect of repeated thermal changes in a component.

At low endurances the S–N and ε–N curves take the form shown in Fig. 21.9. It is usual to consider the S–N curve as starting from a point at a $\frac{1}{4}$ cycle representing a tension test or single pull to fracture. Likewise the ε–N curve is started at a $\frac{1}{4}$ cycle using the true fracture ductility in simple tension as the ε value. Much work has been concentrated on strain cycle testing, and it has been found that, if the plastic strain range ε_p, or

Fig. 21.9 Typical stress-cycle and strain-cycle fatigue curves at low endurance

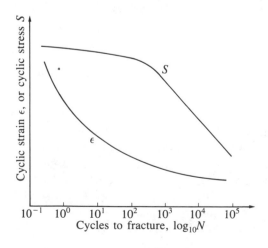

Fig. 21.10 Log ε_p – Log N_f relationship

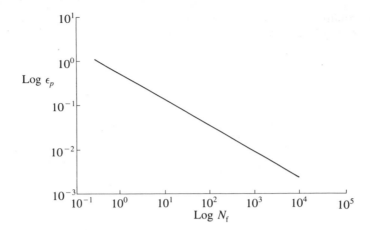

width of the hysteresis loop, is determined during the life of the specimen, then a relationship of the form $\varepsilon_p N^\alpha = $ constant holds for most metals up to 10^5 cycles, and that on a log ε_p–log N graph a straight line results, Fig. 21.10. It has further been demonstrated that α is between 0.5 and 0.6 for many metals at room temperature, and the constant is broadly related to the ductility of the material, i.e. the greater the latter, the larger the constant term.

Thermal strain cycling, that is cyclic strains induced by changes in temperature, appears to yield the same results as if the specimen or component were kept at a constant temperature of a value equal to the upper limit of the previous temperature cycle, and the cyclic strains induced mechanically.

STATISTICAL NATURE OF FATIGUE

Fatigue failure of materials has long been recognized as a random phenomenon. Thus, even in carefully-controlled experiments, at any selected stress amplitude there is often a large scatter in the number of cycles to failure as seen in Fig. 21.7. The reason for this is that, although on a macroscopic scale it is convenient to consider a material as being a homogeneous continuum, on a microscopic scale this is far from being the case. A metal, for example, is known to consist of a random distribution of internal defects such as micro-cracks, dislocations and inclusions contained within a network of grains which have randomly oriented slip planes and grain boundaries. Thus when the same cyclic stress amplitude is applied to nominally identical specimens it is extremely unlikely that each sample will fail after the same number of stress cycles. This is because, on a microscopic scale, the conditions which the fatigue cracks experience as they propagate through each specimen will be quite different.

Thus although fatigue data is usually presented as a single line on an S–N curve (Fig. 21.7) it should be realized that this line does not predict the exact fatigue life at a particular stress amplitude: S–N curves are usually plotted from a large number of fatigue test results which have been analysed by some type of statistical method. In the simplest case the line drawn represents the 50% probability of failure. This means that at

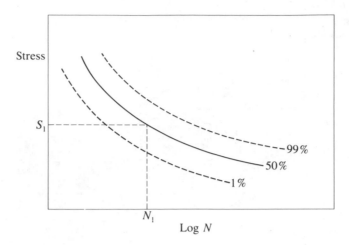

Fig. 21.11 Probability fatigue curve

stress S_1 in Fig. 21.11, 50% of the samples tested will have failed before cycles N_1. In some cases other probability lines will also be drawn. Figure 21.11 shows the 1% and 99% probabilities of failure and this type of graph is sometimes called a *P–S–N* diagram.

USE OF FATIGUE DATA FOR DESIGN

In general engineering, components subjected to cyclical stresses will be designed on an "infinite" life basis. The design stress range will therefore be related to a choice of material having a fatigue or endurance limit, with a suitable safety factor, in excess of that working stress requirement.

There are a number of factors that influence the magnitude of the fatigue limit which will affect design stresses accordingly and these are discussed on pages 552 to 558.

Limited or finite life fatigue relates to endurances principally in the range 10^5–10^7 cycles and it is here that design can benefit from analysis of fatigue crack propagation. The micro-mechanisms of fracture and the fracture mechanics of crack growth in fatigue are introduced in the next two sections.

Finally low-endurance–high-strain fatigue is a rather specialized area of design involving cyclical plasticity which is beyond the scope of this short chapter.

EXAMPLE 21.1

A mild steel shaft has a pulley mounted at each end and is supported symmetrically in between in two bearing housings. When rotating under load the shaft will be subjected to a cyclical bending moment of ±1.2 kN-m. Using a safety factor of 2 and the data in Fig. 21.7, determine a suitable diameter for the shaft for infinite fatigue life.

SOLUTION

From Fig. 21.7 the fatigue limit for infinite life is approximately $\pm 190 \text{ MN/m}^2$ and using a factor of 2 the design stress will be $\pm 95 \text{ MN/m}^2$.

The bending relationship is

$$\sigma = \frac{My}{I}$$

which for a shaft diameter of D gives

$$\pm 95 \times 10^6 = \pm \frac{1.2 \times 10^3 \times 32}{\pi D^3}$$

hence $D^3 = 128.6 \times 10^{-6} \, \text{m}^3$, $D = 50.48$ mm.

MICRO-MECHANISMS OF FATIGUE: INITIATION AND PROPAGATION

A great deal of research has been devoted to a study of the mechanism of fatigue, and yet there is still not a complete understanding of the phenomenon. It is not an easy problem to handle theoretically or experimentally, since the process commences within the atomic structure of the metal crystals and develops from the first few cycles of stress, extending over thousands or millions of subsequent cycles to eventual failure.

The fatigue mechanism has two distinct phases, initiation of a crack and the propagation of this crack to final rupture of the material. One of the earliest (1910) and classic studies of initiation was made by Ewing and Humfrey, who metallurgically examined the polished surface of a rotating bending specimen at intervals during its fatigue life. They observed that above a certain value of cyclic stress (the fatigue limit) some crystals on the surface of the specimen developed bands during cycling. These bands are the result of sliding or shearing of atomic planes within the crystal and are termed *slip bands*. With continued cyclic action these slip bands broaden and intensify to the point where separation occurs within one of the slip bands and a crack is formed as shown in Fig. 21.12(*a*). In the 1950s Forsyth discovered that a process of intrusion and

Fig. 21.12 Stages in the development of a crack along a slip plane

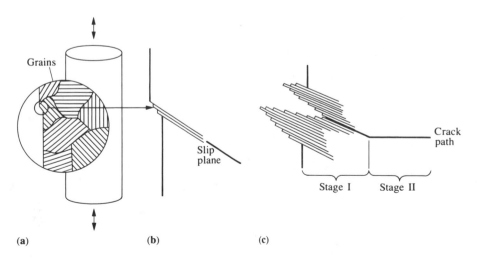

Grains

Slip plane

Crack path

Stage I   Stage II

(a)   (b)   (c)

extrusion at the surface could cause a crack to be formed as illustrated in Fig. 21.12(b) and (c). This crack initially develops along the slip plane of the grain in which it was formed (Stage I growth) but eventually propagates across other grains. This Stage II growth occurs, not as a consequence of any progressive structural damage but as a result of the stress concentration effect at the crack tip as it becomes sharp during unloading.

New theories on the initiation and propagation of fatigue cracks are still being put forward. In recent years it has been suggested that fatigue is initiated by the movement of dislocations. A dislocation is a fault or misplacement in the atomic lattice of the metal. Microscopic plastic deformation allows dislocations or vacancies to "move" through the atomic lattice, and it is thought that coagulation of dislocations forms the beginnings of a crack.

Others have suggested that dislocations are much too small to have any real effect and that it is more likely that cracks develop from intrinsic defects in the material. These may be of the order of 0.5 μm in size and Fig. 21.13 gives an idea of the scale of this relative to microstructural dimensions. It also indicates the rate at which a crack might propagate through a material. Propagation of a fatigue crack is a complex phenomenon depending on the geometry of the component, the material, the type of stressing, and the environment. Propagation can occupy as much as 90% of the total endurance, hence movement is relatively slow. Experimental evidence reveals a discontinuous progress of the crack; it is

Fig. 21.13 Stages of crack growth

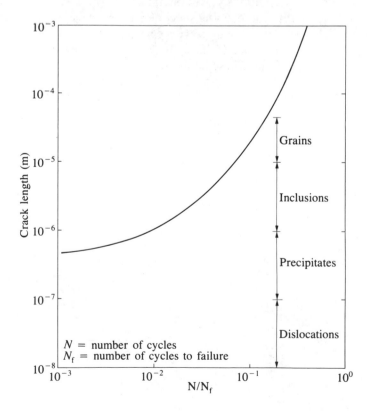

moving for some cycles and stationary for some. Such investigations have also shown that there is a fatigue crack growth threshold below which cracks can exist in a material but will not propagate. This can now be explained in terms of the concept of the stress intensity factor K which was introduced in Chapter 20, and the emergence of fracture mechanics has cast new light on fatigue crack growth phenomena. Fracture mechanics also permits life expectancy to be predicted for a material or structure containing a crack-like defect of known size – something not possible using the S–N curve approach since this does not separate out the initiation and propagation phases.

FRACTURE MECHANICS FOR FATIGUE

Fracture mechanics can only be applicable to fatigue *after* the crack initiation phase to enable crack growth to be predicted. In Chapter 20 it was shown that the state of stress in the vicinity of a crack in an infinite body could be expressed in terms of the stress intensity factor, K where

$$K = Y\sigma\sqrt{\pi a} \tag{21.4}$$

In cyclic loading, K varies over a stress intensity range ΔK where

$$\Delta K = K_{max} - K_{min}$$
$$\Delta K = Y(S_{max} - S_{min})\sqrt{\pi a} \tag{21.5}$$

using S in place of σ.

The simplest and most widely used expression which relates the range of stress intensity factor to the crack growth rate during cyclic loading is the Paris–Erdogan equation. This takes the form

$$\frac{da}{dN} = C(\Delta K)^m \tag{21.6}$$

where C and m are material constants although their values will depend on factors such as the nature of the environment and the level of residual stresses. Typical values of C and m for a range of materials are given in Table 21.1. The way in which the crack growth rate varies during cycling

Table 21.1
Fracture mechanics constants for a range of materials

Material	ΔK_{TH} (MN m$^{-3/2}$)	m	C^* ($\times 10^{-11}$)
Mild steel	3.2–6.6	3.3	0.24
Structural steel	2.0–5.0	3.85–4.2	0.07–0.11
Structural steel in sea-water	1.0–1.5	3.3	1.6
Aluminium	1.0–2.0	2.9	4.56
Aluminium alloy	1.0–2.0	2.6–3.9	3–19
Copper	1.8–2.8	3.9	0.34
Titanium	2.0–3.0	4.4	68.8

* Units of C will give da/dN in m/cycle when ΔK is in MN m$^{-3/2}$.

Fig. 21.14 Crack growth
data from an aluminium
alloy

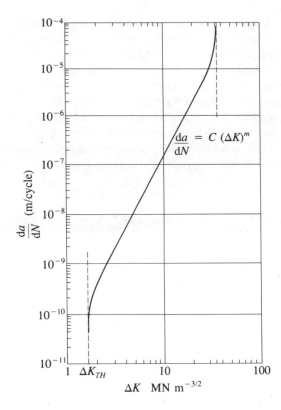

is shown in Fig. 21.14. This illustrates that at the lower end there is a
threshold value of stress intensity range ΔK_{TH} below which the crack will
not propagate. It has been found that for most metals ΔK_{TH} is given by
$(2-3)10^{-5} \times$ modulus. At the upper end, the crack growth rate tends
towards an infinitely large value. As the crack grows, K_{max} increases and
failure will occur when this exceeds the fracture toughness of the material
K_{IC} or when the remaining ligament of material ahead of the crack tip
fails by plastic collapse.

As eqn. (21.6) expresses the rate at which a crack grows under
specified cyclic stress conditions, it is possible to use this to predict
fatigue lifetimes.

From eqns. (21.5) and (21.2)

$$\Delta K = Y \cdot S_R \sqrt{\pi a}$$

where S_R is the range of variation of tensile stress.

Substituting in eqn. (21.6),

$$\frac{da}{dN} = C\{Y \cdot S_R \sqrt{\pi a}\}^m \tag{21.7}$$

Letting the initial crack size in the material be a_0, then the number of
cycles N_f to cause this crack to grow to a size a_f at which failure would

occur in one application of stress, can be obtained by integrating eqn. (21.7)

$$N_f = \frac{2}{C(Y \cdot S_R)^m \pi^{m/2}(2-m)} \{a_f^{1-m/2} - a_0^{1-m/2}\} \tag{21.8}$$

assuming that $m \neq 2$

EXAMPLE 21.2

A support bracket is welded to a backing plate as shown in Fig. 21.15. A fluctuating force in the coupling-rod causes a stress variation of $\pm 50 \text{ MN/m}^2$ at the weld. Using the crack growth data in Fig. 21.14 calculate the maximum size of defect which could be tolerated in the weld.

Fig. 21.15

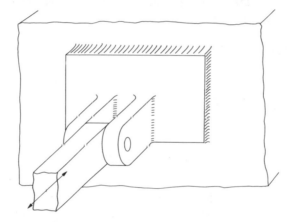

SOLUTION

From Fig. 21.14 the threshold value of the stress intensity factor ΔK_{TH} is $1.65 \text{ MNm}^{-\frac{3}{2}}$.

$$\Delta K_{TH} = Y(\Delta S)\sqrt{\pi a}$$

In this case, assuming an edge crack in the weld then $Y = 1.12$ and ΔS is taken as 50 MN/m^2 since it is only the tensile part of the cycle which causes fatigue crack growth

$$1.65 = 1.12(50)\sqrt{\pi a_0}$$
$$a_0 = 0.27 \text{ mm}$$

EXAMPLE 21.3

The blades of a turbine rotor are fitted into aluminium alloy discs on the rotor as shown in Fig. 21.16. If during the assembly of the system a 0.1 mm deep scratch is made in the surface of the disc as indicated,

Fig. 21.16

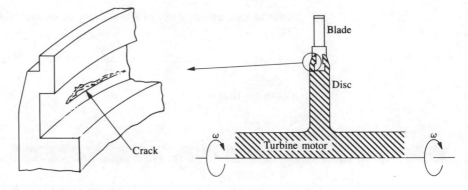

calculate how many stress cycles the disc can withstand before fatigue failure occurs. The rotation of the turbine causes a stress of 350 MN/m² at the plane of the scratch and the crack growth data for the disc material is given in Fig. 21.14.

SOLUTION

From Fig. 21.14 the following information can be obtained:

$$\frac{da}{dN} = 4 \times 10^{-11}(\Delta K)^{3.54}$$

and

$$K_{IC} = 35 \text{ MN m}^{-\frac{3}{2}}$$

From K_{IC} it is possible to calculate the size of crack which would cause failure in a single stress application. This will be the value of a_f in eqn. (21.8)

$$a_f = \frac{1}{\pi}\left(\frac{K_{IC}}{YS}\right)^2 = \frac{1}{\pi}\left(\frac{35}{1.12 \times 350}\right)^2 = 2.54 \text{ mm}$$

It is evident that as the actual crack depth is only 0.1 mm there is no danger of fast fracture during running of the rotor, even at full speed. However, repeated ON/OFF cycles for the rotor will cause the crack to grow. In this case the number of cycles to cause failure can be predicted from eqn. (21.8).

$$N_f = \frac{2}{C(YS_R)^m \pi^{m/2}(2-m)}\{a_f^{1-m/2} - a_0^{1-m/2}\}$$

$$= \frac{2 \times \{(2.54 \times 10^{-3})^{-0.77} - (0.1 \times 10^{-3})^{-0.77}\}}{4 \times 10^{-11}(1.12 \times 350)^{3.54}\pi^{1.77}(-1.54)}$$

$$= 3117 \text{ cycles}$$

INFLUENTIAL FACTORS

MEAN STRESS

It is quite common to think of fatigue in terms of the range of cyclic stress; however, the mean stress in the cycle has an important influence on fatigue behaviour. There are two obvious limiting conditions for the mean and range of stress. One is for the mean stress equal to the static strength in tension or compression, whence the range of stress must be zero. The other condition is for zero mean stress and a stress range equal to twice the static strength. Between these boundaries there is an infinite number of combinations of mean and range of stress. It is obviously impossible to study the problem experimentally completely; consequently, empirical laws have been developed to represent the variation of mean and range of stress, in terms of static strength values and the fatigue curve for reversed stress (zero mean). This latter condition is the most widely used for obtaining experimental data, principally because of the simplicity of the rotating beam test.

If a diagram is plotted of the semi-range of stress as ordinate and the mean stress as abscissa, known as an S_a–S_m diagram, as in Fig. 21.17 then

Fig. 21.17 Diagrams for mean stress v. semi-range of stress

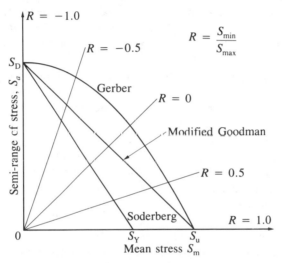

the limiting conditions are $S_m = 0$ and $S_a = S_D$, the fatigue limit for reversed stress, and $S_a = 0$ and $S_m = S_u$ the tensile strength of the material. Between these limits it is required to have a line which represents the locus of all combinations of S_a and S_m which result in the same fatigue endurance. A straight line joining these pairs of co-ordinates represents one empirical law known as the modified Goodman relationship. Algebraically this is given by

$$S_a = S_D\left(1 - \frac{S_m}{S_u}\right)$$

(21.9)

Another relationship is obtained by joining the limiting co-ordinates by a parabola known as the Gerber parabola. This is expressed algebraically in the above symbols as

$$S_a = S_D\left[1 - \left(\frac{S_m}{S_u}\right)^2\right]$$

(21.10)

A more conservative line for design purposes was proposed by Soderberg using a yield stress instead of the tensile strength in the

Goodman relationship:

$$S_a = S_D\left(1 - \frac{S_m}{S_Y}\right) \tag{21.11}$$

The simplest conclusion that can be reached from the foregoing is that an increase in tensile mean stress in the cycle reduces the allowable range of stress for a particular endurance. This applies similarly for direct stress or shear stress (torsional) fatigue.

Compressive mean stresses appear to cause little or no reduction in stress range, some materials even show an increase; consequently the $S_a - S_m$ diagram is not symmetrical about the zero mean stress axis.

GEOMETRICAL FACTORS

1. Stress concentration

Probably the most serious effect in fatigue is that of stress concentration. It is virtually impossible to design any component in a machine without some discontinuity such as a hole, keyway, or change of section. These features are known as *stress raisers* or sources of stress concentration. This concept was introduced in Chapter 13, and it was explained that under static loading in the elastic range the local peak stress at a notch or discontinuity is raised in magnitude above the nominal stress on a cross-section away from the notch. The theoretical elastic stress concentration factor K_t for a notch is defined as the maximum stress at the notch divided by the average stress on the minimum area of cross-section at the notch.

For ductile metals, static stress concentration does not reduce the strength owing to redistribution of stress when the material at the notch enters the plastic range. However, under fatigue loading the position is very different, since a fatigue limit tends to correspond with the static elastic limit of the material. Consequently, the fatigue limit for the material having the peak stress at the discontinuity would correspond to the static elastic limit, and hence the fatigue limit based on the nominal stress would be reduced by a factor dependent on the elastic stress concentration factor. In practice this is generally not so and the notched fatigue strength is rather better than the "plain" fatigue strength divided by K_t. This has led to what is termed the *fatigue strength reduction factor*, which is defined as

$$K_f = \frac{\text{Plain fatigue strength at } N \text{ cycles or fatigue limit}}{\text{Notched fatigue strength at } N \text{ cycles or fatigue limit}}$$

and generally $K_t > K_f > 1$ as shown in Fig. 21.18.

Another way of representing material reaction to stress concentration in fatigue is by means of a *notch sensitivity factor q*, which is defined in terms of the theoretical stress concentration factor and the fatigue strength reduction factor. Thus

$$q = \frac{K_f - 1}{K_t - 1} \tag{21.12}$$

and for a component or specimen which yields K_f in fatigue equal to K_t,

Fig. 21.18 Comparison of plain and notched fatigue curves

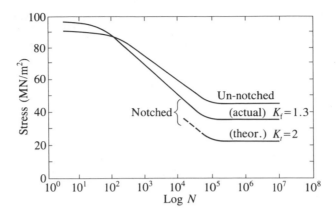

the factor $q = 1$ shows maximum sensitivity. Where there is no strength reduction, $K_f = 1$ and q becomes zero showing no sensitivity.

Discontinuities in form, notches, holes, etc. have little or no strength reduction effect at low endurances for *stress cycling* conditions. When the cycles are low, the material fatigue strength is closer to the static strength. Consequently, if the notched static strength is higher than the plain tensile strength, the notched fatigue strength at the upper end of an *S–N* curve may be higher than the plain fatigue strength as shown in Fig. 21.18.

2. Size It has been found that for geometrically similar components a large size has a slightly lower fatigue strength than a small size in cyclic bending or torsion. As this size effect is not found in uniaxial fatigue, it has been primarily attributed to the different gradients of stress and strain as shown in Fig. 21.19.

Fig. 21.19 Size effect

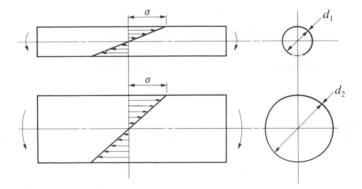

EXAMPLE 21.4

The stepped shaft shown in Fig. 21.20 is subjected to a steady axial pull of 50 kN and a uniform bending moment M. If the yield strength of the rotor material is 300 MN/m^2 and the fatigue limit in reversed bending is 200 MN/m^2, calculate the maximum value of M to avoid fatigue failure in

Fig. 21.20

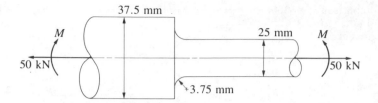

37.5 mm

25 mm

M

M

50 kN

50 kN

3.75 mm

the rotor. K_t for the fillet radius is 1.55 and the notch sensitivity factor $q = 0.9$.

SOLUTION

Mean stress, $\qquad S_m = \dfrac{F}{A} = \dfrac{50 \times 10^3 \times 4}{\pi(25)^2} = 101.8 \text{ MN/m}^2$

Alternating stress, $\quad S_a = \dfrac{My}{I} = \dfrac{M \times 12.5 \times 64}{\pi(25)^4}$

$$= M \times 6.52 \times 10^{-4} \text{ MN/m}^2$$

Here, K_f is obtained from eqn. (21.12),

$$0.9 = \frac{K_f - 1}{1.55 - 1}$$

from which $K_f = 1.5$.

Applying Soderberg's rule to allow for the mean stress,

$$\sigma_a = \frac{\sigma_f}{K_f} \left\{ 1 - \frac{\sigma_m}{\sigma_Y} \right\}$$

$$6.52 \times 10^{-4} M = \frac{200}{1.5} \left\{ 1 - \frac{101.8}{300} \right\}$$

$$M = 135.1 \text{ N-m}$$

ENVIRONMENTAL EFFECTS

1. Temperature

Many components are subjected to fatigue conditions while working at temperatures other than ambient. Components in aeroplanes at high altitude may experience many degrees of frost, while a steam or gas turbine will be running at several hundred degrees Celsius.

Tests on a number of aluminium and steel alloys ranging down to −50 °C, and lower in some cases, have shown that the fatigue strength is as good as and often a little better than at normal temperature (+20 °C).

On the other hand, fatigue tests at higher temperatures show little or no effect up to about 300 °C, after which for steels to about 400 °C there is an increase in fatigue strength to a maximum value, followed by a rapid fall to values well below that at +20 °C. A further interesting feature is that a material having a fatigue limit characteristic at ambient temperature will lose this at high temperatures, and the $S-N$ curve will continue to fall slightly even at high endurances, and an endurance limit has to be quoted.

Above a certain temperature there is interaction between fatigue and creep effects (*see* Ch. 22), and it is found that up to, say, 700 °C for heat-resistant alloy steels, fatigue is the criterion of fracture, whereas at higher temperatures, creep becomes the cause of failure. This has led to the use of combined fatigue–creep diagrams, similar to Fig. 21.17 where the cyclic stress is plotted against the steady stress which produces failure in a specified endurance at a particular temperature.

2. Corrosion

Fatigue tests are generally conducted in air as a reference condition, but in practice many components are subjected to cyclic stress in the presence of a corrosive environment. Corrosion is essentially a process of oxidation, and under static conditions a protective oxide film is formed which tends to retard further corrosion attack. In the presence of cyclic stress the situation is very different, since the partly protective oxide film is ruptured in every cycle allowing further attack. A rather simplified explanation of the corrosion fatigue mechanism is that the microstructure at the surface of the metal is attacked by the corrosive, causing an easier and more rapid initiation of cracks. The stress concentration at the tips of fissures breaks the oxide film and the corrosive in the crack acts as a form of electrolyte with the tip of the crack becoming an anode from which material is removed, thus assisting the propagation under fatigue action. It has been shown that the separate effects of corrosion and fatigue when added do not cause as serious a reduction in strength as the two conditions acting simultaneously.

One of the important aspects of corrosion fatigue is that a metal having a fatigue limit in air no longer possesses one in the corrosive environment, and fractures can be obtained at very low stress after hundreds of millions of cycles.

A particular form of corrosion fatigue may occur in situations which involve the relative movement of contacting surfaces under the action of an alternating load. This is known as fretting corrosion. Some materials are more susceptible to it than others and hence there are preferred combinations of materials in situations where it is likely to arise. The use of anti-fretting compounds, the reduction of surface stresses and surface hardening have also been found to alleviate the problem.

SURFACE FINISH AND TREATMENTS

Fatigue failures in metals almost always develop at a free surface so that the surface condition has a significant effect on fatigue endurance. Immediate improvement can be effected by polishing a machined surface since this reduces the mild stress concentration effect of a lathe turned, milled or ground surface.

Manufacturing processes also may introduce residual stresses and strain hardening. It is important therefore to appreciate the effect which these operations may have on fatigue endurance.

Surface coating of ferrous metals is often done to improve their wear and corrosion characteristics. In general it is found that such coating does not improve the fatigue strength of the substrate and, depending on the plating conditions, may cause a considerable reduction in fatigue life. Again the process of anodizing aluminium to improve its wear and corrosion resistance also has the effect of reducing its fatigue strength.

The introduction of residual compressive stresses on the surface of a metal has been shown to be a very successful method of improving fatigue endurance. Conversely, the presence of residual tensile stresses on the surface has a very detrimental effect. There are a number of metallurgical methods of introducing compressive residual stresses through hardening of the surface layers of the material. The three main methods are induction (or flame) hardening, carburizing and nitriding. It is interesting that the greatest improvement in fatigue strength is observed when stress concentrations are present.

There are also a number of methods of physically introducing compressive residual stresses at the surface. These include shot peening and skin rolling, both of which have the additional advantage that they remove any stress concentration marks left by machining operations.

CUMULATIVE DAMAGE

Although most fatigue tests and some components are subjected to a constant amplitude of cyclic stress during the life to fracture, there are many instances of machine parts and structures which receive a load spectrum, that is, the load and cyclic stress vary in some way under working conditions. To establish any difference between fatigue under varying and constant amplitude conditions, tests are conducted in which a certain number of cycles is done at one stress level followed by a number at a higher or lower stress, and this sequence is repeated till failure occurs. It is suggested that damage by fatigue action accumulates and that a certain total damage line is represented by the $S-N$ curve. One way of representing this algebraically was proposed by Miner. If n_1 cycles are conducted at a stress level S_1, at which the fracture endurance would be N_1, and if this is followed by n_2 cycles out of N_2 at a second stress level S_2, as in Fig. 21.21 and so on, then for h such blocks

$$\frac{n_1}{N_1} + \frac{n}{N_2} + \frac{n_3}{N_3} + \ldots \frac{n_h}{N_h} = 1$$

or

$$\sum_{q=1}^{h} \frac{n_q}{N_q} = 1 \tag{21.13}$$

Test results often show the sum of the cycle ratios (n/N) differing widely from the value of unity, generally covering a range from about 0.6 to 2.0 with, in a few cases, extreme values well outside this range.

The value of unity in eqn. (21.13) tends to be an overall average, but when using this approach in design it is prudent to use a lower value (such as 0.5) in order to allow for uncertainties in the hypothesis.

Fig. 21.21 Diagrammatic representation of cumulative damage

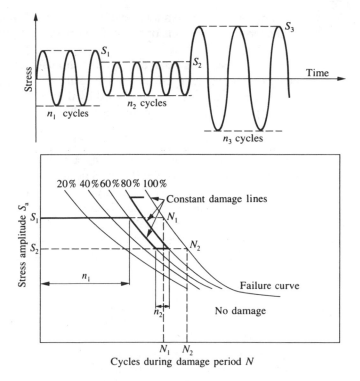

Other models for cumulative damage have been proposed in order to obtain more conservative estimates of the total fatigue life for multi-level sinusoidal stress histories. However, any slight improvements in accuracy are outweighed by the increased complexity of the analysis. The problem is that cumulative damage is dependent on stress history, mean stress, stress concentration, etc. and no simple model can be developed to predict accurately the fatigue life with such a wide range of variables. For example, it is often found that for a two-stress level test, in which one stress is applied for a number of cycles and then run to failure at a second stress, if $S_1 < S_2$, then $\Sigma(n/N) > 1$, and for $S_1 > S_2$, $\Sigma(n/N) < 1$. In addition, the variation from unity is greater for larger differences between S_1 and S_2. It is in spite of these difficulties that Miner's simple approach still provides a useful starting-point to the analysis of fatigue under variable loading spectra and so no other model has received such widespread acceptability and usage.

FAILURE UNDER MULTIAXIAL CYCLIC STRESSES

In the discussion of fatigue so far, we have considered only the effects of cyclical uniaxial stresses on materials. However in practice many components are subjected to biaxial or triaxial stress systems for which laboratory testing would be complex and expensive. Consequently, in a similar manner to the prediction of static yielding under complex stress by the use of a simple tensile yield stress (Chapter 13), attempts have been made to establish a criterion for fatigue failure due to complex cyclic stresses, in terms of the uniaxial stress fatigue limit for the material.

Some success has been achieved in the use of the shear strain energy criterion to predict cyclical failure. Hence from eqn. (13.15) the yield stress in simple tension is replaced by the fatigue limit, S_D, in, say, rotating bending for the material, so that

$$(\sigma_1 - \sigma_2)^2 + (\sigma_2 - \sigma_3)^2 + (\sigma_3 - \sigma_1)^2 = 2S_D^2.$$

The applicability of this criterion may be related to the static yield situation in that micro-yielding will occur at the crack tip as propagation proceeds. Fatigue cracks generally initiate at a free surface so that σ_3 will be zero in the above equation. If mean stresses are present then this must be allowed for by using the above equation as a function of both alternating stresses and mean stresses and using the appropriate values of S_a and S_m, from uniaxial data, on the right hand side.

More recent research has shown that the above criterion has a rather limited range of applicability and that two strain parameters are required instead of the single equivalent uniaxial cyclic stress. The parameters proposed are the largest strain circle created in a fatigue cycle and the position of that strain circle in strain space. These are represented as $\frac{1}{2}(\varepsilon_1 - \varepsilon_3)$ and $\frac{1}{2}(\varepsilon_1 + \varepsilon_3)$ respectively from Mohr's strain circles. The intermediate strain ε_2 has been shown to control the direction in which the crack, once initiated, will grow.

EXAMPLE 21.5

A 25 mm diameter geared shaft, Fig. 21.22, is subjected to a fully reversed bending moment of ± 1094 kN-m as shown. The gear is a shrink fit on to the shaft and sets up radial and tangential stresses of 350 MN/m² on the surface of the shaft. If during operation the gear wheel is subjected to a fluctuating torque of ± 900 N-m estimate the fatigue life of the shaft. The shaft material has a yield stress of 925 MN/m² and its bending fatigue life may be predicted from

$$N_f = 5.2 \times 10^{56} \left(\frac{1}{S_f}\right)^{16.5}$$

where S_f is the stress amplitude in MN/m².

Fig. 21.22

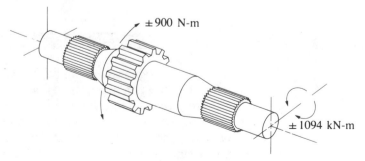

± 900 N-m

± 1094 kN-m

SOLUTION

The bending stress is

$$\sigma_x = \frac{My}{I} = \frac{\pm 1094 \times 10^3 \times 12.5 \times 64}{\pi \times (25)^4} = \pm 713 \text{ MN/m}^2$$

The shear stress is

$$\tau_{xy} = \frac{Tr}{J} = \frac{\pm 900 \times 10^3 \times 12.5 \times 32}{\pi (25)^4} = \pm 293 \text{ MN/m}^2$$

Fig. 21.23 shows the stress systems and from Mohr's circle the principal stresses may be determined for the maximum and minimum limits of cyclic stress

Cyclic maximum (MN/m²)	Cyclic minimum (MN/m²)
$\sigma_1 = 800$	$\sigma_1 = -870$
$\sigma_2 = -420$	$\sigma_2 = -190$
$\sigma_3 = -350$	$\sigma_3 = -350$

Amplitudes (MN/m²)	Mean (MN/m²)
$\sigma_{1a} = 835$	$\sigma_{1m} = -35$
$\sigma_{2a} = 115$	$\sigma_{2m} = -305$
$\sigma_{3a} = 0$	$\sigma_{3m} = -350$

Using the shear strain energy criterion,

$$2S_{Da}^2 = (\sigma_{1a} - \sigma_{2a})^2 + (\sigma_{2a} - \sigma_{3a})^2 + (\sigma_{3a} - \sigma_{1a})^2$$
$$= (835 - 115)^2 + (115 - 0)^2 + (0 - 835)^2$$
$$S_{Da} = 784 \text{ MN/m}^2$$

Similarly, $S_{Dm} = 295 \text{ MN/m}^2$.

Fig. 21.23

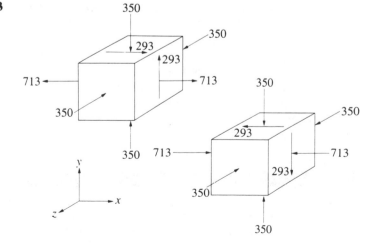

Using the Soderberg rule,

$$S_{Da} = S_f \left\{ 1 - \frac{S_{Dm}}{\sigma_Y} \right\}$$

$$S_f = \frac{784}{1 - \dfrac{295}{925}} = 1151 \text{ MN/m}^2$$

From the equation describing the fatigue life of shaft material

$$N_f = 5.2 \times 10^{56} \left(\frac{1}{1151} \right)^{16.5} = 1.61 \times 10^6 \text{ cycles}$$

FATIGUE OF PLASTICS AND COMPOSITES

Reinforced and unreinforced plastics are susceptible to brittle fatigue failures in much the same way as metals, and the fracture mechanics approach has been used to predict crack growth rates. In addition, however, the high damping and low thermal conductivity of plastics means that under cyclic stressing there is also the possibility of short-term thermal softening failures if precautions are not taken to dissipate the heat generated. Therefore, although in metals the fatigue strength is relatively unaffected by frequencies in the range 3–100 Hz, with plastics the cyclic frequency is important since it has a pronounced effect on the temperature rise in the material.

With non-reinforced moulded plastics, fatigue cracks usually initiate within the bulk of the material because the moulding operation tends to produce a protective skin which inhibits crack growth from the surface. However, plastics are not immune to the effect of stress concentrations and many fatigue cracks in moulded articles are initiated at holes, sharp corners, etc.

In reinforced plastics the failure mechanism appears to take place through a breakdown of the fibre–matrix bond which permits atmospheric corrosion of the fibres. The orientation of the fibres relative to the loading direction has also been found to be a determining factor in regard to fatigue strength. Tests on a wide range of fibre-reinforced plastics has shown that the best performance is achieved with non-woven fibre reinforcement in which the fibre direction is slightly off the loading axis. Woven reinforcement exhibits reduced fatigue strength due to stress concentration effects at the fibre cross-over points.

SUMMARY

It has been shown that the possibility of premature failure of a material by a fatigue mechanism is an extremely important design consideration in situations where fluctuating stresses are either applied directly or transmitted to the material. The majority of all failures which occur in practice can be attributed to fatigue and the greatest percentage of these are caused by bad design, usually ill-considered positioning and/or geometry of stress raisers such as holes, abrupt changes in section, keyways, etc. Other factors which affect fatigue endurance are the level of the mean stress, surface condition and nature of the environment.

Although the crack growth mechanism during fatigue is not completely understood it is known that when a component is subjected to a variety of stress levels the damage incurred is cumulative. It has also been found that fracture mechanics is a useful tool in predicting fatigue life provided crack growth data for the material is available.

BIBLIOGRAPHY

Forrest, P. G. *Fatigue of Metals,* Pergamon, Oxford, 1962.
Osgood, C. G. *Fatigue Design* Pergamon, Oxford, 1982.
Pook, L. P. *The Role of Crack Growth in Metal Fatigue.* The Metal Society, London, 1983.
Duggan, T. V. and Byrne, J. *Fatigue as a Design Criterion,* Macmillan, London, 1977.
Madayag, A. F. *Metal Fatigue*: *Theory and Design,* Wiley, New York, 1969.
Klesnil, M. and Likas, P. *Fatigue of Metallic Materials,* Elsevier, Amsterdam, 1980.

PROBLEMS

21.1 A switching device consists of a rectangular cross-section metal cantilever 200 mm in length and 30 mm in width. The required operating displacement at the free end is ±2.7 mm and the service life is to be 100 000 cycles. To allow for scatter in life performance a factor of 5 is employed on endurance. Using the fatigue curves given in Fig. 21.7 determine the required thickness of the cantilever if made in (a) mild steel, (b) aluminium alloy. $E_{steel} = 208$ GN/m^2, $E_{aluminium} = 79$ GN/m^2.

21.2 (a) A fatigue fracture produced by cyclic uniaxial tension stresses exhibits circular striations which have their centre at the point of crack initiation (usually on the surface). Explain why the striations have this shape. (b) If a shaft is subjected to cyclic torsional stresses, on what planes would you expect the fatigue cracks to grow?

21.3 A pressure vessel support bracket is to be designed so that it can withstand a tensile loading cycle of 0–500 MN/m^2 once every day for 25 years. Which of the following steels would have the greater tolerance to intrinsic defects in this application: (i) a maraging steel ($K_{Ic} = 82$ MN m$^{-\frac{3}{2}}$, $C = 0.15 \times 10^{-11}$, $m = 4.1$), or (ii) a medium-strength steel ($K_{Ic} = 50$ MN m$^{-\frac{3}{2}}$, $C = 0.24 \times 10^{-11}$, $m = 3.3$)? For the loading situation a geometry factor of 1.12 may be assumed.

21.4 A series of crack growth tests on a moulding grade of polymethyl methacrylate gave the following results:

da/dN (m/cycle)	2.25×10^{-7}	4×10^{-7}	6.2×10^{-7}	11×10^{-7}	17×10^{-7}	29×10^{-7}
ΔK(MN m$^{-3/2}$)	0.42	0.53	0.63	0.79	0.94	1.17

If the material has a critical stress intensity factor of 1.8 MN m$^{-3/2}$ and it is known that the moulding process produces defects 40 μm long, estimate the maximum repeated tensile stress which could be applied to this material for at least 10^6 cycles without causing fatigue failure.

21.5 As part of the mechanism of a machine a cam is used to cause a metal beam to oscillate, as shown in Fig. 21.24. If the design of the cam is such that the beam deflection varies between a maximum of 3 mm and a minimum of 1 mm relative to its undeflected position, calculate a suitable beam depth to avoid fatigue failure in the beam material, using the Gerber, modified Goodman and Soderberg methods to allow for the effect of mean stress. A fatigue strength reduction factor of 1.8 should be assumed. The tensile and yield strengths of the beam material are 350 MN/m^2 and 200 MN/m^2 respectively, and its fatigue strength in fully reversed cycling is 100 MN/m^2. Young's modulus for the steel is 207 GN/m^2.

Fig. 21.24

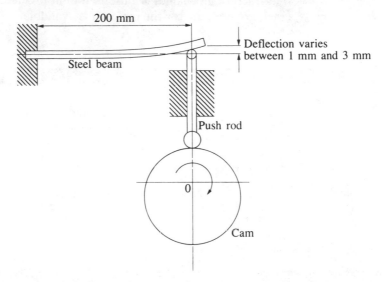

21.6 A shaft of circular cross-section is subjected to a steady bending moment of 1500 N-m and simultaneously to an alternating bending moment of 1000 N-m in the same plane (so that the total moment fluctuates between 2500 N-m and 500 N-m). Calculate the necessary diameter of the shaft if the factor of safety is to be 2.5. The yield stress of the material is 210 MN/m^2 and the fatigue limit in reversed bending is 170 MN/m^2. Calculate also the diameter of the shaft if stress concentrations are to be allowed for with a fatigue strength reduction factor of 2. Assume that the Soderberg rule applies.

Fig. 21.25

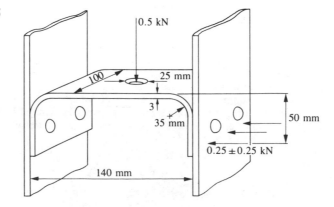

21.7 A connecting-rod of circular cross-section with a diameter of 50 mm is subjected to an eccentric longitudinal load at a distance of 10 mm from the centre of the cross-section. The load varies from a value of $-F/2$ to F. The material used has a tensile strength of 420 MN/m^2 and a fatigue limit in fully reversed loading of 175 MN/m^2. Determine the limiting value of F to avoid fatigue failure if the modified Goodman relation applies.

21.8 Part of the structure of an aircraft is shown in Fig. 21.25. The central hole in the inverted U-section supports a vertical force of 0.5 kN. During flight the upright section is subjected to a cyclical force which varies from 0 to 0.5 kN. If all the parts are made from an aluminium alloy with a yield strength of 392 MN/m^2 and fatigue strength of 270 MN/m^2, estimate where you expect fatigue failure to occur. Use the Soderberg rule to allow for the effect of mean stresses. The stress concentration factor at the curved portion of the U-channel may be taken as 1.85 and K_t at the central hole may be obtained from Chapter 13.

21.9 A series of tensile fatigue tests on stainless steel strips containing a central through hole gave the following values for the fatigue endurance of the steel. If the steel strips were 100 mm wide, comment on the notch sensitivity of the steel.

Hole diameter (mm)	No hole	5	10	20	25
Fatigue endurance (MN/m^2)	600	250	270	320	370

21.10 The fatigue endurances from the S–N curve for a certain steel are:

Stress (MN/m^2)	Fatigue endurance (cycles)
350	2 000 000
380	500 000
410	125 000

If a component manufactured from this steel is subjected to 600 000 cycles at 350 MN/m^2 and 150 000 cycles at 380 MN/m^2, how many cycles can the material be expected to withstand at 410 MN/m^2 before fatigue failure occurs, assuming that Miner's cumulative damage theory applies?

21.11 The analysis of the cyclic stresses on part of the landing gear of an aircraft shows that during each flight it is subjected to the following stress history:

$$100\,000 \text{ cycles at } \pm 50 \text{ MN/m}^2$$
$$10\,000 \text{ cycles at } \pm 100 \text{ MN/m}^2$$
$$500 \text{ cycles at } \pm 150 \text{ MN/m}^2$$
$$200 \text{ cycles at } \pm 180 \text{ MN/m}^2$$
$$100 \text{ cycles at } \pm 185 \text{ MN/m}^2$$
$$50 \text{ cycles at } \pm 200 \text{ MN/m}^2$$
$$10 \text{ cycles at } \pm 210 \text{ MN/m}^2$$

If the S–N curve, for the part material is given by $S = 500\,N^{-0.064}$, where S is in MN/m^2, estimate how many flights this component can withstand before fatigue failure occurs.

21.12 During service a steel cylinder, 320 mm diameter, is to be subjected to an internal pressure which varies from 0 to p. If the wall thickness of the cylinder is 8 mm estimate the maximum permissible value of p to avoid fatigue failure in the cylinder. The tensile strength of the steel is 440 MN/m^2 and the fatigue endurance limit in fully reversed cycling is 210 MN/m^2. A fatigue strength reduction factor of 1.8 may be assumed and the modified Goodman relationship should be used for the mean stress effect.

Fig. 21.26

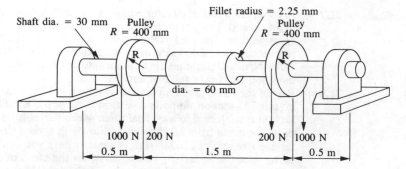

Shaft dia. = 30 mm Pulley R = 400 mm Fillet radius = 2.25 mm Pulley R = 400 mm dia. = 60 mm 1000 N 200 N 200 N 1000 N 0.5 m 1.5 m 0.5 m

21.13 Part of a pulley power transmission system is shown in Fig. 21.26. Assuming a notch sensitivity factor of 1, calculate the required endurance limit in the shaft steel in order to avoid fatigue failure. The stress concentration factors for the shaft should be obtained from Chapter 13. The shaft bearings may be regarded as simple supports and the pulleys each provide additional load of 300 N.

Chapter 22

Creep and viscoelasticity

INTRODUCTION

For the majority of engineering designs the variation of the ambient
temperature is not great and the stiffness and strength of the metal may
be regarded as a constant. However, there are also engineering
applications which occur at high temperature in fields such as steam,
plant, gas turbines, nuclear and chemical processes, kinetic heating of
supersonic aircraft,[1] etc.

It was mentioned briefly in Chapter 19 that in general the effect of
temperatures of up to several hundred degrees Celsius on metals is to
lower their yield and tensile strengths by very considerable amounts
compared with ambient conditions. Another factor which did not arise
during the development of stress–strain solutions of engineering
problems in earlier chapters was the effect of the length of time under
which a component or structure was subjected to stress. It was assumed
that the application of external loading would develop particular values
of stress and strain and these would remain constant until the applied
loading was removed. This is certainly so for the bulk of engineering
alloys in the elastic range at room temperature. However, with increasing
temperature it is possible for a material to have increasing strain with
time even at *constant* applied load. The time dependence of strain is
termed *creep*. Another complementary time-dependent response is
termed *stress-relaxation* which occurs when the strain or deformation of a
component or structure is kept constant for a time period during which a
reducing applied load (stress) is required to maintain the strain. These
are most important properties in the design of any high-temperature
component.

Plastics, both reinforced and unreinforced, have in recent years begun
to take on a developing role among engineering materials. A brief
reference was made in Chapter 3 to their stress–strain–time behaviour
and this is known as *viscoelasticity,* being the interactive results of
Newtonian viscosity and Hookean elasticity. Unreinforced thermoplastics

are particularly susceptible to creep and room temperature is "high" enough for plastics to exhibit the phenomenon, the actual rate of creep being a function of the imposed stress.

The first part of this chapter concentrates on the problem of creep and stress relaxation as a time-dependent–high-temperature phenomenon affecting metals. The second part of the chapter deals with the basic considerations of viscoelastic data and creep design for plastics.

STRESS–STRAIN–TIME–TEMPERATURE RELATIONSHIPS

Creep manifests itself in metals at temperatures above about $0.3T_m$, where T_m is the absolute melting temperature, and around $0.5T_m$ creep strain becomes considerable. Thus Andrade[2] commenced studies of creep behaviour in 1910 using lead, since this metal exhibits creep at room temperature. The majority of creep experiments are carried out under uniaxial constant loading conditions for a particular chosen test temperature which must be very accurately controlled. Measurements of extension are made at frequent intervals of time until the specimen fractures or the experiment is stopped after a sufficiently lengthy period. Typical curves of creep strain against time are plotted in Fig. 22.1 for various constant load (nominal stress) levels at a constant specimen temperature.

There are four principal aspects of each of the curves shown in Fig. 22.1 as follows:

(a) *initial strain,* which is.elastic but may extend marginally into the plastic range due to the first application of load;
(b) *primary stage,* a period of decreasing creep rate during which strain hardening is occurring more rapidly than softening due to the high temperature;
(c) *secondary stage,* in which the creep rate is virtually constant through equilibrium between strain-hardening and thermal softening;

Fig. 22.1 Typical creep curves for various stresses at constant temperature

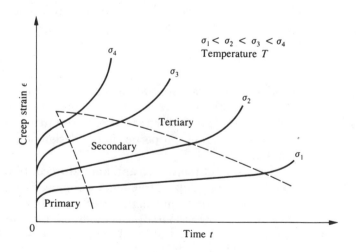

(*d*) *tertiary stage,* an increasing rate of strain, due to microstructural instability from prolonged high temperature and to gradual increase in stress level and stress concentration at cracks in the grain boundaries, which leads to complete fracture of the specimen.

It is important to note that elongation to fracture in creep, even for a ductile metal, is only a fraction of that obtained for continuous loading to fracture at high temperature. The very small gradient of the secondary stage at low stresses suggests that there might be a limiting creep stress, below which $d\varepsilon/dt = 0$, having similar significance as the fatigue limit under cyclic stress. However, it has been shown that there is no reliable criterion of this form for creep, and a "limiting creep stress" is based on a permissible creep strain after a given time.

The family of curves shown in Fig. 22.1 is based on stress as the parameter and temperature constant; however, a similar family of curves would be obtained for a particular constant stress with temperature as the parameter. Thus a complete picture of the creep behaviour of a metal necessitates the construction of several families of creep curves for various stresses and temperatures.

EMPIRICAL REPRESENTATIONS OF CREEP BEHAVIOUR

Typical of the standards of creep strength required for metals are the stresses to give minimum creep rates of 1% strain in 10 000 hours, or 1% strain in 100 000 hours.

A 10 000-hour test occupies approximately 1 year, and it is therefore evident that, although a few creep tests may be conducted over periods of this length or longer, it is a very slow and inconvenient process for obtaining a range of data at different stresses and temperatures. As a result, methods have been sought whereby long-life data can be extrapolated from short-term tests.

Creep strain ε_c, which is a function of stress, σ, time, t, and temperature, T can be represented by three functions as follows:

$$\varepsilon_c = f_1(\sigma) \cdot f_2(t) \cdot f_3(T) \tag{22.1}$$

(*a*) Stress function: the most commonly used functions are

$$\left. \begin{array}{l} f_1(\sigma) = A_1\sigma^\eta; \quad f_1(\sigma) = A_2 \sin h\left\{\dfrac{\sigma}{\sigma_0}\right\}; \\[2ex] f_1(\sigma) = A_3 \exp\left\{\dfrac{\sigma}{\sigma_0'}\right\} \end{array} \right\} \tag{22.2}$$

where A_1, A_2, A_3 are constants and σ_0 and σ_0' are reference stresses.

(*b*) Time function: this is usually expressed as a polynomial, as first suggested by Andrade[3], one reasonably applicable form being

$$\varepsilon_c = \alpha t^{1/3} + \beta t + \gamma t^3 \tag{22.3}$$

in which α, β and γ are material constants, but are functions of stress and temperature, relating to the primary, secondary and tertiary stages respectively.

(c) Temperature function: the most generally used temperature function is

$$f_3(T) = \exp\left\{-\frac{\Delta H}{RT}\right\} \tag{22.4}$$

where ΔH is the activation energy, R is the Universal Gas Constant and T is the absolute temperature. This type of expression is fundamental to all rate processes.

When we come to designing for creep in components, interest centres principally on the secondary stage, where at low stresses the creep rate is constant for very long times producing the major contribution to the total creep strain to fracture. Consequently, in eqn. (22.3) the tertiary term is neglected and the primary is replaced by a constant strain ε_0, being the intercept of the extrapolated secondary stage back on to the strain axis (Fig. 22.2). Thus

$$\varepsilon_c = \varepsilon_0 + \left(\frac{\mathrm{d}\varepsilon}{\mathrm{d}t}\right)t \tag{22.5}$$

where $\mathrm{d}\varepsilon/\mathrm{d}t = \dot{\varepsilon}_c$ is the secondary-stage creep rate. This minimum creep rate has been experimentally related to stress by the empirical expression,

$$\dot{\varepsilon}_c = B\sigma^n \tag{22.6}$$

where B and n are material constants. The dependence on temperature can then be included by writing

$$\dot{\varepsilon}_c = B\sigma^n \exp\left(-\frac{\Delta H}{RT}\right) \tag{22.7}$$

A log-log plot of $\dot{\varepsilon}_c$ against σ yields a straight line. However, extrapolation of the data, particularly at high stress, can be unreliable due to the dependence of n and ΔH on the particular stress/temperature

Fig. 22.2 Simplified creep curve

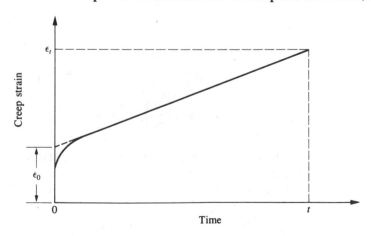

Fig. 22.3 Creep strain v. log time curves, extrapolated to a specified strain

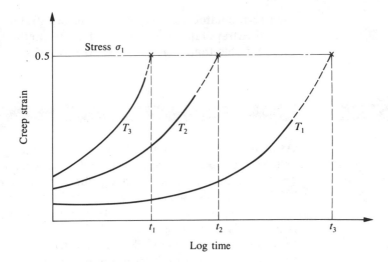

regime. Alternatively, combining eqns (22.5) and (22.6), we can express the time to reach a specified value of total creep strain in the secondary stage as follows:

$$t = \frac{\varepsilon_c - \varepsilon_0}{B\sigma^n} \tag{22.8}$$

However, this approach can be uncertain in that tertiary creep may commence before the predicted value of secondary creep is achieved.

A safer method of predicting the stress and temperature permissible for a specified creep strain of say 0.5% in 100 000 hours is as follows. Creep tests are carried out at various stresses and temperatures to obtain a significant part of the secondary-stage creep in each test. Graphs of creep strain against log time for several temperatures at a particular stress are then extrapolated to the limiting creep strain of say 0.5% as in Fig. 22.3. Next we plot graphs of temperature against log time for various stresses at the strain limit of 0.5% as shown in Fig. 22.4. These curves are

Fig. 22.4 Temperature v. log time curves, extrapolated to specified time

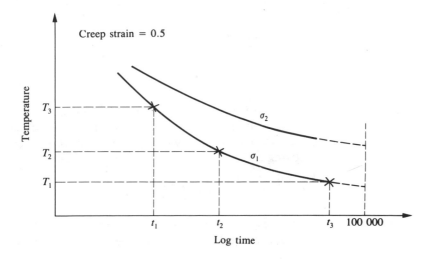

then extrapolated to a required life of say 100 000 hours. Although several extrapolations are involved in this method, they tend to be rather more reliable than a simple extrapolation of a creep–time curve.

CREEP–RUPTURE TESTING

As there is as yet no real substitute for a few long-term tests to ensure reliable creep knowledge, it is useful to have a quick sorting test to enable the best material from a group to be selected for long-term tests.

The creep–rupture test is widely used for the above purpose and also as a guide to the rupture strength at very long endurances. The principle is to apply various values of stress, in successive tests at constant temperature, of a magnitude sufficient to cause rupture in times from a few minutes to several hundred hours. Plotting log stress against log time as in Fig. 22.5 yields a family of straight lines with temperature as parameter. These tend to extrapolate back to the respective hot tensile strengths. Extrapolation forward to longer times is possible, but care has to be exercised in that oxidation owing to the high temperature can cause a marked increase in the slope, and hence reduction in stress for a required life.

Fig. 22.5 Log–rupture-stress/log–time relation for chromium–molybdenum–silicon steel (adapted from Lessels; by courtesy of John Wiley & Sons Inc)

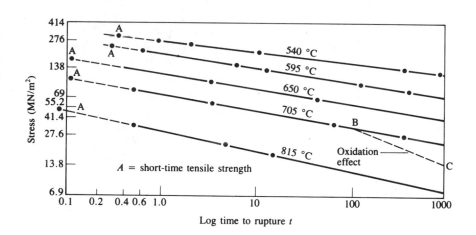

One of the physical approaches to the problem of creep suggests that viscous flow in fluids is analogous to secondary creep in metals and hence a rate-process theory is applicable, relating creep rate and temperature, using eqn. (22.4)

$$\dot{\varepsilon} = \frac{d\varepsilon}{dt} = A\,e^{-\Delta H/RT} \tag{22.9}$$

where A is a material constant.

Larsen and Miller[4] have analysed the above relationship and put it in the form

$$\frac{\Delta H}{R} = T(\log_e A + \log_e t - \log_e \varepsilon) \qquad (22.10)$$

or

$$\left(\frac{\Delta H}{R}\right)_\sigma = T(a + \log_e t) \qquad (22.11)$$

for a given value of strain. Here "a" is a constant for a given strain and $(\Delta H/R)_\sigma$ is a function of the stress level σ. The right-hand side of eqn. (22.11) is known as the Larsen–Miller parameter, $T(a + \log_e t)$, and plotting $\log_e \sigma$ against the parameter often yields a family of straight lines for different creep strains, known as master creep curves, which correlate well over a wide range of times, temperatures and different metals. From the above curves a general relationship may be written in the form

$$\log_e \sigma = C_1 + C_2 T(a + \log_e t) \qquad (22.12)$$

where C_1 and C_2 are constants. Combining the strain-rate/stress equation

$$\dot\varepsilon = B\sigma^n$$

with that above for stress, temperature and time gives

$$\dot\varepsilon = C\sigma^{n'} e^{-\alpha/T} \qquad (22.13)$$

where C, α and n' are material creep constants.

TENSION CREEP TEST EQUIPMENT

The most common form of creep test is conducted in simple tension, and since constant loading on the specimen is required over very long periods a dead-weight loading system is usually employed. A typical arrangement is shown in Fig. 22.6, where the lower end of the test piece is gripped in screwed or colleted chucks attached to a positioning screw. The upper end of the test piece is hung from a beam pivoted on the frame of the machine. The load is suspended from the free end of the beam, and below the weight pan is located a microswitch so that, when the specimen is fractured or has suffered a specified extension, the magnified movement of the load operates the microswitch and cuts out the furnace. This item together with the control circuit is perhaps the most important part of the apparatus. In previous sections it has been stated that slight temperature variation has a great influence on the minimum creep rate; it is therefore essential to have uniform and constant temperature along the length of the specimen. The relevant British Standard specifies a maximum variation along the gauge length of 2 °C and a variation of mean temperature of not more than ±1 °C up to 600 °C and ±2 °C from 600 °C to 1000 °C. If a resistance furnace is used, it is usual to have three separately controlled resistance elements to compensate for non-uniform

Fig. 22.6 Arrangement
of a tension creep
machine (Courtesy of
Electronic and
Mechanical
Engineering Co. Ltd)

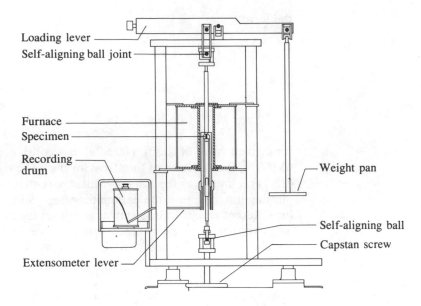

Loading lever
Self-aligning ball joint

Furnace
Specimen

Recording
drum

Weight pan

Self-aligning ball
Capstan screw

Extensometer lever

flow of heat through the furnace. Thermocouples and temperature
indicators are used for measurement and control.

CREEP DURING PURE BENDING OF A BEAM

In order to demonstrate the application of eqn. (22.6), the case of a beam
subjected to pure bending at high temperature will be considered.

The assumptions that will be made in this problem are as follows:

(a) plane sections remain plane under creep deformation;
(b) longitudinal fibres experience only simple axial stress;
(c) creep behaviour is the same in tension as in compression.

If the radius of curvature of the neutral axis is R, then the strain at a
distance y from the axis is

$$\varepsilon = \frac{y}{R}$$

and for creep in the secondary stage,

$$\frac{d\varepsilon}{dt} = B\sigma^n$$

Therefore

$$\frac{d(y/R)}{dt} = B\sigma^n$$

or

$$\frac{y}{R} = B\sigma^n t$$

and

$$\sigma = \left(\frac{y}{RBt}\right)^{1/n} \tag{22.14}$$

If the bar is of width b and depth d, equilibrium of the external and internal moments is given by

$$M = 2\int_0^{d/2} \sigma b y \, dy$$

Substituting for σ, using eqn. (22.14)

$$M = 2\int_0^{d/2} \left(\frac{y}{RBt}\right)^{1/n} by \, dy = \frac{2b}{(RBt)^{1/n}} \int_0^{d/2} y^{1+(1/n)} \, dy$$

Integration gives

$$M = \frac{2b}{(RBt)^{1/n}} \frac{n}{2n+1} \left(\frac{d}{2}\right)^{2+(1/n)}$$

Substituting for $(RBt)^{1/n}$ from eqn. (22.14) gives

$$M = \frac{2b\sigma}{y^{1/n}} \frac{n}{2n+1} \left(\frac{d}{2}\right)^{2+(1/n)}$$

Inserting the second moment of area $I = \frac{1}{12}bd^3$ and rearranging, we have the bending stress at distance y from the neutral axis as

$$\sigma - \frac{My}{I} \frac{(2n+1)}{3n} \left(\frac{2y}{d}\right)^{(1/n)-1} \tag{22.15}$$

It has been put in this form to show the difference from the simple linear elastic distribution of stress, My/I.

The stress distribution across the section is shown in Fig. 22.7 for $n = 1$, which corresponds to no creep and simple elastic conditions, and

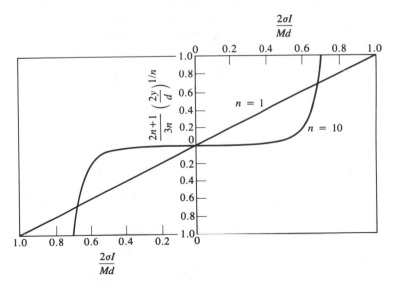

Fig. 22.7 Distribution of bending stress for the cases of zero creep and secondary-stage creep

$n = 10$, where secondary-stage creep is occurring. It is seen that the effect of creep is to relax the outer fibres' stress, but to increase the core bending stress. This is because minimum creep rate must be proportional to the distance from the neutral axis in order to satisfy eqn. (22.14), and this automatically adjusts the stresses to the distribution shown.

CREEP UNDER MULTI-AXIAL STRESSES

Under the above heading we find typical engineering components such as thin-walled tubes subjected to internal pressure and combinations with bending or torsion, thick-walled cylinders under internal pressure, rotating discs and plates subjected to bending. The number of experimental investigations has been limited owing to the intricacies of high-temperature testing and complex loading. However, there have been sufficiently comprehensive studies to verify the analytical approach that is generally adopted.

In order to relate uniaxial creep data to a biaxial problem, some of the laws of plasticity are invoked, namely that (a) the principal strains and stresses are coincident in direction, (b) plastic deformation occurs at constant volume, (c) the maximum shear stresses and shear strains are proportional.

For constant volume the sum of the three principal strains is zero; therefore,

$$\varepsilon_1 + \varepsilon_2 + \varepsilon_3 = 0 \tag{22.16}$$

Expressing maximum shear stress and strain in terms of the difference in principal stress and strains,

$$\frac{\varepsilon_1 - \varepsilon_2}{\sigma_1 - \sigma_2} = \frac{\varepsilon_2 - \varepsilon_3}{\sigma_2 - \sigma_3} = \frac{\varepsilon_3 - \varepsilon_1}{\sigma_3 - \sigma_1} = \beta \tag{22.17}$$

Rearranging the above equations to give the individual principal strains in terms of the principal stresses,

$$\left.\begin{aligned}
\varepsilon_1 &= \frac{2\beta}{3}[\sigma_1 - \tfrac{1}{2}(\sigma_2 + \sigma_3)] \\[2mm]
\varepsilon_2 &= \frac{2\beta}{3}[\sigma_2 - \tfrac{1}{2}(\sigma_3 + \sigma_1)] \\[2mm]
\varepsilon_3 &= \frac{2\beta}{3}[\sigma_3 - \tfrac{1}{2}(\sigma_1 + \sigma_2)]
\end{aligned}\right\} \tag{22.18}$$

These may be expressed as a constant creep rate by writing

$$\left.\begin{aligned}
\dot{\varepsilon}_1 &= \alpha[\sigma_1 - \tfrac{1}{2}(\sigma_2 + \sigma_3)] \\
\dot{\varepsilon}_2 &= \alpha[\sigma_2 - \tfrac{1}{2}(\sigma_3 + \sigma_1)] \\
\dot{\varepsilon}_3 &= \alpha[\sigma_3 - \tfrac{1}{2}(\sigma_1 + \sigma_2)]
\end{aligned}\right\} \tag{22.19}$$

where α is a function relating the three principal stresses to the simple uniaxial stress creep condition.

Using the von Mises yield criterion to obtain the equivalent uniaxial stress σ_e, gives

$$\sigma_e = \frac{1}{\sqrt{2}} \sqrt{[(\sigma_1 - \sigma_2)^2 + (\sigma_2 - \sigma_3)^2 + (\sigma_3 - \sigma_1)^2]} \qquad (22.20)$$

From the simple secondary-stage creep law,

$$\dot{\varepsilon} = B\sigma_e^n$$

and for simple tension, σ_2 and $\sigma_3 = 0$ and $\sigma_e = \sigma_1$; therefore

$$\dot{\varepsilon} = \alpha\sigma_e$$

and hence

$$\alpha = B\sigma_e^{n-1} \qquad (22.21)$$

Therefore the three principal creep rates may be written as

$$\left.\begin{array}{l} \dot{\varepsilon}_1 = B\sigma_e^{n-1}[\sigma_1 - \tfrac{1}{2}(\sigma_2 + \sigma_3)] \\ \dot{\varepsilon}_2 = B\sigma_e^{n-1}[\sigma_2 - \tfrac{1}{2}(\sigma_3 + \sigma_1)] \\ \dot{\varepsilon}_3 = B\sigma_e^{n-1}[\sigma_3 - \tfrac{1}{2}(\sigma_1 + \sigma_2)] \end{array}\right\} \qquad (22.22)$$

EXAMPLE 22.1

A Ni . Cr . Mo alloy steel tube of 100 mm diameter and 3 mm wall thickness is to operate at 400 °C with internal pressure for a service life of 100 000 hours. Determine the allowable pressure for a creep strain limit of 0.5%. The constants in the minimum creep equation at 400 °C are $n = 3$ and $B = 1.45 \times 10^{-23}$ per hour per MN/m^2.

SOLUTION

In the thin tube under internal pressure where σ_1 is the hoop stress and σ_2 the axial stress, $\sigma_1 = 2\sigma_2$ and $\sigma_3 = 0$. Hence from eqns. (22.20) and (22.22)

$$\sigma_e = \frac{\sqrt{3}}{2} \sigma_1$$

and

$$\dot{\varepsilon}_1 = \left(\frac{\sqrt{3}}{2}\right)^{n+1} B\sigma_1^n$$

$$\dot{\varepsilon}_2 = 0$$

$$\dot{\varepsilon}_3 = -\left(\frac{\sqrt{3}}{2}\right)^{n+1} B\sigma_1^n$$

where $\sigma_1 = pr/t$ (eqn. (2.10)).

It is interesting to observe that there is no creep in the axial direction, which has also been verified experimentally.

The allowable internal pressure is controlled by the circumferential or hoop strain rate ε_1, therefore

$$\varepsilon_1 = \left(\frac{\sqrt{3}}{2}\right)^{n+1} B\left(\frac{pr}{t}\right)^n t_h$$

where t_h = time in hours.

Substituting the design values

$$0.005 = \left(\frac{\sqrt{3}}{2}\right)^4 1.45 \times 10^{-23}\left(p \cdot \frac{50}{3}\right)^4 10^5$$

from which

$$p^4 = 794.4 \times 10^8$$

$$p = 530\ \text{N/m}^2$$

If the problem had combined the internal pressure with axial tension or torsion, the solution would only entail the use of a different ratio of σ_1 to σ_2 to give the required expression for σ_e above.

STRESS RELAXATION

The chapter has dealt so far with creep in the form of time-dependent increase in strain at constant stress; an alternative manifestation of creep is a time-dependent decrease in stress at constant strain. A common example of this phenomenon is the relaxation of tightening stress in the bolts of flanged joints in steam and other hot piping, with the resulting possibility of leakage. Another important case is the component subjected to a cycle of thermal strain. The effect is illustrated by means of the stress–strain curve in Fig. 22.8, where thermal expansion has set up compressive stress followed by relaxation with time. On cooling down, a higher residual tensile stress is set up than if there had been no relaxation of compressive stress. This situation can lead to failure eventually in a metal such as a flake cast-iron which is weak in tension. Thermal strain

Fig. 22.8 Hysteresis including stress relaxation

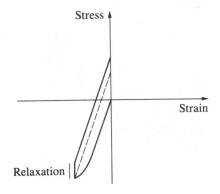

Fig. 22.9 Relaxation of stress with time at constant strain and temperature

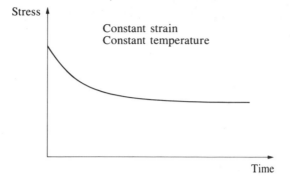

concentration around nozzle openings in pressure vessels is another example where reversal of stress after relaxation could in time lead to a thermal fatigue failure.

A stress-relaxation/time curve is similar to a mirror image of a creep-strain/time curve as shown in Fig. 22.9. Just as much care has to be taken with relaxation testing as with conventional creep testing since results are very sensitive to temperature variation. The procedure usually adopted is to load the specimen to an initial stress which will give a specified strain of, say 0.15%. The stress is then adjusted with time so that the specified strain is maintained.

In the Barr and Bardgett types of test the decrease in stress is noted after 48 hours for various initial stresses at constant temperature. If the decrease in stress is plotted against initial stress, the intercept of the curve on the latter axis gives the initial stress required for "zero" decrease in stress, i.e. no relaxation. However, this is a very short-term test and can only safely be used for sorting materials.

STRESS RELAXATION IN A BOLT

Bolted flanged joints at high temperature represent an example of the problem of stress relaxation. Consider a bolt which is initially tightened to a stress σ_0, producing an elastic strain ε_0, and to simplify the problem it is assumed that the flange is not deformed by the bolt stress. After a period of time t the effect of creep is to induce plastic deformation or creep strain, which allows a relaxation of stress σ and elastic strain. Now, the total strain must remain the same if the flange is rigid; therefore,

$$\varepsilon_0 = \varepsilon_c + \frac{\sigma}{E}$$

or differentiating with respect to time,

$$0 = \frac{\mathrm{d}\varepsilon_c}{\mathrm{d}t} + \frac{1}{E}\frac{\mathrm{d}\sigma}{\mathrm{d}t}$$

579

or

$$\dot{\varepsilon}_c = -\frac{1}{E}\frac{d\sigma}{dt} \tag{22.23}$$

Substituting for $\dot{\varepsilon}_c$ in terms of stress,

$$B\sigma^n = -\frac{1}{E}\frac{d\sigma}{dt}$$

Therefore

$$dt = -\frac{1}{EB}\frac{d\sigma}{\sigma^n} \tag{22.24}$$

The time for relaxation of stress, from σ_0 initially to σ_t at time t, is then obtained by integrating eqn. (22.24), and

$$t = -\frac{1}{EB}\int_{\sigma_0}^{\sigma_t}\frac{d\sigma}{\sigma^n}$$

Therefore

$$t = \frac{1}{EB}\frac{1}{(n-1)}\left(\frac{1}{\sigma_t^{n-1}} - \frac{1}{\sigma_0^{n-1}}\right) \tag{22.25}$$

It is found in practice that relaxation of stress is more rapid than that given above owing to the effects of primary creep, and allowances for this can only be made by using a more complex creep rate–stress–time function.

EXAMPLE 22.2

The bolts holding a flanged joint in steam piping are tightened to an initial stress of $400\ \text{MN/m}^2$. Determine the relaxed stress after $10\,000$ hours, $E = 200\ \text{GN/m}^2$, $n = 3$ and $B = 4.8 \times 10^{-34}$ per hour per MN/m^2.

SOLUTION

From eqn. (22.25)

$$10\,000 = \frac{1}{200 \times 10^9 \times 4.8 \times 10^{-34} \times 2}\left\{\frac{1}{(\sigma_t \times 10^6)^2} - \frac{1}{(400 \times 10^6)^2}\right\}$$

$$19.2 \times 10^{-7} = \frac{1}{\sigma_t^2} - \frac{1}{400^2}$$

and

$$\sigma_t = 349.8\ \text{MN/m}^2$$

CREEP DURING VARIABLE LOAD OR TEMPERATURE

Experiment and analysis has concentrated principally on constant load and constant temperature conditions during creep. There are some engineering applications in which the loading conditions change from time to time at high temperature. These are not necessarily cyclical stress in the fatigue sense, as discussed in the next section, but result in different creep rates during each different load sequence.

Several hypotheses have been proposed for predicting creep strain related to changes of load which are constant before the change and constant after. It is beyond the scope of this chapter to develop these hypotheses, but two which are commonly quoted are *time-hardening* and *strain-hardening* theories. These take the following analytical forms:

Time-hardening: $\quad \dot{\varepsilon}_c = f(\sigma, t)$ (22.26)

Strain-hardening: $\quad \dot{\varepsilon}_c = g(\sigma, \varepsilon_c)$ (22.27)

The former implies that creep rate is a function only of the stress and the current time. The creep curve after the change of stress has the same shape as the constant stress curve from the time of change, i.e. the curve is moved vertically as shown in Fig. 22.10.

Fig. 22.10 Effect of step change in stress on creep strain

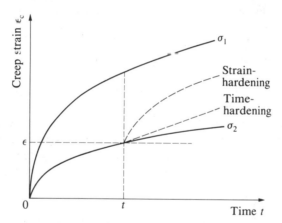

The latter implies that strain rate depends only on the stress and the current plastic strain. Again, as above, the same shape of curve is assumed, but now the appropriate portion of the curve from the time of change is moved horizontally as shown in Fig. 22.10.

These hypotheses can also be applied in a variable strain–stress–relaxation context.

The predictions of total creep strain after load (stress) change are rather better in the case of the strain-hardening approach compared with the time-hardening method both for step-up load change and step-down load change.

Change of temperature in a component or structure, unless extremely slow, will always induce non-uniform thermal strain gradients and hence thermal stresses in addition to the stress due to applied load. It is evident, therefore, that prediction of creep rates or total strain is extremely difficult under variable temperature conditions.

CREEP–FATIGUE INTERACTION

In the last two decades certain engineering developments such as gas and steam turbines, rockets and supersonic aircraft have involved the use of metals not only at very high temperatures but also with dynamic fluctuating stresses. In short, the problem is one in which the mean or steady component of stress can induce creep, and the alternating component of stress may lead to fatigue failure. The earliest investigations into this problem were made between 1936 and 1940 by various German investigators, and since then there have been many interesting studies both in this country and the U.S.A.

The problem of fatigue at high temperature was discussed in Chapter 21, and this phenomenon can be unaccompanied by creep for fully reversed or zero mean stress. Therefore, when considering a material for high-temperature service it is usual to think of the behaviour in terms of a diagram such as Fig. 22.11, in which fatigue failure is the criterion within certain stress and temperature limits, and beyond these creep is the predominant factor. If it is a question not of one or other phenomenon acting on its own, but of both influences operating simultaneously, then the solution becomes rather more involved. Fatigue is essentially a cycle-dependent mechanism, whereas creep is time-dependent. It is therefore both desirable and convenient to express fatigue behaviour at high temperature also in terms of time to rupture as suggested by Tapsell.[5] One of the reasons for this is because of the greater dependence of fatigue on cyclic frequency at high temperature.

The most useful way of presenting data for combined creep and fatigue conditions is in the form of a diagram of alternating stress against steady or mean stress, which is similar in most respects to the S_a–S_m diagram in normal fatigue (Fig. 21.17). Test results are plotted as the combination of alternating and mean stress to produce either rupture or a specified creep strain after a particular number of hours at constant temperature. Points

Fig. 22.11 Strength limitations with increasing temperature

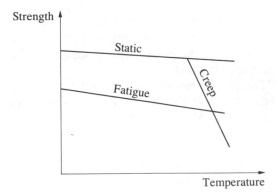

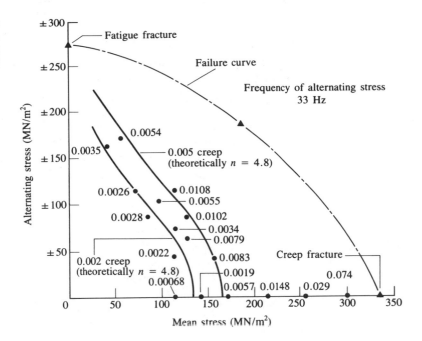

Fig. 22.12 0.26% carbon steel; total creep occurring in 100 hours at 400 °C

along the abscissa axis represent creep conditions only and points along the ordinate axis are for fatigue only. Some results obtained by Tapsell[6] on 0.26% carbon steel are given in Fig. 22.12 for various amounts of total creep strain occurring in 100 hours at 400 °C under different combinations of cyclic and steady stress. Theoretically predicted curves are also shown for creep strains of 0.002 and 0.005.

The influence of alternating stress on the minimum creep rate and time to rupture varies considerably with temperature, material and length of time. At higher temperatures or long life, the alternating stress appears to have little effect on creep rate; in fact, there are cases where creep strengthening has resulted. On the other hand, at lower temperatures or shorter rupture times, fatigue appears to play a more detrimental part, giving a higher creep rate. The rupture strain is also somewhat reduced by the presence of cyclic stress.

VISCOELASTICITY

Because of the increasing use of plastics, both reinforced and unreinforced, in engineering load-bearing applications, it is important that the response of these materials to stress and environmental conditions should be appreciated. A brief mention was made of viscoelastic stress–strain–time behaviour in Chapter 3 and the discussion will be extended somewhat further, particularly in relation to creep, in the remainder of this chapter.

In a viscoelastic material the stress is a function of strain and time and so may be described by an equation of the form

$$\sigma = f(\varepsilon, t) \tag{22.28}$$

This response is known as *non-linear* viscoelasticity, but as it is not amenable to simple analysis it can be reduced to the following form:

$$\sigma = \varepsilon \cdot f(t) \tag{22.29}$$

This response is the basis of *linear* viscoelasticity and simply indicates that, in a tensile test for example, for a fixed value of elapsed time the stress will be directly proportional to the strain. These different stress–strain–time responses are shown schematically in Fig. 22.13.

Fig. 22.13 Stress–strain behaviour of elastic and viscoelastic materials at two values of elapsed time *t*

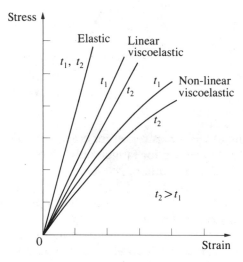

Viscoelastic materials invariably exhibit a time-dependent strain response to a constant strain, which is *relaxation*. In addition, when the applied stress is removed the materials have the ability to recover slowly over a period of time. These effects occur at ambient temperature and, therefore, are a principal design consideration as compared with metals for which creep and relaxation only occur in higher-temperature environments.

In Chapter 3 it was explained that tensile test characteristics for plastics are extremely sensitive to rate of straining. They are equally sensitive to tensile test temperature and, in some materials, the humidity condition. As a result of these special effects in plastics it is not reasonable to quote properties such as modulus, yield strength, etc. as a single value without qualifying these with details of the test condition.

CREEP BEHAVIOUR OF PLASTICS

Plastics exhibit a similar shape of creep curve of creep strain against time for constant stress and temperature as for metals (Fig. 22.1). However, one distinct difference is the ability of plastics to "recover" strain after the removal of the applied load, and this effect is shown in Fig. 22.14. Plastics also have "memory" and the current behaviour in creep or relaxation is dependent on all the past history of stress, strain and time on that sample (assuming constant temperature).

Fig. 22.14 Typical creep and recovery behaviour

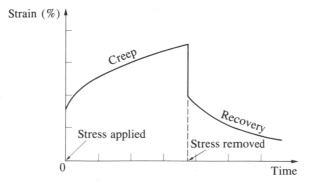

Attempts have been made to simulate polymer structure and its creep and recovery responses by mechanical-type modelling using two principal elements. These are a coil spring, which represents Hookean behaviour (linear load deformation), and a dashpot (a piston in an oil-filled container) which represents viscous Newtonian response (linear load-deformation rate). These elements may be coupled in two ways: (i) with the spring in series with the piston of the dashpot and this is termed a Maxwell model; and (ii) with the spring in a "parallel" location to the dashpot, so that applied load is "shared" between the two elements, and this arrangement is known as the Kelvin–Voigt model. Because these models are individually quite inadequate to represent even linear viscoelasticity, more complex assemblies of the above units have been studied and, although somewhat more representative, they still do not give an adequate prediction of creep response which could be applied in design. This is principally because of the non-linear viscoelastic nature of polymers. The alternative to the empirical methods above is the use of experimental data obtained on the particular plastics for which design exercises have to be carried out.

Creep data is initially presented in the form of graphs of creep strain against log time, since linear time is inconvenient to encompass both short- and long-term tests. A family of creep curves is illustrated in Fig. 22.15(*a*) and two commonly used derivative graphs are shown in Fig. 22.15(*b*) and (*c*). The former is constructed by taking a constant strain section through the curves in Fig. 22.15(*a*) to give what is termed an *isometric* curve. A constant time section through the creep curves gives a stress–strain diagram as shown in Fig. 22.15(*c*) which is known as an *isochronous* curve.

Fig. 22.15 Isometric and isochronous curves from creep curves

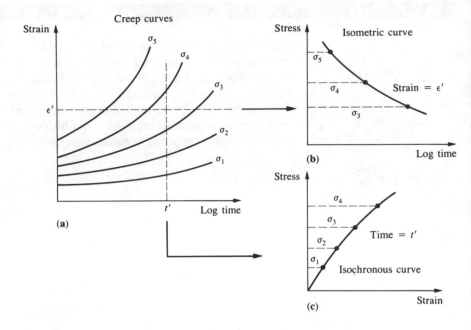

The isometric curve is often used as a good approximation of stress relaxation behaviour since this specific experimental method is less common than creep testing.

Isochronous curves can be developed independently without having to obtain a family of creep curves as above. The method involves a series of mini-creep and recovery tests in tension on a material. A stress is applied to a sample and the strain recorded after a time t (typically 100 sec), the stress is then removed and the specimen allowed to recover for a period of four times the loading time, i.e. $4t$ or 400 sec. A larger stress is then applied to the same specimen and, after recording the strain at time t, this stress is removed and the material allowed to recover. This procedure is repeated until sufficient points have been obtained for the isochronous curve to be plotted. Obtaining isochronous curves by this direct experimental method is less time consuming and more economical than through creep experiments. In fact, one can, of course, derive creep curves from several isochronous experiments for different time intervals, e.g. 10^2, 10^3, 10^4 sec. Isochronous curves plotted on linear scales may not show up the slight non-linearity at low strains so an alternative plot on log-log scales is used so that any non-linearity is demonstrated by the slope of the line being less than unity i.e. less than 45° on the log-log paper.

Another method of representing long-term creep behaviour is by means of curves of modulus against time. These are shown in Fig. 22.16 for three values of constant creep strain. They were derived by taking a constant strain section through a family of creep curves and dividing the stress values by the strain to give relaxation moduli which are plotted against the respective time value intersections.

The effect of temperature on the creep of plastics principally relates to

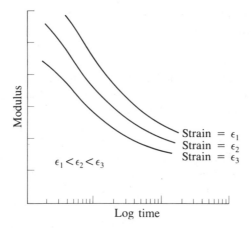

Fig. 22.16 Typical variation of modulus with time

a typical range of atmospheric temperatures such as $-30\,°C$ to $+40\,°C$. However, at the upper end of this range of temperature marked acceleration of creep rates will occur compared to the mid-range value and a family of isochronous curves for one elapsed time and several temperatures will be of the form shown in Fig. 22.17.

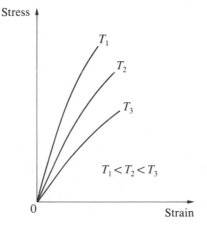

Fig. 22.17 Isochronous curves for different temperatures at one elapsed time

The prediction of creep response to step changes in stress is perhaps even more complex than for metals. It has been tackled by a variety of methods of superposition of parts of individual creep and recovery curves. The principle that is most frequently quoted is the Boltzmann-type superposition but, as with most methods, this only relates to linear viscoelasticity. Solutions for non-linear viscoelastic superposition are too intractable to be of practical use.

DESIGNING FOR CREEP IN PLASTICS

The design of metallic structures and components is generally not temperature or time dependent and is based on linear elastic reversible

587

stress–strain behaviour and small deformations. However, *any* load-bearing component to be made out of plastic has first and foremost to be designed for time-dependent deformation and, secondly, for time-dependent fracture.

Although the basic tenets of equilibrium of forces and compatibility of deformations apply in the design of a plastic omponent, the problem arises in relation to a suitable stress–strain–time law to link the foregoing. The more accurate methods that have been proposed have the drawback of being extremely complex and unattractive to the average designer. Perhaps the most acceptable approach devised in the past two decades has been called the *pseudo-elastic design method*. This involves the use of time-dependent "elastic constants", moduli and contraction (Poisson) ratio substituted into classical equations in place of the true elastic constants. The time-dependent value of modulus must be carefully determined to allow for the service life and limiting strain for the plastic component. The limiting strain value for the particular plastic should generally be decided in consultation with the material manufacturers and, typically, might be of the order of 1–2% strain. From this point the use of published experimental creep data for the material is quite straightforward in developing the component design. The following examples will illustrate the technique and the relevant creep data is given in Fig. 22.18.

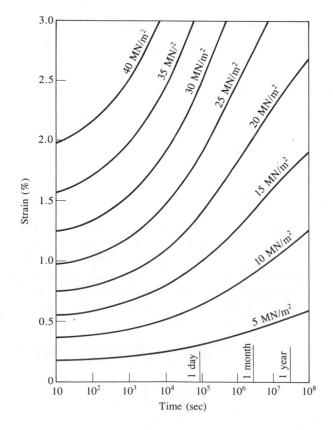

Fig. 22.18 Creep curves for acetal at 20 °C

EXAMPLE 22.3

A solid circular acetal rod, 0.15 m in length, is clamped horizontally at one end and the free end is subjected to a vertical load of 25 N. Determine a suitable diameter for the rod for a limiting strain of 2% in 1 year. What would be the maximum deflection at this time?

SOLUTION

Using the creep curves in Fig. 22.18, a 1-year isochronous curve is plotted as shown in Fig. 22.19 from which an allowable stress of 17.1 MN/m² is obtained at the 2% strain limit.

The maximum bending moment is $25 \times 0.15 = 3.75$ N-m. Using the bending stress relationship,

$$\sigma = \frac{My}{I} = \frac{32M}{\pi d^3}$$

$$d^3 = \frac{32 \times 3.75 \times 10^9}{\pi \times 17.1 \times 10^6} \text{ mm}^3$$

$$d = 13.07 \text{ mm}$$

The maximum deflection at the free end is given by

$$\delta = \frac{WL^3}{3EI}$$

The appropriate value of modulus may be obtained from the isochronous curve at 2% strain, hence the secant modulus

$$E(t) = \frac{17.1}{0.02} = 855 \text{ MN/m}^2$$

Therefore,

$$\delta = \frac{25 \times 0.15^3 \times 64 \times 10^3}{3 \times 855 \times 10^6 \times \pi \times 0.01307^4} = 23 \text{ mm}$$

An alternative way of obtaining the design stress would have been to plot a 2% isometric curve and read off the stress at 1 year.

EXAMPLE 22.4

A circular acetal diaphragm is 2 mm thick and is clamped around its periphery giving a clear diameter of 100 mm. It is to be subjected to uniform pressure for a service life of one year with a material creep strain limitation of 1% and a maximum central deflection of 3 mm. Determine the allowable working pressure.

SOLUTION

The central deflection of a clamped edge circular plate subjected to uniform pressure was given in eqn. (17.22) as

$$w = \frac{12(1 - v^2)pa^4}{64Eh^3}$$

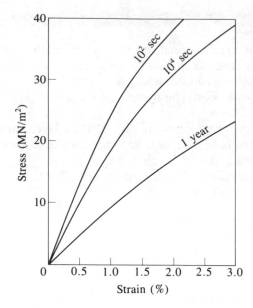

Fig. 22.19 Isochronous curves for acetal at 20 °C

In this problem we shall have two time-dependent functions to consider, the modulus $E(t)$ and the creep or lateral contraction ratio, $v(t)$. The former may be determined from the isochronous curve of Fig. 22.19 for a strain limit of 1%. The secant modulus is given as 920 MN/m². The data available for creep contraction ratio (the time-dependent equivalent of Poisson's ratio) are rather limited and generally lies between 0.3 and 0.4 but can rise to near 0.5 for "rubbery" materials. For this problem we shall take a value of 0.35.

Rewriting the equation above gives

$$p = \frac{64Eh^3w}{12(1-v^2)a^4} = \frac{64 \times 920 \times 2^3 \times 3}{12(1-0.35^2) \times 50^4}$$

$$p = 21.4 \, \text{kN/m}^2$$

We could equally well have obtained the required modulus value by taking a 1% strain section through the creep curves and plotting modulus values against log time and extrapolating to 1 year.

CREEP RUPTURE OF PLASTICS

On page 572 the creep rupture behaviour of metals was described and a similar phenomenon also occurs with plastics. Under the sustained action of a constant load, plastics exhibit a failure mode associated with the creep deformation of the material. This type of failure is sometimes referred to as "static fatigue", but the preferred engineering term is *creep rupture* or, perhaps more generally, *creep failure*. The more general terminology of creep failure is necessary for plastics because, although

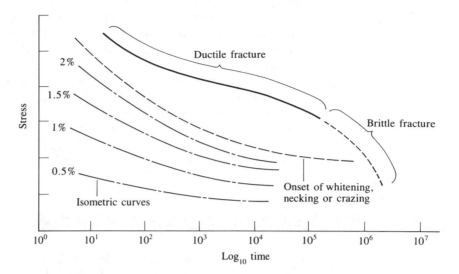

Fig. 22.20 Creep rupture behaviour

rupture of the material will eventually occur due to creep, there may be earlier visible phenomena such as whitening, necking or crazing (crack-like features in glassy plastics) which, as far as the user is concerned, terminate the useful life of the component. Some typical creep failure data for plastics is shown in Fig. 22.20. Two important points should be noted from this. The first is that the creep failure data may be correlated with the isometric data obtained from a constant strain section across the creep curves. A second very important point is that, although the fracture as a result of creep is generally ductile in nature, there is a tendency towards embrittlement in some materials when subjected to constant loads for very long periods of time. This results in a sharp drop-off in the rupture line in Fig. 22.20. Such brittle failures can have serious consequences in practice since there is no prior warning of imminent fracture and there is no ductile tearing of the material to absorb the energy of the fracture. The possible "knee" in the fracture line is something to be wary of when extrapolating short- or medium-term creep failure data to long lifetimes.

SUMMARY

In the case of the designed metal components and structures, creep may be regarded as not a very common occurrence. However, in the case of the design of plastics components, it is of fundamental importance from the start. For either type of material there is no exact analytical method and heavy reliance must be placed on experimental data.

For metals the secondary stage of creep, which yields a constant minimum creep rate for very long periods at low stresses, provides the basis for design using an expression of the form $\varepsilon = B\sigma^n$. This relationship can also be used in the companion time-dependent phenomenon known as stress relaxation.

While situations of constant stress or strain can be handled reasonably, those involving changes in stress or strain at regular intervals of time are very difficult to treat quantitatively. Creep-rupture testing provides very valuable short-term data for the assessment and sorting of materials.

Creep and stress relaxation of plastics, while having many similarities with the behaviour of metals, has the added features of memory and recovery which makes in-depth analysis very difficult. However, the interpretation of basic creep curve data into isometric and isochronous curves allows a convenient and acceptable approach to design by the pseudo-elastic design method as illustrated in the worked examples above.

REFERENCES

1. Pomeroy, C. D., *Creep of Engineering Materials,* Ch. 9, I. Mech. E., 1978.
2. Andrade, E. N. da C., "The viscous flow in metals and allied phenomena", *Proc. Roy. Soc.,* **A84** (1910), 1.
3. Ibid.
4. Larson, F. R. and Miller, J. A., "Time–temperature relationship for rupture and creep stresses", *Trans ASME,* 74, (1952), 765.
5. Tapsell, H. J., *Symposium on High Temp. Steels and Alloys for Gas Turbines,* Iron Steel Inst., 1952. p. 43.
6. Ibid.

BIBLIOGRAPHY

Faupel, J. H. *Engineering Design,* Ch. 12, Wiley, New York 1981.
Penny, R. K., and Marriott, D. L. *Design for Creep,* McGraw-Hill, New York 1971.
Finnie, I. and Heller, W. R. *Creep of Engineering Materials* McGraw-Hill, New York 1959.
Pomeroy, C. D. *Creep of Engineering Materials,* I. Mech. E., 1978.
Crawford, R. J., *Plastics Engineering,* Pergamon Press, Oxford. 1987.

PROBLEMS

22.1 A series of creep tests on an austenitic high-temperature alloy gave the following results:

Stress (MN/m^2)	ε_0(%)	Minimum creep rate (mm/mm/hr)
70	0.041	27×10^{-8}
105	0.061	15.5×10^{-6}
140	0.081	27.5×10^{-5}
210	0.122	15.8×10^{-3}
280	0.162	0.281
350	0.203	2.62

Calculate how long would elapse before a steady stress of 125 MN/m^2 would cause a strain of 1% in this material.

22.2 The following table shows the creep data obtained for a metal using a stress of 100 MN/m^2 at a range of temperatures. If the constant "a" in the Larsen–Miller parameter is 20, calculate the time to failure at a stress of 100 MN/m^2 when the temperature of the material is 700 °C. ($R = 8.314$ J mol^{-1} K)

Temperature (°C)	100	200	250	300
ε_0 (mm/mm/hr)	2.1×10^{-23}	2.77×10^{-17}	4.25×10^{-15}	2.7×10^{-13}
Temperature	400	500	600	
ε_0 (mm/mm/hr)	1.72×10^{-10}	2.06×10^{-8}	8.25×10^{-7}	

22.3 The following creep rupture data was recorded for an alloy steel when it was tested at a range of stresses and temperatures:

Temperature (°C)	Stress (MN/m^2)	Time to failure (hr)
500	300	4724
600	200	570
700	100	297
800	60	95.3
1000	30	8.6

If a component made from this material is required to last at least 10 000 hours at a stress of 150 MN/m^2, what is its maximum permissible service temperature? The constant "a" in the Larsen–Miller parameter for the alloy is 20.

22.4 Figure 22.21 shows a delayed-action contact switch. When the pin is removed the compressed spring causes a tensile stress in the previously unstressed lead rod. Due to creep of the lead the gap between the contact points decreases steadily. Calculate the delay time if the free length of the spring is 40 mm and its stiffness is 10 N/mm. For the lead $\varepsilon_0 = 5 \times 10^{-10}\sigma^{7.5}$ mm/mm/hr

Fig. 22.21

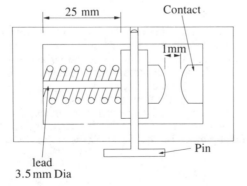

22.5 A sheet of high-temperature alloy is clamped in position using a toggle clamp as illustrated in Fig. 22.22. When the lever A is in the vertical (clamp) position the sheet thickness is reduced by 10 μm. If the sheet is subjected to a pull of 5 kN, calculate how long the clamp could retain the sheet in position. The coefficient of friction between the clamp and the alloy is 0.6. Creep data for the alloy is given in Problem 22.1. $E = 207$ GN/m^2.

Fig. 22.22

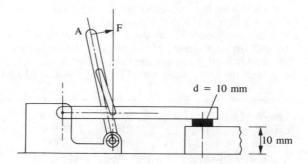

22.6 An aerosol container is to be moulded from an acetal copolymer, for which the creep curves are given in Fig. 22.18. The container diameter is 50 mm and it has a uniform wall thickness of 2 mm. The base of the container is designed with a "skirt" to prevent rocking when the bottom deforms under pressure (see Fig. 22.23). Calculate the depth of the skirt if the container is expected to be subjected to an internal pressure of 120 kN/m² (absolute) for 1 year. Poisson's ratio for acetal may be taken as 0.4.

Fig. 22.23

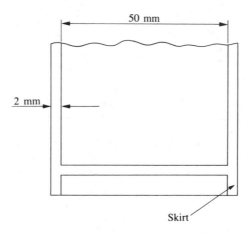

Skirt

22.7 A cylindrical acetal container is subjected to an internal pressure of 0.7 MN/m². For aesthetic reasons the strain in the container is not to exceed 2%. If the diameter and wall thickness of the container are 60 mm and 1 mm respectively calculate how long the container may be regarded as serviceable. Creep curves for the acetal are given in Fig. 22.18.

22.8 A plastic snap-fit connection is shown in Fig. 22.24. If the pin will slip out when the transverse clamping force exerted by the clasp is 33 N calculate (a) the clamping force when the pin is first inserted, and (b) the elapsed time before the pin would slip out. Use the creep curves in Fig. 22.18.

Fig. 22.24

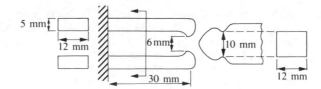

22.9   A long thin-walled pipe constrained by end fittings made of polyvinylchloride is subjected to a steady internal pressure of 700 kN/m^2 at 20 °C. If a tensile stress of 17.5 MN/m^2 is not to be exceeded and the internal radius is 100 mm, determine a suitable wall thickness. What will be the increase in diameter after 1000 hours? The mean creep contraction ratio v, is 0.45, and tensile creep curves provide the following values at 1000 hours:

σ (MN/m^2)	6.9	13.8	20.7	27.6	34.5
ε (%)	0.2	0.48	0.97	1.72	3.38

Appendix A:

Properties of areas

INTRODUCTION

The analysis of stresses developed in symmetrical and unsymmetrical bending of beams (Ch. 6) depends on the shape and area of the cross-section of a beam. The reason for this is because internal *forces* and *moments* derive from stress acting on elements of area. To obtain the *total* shear force or bending-moment effect on a cross-section we must sum up or integrate all the constituent elements and the final expressions involve integrals for the total area (shape) in question. It is these integrals and their solutions which we refer to as "properties" of areas. There are direct comparisons with the "properties" of masses which are used in engineering dynamics or mechanics of machines. The integrals which will need to be evaluated for any cross-sectional shape are described as the first moment, the second moment and the product moment of an area. It may seem incongruous to speak of the "moment" of an area since a moment implies a mass or force multiplied by a distance. However, the integrals do consist of areas multiplied by distances and that is how the expression *moment of area* has become established (this is *not* to be confused with moment of inertia which relates to mass). Co-ordinate axes y and z are used throughout for beam cross-sections, since the x-axis is along the length of the beam.

FIRST MOMENT OF AREA

Referring to the plane figure in Fig. A.1, the first moment of the element of area dA about the z-axis is $y\,dA$; therefore the first moment of the whole figure is $\int_A y\,dA$ about the z-axis, the suffix A indicating summation over the whole area. The first moment of the whole figure about the y-axis is $\int_A z\,dA$.

Fig. A.1

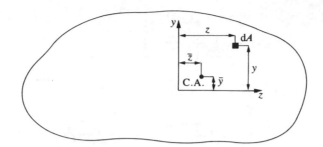

POSITION OF CENTRE OF AREA, OR CENTROID

Let the co-ordinates of the centre of area CA be $\bar{z}$ and $\bar{y}$ as shown in Fig. A.1. Then the moment of the whole area about an axis is the same as the sum of the moments of all the elements of area about that axis, or

$$A\bar{y} = \int_A y \, dA \quad \text{so that} \quad \bar{y} = \frac{1}{A} \int_A y \, dA \qquad \text{(A.1)}$$

Similarly,

$$\bar{z} = \frac{1}{A} \int_A z \, dA \qquad \text{(A.2)}$$

If either or both of the z- and y-axes pass through the centre of area then $\bar{z}$ or y or both $= 0$, and

$$\int_A y \, dA = 0 \quad \text{and/or} \quad \int_A z \, dA = 0$$

If a shape has one axis of symmetry then the centre of area will lie on that axis. If there are two axes of symmetry then their intersection will be the centre of area.

EXAMPLE A.1

Determine the location of the centre of area for the concrete beam cross-section shown in Fig. A.2.

SOLUTION

Since there is a vertical axis of symmetry the centre of area will lie somewhere on that axis as shown by C.

Take as a reference horizontal axis the lower edge of the section, AA, and let the distance of C from AA be $\bar{y}$. The cross-section can be divided into two rectangles by the dotted line. The centre of area of the upper rectangle is at C_2 at a distance of 100 mm from AA. The centre of area of the lower rectangle is at C_1 at a distance of 25 mm from AA.

The areas of the upper and lower rectangles are 5000 and 7500 mm² respectively and the total area of the figure is 12 500 mm².

597

Fig. A.2

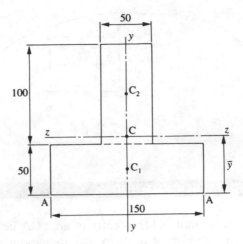

Referring to eqn. (A.1) above we may write

$$12\,500\bar{y} = 5000 \times 100 + 7500 \times 25$$

from which

$$\bar{y} = 55\,\text{mm}$$

SECOND MOMENT OF AREA

If the first moment of an element dA about an axis is multiplied again by its respective co-ordinate we obtain the second moment of area, namely $y^2 dA$, or $z^2 dA$. The second moment of area of the whole figure about the z-axis is

$$\int_A y^2 dA \quad \text{denoted as} \quad I_z \tag{A.3}$$

and about the y-axis is

$$\int_A z^2 dA \quad \text{denoted as} \quad I_y \tag{A.4}$$

SECOND MOMENT OF AREA FOR A RECTANGLE

Common structural cross-sectional shapes are composed of rectangles and the second moment of area of a rectangle is obtained as follows.

Because of the double symmetry the centre of area C is at the centre of the rectangle of width b and depth d shown in Fig. A.3.

Consider an element of area $b\,dy$ as shown.

The second moment of this element about axis zz is $y^2(b\,dy)$. To obtain I_z for the whole section we must integrate between the limits of

Fig. A.3

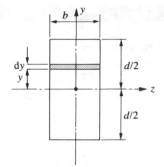

$\pm d/2$ so that

$$I_z = \int_{-d/2}^{+d/2} y^2 b \ \mathrm{d}y = \left[\frac{by^3}{3} \right]_{-d/2}^{+d/2}$$

$$= \frac{bd^3}{12}$$

By a similar analysis we can obtain

$$I_y = \frac{db^3}{12}$$

EXAMPLE A.2

Determine the second moment of area for a solid circular cross-section of 50 mm diameter about an axis through the centre.

SOLUTION

The element of area marked in Fig. A.4 is $\mathrm{d}A = r \ \mathrm{d}\theta \ \mathrm{d}r$ and the second moment of this element about the axis zz is $(r \sin \theta)^2 r \ \mathrm{d}\theta \ \mathrm{d}r$. If we now integrate this expression between limits of 0 to 2π we shall have the second moment of an annular element about zz which is

$$\int_0^{2\pi} (r \sin \theta)^2 r \ \mathrm{d}\theta \ \mathrm{d}r = \pi r^3 \mathrm{d}r$$

and for the solid circle the second moment is

$$I_z = \int_0^R \pi r^3 \mathrm{d}r = \frac{\pi R^4}{4} = \frac{\pi D^4}{64} = \frac{\pi \times 50^4}{64} = 30.7 \times 10^4 \ \mathrm{mm}^4$$

Fig. A.4

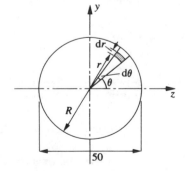

PARALLEL AXES THEOREM

It is sometimes necessary to determine the second moment of area about axes parallel to the centroidal axes.

Referring to Fig. A.5, the second moment of the element dA about the z'-axis is $(y + b)^2 dA$, and for the whole figure

$$I_{z'} = \int_A (y + b)^2 dA$$

$$= \int_A y^2 dA + 2b \int_A y\, dA + \int_A b^2 dA$$

but $\int_A y\, dA = 0$, since it is the first moment about a centroidal axis, therefore

$$I_{z'} = I_z + b^2 A \qquad\qquad\qquad (A.5)$$

and by a similar analysis

$$I_{y'} = I_y + a^2 A \qquad\qquad\qquad (A.6)$$

Fig. A.5

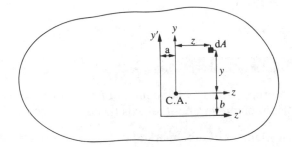

EXAMPLE A.3

Determine the second moments of area of the section in Example A.1 about its centroidal axes.

SOLUTION

Any section composed of rectangles can be broken up for analysis into its separate components. Therefore, in the case of the vertical yy-axis which passes through the centres of area C_1 and C_2, we do not need the parallel axes theorem, so

$$I_y = \frac{100 \times 50^3}{12} + \frac{50 \times 150^3}{12} = 15.1 \times 10^6 \text{ mm}^4$$

In order to calculate the value of I_z we need to apply the parallel axes theorem to both the top and bottom rectangles as follows:

$$I_{z1} = \frac{150 \times 50^3}{12} + (150 \times 50)30^2 = 8.3 \times 10^6 \text{ mm}^4$$

the first term is the I about a horizontal axis through C_1 and the second term is the area of the rectangle multiplied by the square of the distance between C_1 and C.

$$I_{z2} = \frac{50 \times 100^3}{12} + (100 \times 50)45^2 = 14.3 \times 10^6 \text{ mm}^4$$

The total value of I_z is the sum of the two parts above:

$$I_z = 8.3 \times 10^6 + 14.3 \times 10^6 = 22.6 \times 10^6 \text{ mm}^4$$

PRODUCT MOMENT OF AREA

The moment of area requirements for the analysis of *symmetrical* bending of beams have been covered up to this section. However, for *unsymmetrical sections* subjected to bending a further moment of area property is required termed the *product moment of area.* It is defined in relation to the plane figure shown in Fig. A.6 as

$$I_{zy} = \int_A zy \, dA \tag{A.7}$$

where the axes pass through the centre of area C.A. of the figure. Two important differences from the second moments of area I_z and I_y are that I_{zy} can have either *positive or negative* values since z and y values can be positive and negative. Secondly, if either or both of the axes is an axis of symmetry then $I_{zy} = 0$.

Fig. A.6

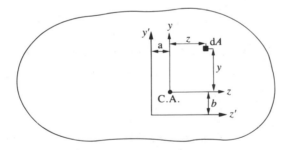

For the case of the product moment of area related to parallel axes z' y', it is straightforward to show that

$$I_{z'y'} = I_{zy} + abA \tag{A.8}$$

EXAMPLE A.4

Determine the product moment of area of the triangular section shown in Fig. A.7 in relation to the centroidal axes z, y.

Fig. A.7

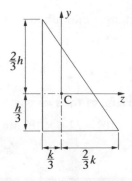

SOLUTION

The equation of the diagonal of the triangle is $y = h/3 - hz/k$. Hence

$$I_{zy} = \int zy \, da = \int_{-k/3}^{2k/3} \int_{-h/3}^{h(k-3z)/3k} zy \, dz \, dy$$

$$= \frac{h^2}{6k^2} \int_{-k/3}^{2k/3} (3z^3 - 2kz^2) \, dz = -\frac{k^2 h^2}{72}$$

TRANSFORMATION OF MOMENTS OF AREA

In unsymmetrical bending of beams it is sometimes necessary to consider the nature of bending about a set of axes rotated through an angle θ with respect to a reference direction of axes as shown in Fig. A.8. Let the moments of area be I_z, I_y and I_{zy} with respect to the reference axes z, y and $I_{z'}$, $I_{y'}$ and $I_{z'y'}$ with respect to different axes z', y' at an anticlockwise angle θ to the former.

$$I_{z'} = \int_A y'^2 dA = \int_A (y \cos \theta - z \sin \theta)^2 dA$$

$$= I_z \cos^2 \theta + I_y \sin^2 \theta - 2I_{zy} \sin \theta \cos \theta$$

$$= \tfrac{1}{2}(I_z + I_y) + \tfrac{1}{2}(I_z - I_y) \cos 2\theta - I_{zy} \sin 2\theta \qquad \text{(A.9)}$$

$$I_{z'y'} = \int_A z'y' dA = \int (z \cos \theta + y \sin \theta)(y \cos \theta - z \sin \theta) \, dA$$

$$= (\cos^2 \theta - \sin^2 \theta)I_{zy} + I_z \sin \theta \cos \theta - I_y \sin \theta \cos \theta$$

$$= \tfrac{1}{2}(I_z - I_y) \sin 2\theta + I_z \cos 2\theta \qquad \text{(A.10)}$$

Fig. A.8

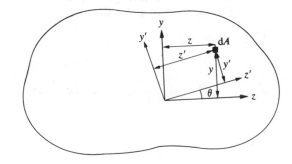

These two equations provide the relationships between moments of area about two sets of rectangular axes with a common origin. It is interesting to note the similarity of form between these equations and those for two-dimensional stress transformation, eqns. (12.13) and (12.14). It does in fact suggest the existence of "principal" second moments of area about axes for which the product moment of area is zero. From eqn. (A.10) putting $I_{z'y'} = 0$ we obtain

$$\tan 2\theta = \frac{2I_{zy}}{(I_z - I_y)} \tag{A.11}$$

and this defines the axes about which maximum and minimum principal second moments of area I_u and I_v occur.

MOHR'S CIRCLE FOR MOMENTS OF AREA

The simplest way of determining the principal second moments of area and indeed the moments of area about any set of axes is to construct a Mohr circle in the same manner as that for stresses or strains.

If eqns. (A.9) and (A.10) are squared and added to each other to eliminate the angle 2θ we obtain the following equation:

$$[I_{z'} - \tfrac{1}{2}(I_z + I_y)]^2 + I_{z'y'}^2 = \tfrac{1}{4}(I_z - I_y)^2 + I_{zy}^2 \tag{A.12}$$

This represents a circle with axes of product moments of area (I_{zy}), as ordinate and second moments of area (I_z, I_y), as abscissa. The centre of the circle is located at

$$\left[\frac{I_z + I_y}{2}, 0\right]$$

Fig. A.9

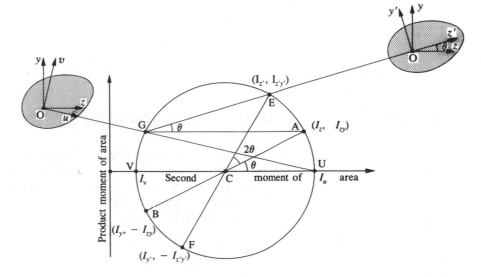

and the radius of the circle is

$$[\tfrac{1}{4}(I_z - I_y)^2 + I_{zy}^2]^{1/2}$$

Since there are no negative values of second moments of area the circle is always to the right of the ordinate.

Figure A.9 shows the circle construction for the shape shown shaded top right and similarly top left. The circle is drawn from known, or calculated, values of I_z, I_y and $\pm I_{zy}$. For required values about axes $z'y'$ at θ to axes zy we draw the diameter ECF at 2θ anticlockwise from ABC. The values at E and F are those required. The principal second moments of area I_u and I_v are the values at U and V, where $I_{zy} = 0$. The directions of the principal axes are given by the chords UG and VG which are shown as axes Ou and Ov on the area (top left).

EXAMPLE A.5

Use the circle construction to determine the principal second moments of area for the angle section of Example 6.14 in which $I_y = 41.3 \times 10^4\,\text{mm}^4$, $I_z = 151.2 \times 10^4\,\text{mm}^4$ and $I_{zy} = 45 \times 10^4\,\text{mm}^4$.

SOLUTION

The centre of the circle is located at $(96.25, 0)$ and with the values of I_z and I_{zy} the circle is drawn as shown in Fig. A.10. The maximum and minimum principal second moments of area are seen to be 168.5×10^4 and $23.5 \times 10^4\,\text{mm}^4$.

Fig. A.10

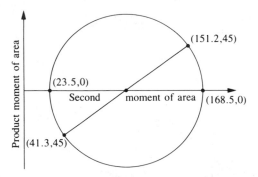

POLAR SECOND MOMENT OF AREA

The second moment of area about an axis perpendicular to the plane of an area is termed the *polar second moment of area* and it is an essential part of the analysis of the shear stresses in the torsion of circular sections (Ch. 5). Referring to Fig. A.11 the polar second moment of area of dA is $r^2 dA$ and for the whole figure

$$J(\text{or } I_p) = \int_A r^2 dA$$

Fig. A.11

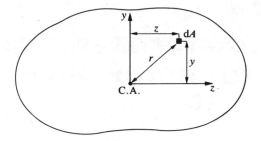

Also, since $r^2 = z^2 + y^2$,

$$J = \int_A z^2 \mathrm{d}A + \int_A y^2 \mathrm{d}A$$

$$= I_y + I_z \qquad (A.13)$$

This is known as the *perpendicular axes theorem*.

Appendix B:

Introduction to matrix algebra

MATRIX DEFINITIONS

A matrix is an array of terms as shown below.

$$[A] = \begin{bmatrix} a_{11} & a_{12} & a_{13} \ldots \ldots a_{1n} \\ a_{21} & a_{22} & a_{23} \ldots \ldots a_{2n} \\ a_{31} & a_{32} & a_{33} \ldots \ldots a_{3n} \\ \vdots \\ a_{m1} & a_{m2} & a_{m3} \ldots \ldots a_{mn} \end{bmatrix}$$

If $n = 1$ then we have a matrix consisting of a single column of terms and this referred to as a *column matrix*. If $m = 1$ then the matrix is called a *row matrix*.

If in the analysis of a problem there is a set of simultaneous equations then the use of matrices can be a very convenient shorthand way of expressing and solving the equations. For example, consider the following set of equations:

$$y_1 = a_{11}x_1 + a_{12}x_2 + a_{13}x_3 + \ldots \ldots a_{1n}x_n$$

$$y_2 = a_{21}x_1 + a_{22}x_2 + a_{23}x_3 + \ldots \ldots a_{2n}x_n$$

$$y_3 = a_{31}x_1 + a_{32}x_2 + a_{33}x_3 + \ldots \ldots a_{3n}x_n$$

$$\vdots$$

$$y_m = a_{m1}x_1 + a_{m2}x_2 + a_{m3}x_3 + \ldots \ldots a_{mn}x_n$$

These may be written in matrix form as follows:

$$\{y\} = [A]\{x\} \tag{B.1}$$

where $\{y\}$ and $\{x\}$ are column matrices.

MATRIX MULTIPLICATION

The matrix equation (B.1) involves the multiplication of the matrices $[A]$ and $\{x\}$. To do this one must apply the simple rules of *matrix multiplication.* These are:

(a) two matrices may only be multiplied if the number of columns in the first is equal to the number of rows in the second;
(b) the terms in the product matrix resulting from the multiplication of matrix $|A|$ with a matrix $|B|$ are given by

$$c_{ij} = \sum_{k=1}^{n} a_{ik}b_{kj} \qquad (B.2)$$

The use of these rules is illustrated in the following example:

$$\begin{bmatrix} a_{11} & a_{12} \\ a_{21} & a_{22} \end{bmatrix}\begin{bmatrix} b_{11} & b_{12} & b_{13} \\ b_{21} & b_{22} & b_{23} \end{bmatrix}$$
$$= \begin{bmatrix} (a_{11}b_{11} + a_{12}b_{21}) & (a_{11}b_{12} + a_{12}b_{22}) & (a_{11}b_{13} + a_{12}b_{23}) \\ (a_{21}b_{11} + a_{22}b_{21}) & (a_{21}b_{12} + a_{22}b_{22}) & (a_{21}b_{13} + a_{22}b_{23}) \end{bmatrix}$$

Suppose

$$[A] = \begin{bmatrix} 2 & 4 \\ 6 & 8 \end{bmatrix}; \qquad [B] = \begin{bmatrix} 1 & 2 & 3 \\ 3 & 2 & -1 \end{bmatrix}$$

Then

$$[C] = [A][B] = \begin{vmatrix} 14 & 12 & 2 \\ 30 & 28 & 10 \end{vmatrix}$$

MATRIX ADDITION AND SUBTRACTION

Matrix algebra also involves the *addition and subtraction* of matrices. The rules for this are as follows:

(a) matrices may only be added or subtracted if they are of the same order, i.e. they each contain the same number of rows and columns.
(b) the terms in the resulting matrix are given by

$$c_{ij} = a_{ij} \pm b_{ij} \qquad (B.3)$$

The following example illustrates the use of these rules:

$$\begin{bmatrix} d_{11} & d_{12} & d_{13} \\ d_{21} & d_{22} & d_{23} \end{bmatrix}\begin{bmatrix} b_{11} & b_{12} & b_{13} \\ b_{21} & b_{22} & b_{23} \end{bmatrix}$$
$$= \begin{bmatrix} (d_{11} \pm b_{11}) & (d_{12} \pm b_{12}) & (d_{13} \pm b_{13}) \\ (d_{21} \pm b_{21}) & (d_{22} \pm b_{22}) & (d_{23} \pm b_{23}) \end{bmatrix}$$

Suppose

$$[D] = \begin{bmatrix} -2 & 4 & 5 \\ 6 & 8 & -3 \end{bmatrix}; \qquad [B] = \begin{bmatrix} 1 & 2 & 3 \\ 3 & 2 & -1 \end{bmatrix}$$

Then

$$[E] = [D] + [B] = \begin{bmatrix} -1 & 6 & 8 \\ 9 & 10 & -4 \end{bmatrix}$$

$$[F] = [D] - [B] = \begin{bmatrix} -3 & 2 & 2 \\ 3 & 6 & -2 \end{bmatrix}$$

INVERSION OF A MATRIX

Referring back to the set of simultaneous equations at the beginning of this appendix, the objective is usually to solve these for the unknown x terms. This is where the use of matrices has a major advantage because referring to eqn. (B.1) we may rewrite this as

$$\{x\} = |A|^{-1} \{y\} \tag{B.4}$$

This equation expresses the solution to the set of simultaneous equations in that each of the unknown x terms is now given by a new matrix $|A|^{-1}$ multiplied by the known y terms. The new matrix is called the *inverse* of matrix $|A|$. The determination of the terms in the inverse matrix is beyond the scope of this brief introduction. Suffice to say that it may be obtained very quickly on a computer and hence the solution to a set of simultaneous equations is determined quickly using eqn. (B.4).

TRANSPOSE OF A MATRIX

The *transpose* of a matrix $|A|$ is denoted by $|A|^T$. It is determined by exchanging the rows and columns in the original matrix. Thus referring to the matrix $|A|$ at the beginning of this appendix, then

$$|A|^T = \begin{bmatrix} a_{11} & a_{21} & a_{31} & \ldots \ldots & a_{m1} \\ a_{12} & a_{22} & a_{32} & \ldots \ldots & a_{m2} \\ a_{13} & a_{23} & a_{33} & \ldots \ldots & a_{m3} \\ \vdots & & & \\ a_{1n} & a_{2n} & a_{3n} & \ldots \ldots & a_{mn} \end{bmatrix}$$

So if

$$|A| = \begin{bmatrix} 2 & 4 \\ 6 & 8 \end{bmatrix} \quad \text{then} \quad |A|^T = \begin{bmatrix} 2 & 6 \\ 4 & 8 \end{bmatrix}$$

SYMMETRIC MATRIX

A square matrix is one in which the number of columns is equal to the number of rows. An important type of square matrix which arises quite often in the finite element method is a *symmetric matrix*. Such matrices possess the property that $a_{ij} = a_{ji}$. An example of such a matrix is given below.

$$\begin{bmatrix} 2 & 4 & 7 & -3 \\ 4 & 5 & 1 & 9 \\ 7 & 1 & 6 & -5 \\ -3 & 9 & -5 & 4 \end{bmatrix} \quad \text{which is often} \quad \begin{bmatrix} 2 & 4 & 7 & -3 \\ & 5 & 1 & 9 \\ sym. & & 6 & -5 \\ & & & 4 \end{bmatrix}$$

written as

Table of mechanical properties of engineering materials

Material	Yield or 0.1% proof stress (MN/m²)	Tensile strength (MN/m²)	Young's modulus (GN/m²)	Shear modulus (GN/m²)
Acrylic	—	50–80	2.7–3.2	—
Aluminium (pure)	40	200	70	26
Aluminium alloy	250–450	320–550	70–72	26–28
Brass	259	427	101	38
Bronze	280	546	122	47
Carbon steel	370	602	208	82
Cast iron (flake)	—	280	175	—
Cast iron (nodular)	—	735	175	—
Concrete*	40	43	18.5	—
Copper (pure)	60	400	110–120	40–46
Douglas fir (dry)	56	125	14	—
Douglas fir (wet)	32	77	11	—
Glass	—	30–1000	50–80	20–35
Magnesium alloy	245	343	45	16.5
Mild steel	280	462	207	81
Nickel steel	1000	1250	207	82.5
Ni–Cr–Mo steel	924	1085	203	77
Nylon	—	65–86	2.0–2.8	—
Polycarbonate	—	56–66	2.0–3.0	—
Polythene	—	8–35	0.2–1.4	—
uPVC	—	30–70	1.0–3.5	—
Red oak (dry)	59	132	12.5	—
Rubber (hard)	—	5–32	0.004	—
Stainless steel	1120	1295	196	87
Titanium (pure)	400	500	110	40
Titanium alloy	750–910	900–1040	106	40
Tungsten	1000	1510	360	150

* Compression.
† On 150 mm.

% Elongation on 50 mm	Endurance or fatigue limit at 10^7 cycles (MN/m^2)	Coeff. of thermal expansion $\times 10^{-6}\,°C^{-1}$	Density (kg/m^3)	Poisson's ratio
2–8	—	0.6	1200	0.4
60	—	23	2710	0.33
8–17	120–140	23	2626–2790	0.33
40	133	18.5	8430	0.34
35	210	17.5	7601	0.34
30	287	12	7850	0.27–0.3
0.6	119	12	7352	0.2–0.3
3.0	315	12	7352	0.2–0.3
0.45†	—	10.8	2400	0.1–0.2
50	—	17	8900	0.33–0.36
—	—	—	560	—
—	—	—	608	—
—	—	5–11	2400–2800	0.2–0.27
12	133	26	1825	0.35
45	224	12	7850	0.27–0.3
14	595	12	7850	0.27–0.3
19	525	12.5	7822	0.27–0.3
60–300	—	0.8–1.0	1150	0.4
100–130	—	0.4–0.7	1100–1250	0.4
100–1200	—	1.3–2.5	914–960	0.4–0.45
10–300	—	0.5–1.0	1300–1500	0.41
—	—	—	691	—
150–7000	—	130–200	860–2000	0.45–0.5
9	616	17.3	7905	0.27–0.3
25	—	8–10	4500	0.33
10–12	490	8–10	4470–4500	0.33
0–4	—	4.3	1900	0.2

Note: This table is only intended to give the reader an indication of the wide range of properties available from various types of material. The numerical values given are fairly typical, but can be varied in most cases very considerably by factors such as composition, heat treatment, temperature, strain rate, etc. For more detailed values of properties and other materials, the reader should consult the relevant British Standard Specification or one of the published handbooks on material properties, e.g. C. J. Smithells, *Metals Reference Book* (5th edn, Butterworths, 1976).

Appendix D:

Answers to problems

1.1   (a) 160.9 N   (b) 201.1 N

1.2   1480 kg

1.3   809 N

1.4   −257.2 kN.

1.5   Yes.

1.6   (a) $\mu \geqslant 0.62$ (b) 1.55 kN.

1.7   2.8 kN; 27.7°.

1.8 $F_A = -24.25$ kN; $F_B = -48.3$ kN; $F_C = -18.8$ kN.

1.9 $F_{AC} = 0.5$ kN; $R_v = 5.0$ kN; $R_h = 7.4$ kN.

1.10 $V_A = 1960$ N; $H_A = 2940$ N; $H_G = 2940$ N.
       $F_{AB} = 2940$ N; $F_{BF} = 0$; $F_{FE} = -3533$ N.

1.11 $V_A = 22.8$ kN; $H_A = 0$; $V_B = 42.8$ kN.
       $F_{AB} = 0$; $F_{BC} = -42.8$; $F_{AC} = -16.1$; $F_{AD} = 36$; $F_{DC} = -22.8$; $F_{DE} = 48.4$.
       $F_{EC} = -20$; $F_{CG} = -48.4$; $F_{EG} = -3.6$; $F_{EF} = 40.3$; $F_{FG} = -72$.
       $F_{GH} = -53.5$; $F_{FH} = 69$ kN.

1.12 $F_{HG} = -29.4$; $F_{HJ} = -39.24$ kN; New $F_{HJ} = -34.33$ kN.

1.13   (i) S.F.: A, 0; B, +1; C, +1/−7; D, −7/+3; F, +3 kN.
          B.M.: A, 0; B, 0; C, +1; D, −6; F, 0 kN-m.
       (ii) S.F.: A, −2; B, −2/+7; E, −11/+8; F, +8 kN.
          B.M.: A, 0; B, −2; E, −8; F, 0 kN-m.
       (iii) S.F.: A, 0; C, −8/−24; F, −36 kN.
          B.M.: A, 0; C, −8; F, −98 kN-m.

1.14   S.F.: D, −29.44; C, −29.44/−22.08; B, −22.08/−7.36; A, −7.36 kN.
       B.M.: D, −147.2; C, −88.32; B, −22.08; A, 0 kN-m.

1.15   At G: $F_z = 1$; $M_x = 0.25$; At F: $F_z = 1$; $M_x = 0.25$; $M_y = 0.25$;
       At E: $F_x = 2.5$; $F_y = 2.5$; At D: $F_x = 2.5$;
       $F_y = 2.5$; At support A: $F_x = -2.5$; $F_y = -2.5$; $F_z = 0$
       At support B: $F_x = +2.5$; $F_y = +2.5$; $F_z = +1$ kN; $M_y = 0.125$, $M_z = 0.05$.

1.16   Torque: AB, +286.5; BC, −668.5; CD, −191 N-m

1.17   B.M.: A, −6; B, +2.8; C, −0.4; D, −10 kN-m.

2.1   7.64 MN/m² in top of rod.

2.2 $A = \dfrac{F}{\sigma} e^{(\rho/\sigma)y}$

Chapter 4

2.3   53 037 mm^2;   10.5 MN.

2.4   1720 mm^2

2.5 $T_{max} = \dfrac{wl}{2 \sin \phi}$

2.6   27.6 m;   3.68 MN/m^2.

2.7   9 mm.

2.8   22.4 MN/m^2

2.9   12 MN/m^2;   211 MN/m^2.

2.10 $\sigma_\theta = 736$ kN/m^2; $\sigma_L = 552$ kN/m^2.

2.11   70 MN/m^2;   138 MN/m^2.

2.12   Key: 131 N-m;   Pin: 138 N-m.

2.13   49.5 kW;   3.15 MN/m^2.

2.14   4.56 mm.

2.15   11.31 kN-m.

CHAPTER 3

3.1   (*a*) Plane stress   (*b*) Plane strain   (*c*) Plane stress.

3.2   0.413 mm

3.3   94.5 MN/m^2.

3.4   0.000 941;   0.0188 rad/m.

3.5 $\sigma_x - \dfrac{E}{(1+v)(1-2v)}[(1-v)\varepsilon_x + v(\varepsilon_y + \varepsilon_z)]$, etc.

$\sigma_x = \dfrac{E}{1-v^2}\{\varepsilon_x + v\varepsilon_y\}$, etc.

$\sigma_x = \dfrac{E}{(1+v)(1-2v)}[(1-v)\varepsilon_x + v\varepsilon_y]$, etc.

3.7   0.000 78;   0.31 mm.

3.8   194 MN/m^2;   162 MN/m^2; -0.0005.

3.9   1.18 MN m/m^3

3.10   37.6 kN/m^2/m^3.

CHAPTER 4

4.1   122 MN/m^2;   19.5 MN/m^2.

4.2   5.069 mm;   5 mm.

4.3   386 MN/m^2 (steel);   214 MN/m^2 (copper).

4.4   Cylinder 89.6 MN/m^2;   Rods 29.6 MN/m^2.

4.5   1.24 m from left end.

4.6   124.2 kN.

4.7   23.9 kN;   76.1 kN;   95.3 °C.

4.8   16.3 MN/m^2 (steel); -65.3 MN/m^2 (copper).

4.9   41 N-m.

4.10   1634 mm^3.

4.11   0.151 mm;   31.4 MN/m^2;   26.2 mm.

4.12   6.1 MN;   34.1 MN.

CHAPTER 5

5.1	$T_h/T_s = 2.35$.
5.2	3.14 kW; 19.7 mm; 11.52 mm.
5.3	163.3 mm.
5.4	CD, 6.52 MN/m^2; AB, 3.7×10^{-3} rads.
5.5	40.3 kW; 20.1 kW.
5.6	5.33 kN-m.
5.7	48.34 mm; 6.85 mm.
5.8	90.3 N-m.
5.9	1 mm; 0.142 rads.
5.10	$3R/t$; $t^2/3R^2$.
5.11	2.75 kN-m; 0.0405 rads.
5.12	324.5 Nm.

CHAPTER 6

6.1	S.F: $+4$, $+4/-1$, $-1/-6$, -6.
	B.M.: 0, $+8$, $+7$, $+12$, 0.
6.2	B, 1101 N-m; C, 1353 Nm
6.3	A, 0; B, 170; F, 70; G, 60; C, 0; H, -10; J, 10; D, 70; E, 0 kN-m.
6.4	$Q = -58.85\,x^2$; $M = -19.62\,x^3$.
6.5	

x	0	1	2	3	4	5	6
Q	$+8$	$+6.8$	$+3.8$	0	-3.8	-6.8	-8
M	0	7.6	13.0	15	13.0	7.6	0

6.6	S.F: $+5.5$, -3.5, $-3.5/-4$, -4 kN
	B.M.: 0, $+7.5$, $+4$, 0 kN-m; $M_{max} = 9.78$ at $x = 2.34$.
6.7	36.41 mm; 91.38 mm.
6.8	5.68 m.
6.9	27.5 MN/m^2; 22 MN/m^2
6.10	$x = \dfrac{d_1 L}{2(d_2 - d_1)}$
6.11	$M_{AB} = 560$ N-m; $T_{AB} = 490$ N-m; $M_{BC} = 501$ N-m; $T_{BC} = 123$ N-m.
	$\sigma_{AB} = 263$ MN/m^2; $\tau_{AB} = 115$ MN/m^2.
6.12	$M_1/M_2 = 1.43$.
6.13	27.9 mm; 7.93 kN-m.
6.14	38 kN m; 89.1 MN/m^2.
6.15	-10 to $+26$ MN/m^2; 45.84 mm.
6.16	51 m; 1.63 MN/m^2.
6.17	66.9 mm.
6.18	$\tau_{max}/\tau_{mean} = 4/3$.
6.19	Sides: 0 at top, 104.6 MN/m^2 at N.A.; 85.2 MN/m^2 at floor; Floor: 17 MN/m^2 to 0.
6.21	1.14 MN/m^2; 1.04 MN/m^2; 25.6 mm.
6.22	4.84 kN; 4.3 kN-m.
6.23	± 136.9 MN/m^2.
6.24	L, -119.2 MN/m^2; M, 70.2 MN/m^2; N, 15.1 MN/m^2; 16.2° to z axis.
6.25	$-13.08°$ to z axis; 54.7 MN/m^2; -33.5 MN/m^2.
6.26	43.4 mm; 155 MN/m^2; 266 MN/m^2.
6.27	$+2.37$ kN.
6.28	78 MN/m^2 in the larger loop.

7.1	-8.54 mm.
7.2	4.92 mm.
7.3	5.62 m; 1.77 mm.
7.4	A, 8.66 mm; B, 2.27 mm; C, 4.98 mm.
7.5	19.19 mm.
7.7	$L/a = 1.24$.
7.8	-0.5 mm; 1.48 mm; 5.14×10^{-3} rads.
7.9	28.1 mm; 32.3° anticlockwise from z axis.
7.10	0.2 mm.
7.11	± 0.0014 rads.
7.12	(a) 2.44 mm; 152.2 MN/m^2 (b) 1.09; 68 MN/m^2.
7.13	3.33 mm.
7.14	1.71 mm.
7.15	1.48 m; 2.9 mm.
7.16	0.715 m; 9.4 mm.
7.17	$\delta = -\dfrac{wa^4}{4EI}$.

8.1	64.8 kN/m.
8.2	$\dfrac{57}{128}$ wL; $\dfrac{7}{128}wL$; $\dfrac{9}{128}wL^2$; $\dfrac{13wL^4}{6144EI}$.
8.3	$\dfrac{\bar{M}L}{16EI}$; $v = \dfrac{\bar{M}}{EI}\left[\dfrac{x^3}{4L} - \dfrac{5x^2}{8} + \dfrac{Lx}{2} - \dfrac{L^2}{8}\right]$.
8.4	15 mm.
8.5	-250 Nm; -187.5; $+2938$; $+2250$ N.
8.6	63.8 mm
8.7	(a) $3\bar{M}/2L$; $\bar{M}/4$. (b) $13wL^4/6144EI$.
8.8	0, 80, 5, 37.5, 0 kN-m.
8.9	S.F.: -4.06; $-4.06/10.62$; $10.62/-9.38$; $-9.38/-2.18$; $-2.18/10$;
	B.M.: 0; -40.63; 65.6; -28.13; -50; 0
8.10	S.F.: 22.1; $-27.9/1.9$; 1.9/9; 9/-11; 11/0
	B.M.: 0; 48.1; -28.8; -9.6; 44.8; 0.
8.11	$L_1 = L_2 = 0.375L$; $L_3 = 0.25L$.

9.1	Long bolts.
9.2	$\sigma_b = 3\sigma_t$.
9.3	$U_u/U_w = 1.6$.
9.4	57.9 mm.
9.5	4 kN/m; 10 coils.
9.6	137 N; 4.1 mm.
9.7	90.4 mm.
9.8	$2W/EI$.
9.10	$F = EI\delta/3\pi R^3$.
9.12	0.412 mm vertical; 0.23 mm horizontal.

CHAPTER 10

10.1 DF, −8; DE, 8; DA, −20; FE, 3; FA, 11.3; FC, −28; EC, −4.2; EA, −11.3; EB, −9 kN.
10.2 AD, 0; AE, 0; BE, 3; ED, 2; EC, −3; DC, 0 kN.
10.3 AD, 0; AF, 100.6; BD, 0; BE, 0; BF, 100.6; CF, −200.2; DE, 50; DF, −70.5; EF, −70.5 kN.
10.4 8.15 mm horizontal; 8.67 mm vertical.
10.5 0.193 mm horizontal; 1.045 mm vertical.
10.6 0.114 mm.
10.7 301.7 kN; 6 mm
10.8 0.785 mm
10.9 AC, 15.4; CE, 8; DE, −6.25; CD, 2.3; AD, 12.3; BD, −15.87 kN.
10.10 AD, 12.06 mm longer than BC
10.11 EC, −56.6 kN; EC(20 °C), −85.7 kN.

CHAPTER 11

11.1 2 kN.
11.2 1.88 kN.
11.3 50 mm.
11.4 1026 kN.
11.5 $1.67\pi^2 EI/L^2$.
11.6 1.33 kN.
11.7 54.65 kN.
11.8 -6.72 MN/m^2.
11.9 (*a*) buckling would occur   (*b*) 2.4
11.10 30 mm.
11.12 37.8 MN/m^2.

CHAPTER 12

12.1 (*a*) $\sigma_x = 100$; $\sigma_y = 200$ MN/m^2 (*b*) Top: $\sigma_x = 240$ MN/m^2; $\tau_{xy} = 0$: Neutral axis: $\sigma_x = 0$; $\tau_{xy} = 6$ MN/m^2 (*c*) $\sigma_x = -10.2$; $\tau_{xy} = 122$ MN/m^2.
12.2 (*a*) $\sigma_n = 199.6$; $\tau_s = 5.98$ MN/m^2 (*b*) $\sigma_n = 30$; $\tau_s = 110$ MN/m^2 (*c*) $\sigma_n = -180.8$; 3.35 MN/m^2.
12.3 56, 21.5 MN/m^2.
12.4 As for 12.2.
12.5 -11; 61 MN/m^2.
12.6 -47; 67 MN/m^2; 30, 17 MN/m^2.
12.7 $+228, -28$; 128 MN/m^2, ± 80 MN/m^2: $-228, +28$; 128 MN/m^2.
12.8 46.3 MN/m^2; 69.3, −23.1 MN/m^2.
12.9 12.5, −0.3 MN/m^2.
12.10 2.25; 6.75 kN.
12.11 132 MN/m^2; 88.14 MN/m^2
12.12 $W = 29.8$ kN; $F = 11.72$ kN.
12.13 7.17×10^{-4}; -6.17×10^{-4}.
12.14 287, −200 MN/m^2; 243 MN/m^2.
12.15 141 MN/m^2; 2.36 kN-m.
12.16 1.13 kN-m.
12.17 $11.35 \times 10^{-4}, 6.15 \times 10^{-4}$; 19° anticlockwise from gauge A.
12.18 153.7; −64.6; 109.2 MN/m^2.

Chapter 15

12.19 $E = 69.93$; $G = 26.9$; $K = 58.3 \, \text{GN/m}^2$; $v = 0.3$.
12.20 $54.74°$.
12.21 $\varepsilon_x = -2.45 \times 10^{-3}$; $\varepsilon_y = 6.85 \times 10^{-3}$.
12.22 $E_x = 15.7$; $E_y = 11.3 \, \text{GN/m}^2$.

CHAPTER 13

13.1 0.96 kN-m; 1.03 kN-m.
13.2 4.45 kN.
13.3 1.037 MN; 1.037 MN.
13.4 143 N.
13.5 45.4 mm.
13.6 15.1 mm.
13.7 $66.7 \, \text{MN/m}^2$; 424.2 kN
13.8 $519.4 \, \text{MN/m}^2$.
13.9 $3.37 \, \text{MN/m}^2$.
13.10 0.265 mm; 61.7; $60.55 \, \text{MN/m}^2$.
13.11 49.7 kN.
13.12 7.53 mm.
13.13 3.35 mm; 6.4.

CHAPTER 14

14.3 Cylinder axially constrained.

14.4 $(a) \ \dfrac{\partial \gamma_{xy}}{\partial z} = \dfrac{\partial \gamma_{xz}}{\partial y} + \dfrac{\partial \gamma_{yz}}{\partial x}$; $(b) \ \dfrac{\partial \varepsilon_z}{\partial \theta} = r \dfrac{\partial \gamma_{\theta z}}{\partial z}$

CHAPTER 15

15.2 1.014; 1.016 MN
15.3 1.915; 110; $30 \, \text{MN/m}^2$.
15.4 65, $66.7 \, \text{MN/m}^2$; 1.05.

15.5 $\sigma_\theta = \dfrac{a^2 - 19b^2}{(b^2 - a^2)}$

15.7 $396 \, \text{MN/m}^2$; 28.7×10^{-4}.
15.9 Inner: -63, $-49.2 \, \text{MN/m}^2$; Outer: 49.3, $35.5 \, \text{MN/m}^2$.
15.10 1.25 mm; 1000 mm.
15.11 (i) -54 (ii) 139.3, $85.6 \, \text{MN/m}^2$.
15.12 $20.3 \, \text{MN/m}^2$.
15.13 Outside, 2.21; Inside, $7.46 \, \text{MN/m}^2$.
15.14 1.61, $13.56 \, \text{MN/m}^2$.
15.16 0.128 mm; $12.75 \, \text{MN/m}^2$.
15.17 7331 rev/min.

CHAPTER 16

16.1   (*a*) 1.7   (*b*) 2   (*c*) 1.272   (*d*) 1.125.
16.2   175 mm;   1.93 m from each end.
16.3   22.44 kN.
16.4   10.42 mm.
16.6 $w_p = 11.67\,M_p/L^2$; $0.414L$ from free end.
16.7 $F_p = M_p$.
16.8   56.5 mm;   3°;   31.4 kN-m.
16.9 5.5×10^{-3};   1.321.
16.10   5328 N-m.
16.12   44.5%
16.13   250, 150 MN/m^2.
16.14   10.57 mm;   3.47 kN.
16.15 -17.4 MN/m^2.
16.16   15 960 rev/min;   10%;   6.85 mm.

CHAPTER 17

17.2   (i) 1.55   (ii) 0.45.
17.3   3.1 mm.
17.4 -1.133 mm.
17.5   0.25 mm;   165.5, 48 MN/m^2.
17.6   68.75, 19.94 MN/m^2.
17.7   0.806 mm.
17.8   0.47 MN.
17.9   6.45 mm.
17.10   0.113 mm.
17.11 -90.64; $+90.64$ kN/m^2.
17.12   1.5 m;   2.94 MN/m^2;   2.25 m, 2.2 MN/m^2.
17.13   8.2, 4.69 MN/m^2.
17.14   487 MN/m^2.
17.15   5.03 MN/m^2.
17.16   4.22 MN/m^2.

CHAPTER 18

18.1 -0.15, -0.325 mm; 65.4, 9.6 kN.
18.2 -0.37 mm; 9.6, -16, -12 kN.
18.3 -1.64 mm.

CHAPTER 19

19.1   198 GN/m^2;   270 MN/m^2;   315 MN/m^2;   451 MN/m^2;   24.4%.
19.2   67.6 GN/m^2;   290 MN/m^2;   335 MN/m^2;   413 MN/m^2.
19.5   16.
19.6   Wrong tempering temperature.

Chapter 22

CHAPTER 20

20.2	$36.3\,\mathrm{MN\text{-}m}^{-3/2}$
20.3	2.38;　261;　200;　5.9;　4.0;　0.75 mm.
20.4	Not acceptable.
20.5	Test is valid.
20.6	1.82 m.
20.7	Yes.
20.8	255 kN.
20.9	0.6 mm;　1.5%.
20.10	$62.8\,\mathrm{MN\text{-}m}^{-3/2}$
20.11	$165.8\,\mathrm{MN\text{-}m}^{-3/2}$;　14.1%.
20.12	$141.6\,\mathrm{MN\text{-}m}^{-3/2}$;　1.07 mm.
20.13	0.106 N-m;　5.08 N-m.

CHAPTER 21

21.1	9.73 mm;　23.28 mm.
21.3	Medium-strength steel.
21.4	$2.13\,\mathrm{MN/m^2}$.
21.5	Goodman, 5.43;　Gerber, 6.6;　Soderberg, 4.6 mm.
21.6	69, 78.4 mm.
21.7	154.7 kN.
21.8	At central hole.
21.9	High notch sensitivity.
21.10	50 000 cycles.
21.11	11 363 flights.
21.12	$10.65\,\mathrm{MN/m^2}$.
21.13	$517.5\,\mathrm{MN/m^2}$.

CHAPTER 22

22.1	104.8 hours.
22.2	262 hours.
22.3	543.5 °C.
22.4	7.1 minutes.
22.5	3.31 hours.
22.6	1.1 mm.
22.7	104 days.
22.8	70.8 N;　13.9 days.
22.9	4 mm;　1.1 mm.

Index